DIFFERENTIAL EQUATIONS
WITH APPLICATIONS AND HISTORICAL NOTES

KU-279-670

International Series in Pure and Applied Mathematics

Churchill/Brown Series

The sole aim of science is the honor of the human mind,
and from this point of view
a question about numbers
is as important
as a question about the system of the world.
—C. G. J. Jacobi

DIFFERENTIAL EQUATIONS WITH APPLICATIONS AND HISTORICAL NOTES

Second Edition

George F. Simmons

Professor of Mathematics
Colorado College

with a new chapter on numerical methods by

John S. Robertson

Department of Mathematical Sciences
United States Military Academy

McGraw-Hill, Inc.

New York St. Louis San Francisco Auckland Bogotá Caracas
Hamburg Lisbon London Madrid Mexico Milan Montreal
New Delhi Paris San Juan São Paulo Singapore Sydney
Tokyo Toronto

DIFFERENTIAL EQUATIONS WITH APPLICATIONS AND HISTORICAL NOTES

International Edition 1991

Exclusive rights by McGraw-Hill Book Co-Singapore for manufacture and export. This book cannot be re-exported from the country to which it is consigned by McGraw-Hill.

2 3 4 5 6 7 8 9 0 KKP PM 9 5 4 3

This book was set in Times Roman.
The editors were Richard Wallis and John M. Morriss;
the production supervisor was Louise Karam.
The cover was designed by Carla Bauer.
Project supervision was done by The Universities Press.

Library of Congress Cataloging-in-Publication Data

Simmons, George Finlay, (date).
  Differential equations with applications and historical notes /
  Geoge F. Simmons.—2nd ed.
     p.   cm.
  ISBN 0-07-057540-1
  1. Differential equations. I. Title.
  QA372.S49   1991
  515 '.35—dc20   90-33686

When ordering this title use ISBN 0-07-112807-7

Printed in Singapore

ABOUT THE AUTHOR

George Simmons has academic degrees from the California Institute of Technology, the University of Chicago, and Yale University. He taught at several colleges and universities before joining the faculty of Colorado College in 1962, where he is Professor of Mathematics. He is also the author of *Introduction to Topology and Modern Analysis* (McGraw-Hill, 1963), *Precalculus Mathematics in a Nutshell* (Janson Publications, 1981), and *Calculus with Analytic Geometry* (McGraw-Hill, 1985).

When not working or talking or eating or drinking or cooking, Professor Simmons is likely to be traveling (Western and Southern Europe, Turkey, Israel, Egypt, Russia, China, Southeast Asia), trout fishing (Rocky Mountain states), playing pocket billiards, or reading (literature, history, biography and autobiography, science, and enough thrillers to achieve enjoyment without guilt).

FOR HOPE AND NANCY
my wife and daughter
who still make it all worthwhile

CONTENTS

PREFACE TO THE
SECOND EDITION

"As correct as a second edition"—so goes the idiom. I certainly hope so, and I also hope that anyone who detects an error will do me the kindness of letting me know, so that repairs can be made. As Confucius said, "A man who makes a mistake and doesn't correct it is making two mistakes."

I now understand why second editions of textbooks are always longer than first editions: as with governments and their budgets, there is always strong pressure from lobbyists to put things in, but rarely pressure to take things out.

The main changes in this new edition are as follows: the number of problems in the first part of the book has been more than doubled; there are two new chapters, on Fourier Series and on Partial Differential Equations; sections on higher order linear equations and operator methods have been added to Chapter 3; and further material on convolutions and engineering applications has been added to the chapter on Laplace Transforms.

Altogether, many different one-semester courses can be built on various parts of this book by using the schematic outline of the chapters given on page xxi. There is even enough material here for a two-semester course, if the appendices are taken into account.

Finally, an entirely new chapter on Numerical Methods (Chapter 14) has been written especially for this edition by Major John S. Robertson of the United States Military Academy. Major Robertson's expertise in these matters is much greater than my own, and I am sure that many users of this new edition will appreciate his contribution, as I do.

McGraw-Hill and I would like to thank the following reviewers for their many helpful comments and suggestions: D. R. Arterburn, New

Mexico Tech; Edward Beckenstein, St. John's University; Harold Carda, South Dakota School of Mines and Technology; Wenxiong Chen, University of Arizona; Jerald P. Dauer, University of Tennessee; Lester B. Fuller, Rochester Institute of Technology; Juan Gatica, University of Iowa; Richard H. Herman, The Pennsylvania State University; Roger H. Marty, Cleveland State University; Jean-Pierre Meyer, The Johns Hopkins University; Krzysztof Ostaszewski, University of Louisville; James L. Rovnyak, University of Virginia; Alan Sharples, New Mexico Tech; Bernard Shiffman, The Johns Hopkins University; and Calvin H. Wilcox, University of Utah.

George F. Simmons

PREFACE TO THE FIRST EDITION

To be worthy of serious attention, a new textbook on an old subject should embody a definite and reasonable point of view which is not represented by books already in print. Such a point of view inevitably reflects the experience, taste, and biases of the author, and should therefore be clearly stated at the beginning so that those who disagree can seek nourishment elsewhere. The structure and contents of this book express my personal opinions in a variety of ways, as follows.

The place of differential equations in mathematics. Analysis has been the dominant branch of mathematics for 300 years, and differential equations are the heart of analysis. This subject is the natural goal of elementary calculus and the most important part of mathematics for understanding the physical sciences. Also, in the deeper questions it generates, it is the source of most of the ideas and theories which constitute higher analysis. Power series, Fourier series, the gamma function and other special functions, integral equations, existence theorems, the need for rigorous justifications of many analytic processes—all these themes arise in our work in their most natural context. And at a later stage they provide the principal motivation behind complex analysis, the theory of Fourier series and more general orthogonal expansions, Lebesgue integration, metric spaces and Hilbert spaces, and a host of other beautiful topics in modern mathematics. I would argue, for example, that one of the main ideas of complex analysis is the liberation of power series from the confining environment of the real number system; and this motive is most clearly felt by those who have tried to use real power series to solve differential equations. In botany, it is obvious that no one can fully appreciate the blossoms of flowering plants without a reasonable understanding of the roots, stems, and leaves which nourish and support them. The same principle is true in mathematics, but is often neglected or forgotten.

Fads are as common in mathematics as in any other human activity, and it is always difficult to separate the enduring from the ephemeral in the achievements of one's own time. At present there is a strong current of abstraction flowing through our graduate schools of mathematics. This current has scoured away many of the individual features of the landscape and replaced them with the smooth, rounded boulders of general theories. When taken in moderation, these general theories are both useful and satisfying; but one unfortunate effect of their predominance is that if a student doesn't learn a little while he is an undergraduate about such colorful and worthwhile topics as the wave equation, Gauss's hypergeometric function, the gamma function, and the basic problems of the calculus of variations—among many others—then he is unlikely to do so later. The natural place for an informal acquaintance with such ideas is a leisurely introductory course on differential equations. Some of our current books on this subject remind me of a sightseeing bus whose driver is so obsessed with speeding along to meet a schedule that his passengers have little or no opportunity to enjoy the scenery. Let us be late occasionally, and take greater pleasure in the journey.

Applications. It is a truism that nothing is permanent except change; and the primary purpose of differential equations is to serve as a tool for the study of change in the physical world. A general book on the subject without a reasonable account of its scientific applications would therefore be as futile and pointless as a treatise on eggs that did not mention their reproductive purpose. This book is constructed so that each chapter except the last has at least one major "payoff"—and often several—in the form of a classic scientific problem which the methods of that chapter render accessible. These applications include

The brachistochrone problem
The Einstein formula $E = mc^2$
Newton's law of gravitation
The wave equation for the vibrating string
The harmonic oscillator in quantum mechanics
Potential theory
The wave equation for the vibrating membrane
The prey–predator equations
Nonlinear mechanics
Hamilton's principle
Abel's mechanical problem

I consider the mathematical treatment of these problems to be among the chief glories of Western civilization, and I hope the reader will agree.

The problem of mathematical rigor. On the heights of pure mathematics, any argument that purports to be a proof must be capable of withstanding the severest criticisms of skeptical experts. This is one of the rules of the game, and if you wish to play you must abide by the rules. But this is not the only game in town.

There are some parts of mathematics—perhaps number theory and abstract algebra—in which high standards of rigorous proof may be appropriate at all levels. But in elementary differential equations a narrow insistence on doctrinaire exactitude tends to squeeze the juice out of the subject, so that only the dry husk remains. My main purpose in this book is to help the student grasp the nature and significance of differential equations; and to this end, I much prefer being occasionally imprecise but understandable to being completely accurate but incomprehensible. I am not at all interested in building a logically impeccable mathematical structure, in which definitions, theorems, and rigorous proofs are welded together into a formidable barrier which the reader is challenged to penetrate.

In spite of these disclaimers, I do attempt a fairly rigorous discussion from time to time, notably in Chapter 13 and Appendices A in Chapters 5, 6 and 7, and B in Chapter 11. I am not saying that the rest of this book is nonrigorous, but only that it leans toward the activist school of mathematics, whose primary aim is to develop methods for solving scientific problems—in contrast to the contemplative school, which analyzes and organizes the ideas and tools generated by the activists.

Some will think that a mathematical argument either is a proof or is not a proof. In the context of elementary analysis I disagree, and believe instead that the proper role of a proof is to carry reasonable conviction to one's intended audience. It seems to me that mathematical rigor is like clothing: in its style it ought to suit the occasion, and it diminishes comfort and restricts freedom of movement if it is either too loose or too tight.

History and biography. There is an old Armenian saying, "He who lacks a sense of the past is condemned to live in the narrow darkness of his own generation." Mathematics without history is mathematics stripped of its greatness: for, like the other arts—and mathematics is one of the supreme arts of civilization—it derives its grandeur from the fact of being a human creation.

In an age increasingly dominated by mass culture and bureaucratic impersonality, I take great pleasure in knowing that the vital ideas of mathematics were not printed out by a computer or voted through by a committee, but instead were created by the solitary labor and individual genius of a few remarkable men. The many biographical notes in this book reflect my desire to convey something of the achievements and personal qualities of these astonishing human beings. Most of the longer notes are placed in the appendices, but each is linked directly to a specific contribution discussed in the text. These notes have as their subjects all but a few of the greatest mathematicians of the past three centuries: Fermat, Newton, the Bernoullis, Euler, Lagrange, Laplace, Fourier, Gauss, Abel, Poisson, Dirichlet, Hamilton, Liouville, Chebyshev, Hermite, Riemann, Minkowski, and Poincaré. As T. S. Eliot wrote in one of his essays, "Someone said: 'The dead writers are remote from us because we *know* so much more than they did.' Precisely, and they are that which we know."

History and biography are very complex, and I am painfully aware that scarcely anything in my notes is actually quite as simple as it may appear. I must also apologize for the many excessively brief allusions to mathematical ideas most student readers have not yet encountered. But with the aid of a good library, sufficiently interested students should be able to unravel most of them for themselves. At the very least, such efforts may help to impart a feeling for the immense diversity of classical mathematics—an aspect of the subject that is almost invisible in the average undergraduate curriculum.

George F. Simmons

SUGGESTIONS FOR THE INSTRUCTOR

The following diagram gives the logical dependence of the chapters and suggests a variety of ways this book can be used, depending on the purposes of the course, the tastes of the instructor, and the backgrounds and needs of the students.

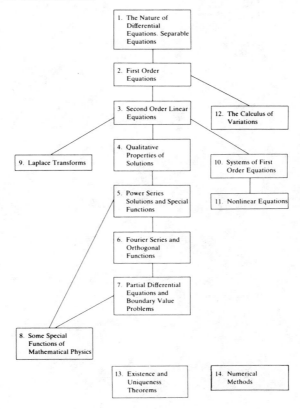

The scientist does not study nature because it is useful; he studies it because he delights in it, and he delights in it because it is beautiful. If nature were not beautiful, it would not be worth knowing, and if nature were not worth knowing, life would not be worth living. Of course I do not here speak of that beauty that strikes the senses, the beauty of qualities and appearances; not that I undervalue such beauty, far from it, but it has nothing to do with science; I mean that profounder beauty which comes from the harmonious order of the parts, and which a pure intelligence can grasp.

—Henri Poincaré

As a mathematical discipline travels far from its empirical source, or still more, if it is a second or third generation only indirectly inspired by ideas coming from "reality," it is beset with very grave dangers. It becomes more and more purely aestheticizing, more and more purely l'art pour l'art. *This need not be bad, if the field is surrounded by correlated subjects, which still have closer empirical connections, or if the discipline is under the influence of men with an exceptionally well-developed taste. But there is a grave danger that the subject will develop along the line of least resistance, that the stream, so far from its source, will separate into a multitude of insignificant branches, and that the discipline will become a disorganized mass of details and complexities. In other words, at a great distance from its empirical source, or after much "abstract" inbreeding, a mathematical subject is in danger of degeneration.*

—John von Neumann

Just as deduction should be supplemented by intuition, so the impulse to progressive generalization must be tempered and balanced by respect and love for colorful detail. The individual problem should not be degraded to the rank of special illustration of lofty general theories. In fact, general theories emerge from consideration of the specific, and they are meaningless if they do not serve to clarify and order the more particularized substance below. The interplay between generality and individuality, deduction and construction, logic and imagination—this is the profound essence of live mathematics. Any one or another of these aspects of mathematics can be at the center of a given achievement. In a far-reaching development all of them will be involved. Generally speaking, such a development will start from the "concrete" ground, then discard ballast by abstraction and rise to the lofty layers of thin air where navigation and observation are easy; after this flight comes the crucial test of landing and reaching specific goals in the newly surveyed low plains of individual "reality." In brief, the flight into abstract generality must start from and return to the concrete and specific.

—Richard Courant

DIFFERENTIAL EQUATIONS
WITH APPLICATIONS AND HISTORICAL NOTES

CHAPTER
1

THE NATURE OF DIFFERENTIAL EQUATIONS. SEPARABLE EQUATIONS

1 INTRODUCTION

An equation involving one dependent variable and its derivatives with respect to one or more independent variables is called a *differential equation*. Many of the general laws of nature—in physics, chemistry, biology, and astronomy—find their most natural expression in the language of differential equations. Applications also abound in mathematics itself, especially in geometry, and in engineering, economics, and many other fields of applied science.

It is easy to understand the reason behind this broad utility of differential equations. The reader will recall that if $y = f(x)$ is a given function, then its derivative dy/dx can be interpreted as the rate of change of y with respect to x. In any natural process, the variables involved and their rates of change are connected with one another by means of the basic scientific principles that govern the process. When this connection is expressed in mathematical symbols, the result is often a differential equation.

The following example may illuminate these remarks. According to Newton's second law of motion, the acceleration a of a body of mass m is proportional to the total force F acting on it, with $1/m$ as the constant of proportionality, so that $a = F/m$ or

$$ma = F. \tag{1}$$

Suppose, for instance, that a body of mass m falls freely under the influence of gravity alone. In this case the only force acting on it is mg, where g is the acceleration due to gravity.[1] If y is the distance down to the body from some fixed height, then its velocity $v = dy/dt$ is the rate of change of position and its acceleration $a = dv/dt = d^2y/dt^2$ is the rate of change of velocity. With this notation, (1) becomes

$$m \frac{d^2y}{dt^2} = mg$$

or

$$\frac{d^2y}{dt^2} = g. \tag{2}$$

If we alter the situation by assuming that air exerts a resisting force proportional to the velocity, then the total force acting on the body is $mg - k(dy/dt)$, and (1) becomes

$$m \frac{d^2y}{dt^2} = mg - k \frac{dy}{dt}. \tag{3}$$

Equations (2) and (3) are the differential equations that express the essential attributes of the physical processes under consideration.

As further examples of differential equations, we list the following:

$$\frac{dy}{dt} = -ky; \tag{4}$$

$$m \frac{d^2y}{dt^2} = -ky; \tag{5}$$

$$\frac{dy}{dx} + 2xy = e^{-x^2}; \tag{6}$$

$$\frac{d^2y}{dx^2} - 5 \frac{dy}{dx} + 6y = 0; \tag{7}$$

$$(1 - x^2) \frac{d^2y}{dx^2} - 2x \frac{dy}{dx} + p(p + 1)y = 0; \tag{8}$$

$$x^2 \frac{d^2y}{dx^2} + x \frac{dy}{dx} + (x^2 - p^2)y = 0. \tag{9}$$

The dependent variable in each of these equations is y, and the independent variable is either t or x. The letters k, m, and p represent

---

[1] g can be considered constant on the surface of the earth in most applications, and is approximately 32 feet per second per second (or 980 centimeters per second per second).

constants. An *ordinary differential equation* is one in which there is only one independent variable, so that all the derivatives occurring in it are ordinary derivatives. Each of these equations is ordinary. The *order* of a differential equation is the order of the highest derivative present. Equations (4) and (6) are first order equations, and the others are second order. Equations (8) and (9) are classical, and are called *Legendre's equation* and *Bessel's equation,* respectively. Each has a vast literature and a history reaching back hundreds of years. We shall study all of these equations in detail later.

A *partial differential equation* is one involving more than one independent variable, so that the derivatives occurring in it are partial derivatives. For example, if $w = f(x,y,z,t)$ is a function of time and the three rectangular coordinates of a point in space, then the following are partial differential equations of the second order:

$$\frac{\partial^2 w}{\partial x^2} + \frac{\partial^2 w}{\partial y^2} + \frac{\partial^2 w}{\partial z^2} = 0;$$

$$a^2\left(\frac{\partial^2 w}{\partial x^2} + \frac{\partial^2 w}{\partial y^2} + \frac{\partial^2 w}{\partial z^2}\right) = \frac{\partial w}{\partial t};$$

$$a^2\left(\frac{\partial^2 w}{\partial x^2} + \frac{\partial^2 w}{\partial y^2} + \frac{\partial^2 w}{\partial z^2}\right) = \frac{\partial^2 w}{\partial t^2}.$$

These equations are also classical, and are called *Laplace's equation,* the *heat equation,* and the *wave equation,* respectively. Each is profoundly significant in theoretical physics, and their study has stimulated the development of many important mathematical ideas. In general, partial differential equations arise in the physics of continuous media—in problems involving electric fields, fluid dynamics, diffusion, and wave motion. Their theory is very different from that of ordinary differential equations, and is much more difficult in almost every respect. For some time to come, we shall confine our attention exclusively to ordinary differential equations.[2]

---

[2] The English biologist J. B. S. Haldane (1892–1964) has a good remark about the one-dimensional special case of the heat equation: "In scientific thought we adopt the simplest theory which will explain all the facts under consideration and enable us to predict new facts of the same kind. The catch in this criterion lies in the word 'simplest.' It is really an aesthetic canon such as we find implicit in our criticism of poetry or painting. The layman finds such a law as

$$a^2\frac{\partial^2 w}{\partial x^2} = \frac{\partial w}{\partial t}$$

much less simple than 'it oozes,' of which it is the mathematical statement. The physicist reverses this judgment, and his statement is certainly the more fruitful of the two, so far as prediction is concerned. It is, however, a statement about something very unfamiliar to the plain man, namely, the rate of change of a rate of change."

2 GENERAL REMARKS ON SOLUTIONS

The general ordinary differential equation of the nth order is

$$F\left(x, y, \frac{dy}{dx}, \frac{d^2y}{dx^2}, \ldots, \frac{d^ny}{dx^n}\right) = 0, \tag{1}$$

or, using the prime notation for derivatives,

$$F(x, y, y', y'', \ldots, y^{(n)}) = 0.$$

Any adequate theoretical discussion of this equation would have to be based on a careful study of explicitly assumed properties of the function F. However, undue emphasis on the fine points of theory often tends to obscure what is really going on. We will therefore try to avoid being overly fussy about such matters—at least for the present.

It is normally a simple task to verify that a given function $y = y(x)$ is a solution of an equation like (1). All that is necessary is to compute the derivatives of $y(x)$ and to show that $y(x)$ and these derivatives, when substituted in the equation, reduce it to an identity in x. In this way we see that

$$y = e^{2x} \quad \text{and} \quad y = e^{3x}$$

are both solutions of the second order equation

$$y'' - 5y' + 6y = 0; \tag{2}$$

and, more generally, that

$$y = c_1 e^{2x} + c_2 e^{3x} \tag{3}$$

is also a solution for every choice of the constants c_1 and c_2. Solutions of differential equations often arise in the form of functions defined implicitly, and sometimes it is difficult or impossible to express the dependent variable explicitly in terms of the independent variable. For instance,

$$xy = \log y + c \tag{4}$$

is a solution of

$$\frac{dy}{dx} = \frac{y^2}{1 - xy} \tag{5}$$

for every value of the constant c, as we can readily verify by differentiating (4) and rearranging the result.[3] These examples also

---

[3] In calculus the notation $\ln x$ is often used for the so-called *natural logarithm*, that is, the function $\log_e x$. In more advanced courses, however, this function is almost always denoted by the symbol $\log x$.

illustrate the fact that a solution of a differential equation usually contains one or more arbitrary constants, equal in number to the order of the equation.

In most cases procedures of this kind are easy to apply to a suspected solution of a given differential equation. The problem of starting with a differential equation and finding a solution is naturally much more difficult. In due course we shall develop systematic methods for solving equations like (2) and (5). For the present, however, we limit ourselves to a few remarks on some of the general aspects of solutions.

The simplest of all differential equations is

$$\frac{dy}{dx} = f(x), \tag{6}$$

and we solve it by writing

$$y = \int f(x)\,dx + c. \tag{7}$$

In some cases the indefinite integral in (7) can be worked out by the methods of calculus. In other cases it may be difficult or impossible to find a formula for this integral. It is known, for instance, that

$$\int e^{-x^2}\,dx \qquad \text{and} \qquad \int \frac{\sin x}{x}\,dx$$

cannot be expressed in terms of a finite number of elementary functions.[4] If we recall, however, that

$$\int f(x)\,dx$$

is merely a symbol for a function (any function) with derivative $f(x)$, then we can almost always give (7) a valid meaning by writing it in the form

$$y = \int_{x_0}^{x} f(t)\,dt + c. \tag{8}$$

The crux of the matter is that this definite integral is a function of the upper limit x (the t under the integral sign is only a dummy variable)

---

[4]Any reader who is curious about the reasons for this should consult D. G. Mead, "Integration," *Am. Math. Monthly*, vol. 68, pp. 152–156 (1961). For additional details, see G. H. Hardy, *The Integration of Functions of a Single Variable*, Cambridge University Press, London, 1916; or J. F. Ritt, *Integration in Finite Terms*, Columbia University Press, New York, 1948.

which always exists when the integrand is continuous over the range of integration, and that its derivative is $f(x)$.[5]

The so-called *separable equations,* or equations with separable variables, are at the same level of simplicity as (6). These are differential equations that can be written in the form

$$\frac{dy}{dx} = f(x)g(y),$$

where the right side is a product of two functions each of which depends on only one of the variables. In such a case we can separate the variables by writing

$$\frac{dy}{g(y)} = f(x)\,dx,$$

and then solve the original equation by integrating:

$$\int \frac{dy}{g(y)} = \int f(x)\,dx + c.$$

These are simple differential equations to deal with in the sense that the problem of solving them can be reduced to the problem of integration, even though the indicated integrations can be difficult or impossible to carry out explicitly.

The general first order equation is the special case of (1) which corresponds to taking $n = 1$:

$$F\left(x,y,\frac{dy}{dx}\right) = 0. \tag{9}$$

We normally expect that an equation like this will have a solution, and that this solution—like (7) and (8)—will contain one arbitrary constant. However,

$$\left(\frac{dy}{dx}\right)^2 + 1 = 0$$

has no real-valued solutions at all, and

$$\left(\frac{dy}{dx}\right)^2 + y^2 = 0$$

has only the single solution $y = 0$ (which contains no arbitrary constants). Situations of this kind raise difficult theoretical questions about

---

[5] This statement is one form of the fundamental theorem of calculus.

the existence and nature of solutions of differential equations. We cannot enter here into a full discussion of these questions, but it may clarify matters if we give an intuitive description of a few of the basic facts.

For the sake of simplicity, let us assume that (9) can be solved for dy/dx:

$$\frac{dy}{dx} = f(x,y). \tag{10}$$

We also assume that $f(x,y)$ is a continuous function throughout some rectangle R in the xy plane. The geometric meaning of a solution of (10) can best be understood as follows (Fig. 1). If $P_0 = (x_0,y_0)$ is a point in R, then the number

$$\left(\frac{dy}{dx}\right)_{P_0} = f(x_0,y_0)$$

determines a direction at P_0. Now let $P_1 = (x_1,y_1)$ be a point near P_0 in this direction, and use

$$\left(\frac{dy}{dx}\right)_{P_1} = f(x_1,y_1)$$

to determine a new direction at P_1. Next, let $P_2 = (x_2,y_2)$ be a point near P_1

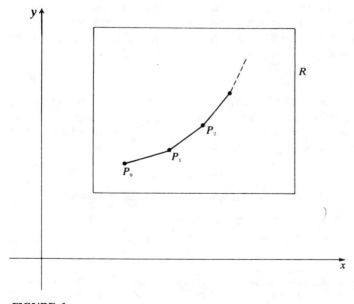

FIGURE 1

in this new direction, and use the number

$$\left(\frac{dy}{dx}\right)_{P_2} = f(x_2, y_2)$$

to determine yet another direction at P_2. If we continue this process, we obtain a broken line with points scattered along it like beads; and if we now imagine that these successive points move closer to one another and become more numerous, then the broken line approaches a smooth curve through the initial point P_0. This curve is a solution $y = y(x)$ of equation (10); for at each point (x,y) on it, the slope is given by $f(x,y)$—and this is precisely the condition required by the differential equation. If we start with a different initial point, then in general we obtain a different curve (or solution). Thus the solutions of (10) form a family of curves, called *integral curves.*[6] Furthermore, it appears to be a reasonable guess that through each point in R there passes just one integral curve of (10). This discussion is intended only to lend plausibility to the following precise statement.

> **Theorem A.   (Picard's theorem.)** *If $f(x,y)$ and $\partial f/\partial y$ are continuous functions on a closed rectangle R, then through each point (x_0, y_0) in the interior of R there passes a unique integral curve of the equation $dy/dx = f(x,y)$.*

If we consider a fixed value of x_0 in this theorem, then the integral curve that passes through (x_0, y_0) is fully determined by the choice of y_0. In this way we see that the integral curves of (10) constitute what is called a *one-parameter family of curves.* The equation of this family can be written in the form

$$y = y(x, c), \tag{11}$$

where different choices of the parameter c yield different curves in the family. The integral curve that passes through (x_0, y_0) corresponds to the value of c for which $y_0 = y(x_0, c)$. If we denote this number by c_0, then (11) is called the *general solution* of (10), and

$$y = y(x, c_0)$$

is called the *particular solution* that satisfies the *initial condition*

$$y = y_0 \quad \text{when} \quad x = x_0.$$

---

[6] Solutions of a differential equation are sometimes called *integrals* of the equation because the problem of finding them is more or less an extension of the ordinary problem of integration.

The essential feature of the general solution (11) is that the constant c in it can be chosen so that an integral curve passes through any given point of the rectangle under consideration.

Picard's theorem is proved in Chapter 13. This proof is quite complicated, and is probably best postponed until the reader has had considerable experience with the more straightforward parts of the subject. The theorem itself can be strengthened in various directions by weakening its hypotheses; it can also be generalized to refer to nth order equations solvable for the nth order derivative. Detailed descriptions of these results would be out of place in the present context, and we content ourselves for the time being with this informal discussion of the main ideas. In the rest of this chapter we explore some of the ways in which differential equations arise in scientific applications.

PROBLEMS

1. Verify that the following functions (explicit or implicit) are solutions of the corresponding differential equations:

   (a) $y = x^2 + c$ $y' = 2x$;
   (b) $y = cx^2$ $xy' = 2y$;
   (c) $y^2 = e^{2x} + c$ $yy' = e^{2x}$;
   (d) $y = ce^{kx}$ $y' = ky$;
   (e) $y = c_1 \sin 2x + c_2 \cos 2x$ $y'' + 4y = 0$;
   (f) $y = c_1 e^{2x} + c_2 e^{-2x}$ $y'' - 4y = 0$;
   (g) $y = c_1 \sinh 2x + c_2 \cosh 2x$ $y'' - 4y = 0$;
   (h) $y = \sin^{-1} xy$ $xy' + y = y'\sqrt{1 - x^2 y^2}$;
   (i) $y = x \tan x$ $xy' = y + x^2 + y^2$;

   (j) $x^2 = 2y^2 \log y$ $y' = \dfrac{xy}{x^2 + y^2}$;

   (k) $y^2 = x^2 - cx$ $2xyy' = x^2 + y^2$;
   (l) $y = c^2 + c/x$ $y + xy' = x^4 (y')^2$;
   (m) $y = ce^{y/x}$ $y' = y^2/(xy - x^2)$;
   (n) $y + \sin y = x$ $(y \cos y - \sin y + x)y' = y$;
   (o) $x + y = \tan^{-1} y$ $1 + y^2 + y^2 y' = 0$.

2. Find the general solution of each of the following differential equations:

   (a) $y' = e^{3x} - x$;   (j) $x^5 y' + y^5 = 0$;
   (b) $xy' = 1$;   (k) $xy' = (1 - 2x^2) \tan y$;
   (c) $y' = xe^{x^2}$;   (l) $y' = 2xy$;
   (d) $y' = \sin^{-1} x$;   (m) $y' \sin y = x^2$;
   (e) $(1 + x)y' = x$;   (n) $y' \sin x = 1$;
   (f) $(1 + x^2)y' = x$;   (o) $y' + y \tan x = 0$;
   (g) $(1 + x^3)y' = x$;   (p) $y' - y \tan x = 0$;
   (h) $(1 + x^2)y' = \tan^{-1} x$;   (q) $(1 + x^2)\, dy + (1 + y^2)\, dx = 0$;
   (i) $xyy' = y - 1$;   (r) $y \log y \, dx - x \, dy = 0$.

3. For each of the following differential equations, find the particular solution

that satisfies the given initial condition:
(a) $y' = xe^x$, $y = 3$ when $x = 1$;
(b) $y' = 2 \sin x \cos x$, $y = 1$ when $x = 0$;
(c) $y' = \log x$, $y = 0$ when $x = e$;
(d) $(x^2 - 1)y' = 1$, $y = 0$ when $x = 2$;
(e) $x(x^2 - 4)y' = 1$, $y = 0$ when $x = 1$;
(f) $(x + 1)(x^2 + 1)y' = 2x^2 + x$, $y = 1$ when $x = 0$.

4. For each of the following differential equations, find the integral curve that passes through the given point:
(a) $y' = e^{3x-2y}$, $(0, 0)$;
(b) $x\,dy = (2x^2 + 1)\,dx$, $(1, 1)$;
(c) $e^{-y}\,dx + (1 + x^2)\,dy = 0$, $(0, 0)$;
(d) $3 \cos 3x \cos 2y\,dx - 2 \sin 3x \sin 2y\,dy = 0$, $(\pi/12, \pi/8)$;
(e) $y' = e^x \cos x$, $(0,0)$;
(f) $xyy' = (x + 1)(y + 1)$, $(1,0)$.

5. Show that $y = e^{x^2}\int_0^x e^{-t^2}\,dt$ is a solution of $y' = 2xy + 1$.

6. For the differential equation (2), namely,

$$y'' - 5y' + 6y = 0,$$

carry out the detailed calculations needed to verify the assertions in the text that
(a) $y = e^{2x}$ and $y = e^{3x}$ are both solutions; and
(b) $y = c_1 e^{2x} + c_2 e^{3x}$ is a solution for every choice of the constants c_1 and c_2.
Remark: In studying a book like this, a student should *never* slide past assertions of this kind—involving such phrases as "we see" or "as we can readily verify"—without personally checking their validity. The mere fact that something is in print does not mean it is necessarily true. Cultivate skepticism as a healthy state of mind, as you would physical fitness; accept nothing on the authority of this writer or any other until you have understood it fully for yourself.

7. In the spirit of Problem 6, verify that (4) is a solution of the differential equation (5) for every value of the constant c.

8. For what values of the constant m will $y = e^{mx}$ be a solution of the differential equation

$$2y''' + y'' - 5y' + 2y = 0?$$

Use the ideas in Problem 6 to find a solution containing three arbitrary constants c_1, c_2, c_3.

3 FAMILIES OF CURVES. ORTHOGONAL TRAJECTORIES

We have seen that the general solution of a first order differential equation normally contains one arbitrary constant, called a *parameter*. When this parameter is assigned various values, we obtain a one-parameter family of curves. Each of these curves is a particular solution, or integral curve, of the given differential equation, and all of them together constitute its general solution.

Conversely, as we might expect, the curves of any one-parameter family are integral curves of some first order differential equation. If the family is

$$f(x,y,c) = 0, \tag{1}$$

then its differential equation can be found by the following steps. First, differentiate (1) implicitly with respect to x to get a relation of the form

$$g\left(x,y,\frac{dy}{dx},c\right) = 0. \tag{2}$$

Next, eliminate the parameter c from (1) and (2) to obtain

$$F\left(x,y,\frac{dy}{dx}\right) = 0 \tag{3}$$

as the desired differential equation. For example,

$$x^2 + y^2 = c^2 \tag{4}$$

is the equation of the family of all circles with centers at the origin (Fig. 2). On differentiation with respect to x this becomes

$$2x + 2y\frac{dy}{dx} = 0;$$

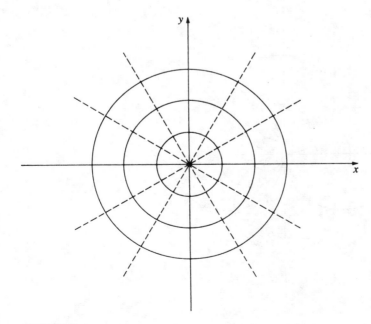

FIGURE 2

and since c is already absent, there is no need to eliminate it and

$$x + y\frac{dy}{dx} = 0 \tag{5}$$

is the differential equation of the given family of circles. Similarly,

$$x^2 + y^2 = 2cx \tag{6}$$

is the equation of the family of all circles tangent to the y-axis at the origin (Fig. 3). When we differentiate this with respect to x, we obtain

$$2x + 2y\frac{dy}{dx} = 2c$$

or

$$x + y\frac{dy}{dx} = c. \tag{7}$$

The parameter c is still present, so it is necessary to eliminate it by combining (6) and (7). This yields

$$\frac{dy}{dx} = \frac{y^2 - x^2}{2xy} \tag{8}$$

as the differential equation of the family (6).

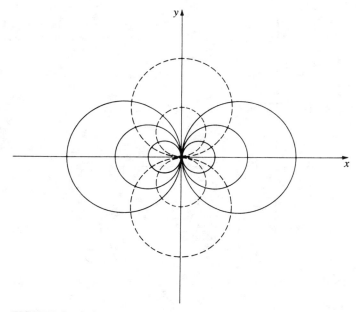

FIGURE 3

As an interesting application of these procedures, we consider the problem of finding orthogonal trajectories. To explain what this problem is, we observe that the family of circles represented by (4) and the family $y = mx$ of straight lines through the origin (the dotted lines in Fig. 2) have the following property: each curve in either family is *orthogonal* (i.e., perpendicular) to every curve in the other family. Whenever two families of curves are related in this way, each is said to be a family of *orthogonal trajectories* of the other. Orthogonal trajectories are of interest in the geometry of plane curves, and also in certain parts of applied mathematics. For instance, if an electric current is flowing in a plane sheet of conducting material, then the lines of equal potential are the orthogonal trajectories of the lines of current flow.

In the example of the circles centered on the origin, it is geometrically obvious that the orthogonal trajectories are the straight lines through the origin, and conversely. In order to cope with more complicated situations, however, we need an analytic method for finding orthogonal trajectories. Suppose that

$$\frac{dy}{dx} = f(x,y) \tag{9}$$

is the differential equation of the family of solid curves in Fig. 4. These curves are characterized by the fact that at any point (x,y) on any one of them the slope is given by $f(x,y)$. The dotted orthogonal trajectory through the same point, being orthogonal to the first curve, has as its slope the negative reciprocal of the first slope. Thus, along any

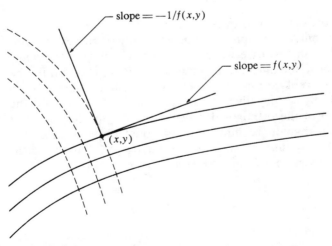

slope $= -1/f(x,y)$

slope $= f(x,y)$

(x,y)

FIGURE 4

orthogonal trajectory, we have $dy/dx = -1/f(x,y)$ or

$$-\frac{dx}{dy} = f(x,y). \tag{10}$$

Our method of finding the orthogonal trajectories of a given family of curves is therefore as follows: first, find the differential equation of the family; next, replace dy/dx by $-dx/dy$ to obtain the differential equation of the orthogonal trajectories; and finally, solve this new differential equation.

If we apply this method to the family of circles (4) with differential equation (5), we get

$$x + y\left(-\frac{dx}{dy}\right) = 0$$

or

$$\frac{dy}{dx} = \frac{y}{x} \tag{11}$$

as the differential equation of the orthogonal trajectories. We can now separate the variables in (11) to obtain

$$\frac{dy}{y} = \frac{dx}{x},$$

which on direct integration yields

$$\log y = \log x + \log c$$

or

$$y = cx$$

as the equation of the orthogonal trajectories.

It is often convenient to express the given family of curves in terms of polar coordinates. In this case we use the fact that if ψ is the angle from the polar radius to the tangent, then $\tan \psi = r\, d\theta/dr$ (Fig. 5). By the above discussion, we replace this expression in the differential equation of the given family by its negative reciprocal, $-dr/r\, d\theta$, to obtain the differential equation of the orthogonal trajectories. As an illustration of the value of this technique, we find the orthogonal trajectories of the family of circles (6). If we use rectangular coordinates, it follows from (8) that the differential equation of the orthogonal trajectories is

$$\frac{dy}{dx} = \frac{2xy}{x^2 - y^2}. \tag{12}$$

Unfortunately, the variables in (12) cannot be separated, so without additional techniques for solving differential equations we can go no

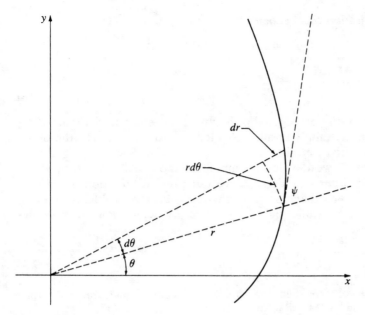

FIGURE 5

further in this direction. However, if we use polar coordinates, the equation of the family (6) can be written as

$$r = 2c \cos \theta. \tag{13}$$

From this we find that

$$\frac{dr}{d\theta} = -2c \sin \theta, \tag{14}$$

and after eliminating c from (13) and (14) we arrive at

$$\frac{r \, d\theta}{dr} = -\frac{\cos \theta}{\sin \theta}$$

as the differential equation of the given family. Accordingly,

$$\frac{r \, d\theta}{dr} = \frac{\sin \theta}{\cos \theta}$$

is the differential equation of the orthogonal trajectories. In this case the variables can be separated, yielding

$$\frac{dr}{r} = \frac{\cos \theta \, d\theta}{\sin \theta};$$

and after integration this becomes

$$\log r = \log (\sin \theta) + \log 2c,$$

so that

$$r = 2c \sin \theta \tag{15}$$

is the equation of the orthogonal trajectories. It will be noted that (15) is the equation of the family of all circles tangent to the x-axis at the origin (see the dotted curves in Fig. 3).

In Chapter 2 we develop a number of more elaborate procedures for solving first order equations. Since our present attention is directed more at applications than formal techniques, all the problems given in this chapter are solvable by the method of separation of variables illustrated above.

PROBLEMS

1. Sketch each of the following families of curves, find the orthogonal trajectories, and add them to the sketch:
   (a) $xy = c$;
   (c) $r = c(1 + \cos \theta)$;
   (b) $y = cx^2$;
   (d) $y = ce^x$.

2. What are the orthogonal trajectories of the family of curves (a) $y = cx^4$; (b) $y = cx^n$ where n is any positive integer? In each case, sketch both families of curves. What is the effect on the orthogonal trajectories of increasing the exponent n?

3. Show that the method for finding orthogonal trajectories in polar coordinates can be expressed as follows. If $dr/d\theta = F(r, \theta)$ is the differential equation of the given family of curves, then $dr/d\theta = -r^2/F(r, \theta)$ is the differential equation of the orthogonal trajectories. Apply this method to the family of circles $r = 2c \sin \theta$.

4. Use polar coordinates to find the orthogonal trajectories of the family of parabolas $r = c/(1 - \cos \theta)$, $c > 0$. Sketch both families of curves.

5. Sketch the family $y^2 = 4c(x + c)$ of all parabolas with axis the x-axis and focus at the origin, and find the differential equation of the family. Show that this differential equation is unaltered when dy/dx is replaced by $-dx/dy$. What conclusion can be drawn from this fact?

6. Find the curves that satisfy each of the following geometric conditions:
   (a) The part of the tangent cut off by the axes is bisected by the point of tangency.
   (b) The projection on the x-axis of the part of the normal between (x,y) and the x-axis has length 1.
   (c) The projection on the x-axis of the part of the tangent between (x,y) and the x-axis has length 1.
   (d) The part of the tangent between (x,y) and the x-axis is bisected by the y-axis.

(e) The part of the normal between (x,y) and the y-axis is bisected by the x-axis.

(f) (x,y) is equidistant from the origin and the point of intersection of the normal with the x-axis.

(g) The polar angle θ equals the angle ψ from the polar radius to the tangent.

(h) The angle ψ from the polar radius to the tangent is constant.

7. A curve rises from the origin in the xy-plane into the first quadrant. The area under the curve from $(0,0)$ to (x,y) is one-third the area of the rectangle with these points as opposite vertices. Find the equation of the curve.

8. Three vertices of a rectangle of area A lie on the x-axis, at the origin, and on the y-axis. If the fourth vertex moves along a curve $y = y(x)$ in the first quadrant in such a way that the rate of change of A with respect to x is proportional to A, find the equation of the curve.

9. A saddle without a saddle-horn (pommel) has the shape of the surface $z = y^2 - x^2$. It is lying outdoors in a rainstorm. Find the paths along which raindrops will run down the saddle. Draw a sketch and use it to convince yourself that your answer is reasonable.

10. Find the differential equation of each of the following one-parameter families of curves:

(a) $y = x \sin(x + c)$;

(b) all circles through $(1,0)$ and $(-1,0)$;

(c) all circles with centers on the line $y = x$ and tangent to both axes;

(d) all lines tangent to the parabola $x^2 = 4y$ (*hint*: the slope of the tangent line at $(2a, a^2)$ is a);

(e) all lines tangent to the unit circle $x^2 + y^2 = 1$.

11. In part (d) of Problem 10, show that the parabola itself is an integral curve of the differential equation of the family of all its tangent lines, and that therefore through each point of this parabola there pass *two* integral curves of this differential equation. Do the same for the unit circle in part (e) of Problem 10.

4 GROWTH, DECAY, CHEMICAL REACTIONS, AND MIXING

We remind the student that the number e is often defined by the limit

$$e = \lim_{n \to \infty} \left(1 + \frac{1}{n}\right)^n,$$

or slightly more generally (put $h = 1/n$), by the limit

$$e = \lim_{h \to 0} (1 + h)^{1/h}. \tag{1}$$

In words, this says that e is the limit of 1 plus a small number, raised to the power of the reciprocal of the small number, as that small number approaches 0.

We recall from calculus that the importance of the number e lies mainly in the fact that the exponential function $y = e^x$ is unchanged by differentiation:

$$\frac{d}{dx} e^x = e^x.$$

An equivalent statement is that $y = e^x$ is a solution of the differential equation

$$\frac{dy}{dx} = y.$$

More generally, if k is any given nonzero constant, then all of the functions $y = ce^{kx}$ are solutions of the differential equation

$$\frac{dy}{dx} = ky. \tag{2}$$

This is easy to verify by differentiation, and can also be discovered by separating the variables and integrating:

$$\frac{dy}{y} = k \, dx, \qquad \log y = kx + c_0, \qquad y = e^{kx + c_0} = e^{c_0}e^{kx} = ce^{kx}.$$

Further, it is not difficult to show that these functions are the *only* solutions of equation (2) [see Problem 1]. In this section we discuss a surprisingly wide variety of applications of these facts to a number of different sciences.

Example 1.   Continuously compounded interest. If P dollars is deposited in a bank that pays an interest rate of 6 percent per year, compounded semiannually, then after t years the accumulated amount is

$$A = P(1 + 0.03)^{2t}.$$

More generally, if the interest rate is $100k$ percent ($k = 0.06$ for 6 percent), and if this interest is compounded n times a year, then after t years the accumulated amount is

$$A = P\left(1 + \frac{k}{n}\right)^{nt}.$$

If n is now increased indefinitely, so that the interest is compounded more and more frequently, then we approach the limiting case of continuously compounded interest.[7] To find the formula for A under these cir-

---

[7] Many banks pay interest daily, which corresponds to $n = 365$. This number is large enough to make continuously compounded interest a very accurate model for what actually happens.

cumstances, we observe that (1) yields

$$\left(1 + \frac{k}{n}\right)^{nt} = \left[\left(1 + \frac{k}{n}\right)^{n/k}\right]^{kt} \to e^{kt},$$

so

$$A = Pe^{kt}. \tag{3}$$

We describe this situation by saying that the amount A *grows exponentially*, or provides an example of *exponential growth*. To understand the meaning of the constant k from a different point of view, we differentiate (3) to obtain

$$\frac{dA}{dt} = Pke^{kt} = kA.$$

If we write this differential equation for A in the form

$$\frac{dA/A}{dt} = k,$$

then we see that k can be thought of as the *fractional change in A per unit time*, and $100k$ is the *percentage change in A per unit time*.

Example 2. Population growth. Suppose that x_0 bacteria are placed in a nutrient solution at time $t = 0$, and that $x = x(t)$ is the population of the colony at a later time t. If food and living space are unlimited, and if as a consequence the population at any moment is increasing at a rate proportional to the population at that moment, find x as a function of t.[8]

Since the rate of increase of x is proportional to x itself, we can write down the differential equation

$$\frac{dx}{dt} = kx.$$

By separating the variables and integrating, we get

$$\frac{dx}{x} = k \, dt, \qquad \log x = kt + c.$$

Since $x = x_0$ when $t = 0$, we have $c = \log x_0$, so $\log x = kt + \log x_0$ and

$$x = x_0 e^{kt}. \tag{4}$$

We therefore have another example of exponential growth.

To make these ideas more concrete, let us assume for the sake of discussion that the total human population of the earth grows in this way. According to the United Nations demographic experts, this population is increasing at an overall rate of approximately 2 percent per year, so

---

[8] Briefly, this assumption about the rate means that we expect twice as many "births" in a given short interval of time when twice as many bacteria are present.

$k = 0.02 = 1/50$ and (4) becomes

$$x = x_0 e^{t/50}. \tag{5}$$

To find the "doubling time" T, that is, the time needed for the total number of people in the world to increase by a factor of 2, we replace (5) by

$$2x_0 = x_0 e^{T/50}.$$

This yields $T/50 = \log 2$, so

$$T = 50 \log 2 \cong 34.65 \text{ years,}$$

since $\log 2 \cong 0.693$.[9]

Example 3. Radioactive decay. If molecules of a certain kind have a tendency to decompose into smaller molecules at a rate unaffected by the presence of other substances, then it is natural to expect that the number of molecules of this kind that will decompose in a unit of time will be proportional to the total number present. A chemical reaction of this type is called a *first order reaction*.

Suppose, for instance, that x_0 grams of matter are present initially, and decompose in a first order reaction. If x is the number of grams present at a later time t, then the principle stated above yields the following differential equation:

$$-\frac{dx}{dt} = kx, \qquad k > 0. \tag{6}$$

[Since dx/dt is the rate of growth of x, $-dx/dt$ is its rate of decay, and (6) says that the rate of decay is proportional to x.] If we separate the variables in (6) and integrate, we obtain

$$\frac{dx}{x} = -k\, dt, \qquad \log x = -kt + c.$$

The initial condition

$$x = x_0 \qquad \text{when} \qquad t = 0 \tag{7}$$

gives $c = \log x_0$, so $\log x = -kt + \log x_0$ and

$$x = x_0 e^{-kt}. \tag{8}$$

---

[9] It is worth mentioning that the population of the industrialized nations is increasing at a rate somewhat less than 2 percent, while that of the third world nations is increasing at a rate greater than 2 percent. From the point of view of the development of the human race and its social and political institutions over the next several centuries, this is perhaps the most important single fact about our contemporary world.

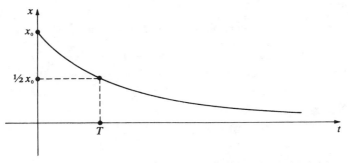

FIGURE 6

This function is therefore the solution of the differential equation (6) that satisfies the initial condition (7). Its graph is given in Fig. 6. The positive constant k is called the *rate constant,* for its value is clearly a measure of the rate at which the reaction proceeds. As we know from Example 1, k can be thought of as the fractional loss of x per unit time.

Very few first order chemical reactions are known, and by far the most important of these is *radioactive decay*. It is convenient to express the rate of decay of a radioactive element in terms of its *half-life*, which is the time required for a given quantity of the element to diminish by a factor of one-half. If we replace x by $x_0/2$ in formula (8), then we get the equation

$$\frac{x_0}{2} = x_0 e^{-kT}$$

for the half-life T, so

$$kT = \log 2.$$

If either k or T is known from observation or experiment, this equation enables us to find the other.

The situation discussed here is an example of *exponential decay*. This phrase refers only to the form of the function (8) and the manner in which the quantity x diminishes, and not necessarily to the idea that something or other is disintegrating.

Example 4. Mixing. A tank contains 50 gallons of brine in which 75 pounds of salt are dissolved. Beginning at time $t = 0$, brine containing 3 pounds of salt per gallon flows in at the rate of 2 gallons per minute, and the mixture (which is kept uniform by stirring) flows out at the same rate. When will there be 125 pounds of dissolved salt in the tank? How much dissolved salt is in the tank after a long time?

If $x = x(t)$ is the number of pounds of dissolved salt in the tank at time $t \geq 0$, then the concentration at that time is $x/50$ pounds per gallon. The rate of change of x is

$$\frac{dx}{dt} = \text{rate at which salt enters tank} - \text{rate at which salt leaves tank.}$$

Since

$$\text{rate of entering} = 3 \cdot 2 = 6 \, \text{lb/min}$$

and

$$\text{rate of leaving} = (x/50) \cdot 2 = \frac{x}{25} \, \text{lb/min},$$

we have

$$\frac{dx}{dt} = 6 - \frac{x}{25} = \frac{150 - x}{25}.$$

Separating variables and integrating give

$$\frac{dx}{150 - x} = \frac{1}{25} dt \quad \text{and} \quad \log(150 - x) = -\frac{1}{25}t + c.$$

Since $x = 75$ when $t = 0$, we see that $c = \log 75$, so

$$\log(150 - x) = -\frac{1}{25}t + \log 75,$$

and therefore

$$150 - x = 75e^{-t/25} \quad \text{or} \quad x = 75(2 - e^{-t/25}).$$

This tells us that $x = 125$ implies $e^{t/25} = 3$ or $t/25 = \log 3$. We conclude that $x = 125$ pounds after

$$t = 25 \log 3 \cong 27.47 \, \text{minutes},$$

since $\log 3 \cong 1.0986$. Also, when t is large we see that x is nearly $75 \cdot 2 = 150$ pounds, as common sense tells us without calculation.

The ideas discussed in Example 3 are the basis for a scientific tool of fairly recent development which has been of great significance for geology and archaeology. In essence, radioactive elements occurring in nature (with known half-lives) can be used to assign dates to events that took place from a few thousand to a few billion years ago. For example, the common isotope of uranium decays through several stages into helium and an isotope of lead, with a half-life of 4.5 billion years. When rock containing uranium is in a molten state, as in lava flowing from the mouth of a volcano, the lead created by this decay process is dispersed by currents in the lava; but after the rock solidifies, the lead is locked in place and steadily accumulates alongside the parent uranium. A piece of granite can be analyzed to determine the ratio of lead to uranium, and this ratio permits an estimate of the time that has elapsed since the critical moment when the granite crystallized. Several methods of age determination involving the decay of thorium and the isotopes of uranium into the various isotopes of lead are in current use. Another method depends on the decay of potassium into argon, with a half-life of

1.3 billion years; and yet another, preferred for dating the oldest rocks, is based on the decay of rubidium into strontium, with a half-life of 50 billion years. These studies are complex and susceptible to errors of many kinds; but they can often be checked against one another, and are capable of yielding reliable dates for many events in geological history linked to the formation of igneous rocks. Rocks tens of millions of years old are quite young, ages ranging into hundreds of millions of years are common, and the oldest rocks yet discovered are upwards of 3 billion years old. This of course is a lower limit for the age of the earth's crust, and so for the age of the earth itself. Other investigations, using various types of astronomical data, age determinations for minerals in meteorites, and so on, have suggested a probable age for the earth of about 4.5 billion years.[10]

The radioactive elements mentioned above decay so slowly that the methods of age determination based on them are not suitable for dating events that took place relatively recently. This gap was filled by Willard Libby's discovery in the late 1940s of *radiocarbon,* a radioactive isotope of carbon with a half-life of about 5600 years. By 1950 Libby and his associates had developed the technique of *radiocarbon dating,* which added a second hand to the slow-moving geological clocks described above and made it possible to date events in the later stages of the Ice Age and some of the movements and activities of prehistoric man. The contributions of this technique to late Pleistocene geology and archaeology have been spectacular.

In brief outline, the facts and principles involved are these. Radiocarbon is produced in the upper atmosphere by the action of cosmic ray neutrons on nitrogen. This radiocarbon is oxidized to carbon dioxide, which in turn is mixed by the winds with the nonradioactive carbon dioxide already present. Since radiocarbon is constantly being formed and constantly decomposing back into nitrogen, its proportion to ordinary carbon in the atmosphere has long since reached an equilibrium state. All air-breathing plants incorporate this proportion of radiocarbon into their tissues, as do the animals that eat these plants. This proportion remains constant as long as a plant or animal lives; but when it dies it ceases to absorb new radiocarbon, while the supply it has at the time of death continues the steady process of decay. Thus, if a piece of old wood has half the radioactivity of a living tree, it lived about 5600 years ago, and if it has only a fourth this radioactivity, it lived about 11,200 years ago. This principle provides a method for dating any ancient object of organic

---

[10] For a full discussion of these matters, as well as many other methods and results of the science of geochronology, see F. E. Zeuner, *Dating the Past,* 4th ed., Methuen, London, 1958.

origin, for instance, wood, charcoal, vegetable fiber, flesh, skin, bone, or horn. The reliability of the method has been verified by applying it to the heartwood of giant sequoia trees whose growth rings record 3000 to 4000 years of life, and to furniture from Egyptian tombs whose age is also known independently. There are technical difficulties, but the method is now felt to be capable of reasonable accuracy as long as the periods of time involved are not too great (up to about 50,000 years).

Radiocarbon dating has been applied to thousands of samples, and laboratories for carrying on this work number in the dozens. Among the more interesting age estimates are these: linen wrappings from the Dead Sea scrolls of the Book of Isaiah, recently found in a cave in Palestine and thought to be first or second century B.C., 1917 ± 200 years; charcoal from the Lascaux cave in southern France, site of the remarkable prehistoric paintings, 15,516 ± 900 years; charcoal from the prehistoric monument at Stonehenge, in southern England, 3798 ± 275 years; charcoal from a tree burned at the time of the volcanic explosion that formed Crater Lake in Oregon, 6453 ± 250 years. Campsites of ancient man throughout the Western Hemisphere have been dated by using pieces of charcoal, fiber sandals, fragments of burned bison bone, and the like. The results suggest that human beings did not arrive in the New World until about the period of the last Ice Age, roughly 25,000 years ago, when the level of the water in the oceans was substantially lower than it now is and they could have walked across the Bering Straits from Siberia to Alaska.[11]

PROBLEMS

1. If k is a given nonzero constant, show that the functions $y = ce^{kx}$ are the only solutions of the differential equation

$$\frac{dy}{dx} = ky.$$

   Hint: Assume that $f(x)$ is a solution of this equation and show that $f(x)/e^{kx}$ is a constant.

2. Suppose that P dollars is deposited in a bank that pays interest at an annual rate of r percent compounded continuously.
   (a) Find the time T required for this investment to double in value as a function of the interest rate r.
   (b) Find the interest rate that must be obtained if the investment is to double in value in 10 years.

---

[11] Libby won the 1960 Nobel Prize for chemistry as a consequence of the work described above. His own account of the method, with its pitfalls and conclusions, can be found in his book *Radiocarbon Dating*, 2d ed., University of Chicago Press, 1955.

3. A bright young executive with foresight but no initial capital makes constant investments of D dollars per year at an annual interest rate of $100k$ percent. Assume that the investments are made continuously and that interest is compounded continuously.
   (a) Find the accumulated amount A at any time t.
   (b) If the interest rate is 6 percent, what must D be if 1 million dollars is to be available for retirement 40 years later?
   (c) If the bright young executive is bright enough to find a safe investment opportunity paying 10 percent, what must D be to achieve the same result of 1 million dollars 40 years later? (It is worth noticing that if this amount of money is simply squirreled away without interest each year for 40 years, the grand total will be less than $80,000.)

4. A newly retired person invests total life savings of P dollars at an interest rate of $100k$ percent per year, compounded continuously. Withdrawals for living expenses are made continuously at a rate of W dollars per year.
   (a) Find the accumulated amount A at any time t.
   (b) Find the withdrawal rate W_0 at which A will remain constant.
   (c) If W is greater than the value W_0 found in part (b), then A will decrease and ultimately disappear. How long will this take?
   (d) Find the time in part (c) if the interest rate is 5 percent and $W = 2W_0$.

5. A certain stock market tycoon has a fortune that increases at a rate proportional to the square of its size at any time. If he had 10 million dollars a year ago, and has 20 million dollars today, how wealthy will he be in 6 months? In a year?

6. A bacterial culture of population x is known to have a growth rate proportional to x itself. Between 6 P.M. and 7 P.M. the population triples. At what time will the population become 100 times what it was at 6 P.M.?

7. The population of a certain mining town is known to increase at a rate proportional to itself. After 2 years the population doubled, and after 1 more year the population was 10,000. What was the original population?

8. It is estimated by experts on agriculture that one-third of an acre of land is needed to provide food for one person on a continuing basis. It is also estimated that there are 10 billion acres of arable land on earth, and that therefore a maximum population of 30 billion people can be sustained if no other sources of food are known. The total world population at the beginning of 1970 was 3.6 billion. Assuming that the population continues to increase at the rate of 2 percent per year, when will the earth be full? What will be the population in the year 2000?

9. A mold grows at a rate proportional to the amount present. At the beginning the amount was 2 grams. In 2 days the amount has increased to 3 grams.
   (a) If $x = x(t)$ is the amount of the mold at time t, show that $x = 2(3/2)^{t/2}$.
   (b) Find the amount at the end of 10 days.

10. In Example 2, assume that living space for the colony of bacteria is limited and food is supplied at a constant rate, so that competition for food and space acts in such a way that ultimately the population will stabilize at a constant level x_1 (x_1 can be thought of as the largest population sustainable by this environment). Assume further that under these conditions the

population grows at a rate proportional to the product of x and the difference $x_1 - x$, and find x as a function of t. Sketch the graph of this function. When is the population increasing most rapidly?

11. Nuclear fission produces neutrons in an atomic pile at a rate proportional to the number of neutrons present at any moment. If n_0 neutrons are present initially, and n_1 and n_2 neutrons are present at times t_1 and t_2, show that

$$\left(\frac{n_1}{n_0}\right)^{t_2} = \left(\frac{n_2}{n_0}\right)^{t_1}.$$

12. If half of a given quantity of radium decomposes in 1600 years, what percentage of the original amount will be left at the end of 2400 years? At the end of 8000 years?

13. If the half-life of a radioactive substance is 20 days, how long will it take for 99 percent of the substance to decay?

14. A field of wheat teeming with grasshoppers is dusted with an insecticide having a kill rate of 200 per 100 per hour. What percentage of the grasshoppers are still alive 1 hour later?

15. Uranium-238 decays at a rate proportional to the amount present. If x_1 and x_2 grams are present at times t_1 and t_2, show that the half-life is

$$\frac{(t_2 - t_1) \log 2}{\log (x_1/x_2)}.$$

16. Suppose that two chemical substances in solution react together to form a compound. If the reaction occurs by means of the collision and interaction of the molecules of the substances, then we expect the rate of formation of the compound to be proportional to the number of collisions per unit time, which in turn is jointly proportional to the amounts of the substances that are untransformed. A chemical reaction that proceeds in this manner is called a *second order reaction*, and this law of reaction is often referred to as the *law of mass action*. Consider a second order reaction in which x grams of the compound contain ax grams of the first substance and bx grams of the second, where $a + b = 1$. If there are aA grams of the first substance present initially, and bB grams of the second, and if $x = 0$ when $t = 0$, find x as a function of the time t.[12]

17. Many chemicals dissolve in water at a rate which is jointly proportional to the amount undissolved and to the difference between the concentration of a saturated solution and the concentration of the actual solution. For a chemical of this kind placed in a tank containing G gallons of water, find the amount x undissolved at time t if $x = x_0$ when $t = 0$ and $x = x_1$ when $t = t_1$, and if S is the amount dissolved in the tank when the solution is saturated.

---

[12] Students who are especially interested in first and second order chemical reactions will find a much more detailed discussion by Linus Pauling, probably the greatest chemist of the twentieth century, in his book *General Chemistry*, 3d ed., W. H. Freeman and Co., San Francisco, 1970. See particularly the chapter "The Rate of Chemical Reactions," which is Chapter 16 in the 3d edition.

18. Suppose that a given population can be divided into two groups: those who have a certain infectious disease, and those who do not have it but can catch it by having contact with an infected person. If x and y are the proportions of infected and uninfected people, then $x + y = 1$. Assume that (1) the disease spreads by the contacts just mentioned between sick people and well people, (2) that the rate of spread dx/dt is proportional to the number of such contacts, and (3) that the two groups mingle freely with each other, so that the number of contacts is jointly proportional to x and y. If $x = x_0$ when $t = 0$, find x as a function of t, sketch the graph, and use this function to show that ultimately the disease will spread through the entire population.

19. A tank contains 100 gallons of brine in which 40 pounds of salt are dissolved. It is desired to reduce the concentration of salt to 0.1 pounds per gallon by pouring in pure water at the rate of 5 gallons per minute and allowing the mixture (which is kept uniform by stirring) to flow out at the same rate. How long will this take?

20. An aquarium contains 10 gallons of polluted water. A filter is attached to this aquarium which drains off the polluted water at the rate of 5 gallons per hour and replaces it at the same rate by pure water. How long does it take to reduce the pollution to half its initial level?

21. A party is being held in a room that contains 1800 cubic feet of air which is originally free of carbon monoxide. Beginning at time $t = 0$ several people start smoking cigarettes. Smoke containing 6 percent carbon monoxide is introduced into the room at the rate of 0.15 cubic feet/min, and the well-circulated mixture leaves at the same rate through a small open window. Extended exposure to a carbon monoxide concentration as low as 0.00018 can be dangerous. When should a prudent person leave this party?

22. According to *Lambert's law of absorption,* the percentage of incident light absorbed by a thin layer of translucent material is proportional to the thickness of the layer.[13] If sunlight falling vertically on ocean water is reduced to one-half its initial intensity at a depth of 10 feet, at what depth is it reduced to one-sixteenth its initial intensity? Solve this problem by merely thinking about it, and also by setting up and solving a suitable differential equation.

23. If sunlight falling vertically on lake water is reduced to three-fifths its initial intensity I_0 at a depth of 15 feet, find its intensity at depths of 30 feet and 60 feet. Find the intensity at a depth of 50 feet.

24. Consider a column of air of cross-sectional area 1 square inch extending from sea level up to "infinity." The atmospheric pressure p at an altitude h above sea level is the weight of the air in this column above the altitude h.

---

[13] Johann Heinrich Lambert (1728–1777) was a Swiss–German astronomer, mathematician, physicist, and man of learning. He was mainly self-educated, and published works on the orbits of comets, the theory of light, and the construction of maps. The Lambert equal-area projection is well known to all cartographers. He is remembered among mathematicians for having given the first proof that π is irrational.

Assuming that the density of the air is proportional to the pressure, show that p satisfies the differential equation

$$\frac{dp}{dh} = -cp, \qquad c > 0,$$

and obtain the formula $p = p_0 e^{-ch}$, where p_0 is the atmospheric pressure at sea level.

25. Assume that the rate at which a hot body cools is proportional to the difference in temperature between it and its surroundings (*Newton's law of cooling*[14]). A body is heated to 110°C and placed in air at 10°C. After 1 hour its temperature is 60°C. How much additional time is required for it to cool to 30°C?

26. A body of unknown temperature is placed in a freezer which is kept at a constant temperature of 0°F. After 15 minutes the temperature of the body is 30°F and after 30 minutes it is 15°F. What was the initial temperature of the body? Solve this problem by merely thinking about it, and also by solving a suitable differential equation.

27. A pot of carrot-and-garlic soup cooling in air at 0°C was initially boiling at 100°C and cooled 20° during the first 30 minutes. How much will it cool during the next 30 minutes?

28. For obvious reasons, the dissecting-room of a certain coroner is kept very cool at a constant temperature of 5°C (= 41°F). While doing an autopsy early one morning on a murder victim, the coroner himself is killed and the victim's body is stolen. At 10 A.M. the coroner's assistant discovers his chief's body and finds its temperature to be 23°C, and at noon the body's temperature is down to 18.5°C. Assuming the coroner had a normal temperature of 37°C (= 98.6°F) when he was alive, when was he murdered?[15]

29. The radiocarbon in living wood decays at the rate of 15.30 disintegrations per minute (dpm) per gram of contained carbon. Using 5600 years as the half-life of radiocarbon, estimate the age of each of the following specimens discovered by archaeologists and tested for radioactivity in 1950:
   (a) a piece of a chair leg from the tomb of King Tutankhamen, 10.14 dpm;
   (b) a piece of a beam of a house built in Babylon during the reign of King Hammurabi, 9.52 dpm;
   (c) dung of a giant sloth found 6 feet 4 inches under the surface of the ground inside Gypsum Cave in Nevada, 4.17 dpm;
   (d) a hardwood atlatl (spear-thrower) found in Leonard Rock Shelter in Nevada, 6.42 dpm.

---

[14] Newton himself applied this rule to estimate the temperature of a red-hot iron ball. So little was known about the laws of heat transfer at that time that his result was only a rough approximation, but it was certainly better than nothing.

[15] The idea for this problem is due to James F. Hurley, "An Application of Newton's Law of Cooling," *The Mathematics Teacher*, vol. 67 (1974), pp. 141–2.

5 FALLING BODIES AND OTHER MOTION PROBLEMS

In this section we study the dynamical problem of determining the motion of a particle along a given path under the action of given forces. We consider only two simple cases: a vertical path, in which the particle is falling either freely under the influence of gravity alone, or with air resistance taken into account; and a circular path, typified by the motion of the bob of a pendulum.

Free fall. The problem of a freely falling body was discussed in Section 1, and we arrived at the differential equation

$$\frac{d^2y}{dt^2} = g \tag{1}$$

for this motion, where y is the distance down to the body from some fixed height. One integration yields the velocity,

$$v = \frac{dy}{dt} = gt + c_1. \tag{2}$$

Since the constant c_1 is clearly the value of v when $t = 0$, it is the initial velocity v_0, and (2) becomes

$$v = \frac{dy}{dt} = gt + v_0. \tag{3}$$

On integrating again we get

$$y = \frac{1}{2}gt^2 + v_0t + c_2.$$

The constant c_2 is the value of y when $t = 0$, or the initial position y_0, so we finally have

$$y = \frac{1}{2}gt^2 + v_0t + y_0 \tag{4}$$

as the general solution of (1). If the body falls from rest starting at $y = 0$, so that $v_0 = y_0 = 0$, then (3) and (4) reduce to

$$v = gt \quad \text{and} \quad y = \frac{1}{2}gt^2.$$

On eliminating t we have the useful equation

$$v = \sqrt{2gy} \tag{5}$$

for the velocity attained in terms of the distance fallen. This result can also be obtained from the *principle of conservation of energy*, which can be stated in the form

kinetic energy + potential energy = a constant.

Since our body falls from rest starting at $y = 0$, the fact that its gain in kinetic energy equals its loss in potential energy gives

$$\frac{1}{2}mv^2 = mgy,$$

and (5) follows at once.

Retarded fall. If we assume that air exerts a resisting force proportional to the velocity of our falling body, then the differential equation of the motion is

$$\frac{d^2y}{dt^2} = g - c\frac{dy}{dt},\tag{6}$$

where $c = k/m$ [see equation 1-(3)]. If dy/dt is replaced by v, this becomes

$$\frac{dv}{dt} = g - cv.\tag{7}$$

On separating variables and integrating, we get

$$\frac{dv}{g - cv} = dt$$

and

$$-\frac{1}{c}\log(g - cv) = t + c_1,$$

so

$$g - cv = c_2 e^{-ct}.$$

The initial condition $v = 0$ when $t = 0$ gives $c_2 = g$, so

$$v = \frac{g}{c}(1 - e^{-ct}).\tag{8}$$

Since c is positive, $v \to g/c$ as $t \to \infty$. This limiting value of v is called the *terminal velocity*. If we wish, we can now replace v by dy/dt in (8) and perform another integration to find y as a function of t.

The motion of a pendulum. Consider a pendulum consisting of a bob of mass m at the end of a rod of negligible mass and length a. If the bob is

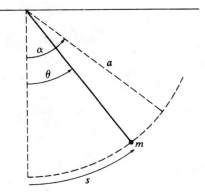

FIGURE 7

pulled to one side through an angle α and released (Fig. 7), then by the principle of conservation of energy we have

$$\tfrac{1}{2}mv^2 = mg(a\cos\theta - a\cos\alpha). \tag{9}$$

Since $s = a\theta$ and $v = ds/dt = a(d\theta/dt)$, this equation gives

$$\tfrac{1}{2}a^2\left(\frac{d\theta}{dt}\right)^2 = ga(\cos\theta - \cos\alpha); \tag{10}$$

and on solving for dt and taking into account the fact that θ decreases as t increases (for small t), we get

$$dt = -\sqrt{\frac{a}{2g}}\frac{d\theta}{\sqrt{\cos\theta - \cos\alpha}}.$$

If T is the *period*, that is, the time required for one complete oscillation, then

$$\frac{T}{4} = -\sqrt{\frac{a}{2g}}\int_\alpha^0 \frac{d\theta}{\sqrt{\cos\theta - \cos\alpha}}$$

or

$$T = 4\sqrt{\frac{a}{2g}}\int_0^\alpha \frac{d\theta}{\sqrt{\cos\theta - \cos\alpha}}. \tag{11}$$

The value of T in this formula clearly depends on α, which is the reason why pendulum clocks vary in their rate of keeping time as the bob swings through a larger or smaller angle.[16] Formula (11) for the period can be expressed more satisfactorily as follows. Since by one of the half-angle

---

[16] This dependence of the period on the amplitude of the swing is what is meant by the "circular error" of pendulum clocks.

formulas of trigonometry we have

$$\cos \theta = 1 - 2 \sin^2 \frac{\theta}{2}$$

and

$$\cos \alpha = 1 - 2 \sin^2 \frac{\alpha}{2},$$

we can write

$$T = 2 \sqrt{\frac{a}{g}} \int_0^\alpha \frac{d\theta}{\sqrt{\sin^2 (\alpha/2) - \sin^2 (\theta/2)}}$$

$$= 2 \sqrt{\frac{a}{g}} \int_0^\alpha \frac{d\theta}{\sqrt{k^2 - \sin^2 (\theta/2)}}, \qquad k = \sin \frac{\alpha}{2}. \tag{12}$$

We now change the variable from θ to ϕ by putting $\sin (\theta/2) = k \sin \phi$, so that ϕ increases from 0 to $\pi/2$ as θ increases from 0 to α, and

$$\frac{1}{2} \cos \frac{\theta}{2} d\theta = k \cos \phi \, d\phi$$

or

$$d\theta = \frac{2k \cos \phi \, d\phi}{\cos (\theta/2)} = \frac{2\sqrt{k^2 - \sin^2 (\theta/2)} \, d\phi}{\sqrt{1 - k^2 \sin^2 \phi}}.$$

This enables us to write (12) in the form

$$T = 4 \sqrt{\frac{a}{g}} \int_0^{\pi/2} \frac{d\phi}{\sqrt{1 - k^2 \sin^2 \phi}} = 4 \sqrt{\frac{a}{g}} F\left(k, \frac{\pi}{2}\right), \tag{13}$$

where

$$F(k, \phi) = \int_0^\phi \frac{d\phi}{\sqrt{1 - k^2 \sin^2 \phi}}$$

is a function of k and ϕ called the *elliptic integral of the first kind*.[17] The *elliptic integral of the second kind*,

$$E(k, \phi) = \int_0^\phi \sqrt{1 - k^2 \sin^2 \phi} \, d\phi,$$

arises in connection with the problem of finding the circumference of an ellipse (see Problem 9). These elliptic integrals cannot be evaluated in terms of elementary functions. Since they occur quite frequently in applications to physics and engineering, their values as numerical functions of k and ϕ are often given in mathematical tables.

---

[17] It is customary in the case of elliptic integrals to violate ordinary usage by allowing the same letter to appear as the upper limit and as the dummy variable of integration.

Our discussion of the pendulum problem up to this point has focused on the first order equation (10). For some purposes it is more convenient to deal with the second order equation obtained by differentiating (10) with respect to t:

$$a\frac{d^2\theta}{dt^2} = -g\sin\theta. \tag{14}$$

If we now recall that $\sin\theta$ is approximately equal to θ for small values of θ, then (14) becomes (approximately)

$$\frac{d^2\theta}{dt^2} + \frac{g}{a}\theta = 0. \tag{15}$$

It will be seen later (in Section 11) that the general solution of the important second order equation

$$\frac{d^2y}{dx^2} + k^2y = 0$$

is

$$y = c_1\sin kx + c_2\cos kx,$$

so (15) yields

$$\theta = c_1\sin\sqrt{\frac{g}{a}}t + c_2\cos\sqrt{\frac{g}{a}}t. \tag{16}$$

The requirement that $\theta = \alpha$ and $d\theta/dt = 0$ when $t = 0$ implies that $c_1 = 0$ and $c_2 = \alpha$, so (16) reduces to

$$\theta = \alpha\cos\sqrt{\frac{g}{a}}t. \tag{17}$$

The period of this approximate solution of (14) is $2\pi\sqrt{a/g}$. It is interesting to note that this is precisely the value of T obtained from (13) when $k = 0$, which is approximately true when the pendulum oscillates through very small angles.

PROBLEMS

1. If the air resistance acting on a falling body of mass m exerts a retarding force proportional to the square of the velocity, then equation (7) becomes

$$\frac{dv}{dt} = g - cv^2,$$

where $c = k/m$. If $v = 0$ when $t = 0$, find v as a function of t. What is the terminal velocity in this case?

2. A torpedo is traveling at a speed of 60 miles/hour at the moment it runs out of fuel. If the water resists its motion with a force proportional to the speed, and if 1 mile of travel reduces its speed to 30 miles/hour, how far will it coast?[18]

3. A rock is thrown upward from the surface of the earth with initial velocity 128 feet/second. Neglecting air resistance and assuming that the only force acting on the rock is a constant gravitational force, find the maximum height it reaches. When does it reach this height, and when does it hit the ground? Answer these questions if the initial velocity is v_0.

4. A mass m is thrown upward from the surface of the earth with initial velocity v_0. If air resistance is assumed to be proportional to velocity, with constant of proportionality k, and if the only other force acting on the mass is a constant gravitational force, show that the maximum height attained is

$$\frac{mv_0}{k} - \frac{m^2g}{k^2} \log\left(1 + \frac{kv_0}{mg}\right).$$

Use l'Hospital's rule to show that this quantity $\rightarrow v_0^2/2g$, in accordance with the result of Problem 3.

5. The force that gravity exerts on a body of mass m at the surface of the earth is mg. In space, however, Newton's law of gravitation asserts that this force varies inversely as the square of the distance to the earth's center. If a projectile fired upward from the surface is to keep traveling indefinitely, and if air resistance is neglected, show that its initial velocity must be at least $\sqrt{2gR}$, where R is the radius of the earth (about 4000 miles). This *escape velocity* is approximately 7 miles/second or 25,000 miles/hour. *Hint:* If x is the distance from the center of the earth to the projectile, and $v = dx/dt$ is its velocity, then

$$\frac{d^2x}{dt^2} = \frac{dv}{dt} = \frac{dv}{dx}\frac{dx}{dt} = v\frac{dv}{dx}.$$

6. In Problem 5, if v_e denotes the escape velocity and $v_0 < v_e$, so that the projectile rises high but does not escape, show that

$$h = \frac{(v_0/v_e)^2}{1 - (v_0/v_e)^2} R$$

is the height it attains before it falls back to earth.

7. Apply the ideas in Problem 5 to find the velocity attained by a body falling freely from rest at an initial altitude $3R$ above the surface of the earth down to the surface. What will be the velocity at the surface if the body falls from an infinite height?

---

[18] In the treatment of dynamical problems by means of vectors, the words *velocity* and *speed* are sharply distinguished from one another. However, in the relatively simple situations we consider, it is permissible (and customary) to use them more or less interchangeably, as we do in everyday speech.

8. Inside the earth, the force of gravity is proportional to the distance from the center. If a hole is drilled through the earth from pole to pole, and a rock is dropped into the hole, with what velocity will it reach the center?

9. (a) Show that the length of the part of the ellipse $x^2/a^2 + y^2/b^2 = 1$ $(a > b)$ that lies in the first quadrant is

$$\int_0^a \sqrt{\frac{a^2 - e^2 x^2}{a^2 - x^2}}\, dx,$$

where e is the eccentricity.

(b) Use the change of variable $x = a \sin \phi$ to transform the integral in (a) into

$$a \int_0^{\pi/2} \sqrt{1 - e^2 \sin^2 \phi}\, d\phi = aE(e, \pi/2),$$

so that the complete circumference of the ellipse is $4aE(e, \pi/2)$.

10. Show that the length of one arch of $y = \sin x$ is $2\sqrt{2}\,E\,(\sqrt{1/2}, \pi/2)$.

11. Show that the total length of the lemniscate $r^2 = a^2 \cos 2\theta$ is $4aF(\sqrt{2}, \pi/4)$.

12. Given the cylinder and sphere whose equations in cylindrical coordinates are $r = a \sin \theta$ and $r^2 + z^2 = b^2$, with $a \le b$, show that:

(a) The area of the part of the cylinder that lies inside the sphere is $4abE(a/b, \pi/2)$.

(b) The area of the part of the sphere that lies inside the cylinder is $2b^2[\pi - 2E(a/b, \pi/2)]$.

13. Establish the following evaluations of definite integrals in terms of elliptic integrals:

(a) $\displaystyle\int_0^{\pi/2} \frac{dx}{\sqrt{\sin x}} = \sqrt{2}F(\sqrt{1/2}, \pi/2)$ [hint: put $x = \pi/2 - y$, then $\cos y$ $= \cos^2 \phi$];

(b) $\displaystyle\int_0^{\pi/2} \sqrt{\cos x}\, dx = 2\sqrt{2}E(\sqrt{1/2}, \pi/2) - \sqrt{2}F(\sqrt{1/2}, \pi/2)$ [hint: put $\cos x = \cos^2 \phi$];

(c) $\displaystyle\int_0^{\pi/2} \sqrt{1 + 4\sin^2 x}\, dx = \sqrt{5}E(\sqrt{4/5}, \pi/2)$ [hint: put $x = \pi/2 - \phi$].

6 THE BRACHISTOCHRONE. FERMAT AND THE BERNOULLIS

Imagine that a point A is joined by a straight wire to a lower point B in the same vertical plane (Fig. 8), and that a bead is allowed to slide without friction down the wire from A to B. We can also consider the case in which the wire is bent into an arc of a circle, so that the motion of the bead is the same as that of the descending bob of a pendulum. Which descent takes the least time, that along the straight path, or that along

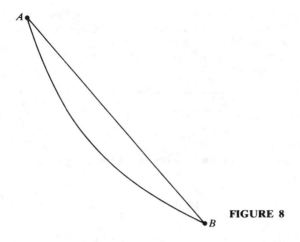

FIGURE 8

the circular path? Since the straight wire joining A and B is clearly the shortest path, we might guess that this wire also yields the shortest time. However, a moment's consideration of the possibilities will make us more skeptical about this conjecture. There might be an advantage in having the bead slide down more steeply at first, thereby increasing its speed more quickly at the beginning of the motion; for with a faster start, it is reasonable to suppose that the bead might reach B in a shorter time, even though it travels over a longer path. For these reasons, Galileo believed that the bead would descend more quickly along the circular path, and probably most people would agree with him.

Many years later, in 1696, John Bernoulli posed a more general problem. He imagined that the wire is bent into the shape of an arbitrary curve, and asked which curve among the infinitely many possibilities will give the shortest possible time of descent. This curve is called the *brachistochrone* (from the Greek *brachistos,* shortest + *chronos,* time). Our purpose in this section is to understand Bernoulli's marvelous solution of this beautiful problem.

We begin by considering an apparently unrelated problem in optics. Figure 9a illustrates a situation in which a ray of light travels from A to P with velocity v_1 and then, entering a denser medium, travels from P to B with a smaller velocity v_2. In terms of the notation in the figure, the total time T required for the journey is given by

$$T = \frac{\sqrt{a^2 + x^2}}{v_1} + \frac{\sqrt{b^2 + (c - x)^2}}{v_2}.$$

If we assume that this ray of light is able to select its path from A to B by way of P in such a way as to minimize T, then $dT/dx = 0$ and by the

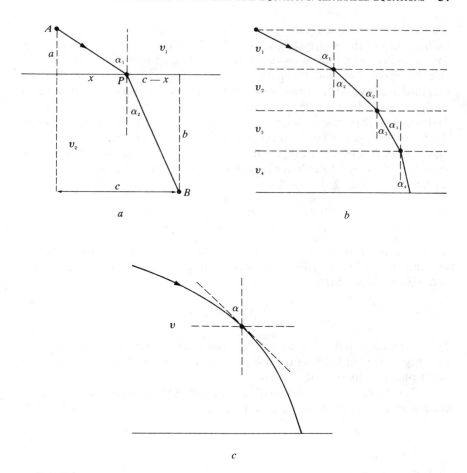

FIGURE 9

methods of elementary calculus we find that

$$\frac{x}{v_1\sqrt{a^2 + x^2}} = \frac{c - x}{v_2\sqrt{b^2 + (c - x)^2}}$$

or

$$\frac{\sin \alpha_1}{v_1} = \frac{\sin \alpha_2}{v_2}.$$

This is *Snell's law of refraction*, which was originally discovered experimentally in the less illuminating form $\sin \alpha_1 / \sin \alpha_2 = $ a constant.[19]

---

[19] Willebrord Snell (1591–1626) was a Dutch astronomer and mathematician. At the age of twenty-two he succeeded his father as professor of mathematics at Leiden. His fame rests mainly on his discovery in 1621 of the law of refraction, which played a significant role in the development of both calculus and the wave theory of light.

The assumption that light travels from one point to another along the path requiring the shortest time is called *Fermat's principle of least time.* This principle not only provides a rational basis for Snell's law, but can also be applied to find the path of a ray of light through a medium of variable density, where in general light will travel along curves instead of straight lines. In Fig. 9b we have a stratified optical medium. In the individual layers the velocity of light is constant, but the velocity decreases from each layer to the one below it. As the descending ray of light passes from layer to layer, it is refracted more and more toward the vertical, and when Snell's law is applied to the boundaries between the layers, we obtain

$$\frac{\sin \alpha_1}{v_1} = \frac{\sin \alpha_2}{v_2} = \frac{\sin \alpha_3}{v_3} = \frac{\sin \alpha_4}{v_4}.$$

If we next allow these layers to grow thinner and more numerous, then in the limit the velocity of light decreases continuously as the ray descends, and we conclude that

$$\frac{\sin \alpha}{v} = \text{a constant.}$$

This situation is indicated in Fig. 9c, and is approximately what happens to a ray of sunlight falling on the earth as it slows in descending through atmosphere of increasing density.

Returning now to Bernoulli's problem, we introduce a coordinate system as in Fig. 10 and imagine that the bead (like the ray of light) is

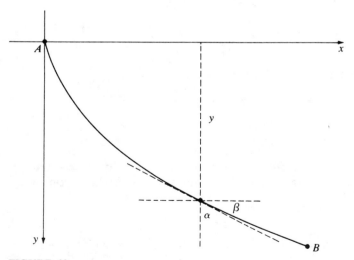

FIGURE 10

capable of selecting the path down which it will slide from A to B in the shortest possible time. The argument given above yields

$$\frac{\sin \alpha}{v} = a \text{ constant.} \tag{1}$$

By the principle of conservation of energy, the velocity attained by the bead at a given level is determined solely by its loss of potential energy in reaching that level, and not at all by the path that brought it there. As in the preceding section, this gives

$$v = \sqrt{2gy}. \tag{2}$$

From the geometry of the situation we also have

$$\sin \alpha = \cos \beta = \frac{1}{\sec \beta} = \frac{1}{\sqrt{1 + \tan^2 \beta}} = \frac{1}{\sqrt{1 + (y')^2}}. \tag{3}$$

On combining equations (1), (2), and (3)—obtained from optics, mechanics, and calculus—we get

$$y[1 + (y')^2] = c \tag{4}$$

as the differential equation of the brachistochrone.

   We now complete our discussion, and discover what curve the brachistochrone actually is, by solving (4). When y' is replaced by dy/dx and the variables are separated, (4) becomes

$$dx = \left(\frac{y}{c - y}\right)^{1/2} dy. \tag{5}$$

At this point we introduce a new variable ϕ by putting

$$\left(\frac{y}{c - y}\right)^{1/2} = \tan \phi, \tag{6}$$

so that $y = c \sin^2 \phi$, $dy = 2c \sin \phi \cos \phi \, d\phi$, and

$$dx = \tan \phi \, dy$$
$$= 2c \sin^2 \phi \, d\phi$$
$$= c(1 - \cos 2\phi) \, d\phi.$$

Integration now yields

$$x = \frac{c}{2}(2\phi - \sin 2\phi) + c_1.$$

Our curve is to pass through the origin, so by (6) we have $x = y = 0$

when $\phi = 0$, and consequently $c_1 = 0$. Thus

$$x = \frac{c}{2}(2\phi - \sin 2\phi) \tag{7}$$

and

$$y = c \sin^2 \phi = \frac{c}{2}(1 - \cos 2\phi). \tag{8}$$

If we now put $a = c/2$ and $\theta = 2\phi$, then (7) and (8) become

$$x = a(\theta - \sin \theta) \quad \text{and} \quad y = a(1 - \cos \theta). \tag{9}$$

These are the standard parameteric equations of the cycloid shown in Fig. 11, which is generated by a point on the circumference of a circle of radius a rolling along the x-axis. We note that there is a single value of a that makes the first arch of this cycloid pass through the point B in Fig. 10; for if a is allowed to increase from 0 to ∞, then the arch inflates, sweeps over the first quadrant of the plane, and clearly passes through B for a single suitably chosen value of a.

Some of the geometric properties of the cycloid are perhaps familiar to the reader from elementary calculus. For example, the length of one arch is 4 times the diameter of the generating circle, and the area under one arch is 3 times the area of this circle. This remarkable curve has many other interesting properties, both geometric and physical, and some of these are described in the problems below.

We hope that the necessary details have not obscured the wonderful imaginative qualities in Bernoulli's brachistochrone problem and his solution of it, for this whole structure of thought is a work of intellectual art of a very high order. In addition to its intrinsic interest, the brachistochrone problem has a larger significance: it was the historical source of the *calculus of variations*—a powerful branch of analysis that in modern times has penetrated deeply into the hidden simplicities at the

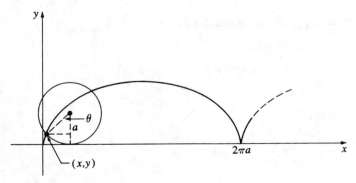

FIGURE 11

heart of the physical world. We shall discuss this subject in Chapter 12, and develop a general method for obtaining equation (4) that is applicable to a wide variety of similar problems.

NOTE ON FERMAT. Pierre de Fermat (1601–1665) was perhaps the greatest mathematician of the seventeenth century, but his influence was limited by his lack of interest in publishing his discoveries, which are known mainly from letters to friends and marginal notes in the books he read. By profession he was a jurist and the king's parliamentary counselor in the French provincial town of Toulouse. However, his hobby and private passion was mathematics. In 1629 he invented analytic geometry, but most of the credit went to Descartes, who hurried into print with his own similar ideas in 1637. At this time—13 years before Newton was born—Fermat also discovered a method for drawing tangents to curves and finding maxima and minima, which amounted to the elements of differential calculus. Newton acknowledged, in a letter that became known only in 1934, that some of his own early ideas on this subject came directly from Fermat. In a series of letters written in 1654, Fermat and Pascal jointly developed the fundamental concepts of the theory of probability. His discovery in 1657 of the principle of least time, and its connection with the refraction of light, was the first step ever taken in the direction of a coherent theory of optics. It was in the theory of numbers, however, that Fermat's genius shone most brilliantly, for it is doubtful whether his insight into the properties of the familiar but mysterious positive integers has ever been equaled. We mention a few of his many discoveries in this field.

1. *Fermat's two squares theorem:* Every prime number of the form $4n + 1$ can be written as the sum of two squares in one and only one way.
2. *Fermat's theorem:* If p is any prime number and n is any positive integer, then p divides $n^p - n$.
3. *Fermat's last theorem:* If $n > 2$, then $x^n + y^n = z^n$ cannot be satisfied by any positive integers x, y, z.

He wrote this last statement in the margin of one of his books, in connection with a passage dealing with the fact that $x^2 + y^2 = z^2$ has many integer solutions. He then added the tantalizing remark, "I have found a truly wonderful proof which this margin is too narrow to contain." Unfortunately no proof has ever been discovered by anyone else, and Fermat's last theorem remains to this day one of the most baffling unsolved problems of mathematics. Finding a proof would confer instant immortality on the finder, but the ambitious student should be warned that many able mathematicians (and some great ones) have tried in vain for hundreds of years.

NOTE ON THE BERNOULLI FAMILY. Most people are aware that Johann Sebastian Bach was one of the greatest composers of all time. However, it is less well known that his prolific family was so consistently talented in this direction that several dozen Bachs were eminent musicians from the sixteenth to the nineteenth centuries. In fact, there were parts of Germany where the very word *bach* meant a musician. What the Bach clan was to music, the Bernoullis were to mathematics and science. In three generations this remarkable Swiss family produced eight mathematicians—three of them outstanding—who in turn had a swarm of descendants who distinguished themselves in many fields.

James Bernoulli (1654–1705) studied theology at the insistence of his father, but abandoned it as soon as possible in favor of his love for science. He taught himself the new calculus of Newton and Leibniz, and was professor of mathematics at Basel from 1687 until his death. He wrote on infinite series, studied many special curves, invented polar coordinates, and introduced the Bernoulli numbers that appear in the power series expansion of the function $\tan x$. In his book *Ars Conjectandi* he formulated the basic principle in the theory of probability known as *Bernoulli's theorem* or the *law of large numbers*: if the probability of a certain event is p, and if n independent trials are made with k successes, then $k/n \to p$ as $n \to \infty$. At first sight this statement may seem to be a triviality, but beneath its surface lies a tangled thicket of philosophical (and mathematical) problems that have been a source of controversy from Bernoulli's time to the present day.

James's younger brother John Bernoulli (1667–1748) also made a false start in his career, by studying medicine and taking a doctor's degree at Basel in 1694 with a thesis on muscle contraction. However, he also became fascinated by calculus, quickly mastered it, and applied it to many problems in geometry, differential equations, and mechanics. In 1695 he was appointed professor of mathematics and physics at Groningen in Holland, and on James's death he succeeded his brother in the professorship at Basel. The Bernoulli brothers sometimes worked on the same problems, which was unfortunate in view of their jealous and touchy dispositions. On occasion the friction between them flared up into a bitter and abusive public feud, as it did over the brachistochrone problem. In 1696 John proposed the problem as a challenge to the mathematicians of Europe. It aroused great interest, and was solved by Newton and Leibniz as well as by the two Bernoullis. John's solution (which we have seen) was the more elegant, while James's— though rather clumsy and laborious—was more general. This situation started an acrimonious quarrel that dragged on for several years and was often conducted in rough language more suited to a street brawl than a scientific discussion. John appears to have been the more cantankerous of the two; for much later, in a fit of jealous rage, he threw his own son out of the house for winning a prize from the French Academy that he coveted for himself.

This son, Daniel Bernoulli (1700–1782), studied medicine like his father and took a degree with a thesis on the action of the lungs; and like his father he soon gave way to his inborn talent and became a professor of mathematics at St. Petersburg. In 1733 he returned to Basel and was successively professor of botany, anatomy, and physics. He won 10 prizes from the French Academy, including the one that infuriated his father, and over the years published many works on physics, probability, calculus, and differential equations. In his famous book *Hydrodynamica* he discussed fluid mechanics and gave the earliest treatment of the kinetic theory of gases. He is considered by many to have been the first genuine mathematical physicist.

PROBLEMS

1. It is stated in the text that the length of one arch of the cycloid (9) is 4 times the diameter of the generating circle (Wren's theorem[20]). Prove this.

2. It is stated in the text that the area under one arch of the cycloid (9) is 3 times the area of the generating circle (Torricelli's theorem[21]). Prove this.

3. Obtain equations (9) for the cycloid by direct integration from the integrated form of equation (5),

$$x = \int \sqrt{\frac{y}{c - y}} \, dy,$$

by starting with the algebraic substitution $u^2 = y/(c - y)$ and continuing with a natural trigonometric substitution.

4. Consider a wire bent into the shape of the cycloid (9), and invert it as in Fig. 10. If a bead is released at the origin and slides down the wire without friction, show that $\pi\sqrt{a/g}$ is the time it takes to reach the point $(\pi a, 2a)$ at the bottom.

5. Show that the number $\pi\sqrt{a/g}$ in Problem 4 is also the time the bead takes to slide to the bottom from any intermediate point, so that the bead will reach the bottom in the same time no matter where it is released. This is known as the *tautochrone* property of the cycloid, from the Greek *tauto*, the same + *chronos*, time.[22]

6. At sunset a man is standing at the base of a dome-shaped hill where it faces the setting sun. He throws a rock straight up in such a manner that the highest

---

[20] Christopher Wren (1632–1723), the greatest of English architects, was an astronomer and mathematician—in fact, Savilian Professor of Astronomy at Oxford—before the Great Fire of London in 1666 gave him his opportunity to build St. Paul's Cathedral, as well as dozens of smaller churches throughout the city.

[21] Evangelista Torricelli (1608–1647) was an Italian physicist and mathematician and a disciple of Galileo, whom he served as secretary. In addition to discovering and proving the theorem stated above, he advanced the first correct ideas—which were narrowly missed by Galileo—about atmospheric pressure and the nature of vacuums, and invented the barometer as an application of his theories. See James B. Conant, *Science and Common Sense,* Yale University Press, New Haven, 1951, pp. 63–71. The geometric theorems of Wren and Torricelli stated in Problems 1 and 2 are straightforward calculus exercises for us. It is interesting to consider how they might have been discovered and proved at a time when the powerful methods of calculus did not exist.

[22] The tautochrone property of the cyloid was discovered by the great Dutch scientist Christiaan Huygens (1629–1695). He published it in 1673 in his treatise on the theory of pendulum clocks, and it was well-known to all European mathematicians at the end of the seventeenth century. When John Bernoulli published his discovery of the brachistochrone in 1696, he expressed himself in the following exuberant language (in Latin, of course): "With justice we admire Huygens because he first discovered that a heavy particle falls down along a common cycloid in the same time no matter from what point on the cycloid it begins its motion. But you will be petrified with astonishment when I say that precisely this cycloid, the tautochrone of Huygens, is our required brachistochrone."

point it reaches is level with the top of the hill. As the rock rises, its shadow moves up the surface of the hill at a constant speed. Show that the profile of the hill is a cycloid.

MISCELLANEOUS PROBLEMS FOR CHAPTER 1

1. It began to snow on a certain morning, and the snow continued to fall steadily throughout the day. At noon a snowplow started to clear a road at a constant rate in terms of the volume of snow removed per hour. The snowplow cleared 2 miles by 2 P.M. and 1 more mile by 4 P.M. When did it start snowing?

2. A mothball whose radius was originally $\frac{1}{4}$ inch is found to have a radius of $\frac{1}{8}$ inch after 1 month. Assuming that it evaporates at a rate proportional to its surface, find the radius as a function of time. After how many more months will it disappear altogether?

3. A tank contains 100 gallons of pure water. Beginning at time $t = 0$, brine containing 1 pound salt/gallon flows in at the rate of 1 gallon/minute, and the mixture (which is kept uniform by stirring) flows out at the same rate. When will there be 50 pounds of dissolved salt in the tank?

4. A large tank contains 100 gallons of brine in which 200 pounds of salt are dissolved. Beginning at time $t = 0$, pure water flows in at the rate of 3 gallons/minute, and the mixture (which is kept uniform by stirring) flows out at the rate of 2 gallons/minute. How long will it take to reduce the amount of salt in the tank to 100 pounds?

5. A smooth football having the shape of an ellipsoid 12 inches long and 6 inches thick is lying outdoors in a rainstorm. Find the paths along which water will run down its sides.

6. If c is a positive constant and a is a positive parameter, then

$$\frac{x^2}{a^2} + \frac{y^2}{a^2 - c^2} = 1$$

is the equation of the family of all ellipses ($a > c$) and hyperbolas ($a < c$) with foci at the points ($\pm c$, 0). Show that this family of *confocal conics* is self-orthogonal (see Problem 3-2).

7. According to *Torricelli's law,* water in an open tank will flow out through a small hole in the bottom with the speed it would acquire in falling freely from the water level to the hole. A hemispherical bowl of radius R is initially full of water, and a small circular hole of radius r is punched in the bottom at time $t = 0$. How long will the bowl take to empty itself?

8. The clepsydra, or ancient water clock, was a bowl from which water was allowed to escape through a small hole in the bottom. It was often used in Greek and Roman courts to time the speeches of lawyers, in order to keep them from talking too much. Find the shape it should have if the water level is to fall at a constant rate.

9. Two open tanks with identical small holes in the bottom drain in the same time. One is a cylinder with a vertical axis and the other is a cone with vertex

down. If they have equal bases and the height of the cylinder is h, what is the height of the cone?

10. A cylindrical can partly filled with water is rotated about its axis with constant angular velocity ω. Show that the surface of the water assumes the shape of a paraboloid of revolution. (Hint: The centripetal force acting on a particle of water of mass m at the free surface is $mx\omega^2$ where x is its distance from the axis, and this is the resultant of the downward gravitational force mg and the normal reaction force R due to other nearby particles of water.)

11. Consider a bead at the highest point of a circle in a vertical plane, and let that point be joined to any lower point on the circle by a straight wire. If the bead slides down the wire without friction, show that it will reach the circle in the same time regardless of the position of the lower point.

12. A chain 4 feet long starts with 1 foot hanging over the edge of a table. Neglect friction, and find the time required for the chain to slide off the table.

13. Experience tells us that a man holding one end of a rope wound around a wooden post can restrain with a small force a much greater force at the other end. Quantitatively, is is not difficult to see that if T and $T + \Delta T$ are the tensions in the rope at angles θ and $\theta + \Delta\theta$ in Fig. 12, then a normal force of approximately $T \Delta\theta$ is exerted by the rope on the post in the region between θ and $\theta + \Delta\theta$. It follows from this that if μ is the coefficient of friction between the rope and the post, then ΔT is approximately $\mu T \Delta\theta$. Use this statement to formulate the differential equation relating T and θ, and solve this equation to find T as a function of θ, μ, and the force T_0 exerted by the man.

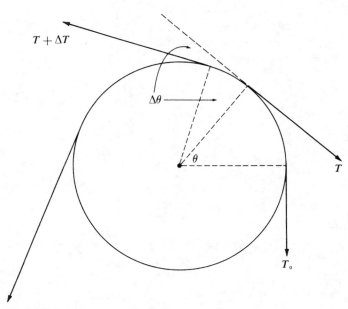

FIGURE 12

14. A load L is supported by a tapered circular column whose material has density a. If the radius of the top of the column is r_0, find the radius r at a distance x below the top if the areas of the horizontal cross sections are proportional to the total loads they bear.

15. The President and the Prime Minister order coffee and receive cups of equal temperature at the same time. The President adds a small amount of cool cream immediately, but does not drink his coffee until 10 minutes later. The Prime Minister waits 10 minutes, and then adds the same amount of cool cream and begins to drink. Who drinks the hotter coffee?

16. A destroyer is hunting a submarine in a dense fog. The fog lifts for a moment, discloses the submarine on the surface 3 miles away, and immediately descends. The speed of the destroyer is twice that of the submarine, and it is known that the latter will at once dive and depart at full speed in a straight course of unknown direction. What path should the destroyer follow to be certain of passing directly over the submarine? Hint: Establish a polar coordinate system with the origin at the point where the submarine was sighted.

17. Four bugs sit at the corners of a square table of side a. At the same instant they all begin to walk with the same speed, each moving steadily toward the bug on its right. If a polar coordinate system is established on the table, with the origin at the center and the polar axis along a diagonal, find the path of the bug that starts on the polar axis and the total distance it walks before all bugs meet at the center.

CHAPTER
2

FIRST
ORDER
EQUATIONS

7 HOMOGENEOUS EQUATIONS

Generally speaking, it is very difficult to solve first order differential equations. Even the apparently simple equation

$$\frac{dy}{dx} = f(x,y)$$

cannot be solved in general, in the sense that no formulas exist for obtaining its solution in all cases. On the other hand, there are certain standard types of first order equations for which routine methods of solution are available. In this chapter we shall briefly discuss a few of the types that have many applications. Since our main purpose is to acquire technical facility, we shall completely disregard questions of continuity, differentiability, the possible vanishing of divisors, and so on. The relevant problems of a purely mathematical nature will be dealt with later, when some of the necessary background has been developed.

The simplest of the standard types is that in which the variables are separable:

$$\frac{dy}{dx} = g(x)h(y).$$

As we know, to solve this we have only to write it in the separated form $dy/h(y) = g(x)\,dx$ and integrate:

$$\int \frac{dy}{h(y)} = \int g(x)\,dx + c.$$

We have seen many examples of this procedure in the preceding chapter.

At the next level of complexity is the homogeneous equation. A function $f(x,y)$ is called *homogeneous of degree n* if

$$f(tx,ty) = t^n f(x,y)$$

for all suitably restricted x, y, and t. This means that if x and y are replaced by tx and ty, t^n factors out of the resulting function, and the remaining factor is the original function. Thus $x^2 + xy$, $\sqrt{x^2 + y^2}$, and $\sin(x/y)$ are homogeneous of degrees 2, 1, and 0. The differential equation

$$M(x,y)\,dx + N(x,y)\,dy = 0$$

is said to be *homogeneous* if M and N are homogeneous functions of the same degree. This equation can then be written in the form

$$\frac{dy}{dx} = f(x, y) \tag{1}$$

where $f(x,y) = -M(x,y)/N(x,y)$ is clearly homogeneous of degree 0. The procedure for solving (1) rests on the fact that it can *always* be changed into an equation with separable variables by means of the substitution $z = y/x$, regardless of the form of the function $f(x,y)$. To see this, we note that the relation

$$f(tx,ty) = t^0 f(x,y) = f(x,y)$$

permits us to set $t = 1/x$ and obtain

$$f(x,y) = f(1,y/x) = f(1,z).$$

Then, since $y = zx$ and

$$\frac{dy}{dx} = z + x\frac{dz}{dx}, \tag{2}$$

equation (1) becomes

$$z + x\frac{dz}{dx} = f(1,z),$$

and the variables can be separated:

$$\frac{dz}{f(1,z) - z} = \frac{dx}{x}.$$

We now complete the solution by integrating and replacing z by y/x.

Example 1. Solve $(x + y) \, dx - (x - y) \, dy = 0$.

We begin by writing the equation in the form suggested by the above discussion:

$$\frac{dy}{dx} = \frac{x + y}{x - y}.$$

Since the function on the right is clearly homogeneous of degree 0, we know that it can be expressed as a function of $z = y/x$. This is easily accomplished by dividing numerator and denominator by x:

$$\frac{dy}{dx} = \frac{1 + y/x}{1 - y/x} = \frac{1 + z}{1 - z}.$$

We next introduce equation (2) and separate the variables, which gives

$$\frac{(1 - z) \, dz}{1 + z^2} = \frac{dx}{x}.$$

On integration this yields

$$\tan^{-1} z - \tfrac{1}{2} \log (1 + z^2) = \log x + c;$$

and when z is replaced by y/x, we obtain

$$\tan^{-1} \frac{y}{x} = \log \sqrt{x^2 + y^2} + c$$

as the desired solution.

PROBLEMS

1. Verify that the following equations are homogeneous, and solve them:

(a) $(x^2 - 2y^2) \, dx + xy \, dy = 0$;   (f) $(x - y) \, dx - (x + y) \, dy = 0$;

(b) $x^2 y' - 3xy - 2y^2 = 0$;   (g) $xy' = 2x + 3y$;

(c) $x^2 y' = 3(x^2 + y^2) \tan^{-1} \dfrac{y}{x} + xy$;   (h) $xy' = \sqrt{x^2 + y^2}$;

(d) $x \sin \dfrac{y}{x} \dfrac{dy}{dx} = y \sin \dfrac{y}{x} + x$;   (i) $x^2 y' = y^2 + 2xy$;

(e) $xy' = y + 2xe^{-y/x}$;   (j) $(x^3 + y^3) \, dx - xy^2 \, dy = 0$.

2. Use rectangular coordinates to find the orthogonal trajectories of the family of all circles tangent to the y-axis at the origin.

3. Show that the substitution $z = ax + by + c$ changes

$$y' = f(ax + by + c)$$

into an equation with separable variables, and apply this method to solve the following equations:

(a) $y' = (x + y)^2$;   (b) $y' = \sin^2 (x - y + 1)$.

4. (a) If $ae \neq bd$, show that constants h and k can be chosen in such a way that the substitutions $x = z - h$, $y = w - k$ reduce

$$\frac{dy}{dx} = F\left(\frac{ax + by + c}{dx + ey + f}\right)$$

to a homogeneous equation.

   (b) If $ae = bd$, discover a substitution that reduces the equation in (a) to one in which the variables are separable.

5. Solve the following equations:

(a) $\dfrac{dy}{dx} = \dfrac{x + y + 4}{x - y - 6}$;

(b) $\dfrac{dy}{dx} = \dfrac{x + y + 4}{x + y - 6}$;

(c) $(2x - 2y)\, dx + (y - 1)\, dy = 0$;

(d) $\dfrac{dy}{dx} = \dfrac{x + y - 1}{x + 4y + 2}$;

(e) $(2x + 3y - 1)\, dx - 4(x + 1)\, dy = 0$.

6. By making the substitution $z = y/x^n$ or $y = zx^n$ and choosing a convenient value of n, show that the following differential equations can be transformed into equations with separable variables, and thereby solve them:

(a) $\dfrac{dy}{dx} = \dfrac{1 - xy^2}{2x^2y}$;

(b) $\dfrac{dy}{dx} = \dfrac{2 + 3xy^2}{4x^2y}$;

(c) $\dfrac{dy}{dx} = \dfrac{y - xy^2}{x + x^2y}$.

7. Show that a straight line through the origin intersects all integral curves of a homogeneous equation at the same angle.

8. Let $y' = f(x,y)$ be a homogeneous differential equation, and prove the following geometric fact about its family of integral curves: If the xy-plane is stretched from (or contracted toward) the origin in such a way that each point (x,y) is moved to a new point (x_1,y_1) which is k times its original distance from the origin, with its direction from the origin unchanged, then every integral curve C is carried into an integral curve C_1. *Hint:* $x_1 = kx$ and $y_1 = ky$.

9. Let $y' = f(x,y)$ be a differential equation whose family of integral curves has the geometric property of invariance under stretching which is stated in Problem 8, and prove that the equation is homogeneous.

10. Let a family of curves be integral curves of a differential equation $y' = f(x,y)$. Let a second family have the property that at each point $P = (x,y)$ the angle from the curve of the first family through P to the curve of the second family through P is α. Show that the curves of the second family are solutions of the differential equation

$$y' = \frac{f(x,y) + \tan \alpha}{1 - f(x,y) \tan \alpha}.$$

11. Use the result of the preceding problem to find the curves that form the angle $\pi/4$ with

(a) all straight lines through the origin;

(b) all circles $x^2 + y^2 = c^2$;

(c) all hyperbolas $x^2 - 2xy - y^2 = c$.

8 EXACT EQUATIONS

If we start with a family of curves $f(x,y) = c$, then its differential equation can be written in the form $df = 0$ or

$$\frac{\partial f}{\partial x} dx + \frac{\partial f}{\partial y} dy = 0.$$

For example, the family $x^2 y^3 = c$ has $2xy^3 dx + 3x^2 y^2 dy = 0$ as its differential equation. Suppose we turn this situation around, and begin with the differential equation

$$M(x,y) dx + N(x,y) dy = 0. \tag{1}$$

If there happens to exist a function $f(x,y)$ such that

$$\frac{\partial f}{\partial x} = M \quad \text{and} \quad \frac{\partial f}{\partial y} = N, \tag{2}$$

then (1) can be written in the form

$$\frac{\partial f}{\partial x} dx + \frac{\partial f}{\partial y} dy = 0 \quad \text{or} \quad df = 0$$

and its general solution is

$$f(x,y) = c.$$

In this case the expression $M\, dx + N\, dy$ is said to be an *exact differential*, and (1) is called an *exact differential equation.*

It is sometimes possible to determine exactness and find the function f by mere inspection. Thus the left sides of

$$y\, dx + x\, dy = 0 \quad \text{and} \quad \frac{1}{y} dx - \frac{x}{y^2} dy = 0$$

are recognizable as the differentials of xy and x/y, respectively, so the general solutions of these equations are $xy = c$ and $x/y = c$. In all but the simplest cases, however, this technique of "solution by insight" is clearly impractical. What is needed is a *test* for exactness and a method for finding the function f. We develop this test and method as follows.

Suppose that (1) is exact, so that there exists a function f satisfying equations (2). We know from elementary calculus that the mixed second partial derivatives of f are equal:

$$\frac{\partial^2 f}{\partial y \, \partial x} = \frac{\partial^2 f}{\partial x \, \partial y} . {}^1 \tag{3}$$

This yields

$$\frac{\partial M}{\partial y} = \frac{\partial N}{\partial x}, \tag{4}$$

so (4) is a necessary condition for the exactness of (1). We shall prove that it is also sufficient by showing that (4) enables us to construct a function f that satisfies equations (2). We begin by integrating the first of equations (2) with respect to x:

$$f = \int M \, dx + g(y). \tag{5}$$

The "constant of integration" occurring here is an arbitrary function of y since it must disappear under differentiation with respect to x. This reduces our problem to that of finding a function $g(y)$ with the property that f as given by (5) satisfies the second of equations (2). On differentiating (5) with respect to y and equating the result to N, we get

$$\frac{\partial}{\partial y} \int M \, dx + g'(y) = N,$$

so

$$g'(y) = N - \frac{\partial}{\partial y} \int M \, dx.$$

This yields

$$g(y) = \int \left(N - \frac{\partial}{\partial y} \int M \, dx \right) dy, \tag{6}$$

provided the integrand here is a function only of y. This will be true if the derivative of the integrand with respect to x is 0; and since the derivative in question is

$$\frac{\partial}{\partial x} \left(N - \frac{\partial}{\partial y} \int M \, dx \right) = \frac{\partial N}{\partial x} - \frac{\partial^2}{\partial x \, \partial y} \int M \, dx$$

$$= \frac{\partial N}{\partial x} - \frac{\partial^2}{\partial y \, \partial x} \int M \, dx$$

$$= \frac{\partial N}{\partial x} - \frac{\partial M}{\partial y},$$

an appeal to our assumption (4) completes the argument.

---

[1] The reader should be aware that equation (3) is true whenever both sides exist and are continuous, and that these conditions are satisfied by almost all functions that are likely to arise in practice. Our blanket hypothesis throughout this chapter (see the first paragraph in Section 7) is that all the functions we discuss are sufficiently continuous and differentiable to guarantee the validity of the operations we perform on them.

In summary, we have proved the following statement: *equation* (1) *is exact if and only if* $\partial M / \partial y = \partial N / \partial x$; *and in this case, its general solution is* $f(x,y) = c$, *where f is given by* (5) *and* (6). Two points deserve emphasis: it is the equation $f(x, y) = c$, and not merely the function f, which is the general solution of (1); and it is the *method* embodied in (5) and (6), not the formulas themselves, which should be learned.

Example 1. Test the equation $e^y \, dx + (xe^y + 2y) \, dy = 0$ for exactness, and solve it if it is exact.

Here we have

$$M = e^y \quad \text{and} \quad N = xe^y + 2y,$$

so

$$\frac{\partial M}{\partial y} = e^y \quad \text{and} \quad \frac{\partial N}{\partial x} = e^y.$$

Thus condition (4) is satisfied, and the equation is exact. This tells us that there exists a function $f(x,y)$ such that

$$\frac{\partial f}{\partial x} = e^y \quad \text{and} \quad \frac{\partial f}{\partial y} = xe^y + 2y.$$

Integrating the first of these equations with respect to x gives

$$f = \int e^y \, dx + g(y) = xe^y + g(y),$$

so

$$\frac{\partial f}{\partial y} = xe^y + g'(y).$$

Since this partial derivative must also equal $xe^y + 2y$, we have $g'(y) = 2y$, so $g(y) = y^2$ and $f = xe^y + y^2$. All that remains is to note that

$$xe^y + y^2 = c$$

is the desired solution of the given differential equation.

PROBLEMS

Determine which of the following equations are exact, and solve the ones that are.

1. $\left(x + \dfrac{2}{y}\right) dy + y \, dx = 0.$
2. $(\sin x \tan y + 1) \, dx + \cos x \sec^2 y \, dy = 0.$
3. $(y - x^3) \, dx + (x + y^3) \, dy = 0.$
4. $(2y^2 - 4x + 5) \, dx = (4 - 2y + 4xy) \, dy.$
5. $(y + y \cos xy) \, dx + (x + x \cos xy) \, dy = 0.$
6. $\cos x \cos^2 y \, dx + 2 \sin x \sin y \cos y \, dy = 0.$
7. $(\sin x \sin y - xe^y) \, dy = (e^y + \cos x \cos y) \, dx.$

8. $-\dfrac{1}{y}\sin\dfrac{x}{y}\,dx + \dfrac{x}{y^2}\sin\dfrac{x}{y}\,dy = 0.$

9. $(1 + y)\,dx + (1 - x)\,dy = 0.$

10. $(2xy^3 + y\cos x)\,dx + (3x^2y^2 + \sin x)\,dy = 0.$

11. $dx = \dfrac{y}{1 - x^2y^2}\,dx + \dfrac{x}{1 - x^2y^2}\,dy.$

12. $(2xy^4 + \sin y)\,dx + (4x^2y^3 + x\cos y)\,dy = 0.$

13. $\dfrac{y\,dx + x\,dy}{1 - x^2y^2} + x\,dx = 0.$

14. $2x(1 + \sqrt{x^2 - y})\,dx = \sqrt{x^2 - y}\,dy.$

15. $(x\log y + xy)\,dx + (y\log x + xy)\,dy = 0.$

16. $(e^{y^2} - \csc y\csc^2 x)\,dx + (2xye^{y^2} - \csc y\cot y\cot x)\,dy = 0.$

17. $(1 + y^2\sin 2x)\,dx - 2y\cos^2 x\,dy = 0.$

18. $\dfrac{x\,dx}{(x^2 + y^2)^{3/2}} + \dfrac{y\,dy}{(x^2 + y^2)^{3/2}} = 0.$

19. $3x^2(1 + \log y)\,dx + \left(\dfrac{x^3}{y} - 2y\right)dy = 0.$

20. Solve

$$\frac{y\,dx - x\,dy}{(x + y)^2} + dy = dx$$

as an exact equation in two ways, and reconcile the results.

21. Solve

$$\frac{4y^2 - 2x^2}{4xy^2 - x^3}\,dx + \frac{8y^2 - x^2}{4y^3 - x^2y}\,dy = 0$$

(a) as an exact equation;

(b) as a homogeneous equation.

22. Find the value of n for which each of the following equations is exact, and solve the equation for that value of n:

(a) $(xy^2 + nx^2y)\,dx + (x^3 + x^2y)\,dy = 0$;

(b) $(x + ye^{2xy})\,dx + nxe^{2xy}\,dy = 0.$

9 INTEGRATING FACTORS

The reader has probably noticed that exact differential equations are comparatively rare, for exactness depends on a precise balance in the form of the equation and is easily destroyed by minor changes in this form. Under these circumstances, it is reasonable to ask whether exact equations are worth discussing at all. In the present section we shall try to convince the reader that they are.

The equation

$$y\,dx + (x^2y - x)\,dy = 0 \tag{1}$$

is easily seen to be nonexact, for $\partial M/\partial y = 1$ and $\partial N/\partial x = 2xy - 1$. However, if we multiply through by the factor $1/x^2$, the equation becomes

$$\frac{y}{x^2}\,dx + \left(y - \frac{1}{x}\right)dy = 0,$$

which is exact. To what extent can other nonexact equations be made exact in this way? In other words, if

$$M(x,y)\,dx + N(x,y)\,dy = 0 \tag{2}$$

is not exact, under what conditions can a function $\mu(x,y)$ be found with the property that

$$\mu(M\,dx + N\,dy) = 0$$

is exact? Any function μ that acts in this way is called an *integrating factor* for (2). Thus $1/x^2$ is an integrating factor for (1). We shall prove that (2) always has an integrating factor if it has a general solution.

Assume then that (2) has a general solution

$$f(x, y) = c,$$

and eliminate c by differentiating:

$$\frac{\partial f}{\partial x}\,dx + \frac{\partial f}{\partial y}\,dy = 0. \tag{3}$$

It follows from (2) and (3) that

$$\frac{dy}{dx} = -\frac{M}{N} = -\frac{\partial f/\partial x}{\partial f/\partial y},$$

so

$$\frac{\partial f/\partial x}{M} = \frac{\partial f/\partial y}{N}. \tag{4}$$

If we denote the common ratio in (4) by $\mu(x,y)$, then

$$\frac{\partial f}{\partial x} = \mu M \quad \text{and} \quad \frac{\partial f}{\partial y} = \mu N.$$

On multiplying (2) by μ, it becomes

$$\mu M\,dx + \mu N\,dy = 0$$

or

$$\frac{\partial f}{\partial x}\,dx + \frac{\partial f}{\partial y}\,dy = 0,$$

which is exact. This argument shows that if (2) has a general solution, then it has at least one integrating factor μ. Actually it has infinitely many integrating factors; for if $F(f)$ is any function of f, then

$$\mu F(f)(M\,dx + N\,dy) = F(f)\,df = d\left[\int F(f)\,df\right],$$

so $\mu F(f)$ is also an integrating factor for (2).

Our discussion so far has not considered the practical problem of finding integrating factors. In general this is quite difficult. There are a few cases, however, in which formal procedures are available. To see how these procedures arise, we consider the condition that μ be an integrating factor for (2):

$$\frac{\partial(\mu M)}{\partial y} = \frac{\partial(\mu N)}{\partial x}.$$

If we write this out, we obtain

$$\mu\frac{\partial M}{\partial y} + M\frac{\partial \mu}{\partial y} = \mu\frac{\partial N}{\partial x} + N\frac{\partial \mu}{\partial x}$$

or

$$\frac{1}{\mu}\left(N\frac{\partial \mu}{\partial x} - M\frac{\partial \mu}{\partial y}\right) = \frac{\partial M}{\partial y} - \frac{\partial N}{\partial x}. \tag{5}$$

It appears that we have "reduced" the problem of solving the ordinary differential equation (2) to the much more difficult problem of solving the partial differential equation (5). On the other hand, we have no need for the general solution of (5) since any particular solution will serve our purpose. And from this point of view, (5) is more fruitful than it looks. Suppose, for instance, that (2) has an integrating factor μ which is a function of x alone. Then $\partial\mu/\partial x = d\mu/dx$ and $\partial\mu/\partial y = 0$, so (5) can be written in the form

$$\frac{1}{\mu}\frac{d\mu}{dx} = \frac{\partial M/\partial y - \partial N/\partial x}{N}. \tag{6}$$

Since the left side of this is a function only of x, the right side is also. If we put

$$\frac{\partial M/\partial y - \partial N/\partial x}{N} = g(x),$$

then (6) becomes

$$\frac{1}{\mu}\frac{d\mu}{dx} = g(x)$$

or

$$\frac{d(\log \mu)}{dx} = g(x),$$

so

$$\log \mu = \int g(x)\, dx$$

and

$$\mu = e^{\int g(x)\, dx}. \tag{7}$$

This reasoning is obviously reversible: if the expression on the right side of (6) is a function only of x, say $g(x)$, then (7) yields a function μ that depends only on x and satisfies equation (5), and is therefore an integrating factor for (2).

Example 1. In the case of equation (1) we have

$$\frac{\partial M / \partial y - \partial N / \partial x}{N} = \frac{1 - (2xy - 1)}{x^2 y - x} = \frac{-2(xy - 1)}{x(xy - 1)} = -\frac{2}{x},$$

which is a function only of x. Accordingly,

$$\mu = e^{\int -(2/x)\, dx} = e^{-2 \log x} = x^{-2}$$

is an integrating factor for (1), as we have already seen.

Similar reasoning gives the following related procedure, which is applicable whenever (2) has an integrating factor depending only on y: if the expression

$$\frac{\partial M / \partial y - \partial N / \partial x}{-M} \tag{8}$$

is a function of y alone, say $h(y)$, then

$$\mu = e^{\int h(y)\, dy} \tag{9}$$

is also a function only of y which satisfies equation (5), and is consequently an integrating factor for (2).

There is another useful technique for converting simple nonexact equations into exact ones. To illustrate it, we again consider equation (1), rearranged as follows:

$$x^2 y\, dy - (x\, dy - y\, dx) = 0. \tag{10}$$

The quantity in parentheses should remind the reader of the differential formula

$$d\left(\frac{y}{x}\right) = \frac{x\, dy - y\, dx}{x^2}, \tag{11}$$

which suggests dividing (10) through by x^2. This transforms the equation into $y \, dy - d(y/x) = 0$, so its general solution is evidently

$$\frac{1}{2}y^2 - \frac{y}{x} = c.$$

In effect, we have found an integrating factor for (1) by noticing in it the combination $x \, dy - y \, dx$ and using (11) to exploit this observation. The following are some other differential formulas that are often useful in similar circumstances:

$$d\left(\frac{x}{y}\right) = \frac{y \, dx - x \, dy}{y^2}; \tag{12}$$

$$d(xy) = x \, dy + y \, dx; \tag{13}$$

$$d(x^2 + y^2) = 2(x \, dx + y \, dy); \tag{14}$$

$$d\left(\tan^{-1}\frac{x}{y}\right) = \frac{y \, dx - x \, dy}{x^2 + y^2}; \tag{15}$$

$$d\left(\log\frac{x}{y}\right) = \frac{y \, dx - x \, dy}{xy}. \tag{16}$$

We see from these formulas that the very simple differential equation $y \, dx - x \, dy = 0$ has $1/x^2$, $1/y^2$, $1/(x^2 + y^2)$, and $1/xy$ as integrating factors, and thus can be solved in this manner in a variety of ways.

Example 2. Find the shape of a curved mirror such that light from a source at the origin will be reflected in a beam of rays parallel to the x-axis.

By symmetry, the mirror will have the shape of the surface of revolution generated by revolving a curve APB (Fig. 13) about the x-axis.

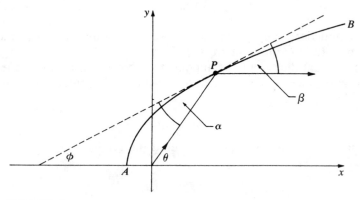

FIGURE 13

It follows from the law of reflection that $\alpha = \beta$. By the geometry of the situation, $\phi = \beta$ and $\theta = \alpha + \phi = 2\beta$. Since $\tan \theta = y/x$ and

$$\tan \theta = \tan 2\beta = \frac{2 \tan \beta}{1 - \tan^2 \beta},$$

we have

$$\frac{y}{x} = \frac{2\, dy/dx}{1 - (dy/dx)^2}.$$

Solving this quadratic equation for dy/dx gives

$$\frac{dy}{dx} = \frac{-x \pm \sqrt{x^2 + y^2}}{y}$$

or

$$x\, dx + y\, dy = \pm\sqrt{x^2 + y^2}\, dx.$$

By using (14), we get

$$\pm \frac{d(x^2 + y^2)}{2\sqrt{x^2 + y^2}} = dx,$$

so

$$\pm\sqrt{x^2 + y^2} = x + c.$$

On simplification this yields

$$y^2 = 2cx + c^2,$$

which is the equation of the family of all parabolas with focus at the origin and axis the x-axis. It is often shown in elementary calculus that all parabolas have this so-called *focal property*. The conclusion of this example is the converse: parabolas are the only curves with this property.

PROBLEMS

1. Show that if $(\partial M/\partial y - \partial N/\partial x)/(Ny - Mx)$ is a function $g(z)$ of the product $z = xy$, then

$$\mu = e^{\int g(z)\, dz}$$

is an integrating factor for equation (2).
2. Sove each of the following equations by finding an integrating factor:
   (a) $(3x^2 - y^2)\, dy - 2xy\, dx = 0$;
   (b) $(xy - 1)\, dx + (x^2 - xy)\, dy = 0$;
   (c) $x\, dy + y\, dx + 3x^3 y^4\, dy = 0$;
   (d) $e^x\, dx + (e^x \cot y + 2y \csc y)\, dy = 0$;
   (e) $(x + 2)\sin y\, dx + x \cos y\, dy = 0$;
   (f) $y\, dx + (x - 2x^2 y^3)\, dy = 0$;
   (g) $(x + 3y^2)\, dx + 2xy\, dy = 0$;
   (h) $y\, dx + (2x - ye^y)\, dy = 0$;
   (i) $(y \log y - 2xy)\, dx + (x + y)\, dy = 0$;
   (j) $(y^2 + xy + 1)\, dx + (x^2 + xy + 1)\, dy = 0$;
   (k) $(x^3 + xy^3)\, dx + 3y^2\, dy = 0$.

3. Under what circumstances will equation (2) have an integrating factor that is a function of the sum $z = x + y$?

4. Solve the following equations by using the differential formulas (12)–(16):

(a) $x \, dy - y \, dx = (1 + y^2) \, dy$;

(b) $y \, dx - x \, dy = xy^3 \, dy$;

(c) $x \, dy = (x^5 + x^3 y^2 + y) \, dx$;

(d) $(y + x) \, dy = (y - x) \, dx$;

(e) $x \, dy = (y + x^2 + 9y^2) \, dx$;

(f) $(y^2 - y) \, dx + x \, dy = 0$;

(g) $x \, dy - y \, dx = (2x^2 - 3) \, dx$;

(h) $x \, dy + y \, dx = \sqrt{xy} \, dy$;

(i) $(y - xy^2) \, dx + (x + x^2 y^2) \, dy = 0$;

(j) $x \, dy - y \, dx = x^2 y^4 (x \, dy + y \, dx)$;

(k) $x \, dy + y \, dx + x^2 y^5 \, dy = 0$;

(l) $(2xy^2 - y) \, dx + x \, dy = 0$;

(m) $dy + \dfrac{y}{x} \, dx = \sin x \, dx$.

5. Solve the following equation by making the substitution $z = y/x^n$ or $y = xz^n$ and choosing a convenient value for n:

$$\frac{dy}{dx} = \frac{2y}{x} + \frac{x^3}{y} + x \tan \frac{y}{x^2}.$$

6. Find the curve APB in Example 2 by using polar coordinates instead of rectangular coordinates. *Hint:* $\psi + \alpha = \pi$.

10 LINEAR EQUATIONS

The most important type of differential equation is the *linear equation,* in which the derivative of highest order is a linear function of the lower order derivatives. Thus the general first order linear equation is

$$\frac{dy}{dx} = p(x)y + q(x),$$

the general second order linear equation is

$$\frac{d^2y}{dx^2} = p(x)\frac{dy}{dx} + q(x)y + r(x),$$

and so on. It is understood that the coefficients on the right in these expressions, namely, $p(x)$, $q(x)$, $r(x)$, etc., are functions of x alone.

Our present concern is with the general first order linear equation, which we write in the standard form

$$\frac{dy}{dx} + P(x)y = Q(x). \tag{1}$$

The simplest method of solving this depends on the observation that

$$\frac{d}{dx}(e^{\int P\,dx}y) = e^{\int P\,dx}\frac{dy}{dx} + yPe^{\int P\,dx} = e^{\int P\,dx}\left(\frac{dy}{dx} + Py\right). \tag{2}$$

Accordingly, if (1) is multiplied through by $e^{\int P\,dx}$, it becomes

$$\frac{d}{dx}(e^{\int P\,dx}y) = Qe^{\int P\,dx}. \tag{3}$$

Integration now yields

$$e^{\int P\,dx}y = \int Qe^{\int P\,dx}\,dx + c,$$

so

$$y = e^{-\int P\,dx}\left(\int Qe^{\int P\,dx}\,dx + c\right) \tag{4}$$

is the general solution of (1).

Example 1. Solve $\dfrac{dy}{dx} + \dfrac{1}{x}y = 3x$.

This equation is obviously linear with $P = 1/x$, so we have

$$\int P\,dx = \int \frac{1}{x}\,dx = \log x \quad \text{and} \quad e^{\int P\,dx} = e^{\log x} = x.$$

On multiplying through by x and remembering (3), we obtain

$$\frac{d}{dx}(xy) = 3x^2,$$

so

$$xy = x^3 + c \quad \text{or} \quad y = x^2 + cx^{-1}.$$

As the method of this example indicates, one should not try to learn the complicated formula (4) and apply it mechanically in solving linear equations. Instead, it is much better to remember and use the procedure by which (4) was derived: *multiply by $e^{\int P\,dx}$ and integrate.* One drawback to the above discussion is that everything hinges on noticing the fact stated in (2). In other words, the integrating factor $e^{\int P\,dx}$ seems to have been plucked mysteriously out of thin air. In Problem 1 below we ask the reader to discover it for himself by the methods of Section 9.

PROBLEMS

1. Write equation (1) in the form $M\,dx + N\,dy = 0$ and use the ideas of Section 9 to show that this equation has an integrating factor μ that is a function of x alone. Find μ and obtain (4) by solving $\mu M\,dx + \mu N\,dy = 0$ as an exact equation.

2. Solve the following as linear equations:

(a) $x\dfrac{dy}{dx} - 3y = x^4$;

(f) $(2y - x^3)\,dx = x\,dy$;

(b) $y' + y = \dfrac{1}{1 + e^{2x}}$;

(g) $y - x + xy\cot x + xy' = 0$;

(c) $(1 + x^2)\,dy + 2xy\,dx = \cot x\,dx$;

(h) $\dfrac{dy}{dx} - 2xy = 6xe^{x^2}$;

(d) $y' + y = 2xe^{-x} + x^2$;

(i) $(x\log x)y' + y = 3x^3$;

(e) $y' + y\cot x = 2x\csc x$;

(j) $(y - 2xy - x^2)\,dx + x^2\,dy = 0$.

3. The equation

$$\frac{dy}{dx} + P(x)y = Q(x)y^n,$$

which is known as *Bernoulli's equation*, is linear when $n = 0$ or 1. Show that it can be reduced to a linear equation for any other value of n by the change of variable $z = y^{1-n}$, and apply this method to solve the following equations:
(a) $xy' + y = x^4y^3$; (c) $x\,dy + y\,dx = xy^2\,dx$.
(b) $xy^2y' + y^3 = x\cos x$;

4. The usual notation dy/dx implies that x is the independent variable and y is the dependent variable. In trying to solve a differential equation, it is sometimes helpful to replace x by y and y by x and work on the resulting equation. Apply this method to the following equations:
(a) $(e^y - 2xy)y' = y^2$; (c) $xy' + 2 = x^3(y - 1)y'$;

(b) $y - xy' = y'y^2e^y$; (d) $f(y)^2\dfrac{dx}{dy} + 3f(y)f'(y)x = f'(y)$.

5. Find the orthogonal trajectories of the family of curves
   (a) $y = x + ce^{-x}$;
   (b) $y^2 = ce^x + x + 1$.

6. We know from (4) that the general solution of a first order linear equation is a family of curves of the form

$$y = cf(x) + g(x).$$

Show, conversely, that the differential equation of any such family is linear.

7. Show that $y' + Py = Qy\log y$ can be solved by the change of variable $z = \log y$, and apply this method to solve $xy' = 2x^2y + y\log y$.

8. One solution of $y'\sin 2x = 2y + 2\cos x$ remains bounded as $x \to \pi/2$. Find it.

9. A tank contains 10 gallons of brine in which 2 pounds of salt are dissolved. Brine containing 1 pound of salt per gallon is pumped into the tank at the rate of 3 gallons/minute, and the stirred mixture is drained off at the rate of 4 gallons/minute. Find the amount $x = x(t)$ of salt in the tank at any time t.

10. A tank contains 40 gallons of pure water. Brine with 3 pounds of salt per gallon flows in at the rate of 2 gallons/minute, and the stirred mixture flows out at 3 gallons/minute.

(a) Find the amount of salt in the tank when the brine in it has been reduced to 20 gallons.
(b) When is the amount of salt in the tank largest?

11. (a) Suppose that a given radioactive element A decomposes into a second radioactive element B, and that B in turn decomposes into a third element C. If the amount of A present initially is x_0, if the amounts of A and B present at a later time t are x and y, respectively, and if k_1 and k_2 are the rate constants of these two reactions, find y as a function of t.

(b) Radon (with a half-life of 3.8 days) is an intensely radioactive gas that is produced as the immediate product of the decay of radium (with a half-life of 1600 years). The atmosphere contains traces of radon near the ground as a result of seepage from soil and rocks, all of which contain minute quantities of radium. There is concern in some parts of the American West about possibly dangerous accumulations of radon in the enclosed basements of houses whose concrete foundations and underlying ground contain appreciably greater quantities of radium than normal because of nearby uranium mining. If the rate constants (fractional losses per unit time, in years) for the decay of radium and radon are $k_1 = 0.00043$ and $k_2 = 66$, use the result of part (a) to determine how long after the completion of a basement the amount of radon will be at a maximum.

11 REDUCTION OF ORDER

As we have seen, the general second order differential equation has the form

$$F(x,y,y',y'') = 0.$$

In this section we consider two special types of second order equations that can be solved by first order methods.

Dependent variable missing. If y is not explicitly present, our equation can be written

$$f(x,y',y'') = 0. \tag{1}$$

In this case we introduce a new dependent variable p by putting

$$y' = p \quad \text{and} \quad y'' = \frac{dp}{dx}. \tag{2}$$

This substitution transforms (1) into the first order equation

$$f\left(x,p,\frac{dp}{dx}\right) = 0. \tag{3}$$

If we can find a solution for (3), we can replace p in this solution by dy/dx and attempt to solve the result. This procedure reduces the problem of solving the second order equation (1) to that of solving two first order equations in succession.

Example 1. Solve $xy'' - y' = 3x^2$.

The variable y is missing from this equation, so (2) reduces it to

$$x\frac{dp}{dx} - p = 3x^2$$

or

$$\frac{dp}{dx} - \frac{1}{x}p = 3x,$$

which is linear. On solving this by the method of Section 10, we obtain

$$p = \frac{dy}{dx} = 3x^2 + c_1 x,$$

so

$$y = x^3 + \frac{1}{2}c_1 x^2 + c_2$$

is the desired solution.

Independent variable missing. If x is not explicitly present, our second order equation can be written

$$g(y, y', y'') = 0. \tag{4}$$

Here we introduce our new dependent variable p in the same way, but this time we express y'' in terms of a derivative with respect to y:

$$y' = p \quad \text{and} \quad y'' = \frac{dp}{dx} = \frac{dp}{dy}\frac{dy}{dx} = p\frac{dp}{dy}. \tag{5}$$

This enables us to write (4) in the form

$$g\left(y, p, p\frac{dp}{dy}\right) = 0; \tag{6}$$

and from this point on we proceed as above, solving two first order equations in succession.

Example 2. Solve $y'' + k^2 y = 0$.

With the aid of (5), we can write this in the form

$$p\frac{dp}{dy} + k^2 y = 0 \quad \text{or} \quad p\,dp + k^2 y\,dy = 0.$$

Integration yields

$$p^2 + k^2 y^2 = k^2 a^2,$$

so

$$p = \frac{dy}{dx} = \pm k\sqrt{a^2 - y^2}$$

or

$$\frac{dy}{\sqrt{a^2 - y^2}} = \pm k\,dx.$$

A second integration gives

$$\sin^{-1}\frac{y}{a} = \pm kx + b,$$

so

$$y = a \sin (\pm kx + b) \qquad \text{or} \qquad y = A \sin (kx + B).$$

This general solution can also be written as

$$y = c_1 \sin kx + c_2 \cos kx, \tag{7}$$

by expanding $\sin (kx + B)$ and changing the form of the constants.

The equation solved in Example 2 occurs quite often in applications (see Section 5). It is linear, and its solution (7) will be fitted into the general theory of second order linear equations in the next chapter.

PROBLEMS

1. Solve the following equations:
   (a) $yy'' + (y')^2 = 0$;
   (b) $xy'' = y' + (y')^3$;
   (c) $y'' - k^2 y = 0$;
   (d) $x^2 y'' = 2xy' + (y')^2$;
   (e) $2yy'' = 1 + (y')^2$;
   (f) $yy'' - (y')^2 = 0$;
   (g) $xy'' + y' = 4x$.

2. Find the specified particular solution of each of the following equations:
   (a) $(x^2 + 2y')y'' + 2xy' = 0$, $y = 1$ and $y' = 0$ when $x = 0$;
   (b) $yy'' = y^2 y' + (y')^2$, $y = -\frac{1}{2}$ and $y' = 1$ when $x = 0$;
   (c) $y'' = y'e^y$, $y = 0$ and $y' = 2$ when $x = 0$.

3. Solve each of the following equations by both methods of this section, and reconcile the results:
   (a) $y'' = 1 + (y')^2$;
   (b) $y'' + (y')^2 = 1$.

4. In Problem 5-8 we considered a hole drilled through the earth from pole to pole and a rock dropped into the hole. This rock will fall through the hole, pause at the other end, and return to its starting point. How long will this complete round trip take?

5. Consider a wire bent into the shape of the cycloid whose parametric equations are $x = a(\theta - \sin \theta)$ and $y = a(1 - \cos \theta)$, and invert it as in Fig. 10. If a bead is released on the wire and slides without friction and under the influence of gravity alone, show that its velocity v satisfies the equation

$$4av^2 = g(s_0^2 - s^2),$$

where s_0 and s are the arc lengths from the bead's lowest point to the bead's initial position and its position at any later time, respectively. By differentiation obtain the equation

$$\frac{d^2 s}{dt^2} + \frac{g}{4a} s = 0,$$

and from this find s as a function of t and determine the period of the motion. Note that these results establish once again the tautochrone property of the cycloid discussed in Problem 6-5.

12 THE HANGING CHAIN. PURSUIT CURVES

We now discuss several applications leading to differential equations that can be solved by the methods of this chapter.

Example 1. Find the shape assumed by a flexible chain suspended between two points and hanging under its own weight.

Let the y-axis pass through the lowest point of the chain (Fig. 14), let s be the arc length from this point to a variable point (x,y), and let $w(s)$ be the linear density of the chain. We obtain the equation of the curve from the fact that the portion of the chain between the lowest point and (x,y) is in equilibrium under the action of three forces: the horizontal tension T_0 at the lowest point; the variable tension T at (x, y), which acts along the tangent because of the flexibility of the chain; and a downward force equal to the weight of the chain between these two points. Equating the horizontal component of T to T_0 and the vertical component of T to the weight of the chain gives

$$T \cos \theta = T_0 \quad \text{and} \quad T \sin \theta = \int_0^s w(s)\, ds.$$

It follows from the first of these equations that

$$T \sin \theta = T_0 \tan \theta = T_0 \frac{dy}{dx},$$

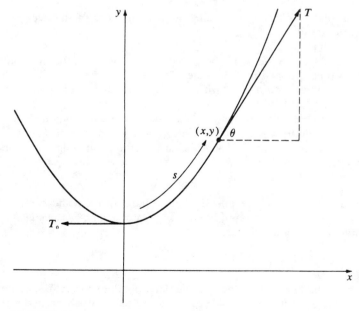

FIGURE 14

so

$$T_0 y' = \int_0^s w(s)\, ds.$$

We eliminate the integral here by differentiating with respect to x:

$$T_0 y'' = \frac{d}{dx} \int_0^s w(s)\, ds = \frac{d}{ds} \int_0^s w(s)\, ds\, \frac{ds}{dx}$$

$$= w(s)\sqrt{1 + (y')^2}.$$

Thus

$$T_0 y'' = w(s)\sqrt{1 + (y')^2} \tag{1}$$

is the differential equation of the desired curve, and the curve itself is found by solving this equation. To proceed further, we must have definite information about the function $w(s)$. We shall solve (1) for the case in which $w(s)$ is a constant w_0, so that

$$y'' = a\sqrt{1 + (y')^2}, \qquad a = \frac{w_0}{T_0}. \tag{2}$$

On substituting $y' = p$ and $y'' = dp/dx$, as in Section 11, equation (2) reduces to

$$\frac{dp}{\sqrt{1 + p^2}} = a\, dx. \tag{3}$$

We now integrate (3) and use the fact that $p = 0$ when $x = 0$ to obtain

$$\log{(p + \sqrt{1 + p^2})} = ax.$$

Solving for p yields

$$p = \frac{dy}{dx} = \frac{1}{2}(e^{ax} - e^{-ax}).$$

If we place the x-axis at the proper height, so that $y = 1/a$ when $x = 0$, we get

$$y = \frac{1}{2a}(e^{ax} + e^{-ax}) = \frac{1}{a}\cosh ax$$

as the equation of the curve assumed by a uniform flexible chain hanging under its own weight. This curve is called a *catenary*, from the Latin word for chain, *catena*. Catenaries also arise in other interesting problems. For instance, it will be shown in Chapter 12 that if an arc joining two given points and lying above the x-axis is revolved about this axis, then the area of the resulting surface of revolution is smallest when the arc is part of a catenary.

Example 2. A point P is dragged along the xy-plane by a string PT of length a. If T starts at the origin and moves along the positive y-axis, and if P starts at $(a,0)$, what is the path of P? This curve is called a *tractrix* (from the Latin *tractum*, meaning drag).

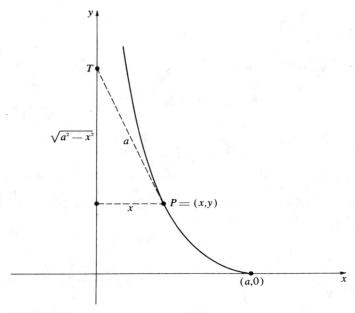

FIGURE 15

It is easy to see from Fig. 15 that the differential equation of the path is

$$\frac{dy}{dx} = -\frac{\sqrt{a^2 - x^2}}{x}.$$

On separating variables and integrating, and using the fact that $y = 0$ when $x = a$, we find that

$$y = a \log\left(\frac{a + \sqrt{a^2 - x^2}}{x}\right) - \sqrt{a^2 - x^2}$$

is the equation of the tractrix. This curve is of considerable importance in geometry, because the trumpet-shaped surface obtained by revolving it about the y-axis is a model for Lobachevsky's version of non-Euclidean geometry, since the sum of the angles of any triangle drawn on the surface is less than 360°. Also, in the context of differential geometry this surface is called a *pseudosphere,* because it has constant negative curvature as opposed to the constant positive curvature of a sphere.

Example 3. A rabbit starts at the origin and runs up the y-axis with speed a. At the same time a dog, running with speed b, starts at the point $(c,0)$ and pursues the rabbit. What is the path of the dog?

At time t, measured from the instant both start, the rabbit will be at the point $R = (0,at)$ and the dog at $D = (x,y)$ (Fig. 16). Since the line DR

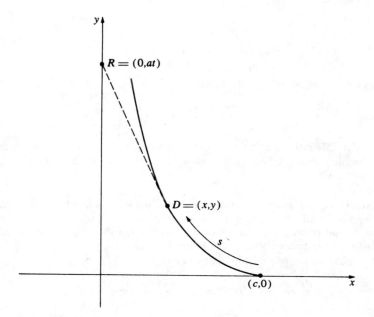

FIGURE 16

is tangent to the path, we have

$$\frac{dy}{dx} = \frac{y - at}{x} \qquad \text{or} \qquad xy' - y = -at. \tag{4}$$

To eliminate t, we begin by differentiating (4) with respect to x, which gives

$$xy'' = -a\frac{dt}{dx}. \tag{5}$$

Since $ds/dt = b$, we have

$$\frac{dt}{dx} = \frac{dt}{ds}\frac{ds}{dx} = -\frac{1}{b}\sqrt{1 + (y')^2}, \tag{6}$$

where the minus sign appears because s increases as x decreases. When (5) and (6) are combined, we obtain the differential equation of the path:

$$xy'' = k\sqrt{1 + (y')^2}, \qquad k = \frac{a}{b}. \tag{7}$$

The substitution $y' = p$ and $y'' = dp/dx$ reduces (7) to

$$\frac{dp}{\sqrt{1 + p^2}} = k\frac{dx}{x};$$

and on integrating and using the initial condition $p = 0$ when $x = c$, we find that

$$\log (p + \sqrt{1 + p^2}) = \log \left(\frac{x}{c}\right)^k.$$

This can readily be solved for p, yielding

$$p = \frac{dy}{dx} = \frac{1}{2}\left[\left(\frac{x}{c}\right)^k - \left(\frac{c}{x}\right)^k\right].$$

In order to continue and find y as a function of x, we must have further information about k. We ask the reader to explore some of the possibilities in Problem 8.

Example 4. The y-axis and the line $x = c$ are the banks of a river whose current has uniform speed a in the negative y-direction. A boat enters the river at the point $(c,0)$ and heads directly toward the origin with speed b relative to the water. What is the path of the boat?

The components of the boat's velocity (Fig. 17) are

$$\frac{dx}{dt} = -b \cos \theta \quad \text{and} \quad \frac{dy}{dt} = -a + b \sin \theta,$$

so

$$\frac{dy}{dx} = \frac{-a + b \sin \theta}{-b \cos \theta} = \frac{-a + b(-y/\sqrt{x^2 + y^2})}{-b(x/\sqrt{x^2 + y^2})}$$

$$= \frac{a\sqrt{x^2 + y^2} + by}{bx}.$$

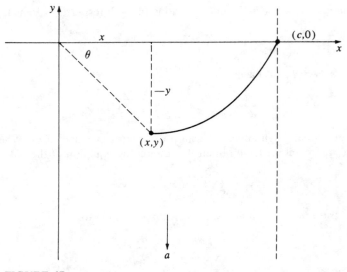

FIGURE 17

This equation is homogeneous, and its solution as found by the method of Section 7 is

$$c^k(y + \sqrt{x^2 + y^2}) = x^{k+1},$$

where $k = a/b$. It is clear that the fate of the boat depends on the relation between a and b. In Problem 9 we ask the reader to discover under what circumstances the boat will be able to land, and where.

PROBLEMS

1. In Example 1, show that the tension T at an arbitrary point (x,y) on the chain is given by $w_0 y$.
2. If the chain in Example 1 supports a load of horizontal density $L(x)$, what differential equation should be used in place of (1)?
3. What is the shape of a cable of negligible density [so that $w(s) = 0$] that supports a bridge of constant horizontal density given by $L(x) = L_0$?
4. If the length of any small portion of an elastic cable of uniform density is proportional to the tension in it, show that it assumes the shape of a parabola when hanging under its own weight.
5. A curtain is made by hanging thin rods from a cord of negligible density. If the rods are close together and equally spaced horizontally, and if the bottom of the curtain is trimmed to be horizontal, what is the shape of the cord?
6. What curve lying above the x-axis has the property that the length of the arc joining any two points on it is proportional to the area under that arc?
7. Show that the tractrix in Example 2 is orthogonal to the lower half of each circle with radius a and center on the positive y-axis.
8. (a) In Example 3, assume that $a < b$ (so that $k < 1$) and find y as a function of x. How far does the rabbit run before the dog catches him?
   (b) Assume that $a = b$ and find y as a function of x. How close does the dog come to the rabbit?
9. In Example 4, solve the equation of the path for y and determine conditions on a and b that will allow the boat to reach the opposite bank. Where will it land?

13 SIMPLE ELECTRIC CIRCUITS

In the present section we consider the linear differential equations that govern the flow of electricity in the simple circuit shown in Fig. 18. This circuit consists of four elements whose action can be understood quite easily without any special knowledge of electricity.

A. A source of electromotive force (emf) E—perhaps a battery or generator—which drives electric charge and produces a current I. Depending on the nature of the source, E may be a constant or a function of time.

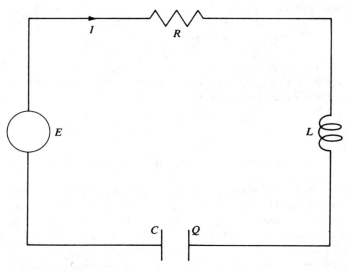

FIGURE 18

B. A resistor of resistance R, which opposes the current by producing a drop in emf of magnitude

$$E_R = RI.$$

This equation is called *Ohm's law.*[2]

C. An inductor of inductance L, which opposes any change in the current by producing a drop in emf of magnitude

$$E_L = L\frac{dI}{dt}.$$

D. A capacitor (or condenser) of capacitance C, which stores the charge Q. The charge accumulated by the capacitor resists the inflow of additional charge, and the drop in emf arising in this way is

$$E_C = \frac{1}{C}Q.$$

---

[2] Georg Simon Ohm (1787–1854) was a German physicist whose only significant contribution to science was his discovery of the law stated above. When he announced it in 1827 it seemed too good to be true, and was not believed. Ohm was considered unreliable because of this, and was so badly treated that he resigned his professorship at Cologne and lived for several years in obscurity and poverty before it was recognized that he was right. One of his pupils in Cologne was Peter Dirichlet, who later became one of the most eminent German mathematicians of the nineteenth century.

Furthermore, since the current is the rate of flow of charge, and hence the rate at which charge builds up on the capacitor, we have

$$I = \frac{dQ}{dt}.$$

Students who are unfamiliar with electric circuits may find it helpful to think of the current I as analogous to the rate of flow of water in a pipe. The electromotive force E plays the role of a pump producing pressure (voltage) that causes the water to flow. The resistance R is analogous to friction in the pipe, which opposes the flow by producing a drop in the pressure. The inductance L is a kind of inertia that opposes any change in the flow by producing a drop in pressure if the flow is increasing and an increase in pressure if the flow is decreasing. The best way to think of the capacitor is to visualize a cylindrical storage tank that the water enters through a hole in the bottom: the deeper the water is in the tank (Q), the harder it is to pump more water in; and the larger the base of the tank is (C) for a given quantity of stored water, the shallower the water is in the tank and the easier it is to pump more water in.

These circuit elements act together in accordance with *Kirchhoff's law,* which states that the algebraic sum of the electromotive forces around a closed circuit is zero.[3] This principle yields

$$E - E_R - E_L - E_C = 0$$

or

$$E - RI - L\frac{dI}{dt} - \frac{1}{C}Q = 0,$$

which we rewrite in the form

$$L\frac{dI}{dt} + RI + \frac{1}{C}Q = E. \tag{1}$$

Depending on the circumstances, we may wish to regard either I or Q as the dependent variable. In the first case, we eliminate Q by differentiating (1) with respect to t and replacing dQ/dt by I:

$$L\frac{d^2I}{dt^2} + R\frac{dI}{dt} + \frac{1}{C}I = \frac{dE}{dt}. \tag{2}$$

---

[3] Gustav Robert Kirchhoff (1824–1887) was another German scientist whose work on electric circuits is familiar to every student of elementary physics. He also established the principles of spectrum analysis and paved the way for the applications of spectroscopy in determining the chemical constitution of the stars.

In the second case, we simply replace I by dQ/dt:

$$L\frac{d^2Q}{dt^2} + R\frac{dQ}{dt} + \frac{1}{C}Q = E. \tag{3}$$

We shall consider these second order linear equations in more detail later. Our concern in this section is primarily with the first order linear equation

$$L\frac{dI}{dt} + RI = E \tag{4}$$

obtained from (1) when no capacitor is present.

Example 1. Solve equation (4) for the case in which an initial current I_0 is flowing and a constant emf E_0 is impressed on the circuit at time $t = 0$.
   For $t \geq 0$, our equation is

$$L\frac{dI}{dt} + RI = E_0.$$

The variables can be separated, yielding

$$\frac{dI}{E_0 - RI} = \frac{1}{L}dt.$$

On integrating and using the initial condition $I = I_0$ when $t = 0$, we get

$$\log(E_0 - RI) = -\frac{R}{L}t + \log(E_0 - RI_0),$$

so

$$I = \frac{E_0}{R} + \left(I_0 - \frac{E_0}{R}\right)e^{-Rt/L}.$$

Note that the current I consists of a *steady-state* part E_0/R and a *transient* part $(I_0 - E_0/R)e^{-Rt/L}$ that approaches zero as t increases. Consequently, Ohm's law $E_0 = RI$ is nearly true for large t. We also observe that if $I_0 = 0$, then

$$I = \frac{E_0}{R}(1 - e^{-Rt/L}),$$

and if $E_0 = 0$, then $I = I_0 e^{-Rt/L}$.

PROBLEMS

1. In Example 1, with $I_0 = 0$ and $E_0 \neq 0$, show that the current in the circuit builds up to half its theoretical maximum in $(L \log 2)/R$ seconds.

2. Solve equation (4) for the case in which the circuit has an initial current I_0 and the emf impressed at time $t = 0$ is given by
   (a) $E = E_0 e^{-kt}$;
   (b) $E = E_0 \sin \omega t$.
3. Consider a circuit described by equation (4) and show that:
   (a) Ohm's law is satisfied whenever the current is at a maximum or minimum.
   (b) The emf is increasing when the current is at a minimum and decreasing when it is at a maximum.
4. If $L = 0$ in equation (3), and if $Q = 0$ when $t = 0$, find the charge buildup $Q = Q(t)$ on the capacitor in each of the following cases:
   (a) E is a constant E_0;
   (b) $E = E_0 e^{-t}$;
   (c) $E = E_0 \cos \omega t$.
5. Use equation (1) with $R = 0$ and $E = 0$ to find $Q = Q(t)$ and $I = I(t)$ for the discharge of a capacitor through an inductor of inductance L, with initial conditions $Q = Q_0$ and $I = 0$ when $t = 0$.

MISCELLANEOUS PROBLEMS FOR CHAPTER 2

Among the following 50 differential equations are representatives of all the types discussed in this chapter, in random order. Many are solvable by several methods. They are presented for the use of students who wish to practice identifying the method or methods applicable to a given equation, without having the hint provided by the title of the section in which the equation occurs.

1. $yy'' = (y')^2$.
2. $(1 - xy)y' = y^2$.
3. $(2x + 3y + 1) \, dx + (2y - 3x + 5) \, dy = 0$.
4. $xy' = \sqrt{x^2 + y^2}$.
5. $y^2 \, dx = (x^3 - xy) \, dy$.
6. $(x^2y^3 + y) \, dx = (x^3y^2 - x) \, dy$.
7. $yy'' + (y')^2 - 2yy' = 0$.
8. $x \, dy + y \, dx = x \cos x \, dx$.
9. $xy \, dy = x^2 \, dy + y^2 \, dx$.
10. $(e^x - 3x^2y^2)y' + ye^x = 2xy^3$.
11. $y'' + 2x(y')^2 = 0$.
12. $(x^2 + y) \, dx = x \, dy$.
13. $xy' + y = x^2 \cos x$.
14. $(6x + 4y + 3) \, dx + (3x + 2y + 2) \, dy = 0$.
15. $\cos (x + y) \, dx = x \sin (x + y) \, dx + x \sin (x + y) \, dy$.
16. $x^2y'' + xy' = 1$.
17. $(y^2e^{xy} + \cos x) \, dx + (e^{xy} + xye^{xy}) \, dy = 0$.
18. $y' \log (x - y) = 1 + \log (x - y)$.
19. $y' + 2xy = e^{-x^2}$.
20. $(y^2 - 3xy - 2x^2) \, dx = (x^2 - xy) \, dy$.

21. $(1 + x^2)y' + 2xy = 4x^3$.

22. $e^x \sin y \, dx + e^x \cos y \, dy = y \sin xy \, dx + x \sin xy \, dy$.

23. $(1 + x^2)y'' + xy' = 0$.

24. $(xe^y + y - x^2) \, dy = (2xy - e^y - x) \, dx$.

25. $e^x(1 + x) \, dx = (xe^x - ye^y) \, dy$.

26. $(x^2y^4 + x^6) \, dx - x^3y^3 \, dy = 0$.

27. $y' = 1 + 3y \tan x$.

28. $\dfrac{dy}{dx} = 1 + \dfrac{y}{x} - \left(\dfrac{y}{x}\right)^2$.

29. $\dfrac{dy}{dx} = \dfrac{2xye^{(x/y)^2}}{y^2 + y^2e^{(x/y)^2} + 2x^2e^{(x/y)^2}}$.

30. $\dfrac{dy}{dx} = \dfrac{x + 2y + 2}{-2x + y}$.

31. $3x^2 \log y \, dx + \dfrac{x^3}{y} \, dy = 0$.

32. $\dfrac{3y^2}{x^2 + 3x} \, dx + \left(2y \log \dfrac{5x}{x + 3} + 3 \sin y\right) dy = 0$.

33. $\dfrac{y - x}{(x + y)^3} \, dx - \dfrac{2x}{(x + y)^3} \, dy = 0$.

34. $(xy^2 + y) \, dx + x \, dy = 0$.

35. $x^2y'' = y'(3x - 2y')$.

36. $(3x^2y - y^3) \, dx - (3xy^2 - x^3) \, dy = 0$.

37. $x(x^2 + 1)y' + 2y = (x^2 + 1)^3$.

38. $\dfrac{dy}{dx} = \dfrac{-3x - 2y - 1}{2x + 3y - 1}$.

39. $e^{x^2y}(1 + 2x^2y) \, dx + x^3e^{x^2y} \, dy = 0$.

40. $(3x^2e^y - 2x) \, dx + (x^3e^y - \sin y) \, dy = 0$.

41. $y^2y'' + (y')^3 = 0$.

42. $(3xy + y^2) \, dx + (3xy + x^2) \, dy = 0$.

43. $x^2y' = x^2 + xy + y^2$.

44. $xy' + y = y^2 \log x$.

45. $\dfrac{\cos y}{x + 3} \, dx - \left(\sin y \log (5x + 15) - \dfrac{1}{y}\right) dy = 0$.

46. $x^2y'' + (y')^2 = 0$.

47. $(xy + y - 1) \, dx + x \, dy = 0$.

48. $x^2y' - y^2 = 2xy$.

49. $y'' = 2y(y')^3$.

50. $\dfrac{dx}{dy} + x \cot y = \sec y$.

51. A tank contains 50 gallons of brine in which 25 pounds of salt are dissolved. Beginning at time $t = 0$, water runs into this tank at the rate of 2 gallons/minute, and the mixture flows out at the same rate through a second tank initially containing 50 gallons of pure water. When will the second tank contain the greatest amount of salt?

52. A natural extension of the first order linear equation

$$y' = p(x) + q(x)y$$

is the *Riccati equation*[4]

$$y' = p(x) + q(x)y + r(x)y^2.$$

In general, this equation cannot be solved by elementary methods. However, if a particular solution $y_1(x)$ is known, then the general solution has the form

$$y(x) = y_1(x) + z(x)$$

where $z(x)$ is the general solution of the Bernoulli equation

$$z' - (q + 2ry_1)z = rz^2.$$

Prove this, and find the general solution of the equation

$$y' = \frac{y}{x} + x^3y^2 - x^5,$$

which has $y_1(x) = x$ as an obvious particular solution.

53. The propagation of a single act in a large population (for example, buying a Japanese- or German-made car) often depends partly on external circumstances (price, quality, and frequency-of-repair records) and partly on a human tendency to imitate other people who have already performed the same act. In this case the rate of increase of the proportion $y(t)$ of people who have performed the act can be expressed by the formula

$$\frac{dy}{dt} = (1 - y)[s(t) + Iy], \qquad (*)$$

---

[4] Count Jacopo Francesco Riccati (1676–1754) was an Italian savant who wrote on mathematics, physics, and philosophy. He was chiefly responsible for introducing the ideas of Newton to Italy. At one point he was offered the presidency of the St. Petersburg Academy of Sciences, but understandably he preferred the leisure and comfort of his aristocratic life in Italy to administrative responsibilities in Russia. Though widely known in scientific circles of this time, he now survives only through the differential equation bearing his name. Even this was an accident of history, for Riccati merely discussed special cases of this equation without offering any solutions, and most of these special cases were successfully treated by various members of the Bernoulli family. The details of this complex story can be found in G. N. Watson, *A Treatise on the Theory of Bessel Functions*, 2d ed., pp. 1–3, Cambridge University Press, London, 1944. The *special Riccati equation* $y' + by^2 = cx^m$ is known to be solvable in finite terms if and only if the exponent m is -2 or of the form $-4k/(2k + 1)$ for some integer k (see Problem 47-8).

where $s(t)$ measures the external stimulus and I is a constant called the imitation coefficient.[5]

(a) Notice that (∗) is a Riccati equation and that $y = 1$ is an obvious solution, and use the result of Problem 52 to find the Bernoulli equation satisfied by $z(t)$.

(b) Find $y(t)$ for the case in which the external stimulus increases steadily with time, so that $s(t) = at$ for a positive constant a. Leave your answer in the form of an integral.

54. (a) If Riccati's equation in Problem 52 has a known solution $y_1(x)$, show that the general solution has the form of the one-parameter family of curves

$$y = \frac{cf(x) + g(x)}{cF(x) + G(x)}.$$

(b) Show, conversely, that the differential equation of any one-parameter family of this form is a Riccati equation.

Dynamical problems with variable mass. In the preceding pages, we have considered many applications of Newton's second law of motion in the form given in Section 1:

$$F = ma,$$

where F is the force acting on a body of mass m whose acceleration is a. It should be realized, however, that this formulation applies only to situations in which the mass is constant. Newton's law is actually somewhat more general, and states that when a force F acts on a body of mass m, it produces momentum (mv, where v is the velocity) at a rate equal to the force:

$$F = \frac{d}{dt}(mv).$$

This equation reduces to $F = ma$ when m is constant. In applying this form of the law to a moving body with variable mass, it is necessary to distinguish momentum produced by F from momentum produced by mass joining the body from an outside source. Thus, if mass with velocity $v + w$ (so that w is its velocity relative to m) is being added to m at the rate dm/dt, the effect of F in increasing momentum must be supplemented by $(v + w) \, dm/dt$, giving

$$(v + w)\frac{dm}{dt} + F = \frac{d}{dt}(mv),$$

---

[5] See Anatol Rapoport, "Contribution to the Mathematical Theory of Mass Behavior: I. The Propagation of Single Acts," *Bulletin of Mathematical Biophysics*, Vol. 14, pp. 159–169 (1952).

which simplifies to

$$w\frac{dm}{dt} + F = m\frac{dv}{dt}.$$

We note that dm/dt is positive or negative according as the body is gaining or losing mass, and that w is positive or negative depending on the motion of the mass gained or lost relative to m. The following problems provide several illustrations of these ideas.

55. A rocket of structural mass m_1 contains fuel of initial mass m_2. It is fired straight up from the surface of the earth by burning fuel at a constant rate a (so that $dm/dt = -a$ where m is the variable total mass of the rocket) and expelling the exhaust products backward at a constant velocity b relative to the rocket. Neglecting all external forces except a gravitational force mg, where g is assumed constant, find the velocity and height attained at the moment when the fuel is exhausted (the *burnout velocity* and *burnout height*).[6]

56. A spherical raindrop, starting from rest, falls under the influence of gravity. If it gathers in water vapor (assumed at rest) at a rate proportional to its surface, and if its initial radius is 0, show that it falls with constant acceleration $g/4$.

57. If the initial radius of the raindrop in Problem 56 is r_0 and r is its radius at time t, show that its acceleration at time t is

$$\frac{g}{4}\left(1 + \frac{3r_0^4}{r^4}\right).$$

Thus the acceleration is constant—with value $g/4$—if and only if the raindrop has zero initial radius.

58. A spherical raindrop, starting from rest, falls through a uniform mist. If it gathers in water droplets in its path (assumed at rest) as it moves, and if its initial radius is 0, show that it falls with constant acceleration $g/7$.

59. Einstein's special theory of relativity asserts that the mass m of a particle moving with velocity v is given by the formula

$$m = \frac{m_0}{\sqrt{1 - v^2/c^2}}, \qquad (*)$$

where c is the velocity of light and m_0 is the rest mass.

---

[6] The experience of engineering experts strongly suggests that no foreseeable combination of fuel and rocket design will enable a rocket, starting from rest, to acquire a burnout velocity as large as the escape velocity $\sqrt{2gR}$. This means that single-stage rockets of this kind cannot be used for journeys into space from the surface of the earth, and all such journeys will continue to require the multistage rockets familiar to us from recent decades.

(a) If the particle starts from rest in empty space and moves for a long time under the influence of a constant gravitational field, find v as a function of time by taking $w = -v$, and show that $v \to c$ as $t \to \infty$.[7]

(b) Let $M = m - m_0$ be the increase in the mass of the particle. If the corresponding increase E in its energy is taken to be the work done on it by the prevailing force F, so that

$$E = \int_0^v F \, dx = \int_0^v \frac{d}{dt}(mv) \, dx = \int_0^v v \, d(mv),$$

verify that

$$E = Mc^2. \tag{**}$$

(c) Deduce (*) from (**).

---

[7] Enrico Fermi has suggested that the phenomenon described here, transferred to the case of charged particles of interstellar dust accelerated by the magnetic fields of stars, can account in part for the origin of primary cosmic rays.

CHAPTER 3

SECOND
ORDER
LINEAR
EQUATIONS

14 INTRODUCTION

In the preceding chapters we studied a few restricted types of differential equations that can be solved in terms of familiar elementary functions. The methods we developed require considerable skill in the techniques of integration, and their many interesting applications have a tasty flavor of practicality. Unfortunately, however, it must be admitted that this part of the subject tends to be a miscellaneous bag of tricks, and conveys little insight into the general nature of differential equations and their solutions. In the present chapter we discuss an important class of equations with a rich and far-reaching theory. We shall see that this theory can be given a coherent and satisfying structure based on a few simple principles.

The general second order linear differential equation is

$$\frac{d^2y}{dx^2} + P(x)\frac{dy}{dx} + Q(x)y = R(x),$$

or, more simply,

$$y'' + P(x)y' + Q(x)y = R(x). \tag{1}$$

As the notation indicates, it is understood that $P(x)$, $Q(x)$, and $R(x)$ are functions of x alone (or perhaps constants). It is clear that no loss of generality results from taking the coefficient of y'' to be 1, since this can always be accomplished by division. Equations of this kind are of great significance in physics, especially in connection with vibrations in mechanics and the theory of electric circuits. In addition—as we shall see in later chapters—many profound and beautiful ideas in pure mathematics have grown out of the study of these equations.

We should not be misled by the fact that first order linear equations are easily solved by means of formulas. In general, (1) cannot be solved explicitly in terms of known elementary functions, or even in terms of indicated integrations. To find solutions, it is commonly necessary to resort to infinite processes of one kind or another, usually infinite series. Many special equations of particular importance in applications, for instance those of Legendre and Bessel mentioned in Section 1, have been studied at great length; and the theory of a single such equation has often been found so complicated as to constitute by itself an entire department of analysis. We shall discuss these matters in Chapters 5 and 8.

In this chapter our detailed consideration of actual methods for solving (1) will be restricted, for the most part, to the special case in which the coefficients $P(x)$ and $Q(x)$ are constants. It should also be emphasized that most of the ideas and procedures we discuss can be generalized at once to linear equations of higher order, with no change in the underlying principles but only an increasing complexity in the surrounding details. By restricting ourselves for the most part to second order equations, we attain as much simplicity as possible without distorting the main ideas, and yet we still have enough generality to include all the linear equations of greatest interest in mathematics and physics.

Since in general it is not possible to produce an explicit solution of (1) for inspection, our first order of business is to assure ourselves that this equation really has a solution. The following existence and uniqueness theorem is proved in Chapter 13.

Theorem A. *Let $P(x)$, $Q(x)$, and $R(x)$ be continuous functions on a closed interval $[a,b]$.*[1] *If x_0 is any point in $[a,b]$, and if y_0 and y_0' are any numbers whatever, then equation (1) has one and only one solution $y(x)$ on the entire interval such that $y(x_0) = y_0$ and $y'(x_0) = y_0'$.*

---

[1] If a and b are real numbers such that $a < b$, then the symbol $[a,b]$ denotes the interval consisting of all real numbers x that satisfy the inequalities $a \leq x \leq b$. This interval is called *closed* because it contains its endpoints. The *open* interval resulting from the exclusion of the endpoints is denoted by (a,b) and is defined by the inequalities $a < x < b$.

Thus, under these hypotheses, at any given point x_0 in $[a,b]$ we can arbitrarily prescribe the values of $y(x)$ and $y'(x)$, and there will then exist precisely one solution of (1) on $[a,b]$ that assumes the prescribed values at the given point; or, more geometrically, (1) has a unique solution on $[a,b]$ that passes through a specified point (x_0,y_0) with a specified slope y_0'. In our general discussions through the remainder of this chapter, we shall always assume (without necessarily saying so explicitly) that the hypotheses of Theorem A are satisfied.

Example 1. Find the solution of the initial value problem

$$y'' + y = 0, \qquad y(0) = 0 \text{ and } y'(0) = 1.$$

We know that $y = \sin x$, $y = \cos x$, and more generally $y = c_1 \sin x + c_2 \cos x$ for any constants c_1 and c_2, are all solutions of the differential equation. Also, $y = \sin x$ clearly satisfies the initial conditions, because $\sin 0 = 0$ and $\cos 0 = 1$. By Theorem A, $y = \sin x$ is the *only* solution of the given initial value problem, and is therefore completely characterized as a function by this problem. In just the same way, the function $y = \cos x$ is easily seen to be a solution, and therefore the only solution, of the corresponding initial value problem

$$y'' + y = 0, \qquad y(0) = 1 \text{ and } y'(0) = 0.$$

Since all of trigonometry can be regarded as the development of the properties of these two functions, it follows that all of trigonometry is contained by implication (as the acorn contains the oak tree) within the two initial value problems stated above. We shall examine this remarkable idea in greater detail in Chapter 4.

We emphasize again that in Theorem A the initial conditions that determine a unique solution of equation (1) are conditions on the value of the solution and its first derivative at a single fixed point x_0 in the interval $[a,b]$. In contrast to this, the problem of finding a solution of equation (1) that satisfies conditions of the form $y(x_0) = y_0$ and $y(x_1) = y_1$, where x_0 and x_1 are different points in the interval, is not covered by Theorem A. Problems of this kind are called *boundary value problems*, and are discussed in Chapter 7.

The term $R(x)$ in equation (1) is isolated from the others and written on the right because it does not contain the dependent variable y or any of its derivatives. If $R(x)$ is identically zero, then (1) reduces to the *homogeneous equation*

$$y'' + P(x)y' + Q(x)y = 0. \tag{2}$$

(This traditional use of the word *homogeneous* should not be confused with the equally traditional but totally different use given in Section 7.) If $R(x)$ is not identically zero, then (1) is said to be *nonhomogeneous*.

In studying the nonhomogeneous equation (1) it is necessary to consider along with it the homogeneous equation (2) obtained from it by replacing $R(x)$ by 0. Under these circumstances (1) is often called the *complete equation,* and (2) the *reduced equation* associated with it. The reason for this linkage between (1) and (2) is easy to understand, as follows.

Suppose that in some way we know that $y_g(x,c_1,c_2)$ is the general solution of (2)—we expect it to contain two arbitrary constants since the equation is of the second order—and that $y_p(x)$ is a fixed particular solution of (1). If $y(x)$ is any solution whatever of (1), then an easy calculation shows that $y(x) - y_p(x)$ is a solution of (2):

$$
\begin{aligned}
(y - y_p)'' &+ P(x)(y - y_p)' + Q(x)(y - y_p) \\
&= [y'' + P(x)y' + Q(x)y] - [y_p'' + P(x)y_p' + Q(x)y_p] \\
&= R(x) - R(x) = 0. \tag{3}
\end{aligned}
$$

Since $y_g(x,c_1,c_2)$ is the general solution of (2), it follows that $y(x) - y_p(x) = y_g(x,c_1,c_2)$ or

$$
y(x) = y_g(x,c_1,c_2) + y_p(x)
$$

for a suitable choice of the constants c_1 and c_2. This argument proves the following theorem.

> **Theorem B.** *If y_g is the general solution of the reduced equation (2) and y_p is any particular solution of the complete equation (1), then $y_g + y_p$ is the general solution of (1).*

We shall see in Section 19 that if y_g is known, then a formal procedure is available for finding y_p. This shows that the central problem in the theory of linear equations is that of solving the homogeneous equation. Accordingly, most of our attention will be devoted to studying the structure of y_g and investigating various ways of determining its explicit form—none of which is effective in all cases.

The first thing we should notice about the homogeneous equation (2) is that the function $y(x)$ which is identically zero—that is, $y(x) = 0$ for all x—is always a solution. This is called the *trivial solution,* and is usually of no interest. The basic structural fact about solutions of (2) is given in the following theorem.

> **Theorem C.** *If $y_1(x)$ and $y_2(x)$ are any two solutions of (2), then*
>
> $$
> c_1 y_1(x) + c_2 y_2(x) \tag{4}
> $$
>
> *is also a solution for any constants c_1 and c_2.*

Proof. The statement follows immediately from the fact that

$$(c_1 y_1 + c_2 y_2)'' + P(x)(c_1 y_1 + c_2 y_2)' + Q(x)(c_1 y_1 + c_2 y_2)$$
$$= (c_1 y_1'' + c_2 y_2'') + P(x)(c_1 y_1' + c_2 y_2') + Q(x)(c_1 y_1 + c_2 y_2)$$
$$= c_1[y_1'' + P(x)y_1' + Q(x)y_1] + c_2[y_2'' + P(x)y_2' + Q(x)y_2]$$
$$= c_1 \cdot 0 + c_2 \cdot 0 = 0, \tag{5}$$

where the multipliers of c_1 and c_2 are zero because, by assumption, y_1 and y_2 are solutions of (2).

For reasons connected with the elementary algebra of vectors, the solution (4) is commonly called a *linear combination* of the solutions $y_1(x)$ and $y_2(x)$. If we use this terminology, Theorem C can be restated as follows: *any linear combination of two solutions of the homogeneous equation* (2) *is also a solution.*

Suppose that by some means or other we have managed to find two solutions of equation (2). Then this theorem provides us with another which involves two arbitrary constants, and which therefore may be the general solution of (2). There is one difficulty: if either y_1 or y_2 is a constant multiple of the other, say $y_1 = ky_2$, then

$$c_1 y_1 + c_2 y_2 = c_1 k y_2 + c_2 y_2 = (c_1 k + c_2)y_2 = cy_2,$$

and only one essential constant is present. On this basis we have reasonable grounds for hoping that if neither y_1 nor y_2 is a constant multiple of the other, then

$$c_1 y_1(x) + c_2 y_2(x)$$

will be the general solution of (2). We shall prove this in the next section.

Occasionally the special form of a linear equation enables us to find simple particular solutions by inspection or by experimenting with power, exponential, or trigonometric functions.

Example 2. Solve

$$y'' + y' = 0.$$

By inspection we see that $y_1 = 1$ and $y_2 = e^{-x}$ are solutions. It is obvious that neither function is a constant multiple of the other, so (assuming the theorem stated above, but not yet proved) we conclude that

$$y = c_1 + c_2 e^{-x}$$

is the general solution.

Example 3. Solve

$$x^2 y'' + 2xy' - 2y = 0.$$

Since differentiating a power pushes down the exponent by one unit, the form of this equation suggests that we look for possible solutions of the type $y = x^n$. On substituting this in the differential equation and dividing by the common factor x^n, we obtain the quadratic equation $n(n - 1) + 2n - 2 = 0$ or $n^2 + n - 2 = 0$. This has roots $n = 1$, -2, so $y_1 = x$ and $y_2 = x^{-2}$ are solutions and

$$y = c_1 x + c_2 x^{-2}$$

is the general solution on any interval not containing the origin.

It is worth remarking at this point that a large part of the theory of linear equations rests on the fundamental properties stated in Theorems B and C. An inspection of the calculations (3) and (5) will show at once that these properties in turn depend on the *linearity of differentiation*, that is, on the fact that

$$[\alpha f(x) + \beta g(x)]' = \alpha f'(x) + \beta g'(x)$$

for all constants α and β and all differentiable functions $f(x)$ and $g(x)$.

PROBLEMS

In the following problems, assume the fact stated above (but not yet proved), that if $y_1(x)$ and $y_2(x)$ are two solutions of (2) and neither is a constant multiple of the other, then $c_1 y_1(x) + c_2 y_2(x)$ is the general solution.

1. (a) Verify that $y_1 = 1$ and $y_2 = x^2$ are solutions of the reduced equation $xy'' - y' = 0$, and write down the general solution.
   (b) Determine the value of a for which $y_p = ax^3$ is a particular solution of the complete equation $xy'' - y' = 3x^2$. Use this solution and the result of part (a) to write down the general solution of this equation. (Compare with Example 1 in Section 11.)
   (c) Can you discover y_1, y_2, and y_p by inspection?

2. Verify that $y_1 = 1$ and $y_2 = \log x$ are solutions of the equation $xy'' + y' = 0$, and write down the general solution. Can you discover y_1 and y_2 by inspection?

3. (a) Show that $y_1 = e^{-x}$ and $y_2 = e^{2x}$ are solutions of the reduced equation $y'' - y' - 2y = 0$. What is the general solution?
   (b) Find a and b so that $y_p = ax + b$ is a particular solution of the complete equation $y'' - y' - 2y = 4x$. Use this solution and the result of part (a) to write down the general solution of this equation.

4. Use inspection or experiment to find a particular solution for each of the following equations:
   (a) $x^3 y'' + x^2 y' + xy = 1$;   (c) $y'' - 2y = \sin x$.
   (b) $y'' - 2y' = 6$;

5. In each of the following cases, use inspection or experiment to find particular solutions of the reduced and complete equations and write down the general

solution:

(a) $y'' = e^x$;

(b) $y'' - 2y' = 4$;

(c) $y'' - y = \sin x$;

(d) $(x - 1)y'' - xy' + y = 0$;

(e) $y'' + 2y' = 6e^x$.

6. By eliminating the constants c_1 and c_2, find the differential equation of each of the following families of curves:

(a) $y = c_1x + c_2x^2$;

(b) $y = c_1e^{kx} + c_2e^{-kx}$;

(c) $y = c_1 \sin kx + c_2 \cos kx$;

(d) $y = c_1 + c_2e^{-2x}$;

(e) $y = c_1x + c_2 \sin x$;

(f) $y = c_1e^x + c_2xe^x$;

(g) $y = c_1e^x + c_2e^{-3x}$;

(h) $y = c_1x + c_2x^{-1}$.

7. Verify that $y = c_1x^{-1} + c_2x^5$ is a solution of

$$x^2y'' - 3xy - 5y = 0$$

on any interval $[a,b]$ that does not contain the origin. If $x_0 \neq 0$, and if y_0 and y_0' are arbitrary, show directly that c_1 and c_2 can be chosen in one and only one way so that $y(x_0) = y_0$ and $y'(x_0) = y_0'$.

8. Show that $y = x^2 \sin x$ and $y = 0$ are both solutions of

$$x^2y'' - 4xy' + (x^2 + 6)y = 0,$$

and that both satisfy the conditions $y(0) = 0$ and $y'(0) = 0$. Does this contradict Theorem A? If not, why not?

9. If a solution of equation (2) on an interval $[a,b]$ is tangent to the x-axis at any point of this interval, then it must be identically zero. Why?

10. If $y_1(x)$ and $y_2(x)$ are two solutions of equation (2) on an interval $[a,b]$, and have a common zero in this interval, show that one is a constant multiple of the other. [Recall that a point x_0 is said to be a *zero* of a function $f(x)$ if $f(x_0) = 0$.]

15 THE GENERAL SOLUTION OF THE HOMOGENEOUS EQUATION

If two functions $f(x)$ and $g(x)$ are defined on an interval $[a,b]$ and have the property that one is a constant multiple of the other, then they are said to be *linearly dependent* on $[a,b]$. Otherwise—that is, if neither is a constant multiple of the other—they are called *linearly independent*. It is worth noting that if $f(x)$ is identically zero, then $f(x)$ and $g(x)$ are linearly dependent for every function $g(x)$, since $f(x) = 0 \cdot g(x)$.

Our purpose in this section is to prove the following theorem.

Theorem A. *Let $y_1(x)$ and $y_2(x)$ be linearly independent solutions of the homogeneous equation*

$$y'' + P(x)y' + Q(x)y = 0 \tag{1}$$

on the interval $[a,b]$. Then

$$c_1y_1(x) + c_2y_2(x) \tag{2}$$

is the general solution of equation (1) *on* $[a,b]$, *in the sense that every solution of* (1) *on this interval can be obtained from* (2) *by a suitable choice of the arbitrary constants c_1 and c_2.*

The proof will be given in stages, by means of several lemmas and auxiliary ideas.

Let $y(x)$ be any solution of (1) on $[a,b]$. We must show that constants c_1 and c_2 can be found so that

$$y(x) = c_1 y_1(x) + c_2 y_2(x)$$

for all x in $[a,b]$. By Theorem 14-A, a solution of (1) over all of $[a,b]$ is completely determined by its value and the value of its derivative at a single point. Consequently, since $c_1 y_1(x) + c_2 y_2(x)$ and $y(x)$ are both solutions of (1) on $[a,b]$, it suffices to show that for some point x_0 in $[a,b]$ we can find c_1 and c_2 so that

$$c_1 y_1(x_0) + c_2 y_2(x_0) = y(x_0)$$

and

$$c_1 y_1'(x_0) + c_2 y_2'(x_0) = y'(x_0).$$

For this system to be solvable for c_1 and c_2, it suffices that the determinant

$$\begin{vmatrix} y_1(x_0) & y_2(x_0) \\ y_1'(x_0) & y_2'(x_0) \end{vmatrix} = y_1(x_0)y_2'(x_0) - y_2(x_0)y_1'(x_0)$$

have a value different from zero. This leads us to investigate the function of x defined by

$$W(y_1, y_2) = y_1 y_2' - y_2 y_1',$$

which is known as the *Wronskian*[2] of y_1 and y_2, with special reference to whether it vanishes at x_0. Our first lemma simplifies this problem by showing that the location of the point x_0 is of no consequence.

Lemma 1. *If $y_1(x)$ and $y_2(x)$ are any two solutions of equation* (1) *on* $[a,b]$, *then their Wronskian $W = W(y_1,y_2)$ is either identically zero or never zero on* $[a,b]$.

---

[2] Hoëné Wronski (1778–1853) was an impecunious Pole of erratic personality who spent most of his life in France. The Wronskian determinant mentioned above was his sole contribution to mathematics. He was the only Polish mathematician of the nineteenth century whose name is remembered today, which is a little surprising in view of the many eminent men in this field whom Poland has given to the twentieth century.

Proof. We begin by observing that

$$W' = y_1 y_2'' + y_1' y_2' - y_2 y_1'' - y_2' y_1'$$
$$= y_1 y_2'' - y_2 y_1''.$$

Next, since y_1 and y_2 are both solutions of (1), we have

$$y_1'' + P y_1' + Q y_1 = 0$$

and

$$y_2'' + P y_2' + Q y_2 = 0.$$

On multiplying the first of these equations by y_2 and the second by y_1, and subtracting the first from the second, we obtain

$$(y_1 y_2'' - y_2 y_1'') + P(y_1 y_2' - y_2 y_1') = 0$$

or

$$\frac{dW}{dx} + PW = 0.$$

The general solution of this first order equation is

$$W = ce^{-\int P\,dx};$$

and since the exponential factor is never zero we see that W is identically zero if the constant $c = 0$, and never zero if $c \neq 0$, and the proof is complete.[3]

This result reduces our overall task of proving the theorem to that of showing that the Wronskian of any two linearly independent solutions of (1) is not identically zero. We accomplish this in our next lemma, which actually yields a bit more than is needed.

Lemma 2. *If $y_1(x)$ and $y_2(x)$ are two solutions of equation (1) on $[a,b]$, then they are linearly dependent on this interval if and only if their Wronskian $W(y_1,y_2) = y_1 y_2' - y_2 y_1'$ is identically zero.*

Proof. We begin by assuming that y_1 and y_2 are linearly dependent, and we show as a consequence of this that $y_1 y_2' - y_2 y_1' = 0$. First, if either function is identically zero on $[a,b]$, then the conclusion is clear. We may therefore assume, without loss of generality, that neither is identically zero; and it follows from this and their linear dependence that each is a constant multiple of the other. Accordingly we have $y_2 = cy_1$ for some constant c, so $y_2' = cy_1'$. These equations enable us to write

$$y_1 y_2' - y_2 y_1' = y_1(cy_1') - (cy_1)y_1'$$
$$= 0,$$

which proves this half of the lemma.

---

[3] Formula (3) is due to the great Norwegian mathematician Niels Henrik Abel (see Appendix B in Chapter 9), and is called *Abel's formula*.

We now assume that the Wronskian is identically zero and prove linear dependence. If y_1 is identically zero on $[a,b]$, then (as we remarked at the beginning of the section) the functions are linearly dependent. We may therefore assume that y_1 does not vanish identically on $[a,b]$, from which it follows by continuity that y_1 does not vanish at all on some subinterval $[c,d]$ of $[a,b]$. Since the Wronskian is identically zero on $[a,b]$, we can divide it by y_1^2 to get

$$\frac{y_1 y_2' - y_2 y_1'}{y_1^2} = 0$$

on $[c,d]$. This can be written in the form $(y_2/y_1)' = 0$, and by integrating we obtain $y_2/y_1 = k$ or $y_2(x) = ky_1(x)$ for some constant k and all x in $[c,d]$. Finally, since $y_2(x)$ and $ky_1(x)$ have equal values in $[c,d]$, they have equal derivatives there as well; and Theorem 14-A allows us to infer that

$$y_2(x) = ky_1(x)$$

for all x in $[a,b]$, which concludes the argument.

With this lemma, the proof of Theorem A is complete.

Ordinarily, the simplest way of showing that two solutions of (1) are linearly independent over an interval is to show that their ratio is not constant there, and in most cases this is easily determined by inspection. On occasion, however, it is convenient to employ the formal test embodied in Lemma 2: compute the Wronskian, and show that it does not vanish. Both procedures are illustrated in the following example.

Example 1. Show that $y = c_1 \sin x + c_2 \cos x$ is the general solution of $y'' + y = 0$ on any interval, and find the particular solution for which $y(0) = 2$ and $y'(0) = 3$.

The fact that $y_1 = \sin x$ and $y_2 = \cos x$ are solutions is easily verified by substitution. Their linear independence on any interval $[a,b]$ follows from the observation that $y_1/y_2 = \tan x$ is not constant, and also from the fact that their Wronskian never vanishes:

$$W(y_1, y_2) = \begin{vmatrix} \sin x & \cos x \\ \cos x & -\sin x \end{vmatrix} = -\sin^2 x - \cos^2 x = -1.$$

Since $P(x) = 0$ and $Q(x) = 1$ are continuous on $[a,b]$, it now follows from Theorem A that $y = c_1 \sin x + c_2 \cos x$ is the general solution of the given equation on $[a,b]$. Furthermore, since the interval $[a,b]$ can be expanded indefinitely without introducing points at which $P(x)$ or $Q(x)$ is discontinuous, this general solution is valid for all x. To find the required particular solution, we solve the system

$$c_1 \sin 0 + c_2 \cos 0 = 2,$$
$$c_1 \cos 0 - c_2 \sin 0 = 3.$$

This yields $c_2 = 2$ and $c_1 = 3$, so $y = 3 \sin x + 2 \cos x$ is the particular solution that satisfies the given conditions.

The concepts of linear dependence and independence are significant in a much wider context than appears here. As the reader is perhaps already aware, the important branch of mathematics known as *linear algebra* is in essence little more than an abstract treatment of these concepts, with many applications to algebra, geometry, and analysis.

PROBLEMS

In Problems 1 to 7, use Wronskians to establish linear independence.

1. Show that e^x and e^{-x} are linearly independent solutions of $y'' - y = 0$ on any interval.

2. Show that $y = c_1 x + c_2 x^2$ is the general solution of

$$x^2 y'' - 2xy' + 2y = 0$$

on any interval not containing 0, and find the particular solution for which $y(1) = 3$ and $y'(1) = 5$.

3. Show that $y = c_1 e^x + c_2 e^{2x}$ is the general solution of

$$y'' - 3y' + 2y = 0$$

on any interval, and find the particular solution for which $y(0) = -1$ and $y'(0) = 1$.

4. Show that $y = c_1 e^{2x} + c_2 x e^{2x}$ is the general solution of

$$y'' - 4y' + 4y = 0$$

on any interval.

5. By inspection or experiment, find two linearly independent solutions of $x^2 y'' - 2y = 0$ on the interval $[1,2]$, and determine the particular solution satisfying the initial conditions $y(1) = 1$, $y'(1) = 8$.

6. In each of the following, verify that the functions $y_1(x)$ and $y_2(x)$ are linearly independent solutions of the given differential equation on the interval $[0,2]$, and find the solution satisfying the stated initial conditions:
   (a) $y'' + y' - 2y = 0$, $y_1 = e^x$ and $y_2 = e^{-2x}$, $y(0) = 8$ and $y'(0) = 2$;
   (b) $y'' + y' - 2y = 0$, $y_1 = e^x$ and $y_2 = e^{-2x}$, $y(1) = 0$ and $y'(1) = 0$;
   (c) $y'' + 5y' + 6y = 0$, $y_1 = e^{-2x}$ and $y_2 = e^{-3x}$, $y(0) = 1$ and $y'(0) = 1$;
   (d) $y'' + y' = 0$, $y_1 = 1$ and $y_2 = e^{-x}$, $y(2) = 0$ and $y'(2) = e^{-2}$.

7. (a) Use one (or both) of the methods described in Section 11 to find all solutions of $y'' + (y')^2 = 0$.
   (b) Verify that $y_1 = 1$ and $y_2 = \log x$ are linearly independent solutions of the equation in part (a) on any interval to the right of the origin. Is $y = c_1 + c_2 \log x$ the general solution? If not, why not?

8. Use the Wronskian to prove that two solutions of the homogeneous equation (1) on an interval $[a,b]$ are linearly dependent if
   (a) they have a common zero x_0 in the interval (Problem 14-10);
   (b) they have maxima or minima at the same point x_0 in the interval.

9. Consider the two functions $f(x) = x^3$ and $g(x) = x^2|x|$ on the interval $[-1, 1]$.
   (a) Show that their Wronskian $W(f, g)$ vanishes identically.
   (b) Show that f and g are not linearly dependent.
   (c) Do (a) and (b) contradict Lemma 2? If not, why not?

10. It is clear that $\sin x$, $\cos x$ and $\sin x$, $\sin x - \cos x$ are two distinct pairs of linearly independent solutions of $y'' + y = 0$. Thus, if y_1 and y_2 are linearly independent solutions of the homogeneous equation

$$y'' + P(x)y' + Q(x)y = 0,$$

   we see that y_1 and y_2 are *not* uniquely determined by the equation.
   (a) Show that

$$P(x) = -\frac{y_1 y_2'' - y_2 y_1''}{W(y_1, y_2)}$$

   and

$$Q(x) = \frac{y_1' y_2'' - y_2' y_1''}{W(y_1, y_2)},$$

   so that the equation *is* uniquely determined by any given pair of linearly independent solutions.
   (b) Use (a) to reconstruct the equation $y'' + y = 0$ from each of the two pairs of linearly independent solutions mentioned above.
   (c) Use (a) to reconstruct the equation in Problem 4 from the pair of linearly independent solutions e^{2x}, xe^{2x}.

11. (a) Show that by applying the substitution $y = uv$ to the homogeneous equation (1) it is possible to obtain a homogeneous second order linear equation for v with no v' term present. Find u and the equation for v in terms of the original coefficients $P(x)$ and $Q(x)$.
    (b) Use the method of part (a) to find the general solution of $y'' + 2xy' + (1 + x^2)y = 0$.

16   THE USE OF A KNOWN SOLUTION TO FIND ANOTHER

As we have seen, it is easy to write down the general solution of the homogeneous equation

$$y'' + P(x)y' + Q(x)y = 0 \tag{1}$$

whenever we know two linearly independent solutions $y_1(x)$ and $y_2(x)$. But how do we find y_1 and y_2? Unfortunately there is no general method for doing this. However, there does exist a standard procedure for determining y_2 when y_1 is known. This is of considerable importance, for in many cases a single solution of (1) can be found by inspection or some other device.

To develop this procedure, we assume that $y_1(x)$ is a known nonzero solution of (1), so that $cy_1(x)$ is also a solution for any constant *c.* The basic idea is to replace the constant c by an unknown function $v(x)$, and then to attempt to determine v in such a manner that $y_2 = vy_1$ will be a solution of (1). It isn't at all clear in advance that this approach will work, but it does. To see how we might think of trying it, recall that the linear independence of the two solutions y_1 and y_2 requires that the ratio y_2/y_1 must be a nonconstant function of x, say v; and if we can find v, then since we know y_1 we have y_2 and our problem is solved.

We assume, then, that $y_2 = vy_1$ is a solution of (1), so that

$$y_2'' + Py_2' + Qy_2 = 0, \tag{2}$$

and we try to discover the unknown function $v(x)$. On substituting $y_2 = vy_1$ and the expressions

$$y_2' = vy_1' + v'y_1 \quad \text{and} \quad y_2'' = vy_1'' + 2v'y_1' + v''y_1$$

into (2) and rearranging, we get

$$v(y_1'' + Py_1' + Qy_1) + v''y_1 + v'(2y_1' + Py_1) = 0.$$

Since y_1 is a solution of (1), this reduces to

$$v''y_1 + v'(2y_1' + Py_1) = 0$$

or

$$\frac{v''}{v'} = -2\frac{y_1'}{y_1} - P.$$

An integration now gives

$$\log v' = -2 \log y_1 - \int P \, dx,$$

so

$$v' = \frac{1}{y_1^2} e^{-\int P \, dx}$$

and

$$v = \int \frac{1}{y_1^2} e^{-\int P \, dx} \, dx. \tag{3}$$

All that remains is to show that y_1 and $y_2 = vy_1$, where v is given by (3), actually are linearly independent as claimed; and this we leave to the reader in Problem 1.[4]

---

[4] Formula (3) is due to the eminent French mathematician Joseph Liouville (see the note at the end of Section 43).

Example 1. $y_1 = x$ is a solution of $x^2y'' + xy' - y = 0$ which is simple enough to be discovered by inspection. Find the general solution.

We begin by writing the given equation in the form of (1):

$$y'' + \frac{1}{x}y' - \frac{1}{x^2}y = 0.$$

Since $P(x) = 1/x$, a second linearly independent solution is given by $y_2 = vy_1$, where

$$v = \int \frac{1}{x^2} e^{-\int (1/x)\,dx}\,dx = \int \frac{1}{x^2} e^{-\log x}\,dx = \int x^{-3}\,dx = \frac{x^{-2}}{-2}.$$

This yields $y_2 = (-1/2)x^{-1}$, so the general solution is $y = c_1x + c_2x^{-1}$.

PROBLEMS

1. If y_1 is a nonzero solution of equation (1) and $y_2 = vy_1$, where v is given by formula (3), is the second solution found in the text, show by computing the Wronskian that y_1 and y_2 are linearly independent.
2. Use the method of this section to find y_2 and the general solution of each of the following equations from the given solution y_1:
   (a) $y'' + y = 0$, $y_1 = \sin x$;   (b) $y'' - y = 0$, $y_1 = e^x$.
3. The equation $xy'' + 3y' = 0$ has the obvious solution $y_1 = 1$. Find y_2 and the general solution.
4. Verify that $y_1 = x^2$ is one solution of $x^2y'' + xy' - 4y = 0$, and find y_2 and the general solution.
5. The equation $(1 - x^2)y'' - 2xy' + 2y = 0$ is the special case of Legendre's equation

$$(1 - x^2)y'' - 2xy' + p(p + 1)y = 0$$

   corresponding to $p = 1$. It has $y_1 = x$ as an obvious solution. Find the general solution.
6. The equation $x^2y'' + xy' + (x^2 - \frac{1}{4})y = 0$ is the special case of Bessel's equation

$$x^2y'' + xy' + (x^2 - p^2)y = 0$$

   corresponding to $p = \frac{1}{2}$. Verify that $y_1 = x^{-1/2} \sin x$ is one solution over any interval including only positive values of x, and find the general solution.
7. Use the fact that $y_1 = x$ is an obvious solution of each of the following equations to find their general solutions:
   (a) $y'' - \dfrac{x}{x - 1}y' + \dfrac{1}{x - 1}y = 0$;
   (b) $x^2y'' + 2xy' - 2y = 0$;
   (c) $x^2y'' - x(x + 2)y' + (x + 2)y = 0$.
8. Find the general solution of $y'' - xf(x)y' + f(x)y = 0$.
9. Verify that one solution of $xy'' - (2x + 1)y' + (x + 1)y = 0$ is given by $y_1 = e^x$, and find the general solution.

10. (a) If n is a positive integer, find two linearly independent solutions of

$$xy'' - (x + n)y' + ny = 0.$$

(b) Find the general solution of the equation in part (a) for the cases $n = 1$, 2, 3.

11. Find the general solution of $y'' - f(x)y' + [f(x) - 1]y = 0$.

12. For another, faster approach to formula (3), show that $v' = (y_2/y_1)' = W(y_1,y_2)/y_1^2$ and use Abel's formula in Section 15 to obtain v.

17 THE HOMOGENEOUS EQUATION WITH CONSTANT COEFFICIENTS

We are now in a position to give a complete discussion of the homogeneous equation $y'' + P(x)y' + Q(x)y = 0$ for the special case in which $P(x)$ and $Q(x)$ are constants p and q:

$$y'' + py' + qy = 0. \tag{1}$$

Our starting point is the fact that the exponential function e^{mx} has the property that its derivatives are all constant multiples of the function itself. This leads us to consider

$$y = e^{mx} \tag{2}$$

as a possible solution for (1) if the constant m is suitably chosen. Since $y' = me^{mx}$ and $y'' = m^2 e^{mx}$, substitution in (1) yields

$$(m^2 + pm + q)e^{mx} = 0; \tag{3}$$

and since e^{mx} is never zero, (3) holds if and only if m satisfies the *auxiliary equation*

$$m^2 + pm + q = 0. \tag{4}$$

The two roots m_1 and m_2 of this equation, that is, the values of m for which (2) is a solution of (1), are given by the quadratic formula:

$$m_1, m_2 = \frac{-p \pm \sqrt{p^2 - 4q}}{2}. \tag{5}$$

Further development of this situation requires separate treatment of the three possibilities inherent in (5).

Distinct real roots. It is clear that the roots m_1 and m_2 are distinct real numbers if and only if $p^2 - 4q > 0$. In this case we get the two solutions

$$e^{m_1 x} \qquad \text{and} \qquad e^{m_2 x}.$$

Since the ratio

$$\frac{e^{m_1 x}}{e^{m_2 x}} = e^{(m_1 - m_2)x}$$

is not constant, these solutions are linearly independent and

$$y = c_1 e^{m_1 x} + c_2 e^{m_2 x} \tag{6}$$

is the general solution of (1).

Distinct complex roots.[5] The roots m_1 and m_2 are distinct complex numbers if and only if $p^2 - 4q < 0$. In this case m_1 and m_2 can be written in the form $a \pm ib$; and by *Euler's formula*

$$e^{i\theta} = \cos \theta + i \sin \theta, \tag{7}$$

our two solutions of (1) are

$$e^{m_1 x} = e^{(a+ib)x} = e^{ax} e^{ibx} = e^{ax}(\cos bx + i \sin bx) \tag{8}$$

and

$$e^{m_2 x} = e^{(a-ib)x} = e^{ax} e^{-ibx} = e^{ax}(\cos bx - i \sin bx). \tag{9}$$

Since we are interested only in solutions that are real-valued functions, we can add (8) and (9) and divide by 2, and subtract and divide by $2i$, to obtain

$$e^{ax} \cos bx \qquad \text{and} \qquad e^{ax} \sin bx. \tag{10}$$

These solutions are linearly independent, so the general solution of (1) in this case is

$$y = e^{ax}(c_1 \cos bx + c_2 \sin bx). \tag{11}$$

We can look at this matter from another point of view. A complex-valued function $w(x) = u(x) + iv(x)$ satisfies equation (1), in which p and q are *real* numbers, if and only if $u(x)$ and $v(x)$ satisfy (1) separately. Accordingly, a complex solution of (1) always contains two real solutions, and (8) yields the two solutions (10) at once.

Equal real roots. It is evident that the roots m_1 and m_2 are equal real numbers if and only if $p^2 - 4q = 0$. Here we obtain only one solution $y = e^{mx}$ with $m = -p/2$. However, we can easily find a second linearly independent solution by the method of the preceding section: if we take $y_1 = e^{(-p/2)x}$, then

$$v = \int \frac{1}{y_1^2} e^{-\int p \, dx} \, dx = \int \frac{1}{e^{-px}} e^{-px} \, dx = x$$

---

[5] We take it for granted that the reader is acquainted with the elementary algebra of complex numbers. Euler's formula (7) is—or ought to be—a standard part of any reasonably satisfactory course in calculus.

and $y_2 = vy_1 = xe^{mx}$. In this case (1) has

$$y = c_1 e^{mx} + c_2 xe^{mx} \tag{12}$$

as its general solution.

　　In summary, we have three possible forms—given by formulas (6), (11), and (12)—for the general solution of the homogeneous equation (1) with constant coefficients, depending on the nature of the roots m_1 and m_2 of the auxiliary equation (4). It is clear that the qualitative nature of this general solution is fully determined by the signs and relative magnitudes of the coefficients p and q, and can be radically changed by altering their numerical values. This matter is important for physicists concerned with the detailed analysis of mechanical systems or electric circuits described by equations of the form (1). For instance, if $p^2 < 4q$, the graph of the solution is a wave whose amplitude increases or decreases exponentially according as p is negative or positive. This statement and others like it are obvious consequences of the above discussion, and are given exhaustive treatment in books dealing more fully with the elementary physical applications of differential equations.

　　The ideas of this section are primarily due to Euler. A brief sketch of a few of the many achievements of this great scientific genius is given in Appendix A.

PROBLEMS

1. Find the general solution of each of the following equations:

(a) $y'' + y' - 6y = 0$;

(b) $y'' + 2y' + y = 0$;

(c) $y'' + 8y = 0$;

(d) $2y'' - 4y' + 8y = 0$;

(e) $y'' - 4y' + 4y = 0$;

(f) $y'' - 9y' + 20y = 0$;

(g) $2y'' + 2y' + 3y = 0$;

(h) $4y'' - 12y' + 9y = 0$;

(i) $y'' + y' = 0$;

(j) $y'' - 6y' + 25y = 0$;

(k) $4y'' + 20y' + 25y = 0$;

(l) $y'' + 2y' + 3y = 0$;

(m) $y'' = 4y$;

(n) $4y'' - 8y' + 7y = 0$;

(o) $2y'' + y' - y = 0$;

(p) $16y'' - 8y' + y = 0$;

(q) $v'' + 4y' + 5y = 0$;

(r) $y'' + 4y' - 5y = 0$.

2. Find the solutions of the following initial value problems:

(a) $y'' - 5y' + 6y = 0$, $\quad y(1) = e^2$ and $y'(1) = 3e^2$;

(b) $y'' - 6y' + 5y = 0$, $\quad y(0) = 3$ and $y'(0) = 11$;

(c) $y'' - 6y' + 9y = 0$, $\quad y(0) = 0$ and $y'(0) = 5$;

(d) $y'' + 4y' + 5y = 0$, $\quad y(0) = 1$ and $y'(0) = 0$;

(e) $y'' + 4y' + 2y = 0$, $\quad y(0) = -1$ and $y'(0) = 2 + 3\sqrt{2}$;

(f) $y'' + 8y' - 9y = 0$, $\quad y(1) = 2$ and $y'(1) = 0$.

3. Show that the general solution of equation (1) approaches 0 as $x \to \infty$ if and only if p and q are both positive.

4. Without using the formulas obtained in this section, show that the derivative of any solution of equation (1) is also a solution.

5. The equation

$$x^2y'' + pxy' + qy = 0,$$

where p and q are constants, is called *Euler's equidimensional equation.*[6] Show that the change of independent variable given by $x = e^z$ transforms it into an equation with constant coefficients, and apply this technique to find the general solution of each of the following equations:

(a) $x^2y'' + 3xy' + 10y = 0$; (f) $x^2y'' + 2xy' - 6y = 0$;
(b) $2x^2y'' + 10xy' + 8y = 0$; (g) $x^2y'' + 2xy' + 3y = 0$;
(c) $x^2y'' + 2xy' - 12y = 0$; (h) $x^2y'' + xy' - 2y = 0$;
(d) $4x^2y'' - 3y = 0$; (i) $x^2y'' + xy' - 16y = 0$.
(e) $x^2y'' - 3xy' + 4y = 0$;

6. In Problem 5 certain homogeneous equations with variable coefficients were transformed into equations with constant coefficients by changing the independent variable from x to $z = \log x$. Consider the general homogeneous equation

$$y'' + P(x)y' + Q(x)y = 0, \tag{*}$$

and change the independent variable from x to $z = z(x)$, where $z(x)$ is an unspecified function of x. Show that equation (*) can be transformed in this way into an equation with constant coefficients if and only if $(Q' + 2PQ)/Q^{3/2}$ is constant, in which case $z = \int \sqrt{Q(x)}\, dx$ will effect the desired result.

7. Use the result of Problem 6 to discover whether each of the following equations can be transformed into an equation with constant coefficients by changing the independent variable, and solve it if this is possible:

(a) $xy'' + (x^2 - 1)y' + x^3y = 0$;
(b) $y'' + 3xy' + x^2y = 0$.

8. In this problem we present another way of discovering the second linearly independent solution of (1) when the roots of the auxiliary equation are real and equal.

(a) If $m_1 \neq m_2$, verify that the differential equation

$$y'' - (m_1 + m_2)y' + m_1m_2 y = 0$$

has

$$y = \frac{e^{m_1x} - e^{m_2x}}{m_1 - m_2}$$

as a solution.

(b) Think of m_2 as fixed and use l'Hospital's rule to find the limit of the solution in part (a) as $m_1 \to m_2$.

(c) Verify that the limit in part (b) satisfies the differential equation obtained from the equation in part (a) by replacing m_1 by m_2.

---

[6] It is also known as *Cauchy's equidimensional equation.* Euler's researches were so extensive that many mathematicians try to avoid confusion by naming equations, formulas, theorems, etc., for the person who first studied them after Euler.

18 THE METHOD OF UNDETERMINED COEFFICIENTS

In the preceding two sections we considered several ways of finding the general solution of the homogeneous equation

$$y'' + P(x)y' + Q(x)y = 0. \tag{1}$$

As we saw, these methods are effective in only a few special cases: when the coefficients $P(x)$ and $Q(x)$ are constants, and when they are not constants but are still simple enough to enable us to discover one nonzero solution by inspection. Fortunately these categories are sufficiently broad to cover a number of significant applications. However, it should be clearly understood that many homogeneous equations of great importance in mathematics and physics are beyond the reach of these procedures, and can only be solved by the method of power series developed in Chapter 5.

   In this and the next section we turn to the problem of solving the nonhomogeneous equation

$$y'' + P(x)y' + Q(x)y = R(X) \tag{2}$$

for those cases in which the general solution $y_g(x)$ of the corresponding homogeneous equation (1) is already known. By Theorem 14-B, if $y_p(x)$ is any particular solution of (2), then

$$y(x) = y_g(x) + y_p(x)$$

is the general solution of (2). But how do we find y_p? This is the practical problem that we now consider.

   The method of undetermined coefficients is a procedure for finding y_p when (2) has the form

$$y'' + py' + qy = R(x), \tag{3}$$

where p and q are constants and $R(x)$ is an exponential, a sine or cosine, a polynomial, or some combination of such functions. As an example, we study the equation

$$y'' + py' + qy = e^{ax}. \tag{4}$$

Since differentiating an exponential such as e^{ax} merely reproduces the function with a possible change in the numerical coefficient, it is natural to guess that

$$y_p = Ae^{ax} \tag{5}$$

might be a particular solution of (4). Here A is the *undetermined coefficient* that we want to determine in such a way that (5) will actually satisfy (4). On substituting (5) into (4), we get

$$A(a^2 + pa + q)e^{ax} = e^{ax},$$

so

$$A = \frac{1}{a^2 + pa + q}. \tag{6}$$

This value of A will make (5) a solution of (4) *except when the denominator on the right of* (6) *is zero.* The source of this difficulty is easy to understand, for the exception arises when a is a root of the auxiliary equation

$$m^2 + pm + q = 0, \tag{7}$$

and in this case we know that (5) reduces the left side of (4) to zero and cannot possibly satisfy (4) as it stands, with the right side different from zero.

What can be done to continue the procedure in this exceptional case? We saw in the previous section that when the auxiliary equation has a double root, the second linearly independent solution of the homogeneous equation is obtained by multiplying by x. With this as a hint, we take

$$y_p = Axe^{ax} \tag{8}$$

as a substitute trial solution. On inserting (8) into (4), we get

$$A(a^2 + pa + q)xe^{ax} + A(2a + p)e^{ax} = e^{ax}.$$

The first expression in parentheses is zero because of our assumption that a is a root of (7), so

$$A = \frac{1}{2a + p}. \tag{9}$$

This gives a valid coefficient for (8) except when $a = -p/2$, that is, except when a is a double root of (7). In this case we hopefully continue the successful pattern indicated above and try

$$y_p = Ax^2 e^{ax}. \tag{10}$$

Substitution of (10) into (4) yields

$$A(a^2 + pa + q)x^2 e^{ax} + 2A(2a + p)xe^{ax} + 2Ae^{ax} = e^{ax}.$$

Since a is now assumed to be a double root of (7), both expressions in parentheses are zero and

$$A = \frac{1}{2}. \tag{11}$$

To summarize: If a is not a root of the auxiliary equation (7), then (4) has a particular solution of the form Ae^{ax}; if a is a simple root of (7), then (4) has no solution of the form Ae^{ax} but does have one of the form Axe^{ax}; and if a is a double root, then (4) has no solution of the form Axe^{ax} but does have one of the form Ax^2e^{ax}. In each case we have given a formula for A, but only for the purpose of clarifying the reasons behind the events. In practice it is easier to find A by direct substitution in the equation at hand.

Another important case where the method of undetermined coefficients can be applied is that in which the right side of equation (4) is replaced by $\sin bx$:

$$y'' + py' + qy = \sin bx. \tag{12}$$

Since the derivatives of $\sin bx$ are constant multiples of $\sin bx$ and $\cos bx$, we take a trial solution of the form

$$y_p = A \sin bx + B \cos bx. \tag{13}$$

The undetermined coefficients A and B can now be computed by substituting (13) into (12) and equating the resulting coefficients of $\sin bx$ and $\cos bx$ on the left and right. These steps work just as well if the right side of equation (12) is replaced by $\cos bx$ or any linear combination of $\sin bx$ and $\cos bx$, that is, any function of the form $\alpha \sin bx + \beta \cos bx$. As before, the method breaks down if (13) satisfies the homogeneous equation corresponding to (12). When this happens, the procedure can be carried through by using

$$y_p = x(A \sin bx + B \cos bx) \tag{14}$$

as our trial solution instead of (13).

Example 1. Find a particular solution of

$$y'' + y = \sin x. \tag{15}$$

The reduced homogeneous equation $y'' + y = 0$ has $y = c_1 \sin x + c_2 \cos x$ as its general solution, so it is useless to take $y_p = A \sin x + B \cos x$ as a trial solution for the complete equation (15). We therefore try $y_p = x(A \sin x + B \cos x)$. This yields

$$y_p' = A \sin x + B \cos x + x(A \cos x - B \sin x)$$

and

$$y_p'' = 2A \cos x - 2B \sin x + x(-A \sin x - B \cos x),$$

and by substituting in (15) we obtain

$$2A \cos x - 2B \sin x = \sin x.$$

This tells us that the choice $A = 0$ and $B - \frac{1}{2}$ satisfies our requirement, so $y_p = -\frac{1}{2}x \cos x$ is the desired particular solution.

Finally, we consider the case in which the right side of equation (4) is replaced by a polynomial:

$$y'' + py' + qy = a_0 + a_1 x + \cdots + a_n x^n. \tag{16}$$

Since the derivative of a polynomial is again a polynomial, we are led to seek a particular solution of the form

$$y_p = A_0 + A_1 x + \cdots + A_n x^n. \tag{17}$$

When (17) is substituted into (16), we have only to equate the coefficients of like powers of x to find the values of the undetermined coefficients A_0, $A_1, \ldots, A_n$. If the constant q happens to be zero, then this procedure gives x^{n-1} as the highest power of x on the left of (16), so in this case we take our trial solution in the form

$$\begin{aligned} y_p &= x(A_0 + A_1 x + \cdots + A_n x^n) \\ &= A_0 x + A_1 x^2 + \cdots + A_n x^{n+1}. \end{aligned} \tag{18}$$

If p and q are both zero, then (16) can be solved at once by direct integration.

Example 2. Find the general solution of

$$y'' - y' - 2y = 4x^2. \tag{19}$$

The reduced homogeneous equation $y'' - y' - 2y = 0$ has $m^2 - m - 2 = 0$ or $(m - 2)(m + 1) = 0$ as its auxiliary equation, so the general solution of the reduced equation is $y_g = c_1 e^{2x} + c_2 e^{-x}$.

Since the right side of the complete equation (19) is a polynomial of the second degree, we take a trial solution of the form $y_p = A + Bx + Cx^2$ and substitute it into (19):

$$2C - (B + 2Cx) - 2(A + Bx + Cx^2) = 4x^2.$$

Equating coefficients of like powers of x gives the system of linear equations

$$2C - B - 2A = 0,$$
$$-2C - 2B = 0,$$
$$-2C = 4.$$

We now easily see that $C = -2$, $B = 2$, and $A = -3$, so our particular solution is $y_p = -3 + 2x - 2x^2$ and

$$y = c_1 e^{2x} + c_2 e^{-x} - 3 + 2x - 2x^2$$

is the general solution of the complete equation (19).

The above discussions show that the form of a particular solution of equation (3) can often be inferred from the form of the right-hand member $R(x)$. In general this is true whenever $R(x)$ is a function with only a finite number of essentially different derivatives. We have seen how this works for exponentials, sines and cosines, and polynomials. In Problem 3 we indicate a course of action for the case in which $R(x)$ is a sum of such functions. It is also possible to develop slightly more elaborate techniques for handling various products of these elementary functions, but for most practical purposes this is unnecessary. In essence, the whole matter is simply a question of intelligent guesswork involving a sufficient number of undetermined coefficients that can be tailored to fit the circumstances.

PROBLEMS

1. Find the general solution of each of the following equations:
   (a) $y'' + 3y' - 10y = 6e^{4x}$;
   (b) $y + 4y = 3 \sin x$;
   (c) $y'' + 10y' + 25y = 14e^{-5x}$;
   (d) $y'' - 2y' + 5y = 25x^2 + 12$;
   (e) $y'' - y' - 6y = 20e^{-2x}$;
   (f) $y'' - 3y' + 2y = 14 \sin 2x - 18 \cos 2x$;
   (g) $y'' + y = 2 \cos x$;
   (h) $y'' - 2y' = 12x - 10$;
   (i) $y'' - 2y' + y = 6e^x$;
   (j) $y'' - 2y' + 2y = e^x \sin x$;
   (k) $y'' + y' = 10x^4 + 2$.

2. If k and b are positive constants, find the general solution of

$$y'' + k^2 y = \sin bx.$$

3. If $y_1(x)$ and $y_2(x)$ are solutions of

$$y'' + P(x)y' + Q(x)y = R_1(x)$$

and

$$y'' + P(x)y' + Q(x)y = R_2(x),$$

show that $y(x) = y_1(x) + y_2(x)$ is a solution of

$$y'' + P(x)y' + Q(x)y = R_1(x) + R_2(x).$$

This is called the *principle of superposition*. Use this principle to find the general solution of
   (a) $y'' + 4y = 4 \cos 2x + 6 \cos x + 8x^2 - 4x$;
   (b) $y'' + 9y = 2 \sin 3x + 4 \sin x - 26e^{-2x} + 27x^3$.

19 THE METHOD OF VARIATION OF PARAMETERS

The technique described in Section 18 for determining a particular solution of the nonhomogeneous equation

$$y'' + P(x)y' + Q(x)y = R(x) \tag{1}$$

has two severe limitations: it can be used only when the coefficients $P(x)$ and $Q(x)$ are constants, and even then it works only when the right-hand term $R(x)$ has a particularly simple form. Within these limitations, however, this procedure is usually the easiest to apply.

We now develop a more powerful method that always works—regardless of the nature of P, Q, and R—provided only that the general solution of the corresponding homogeneous equation

$$y'' + P(x)y' + Q(x)y = 0 \tag{2}$$

is already known. We assume, then, that in some way the general solution

$$y(x) = c_1 y_1(x) + c_2 y_2(x) \tag{3}$$

of (2) has been found. The method is similar to that discussed in Section 16; that is, we replace the constants c_1 and c_2 by unknown functions $v_1(x)$ and $v_2(x)$, and attempt to determine v_1 and v_2 in such a manner that

$$y = v_1 y_1 + v_2 y_2 \tag{4}$$

will be a solution of (1).[7] With two unknown functions to find, it will be necessary to have two equations relating these functions. We obtain one of these by requiring that (4) be a solution of (1). It will soon be clear what the second equation should be. We begin by computing the derivative of (4), arranged as follows:

$$y' = (v_1 y_1' + v_2 y_2') + (v_1' y_1 + v_2' y_2). \tag{5}$$

Another differentiation will introduce second derivatives of the unknowns v_1 and v_2. We avoid this complication by requiring the second expression in parentheses to vanish:

$$v_1' y_1 + v_2' y_2 = 0. \tag{6}$$

This gives

$$y' = v_1 y_1' + v_2 y_2', \tag{7}$$

so

$$y'' = v_1 y_1'' + v_1' y_1' + v_2 y_2'' + v_2' y_2'. \tag{8}$$

On substituting (4), (7), and (8) into (1), and rearranging, we get

$$v_1(y_1'' + Py_1' + Qy_1) + v_2(y_2'' + Py_2' + Qy_2)$$
$$+ v_1' y_1' + v_2' y_2' = R(x). \tag{9}$$

Since y_1 and y_2 are solutions of (2), the two expressions in parentheses are equal to 0, and (9) collapses to

$$v_1' y_1' + v_2' y_2' = R(x). \tag{10}$$

---

[7] This is the source of the name *variation of parameters*: we vary the parameters c_1 and c_2.

Taking (6) and (10) together, we have two equations in the two unknowns v_1' and v_2':

$$v_1'y_1 + v_2'y_2 = 0,$$
$$v_1'y_1' + v_2'y_2' = R(x).$$

These can be solved at once, giving

$$v_1' = \frac{-y_2 R(x)}{W(y_1, y_2)} \quad \text{and} \quad v_2' = \frac{y_1 R(x)}{W(y_1, y_2)}. \tag{11}$$

It should be noted that these formulas are legitimate, for the Wronskian in the denominators is nonzero by the linear independence of y_1 and y_2. All that remains is to integrate formulas (11) to find v_1 and v_2:

$$v_1 = \int \frac{-y_2 R(x)}{W(y_1, y_2)} \, dx \quad \text{and} \quad v_2 = \int \frac{y_1 R(x)}{W(y_1, y_2)} \, dx. \tag{12}$$

We can now put everything together and assert that

$$y = y_1 \int \frac{-y_2 R(x)}{W(y_1, y_2)} \, dx + y_2 \int \frac{y_1 R(x)}{W(y_1, y_2)} \, dx \tag{13}$$

is the particular solution of (1) we are seeking.

The reader will see that this method has disadvantages of its own. In particular, the integrals in (12) may be difficult or impossible to work out. Also, of course, it is necessary to know the general solution of (2) before the process can even be started; but this objection is really immaterial because we are unlikely to care about finding a particular solution of (1) unless the general solution of (2) is already at hand.

The method of variation of parameters was invented by the French mathematician Lagrange in connection with his epoch-making work in analytical mechanics (see Appendix A in Chapter 12).

Example 1. Find a particular solution of $y'' + y = \csc x$.

The corresponding homogeneous equation $y'' + y = 0$ has $y(x) = c_1 \sin x + c_2 \cos x$ as its general solution, so $y_1 = \sin x$, $y_1' = \cos x$, $y_2 = \cos x$, and $y_2' = -\sin x$. The Wronskian of y_1 and y_2 is

$$W(y_1, y_2) = y_1 y_2' - y_2 y_1' = -\sin^2 x - \cos^2 x = -1,$$

so by (12) we have

$$v_1 = \int \frac{-\cos x \csc x}{-1} \, dx = \int \frac{\cos x}{\sin x} \, dx = \log (\sin x)$$

and

$$v_2 = \int \frac{\sin x \csc x}{-1} \, dx = -x.$$

Accordingly,

$$y = \sin x \log (\sin x) - x \cos x$$

is the desired particular solution.

PROBLEMS

1. Find a particular solution of

$$y'' - 2y' + y = 2x,$$

first by inspection and then by variation of parameters.

2. Find a particular solution of

$$y'' - y' - 6y = e^{-x},$$

first by undetermined coefficients and then by variation of parameters.

3. Find a particular solution of each of the following equations:
(a) $y'' + 4y = \tan 2x$;
(b) $y'' + 2y' + y = e^{-x} \log x$;
(c) $y'' - 2y' - 3y = 64xe^{-x}$;
(d) $y'' + 2y' + 5y = e^{-x} \sec 2x$;
(e) $2y'' + 3y' + y = e^{-3x}$;
(f) $y'' - 3y' + 2y = (1 + e^{-x})^{-1}$.

4. Find a particular solution of each of the following equations:
(a) $y'' + y = \sec x$;
(b) $y'' + y = \cot^2 x$;
(c) $y'' + y = \cot 2x$;
(d) $y'' + y = x \cos x$;
(e) $y'' + y = \tan x$;
(f) $y'' + y = \sec x \tan x$;
(g) $y'' + y = \sec x \csc x$.

5. (a) Show that the method of variation of parameters applied to the equation $y'' + y = f(x)$ leads to the particular solution

$$y_p(x) = \int_0^x f(t) \sin (x - t) \, dt.$$

(b) Find a similar formula for a particular solution of the equation $y'' + k^2 y = f(x)$, where k is a positive constant.

6. Find the general solution of each of the following equations:
(a) $(x^2 - 1)y'' - 2xy' + 2y = (x^2 - 1)^2$;
(b) $(x^2 + x)y'' + (2 - x^2)y' - (2 + x)y = x(x + 1)^2$;
(c) $(1 - x)y'' + xy' - y = (1 - x)^2$;
(d) $xy'' - (1 + x)y' + y = x^2 e^{2x}$;
(e) $x^2 y'' - 2xy' + 2y = xe^{-x}$.

20 VIBRATIONS IN MECHANICAL AND ELECTRICAL SYSTEMS

Generally speaking, vibrations occur whenever a physical system in stable equilibrium is disturbed, for then it is subject to forces tending to restore its equilibrium. In the present section we shall see how situations of this kind can lead to differential equations of the form

$$\frac{d^2x}{dt^2} + p\frac{dx}{dt} + qx = R(t),$$

and also how the study of these equations sheds light on the physical circumstances.

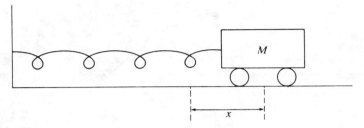

FIGURE 19

Undamped simple harmonic vibrations. As a continuing example, we consider a cart of mass M attached to a nearby wall by means of a spring (Fig. 19). The spring exerts no force when the cart is at its equilibrium position $x = 0$. If the cart is displaced by a distance x, then the spring exerts a restoring force $F_s = -kx$, where k is a positive constant whose magnitude is a measure of the stiffness of the spring. By Newton's second law of motion, which says that the mass of the cart times its acceleration equals the total force acting on it, we have

$$M\frac{d^2x}{dt^2} = F_s \tag{1}$$

or

$$\frac{d^2x}{dt^2} + \frac{k}{M}x = 0. \tag{2}$$

It will be convenient to write this equation of motion in the form

$$\frac{d^2x}{dt^2} + a^2x = 0, \tag{3}$$

where $a = \sqrt{k/M}$, and its general solution can be written down at once:

$$x = c_1 \sin at + c_2 \cos at. \tag{4}$$

If the cart is pulled aside to the position $x = x_0$ and released without any initial velocity at time $t = 0$, so that our initial conditions are

$$x = x_0 \quad \text{and} \quad v = \frac{dx}{dt} = 0 \quad \text{when } t = 0, \tag{5}$$

then it is easily seen that $c_1 = 0$ and $c_2 = x_0$, so (4) becomes

$$x = x_0 \cos at. \tag{6}$$

The graph of (6) is shown in Fig. 20. The *amplitude* of this simple harmonic vibration is x_0; and since its *period* T is the time required for

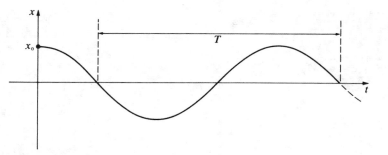

FIGURE 20

one complete cycle, we have $aT = 2\pi$ and

$$T = \frac{2\pi}{a} = 2\pi\sqrt{\frac{M}{k}}. \tag{7}$$

Its *frequency* f is the number of cycles per unit time, so $fT = 1$ and

$$f = \frac{1}{T} = \frac{a}{2\pi} = \frac{1}{2\pi}\sqrt{\frac{k}{M}}. \tag{8}$$

It is clear from (8) that the frequency of this vibration increases if the stiffness of the spring is increased or if the mass of the cart is decreased, as our common sense would have led us to predict.

Damped vibrations. As our next step in developing this physical problem, we consider the additional effect of a *damping force* F_d due to the viscosity of the medium through which the cart moves (air, water, oil, etc.). We make the specific assumption that this force opposes the motion and has magnitude proportional to the velocity, that is, that $F_d = -c(dx/dt)$, where c is a positive constant measuring the resistance of the medium. Equation (1) now becomes

$$M\frac{d^2x}{dt^2} = F_s + F_d, \tag{9}$$

so

$$\frac{d^2x}{dt^2} + \frac{c}{M}\frac{dx}{dt} + \frac{k}{M}x = 0. \tag{10}$$

Again for the sake of convenience, we write this in the form

$$\frac{d^2x}{dt^2} + 2b\frac{dx}{dt} + a^2x = 0, \tag{11}$$

where $b = c/2M$ and $a = \sqrt{k/M}$. The auxiliary equation is

$$m^2 + 2bm + a^2 = 0, \tag{12}$$

and its roots m_1 and m_2 are given by

$$m_1, m_2 = \frac{-2b \pm \sqrt{4b^2 - 4a^2}}{2} = -b \pm \sqrt{b^2 - a^2}. \tag{13}$$

The general solution of (11) is of course determined by the nature of the numbers m_1 and m_2. As we know, there are three cases, which we consider separately.

CASE A. $b^2 - a^2 > 0$ or $b > a$. In loose terms this amounts to assuming that the frictional force due to the viscosity is large compared to the stiffness of the spring. It follows that m_1 and m_2 are distinct negative numbers, and the general solution of (11) is

$$x = c_1 e^{m_1 t} + c_2 e^{m_2 t}. \tag{14}$$

If we apply the initial conditions (5) to evaluate c_1 and c_2, (14) becomes

$$x = \frac{x_0}{m_1 - m_2} (m_1 e^{m_2 t} - m_2 e^{m_1 t}). \tag{15}$$

The graph of this function is given in Fig. 21. It is clear that no vibration occurs, and that the cart merely subsides to its equilibrium position. This type of motion is called *overdamped*. We now imagine that the viscosity is decreased until we reach the condition of the next case.

CASE B. $b^2 - a^2 = 0$ or $b = a$. Here we have $m_1 = m_2 = -b = -a$, and the general solution of (11) is

$$x = c_1 e^{-at} + c_2 t e^{-at}. \tag{16}$$

When the initial conditions (5) are imposed, we obtain

$$x = x_0 e^{-at}(1 + at). \tag{17}$$

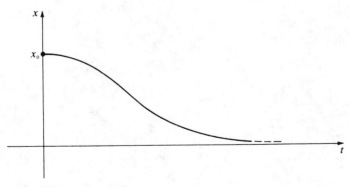

FIGURE 21

This function has a graph similar to that of (15), and again we have no vibration. Any motion of this kind is said to be *critically damped*. If the viscosity is now decreased by any amount, however small, then the motion becomes vibratory, and is called *underdamped*. This is the really interesting situation, which we discuss as follows.

CASE C. $b^2 - a^2 < 0$ or $b < a$. Here m_1 and m_2 are conjugate complex numbers $-b \pm i\alpha$, where

$$\alpha = \sqrt{a^2 - b^2},$$

and the general solution of (11) is

$$x = e^{-bt}(c_1 \cos \alpha t + c_2 \sin \alpha t). \tag{18}$$

When c_1 and c_2 are evaluated in accordance with the initial conditions (5), this becomes

$$x = \frac{x_0}{\alpha} e^{-bt}(\alpha \cos \alpha t + b \sin \alpha t). \tag{19}$$

If we introduce $\theta = \tan^{-1}(b/\alpha)$, then (19) can be expressed in the more revealing form

$$x = \frac{x_0 \sqrt{\alpha^2 + b^2}}{\alpha} e^{-bt} \cos(\alpha t - \theta). \tag{20}$$

This function oscillates with an amplitude that falls off exponentially, as Fig. 22 shows. It is not periodic in the strict sense, but its graph crosses the equilibrium position $x = 0$ at regular intervals. If we consider its "period" T as the time required for one complete "cycle," then $\alpha T = 2\pi$ and

$$T = \frac{2\pi}{\alpha} = \frac{2\pi}{\sqrt{a^2 - b^2}} = \frac{2\pi}{\sqrt{k/M - c^2/4M^2}}. \tag{21}$$

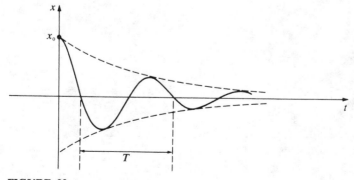

FIGURE 22

Also, its "frequency" f is given by

$$f = \frac{1}{T} = \frac{1}{2\pi}\sqrt{a^2 - b^2} = \frac{1}{2\pi}\sqrt{\frac{k}{M} - \frac{c^2}{4M^2}}. \tag{22}$$

This number is usually called the *natural frequency* of the system. When the viscosity vanishes, so that $c = 0$, it is clear that (21) and (22) reduce to (7) and (8). Furthermore, on comparing (8) and (22) we see that the frequency of the vibration is decreased by damping, as we might expect.

Forced vibrations. The vibrations discussed above are known as *free* vibrations because all the forces acting on the system are internal to the system itself. We now extend our analysis to cover the case in which an impressed external force $F_e = f(t)$ acts on the cart. Such a force might arise in many ways: for example, from vibrations of the wall to which the spring is attached, or from the effect on the cart of an external magnetic field (if the cart is made of iron). In place of (9) we now have

$$M\frac{d^2x}{dt^2} = F_s + F_d + F_e, \tag{23}$$

so

$$M\frac{d^2x}{dt^2} + c\frac{dx}{dt} + kx = f(t). \tag{24}$$

The most important case is that in which the impressed force is periodic and has the form $f(t) = F_0 \cos \omega t$, so that (24) becomes

$$M\frac{d^2x}{dt^2} + c\frac{dx}{dt} + kx = F_0 \cos \omega t. \tag{25}$$

We have already solved the corresponding homogeneous equation (10), so in seeking the general solution of (25) all that remains is to find a particular solution. This is most readily accomplished by the method of undetermined coefficients. Accordingly, we take $x = A \sin \omega t + B \cos \omega t$ as a trial solution. On substituting this into (25), we obtain the following pair of equations for A and B:

$$\omega c A + (k - \omega^2 M)B = F_0,$$
$$(k - \omega^2 M)A - \omega c B = 0.$$

The solution of this system is

$$A = \frac{\omega c F_0}{(k - \omega^2 M)^2 + \omega^2 c^2} \quad \text{and} \quad B = \frac{(k - \omega^2 M)F_0}{(k - \omega^2 M)^2 + \omega^2 c^2}.$$

Our desired particular solution is therefore

$$x = \frac{F_0}{(k - \omega^2 M)^2 + \omega^2 c^2}[\omega c \sin \omega t + (k - \omega^2 M) \cos \omega t]. \tag{26}$$

By introducing $\phi = \tan^{-1}[\omega c/(k - \omega^2 M)]$, we can write (26) in the more useful form

$$x = \frac{F_0}{\sqrt{(k - \omega^2 M)^2 + \omega^2 c^2}} \cos(\omega t - \phi). \tag{27}$$

If we now assume that we are dealing with the underdamped motion discussed above, then the general solution of (25) is

$$x = e^{-bt}(c_1 \cos \alpha t + c_2 \sin \alpha t) + \frac{F_0}{\sqrt{(k - \omega^2 M)^2 + \omega^2 c^2}} \cos(\omega t - \phi). \tag{28}$$

The first term here is clearly *transient* in the sense that it approaches 0 as $t \to \infty$. As a matter of fact, this is true whether the motion is underdamped or not, as long as some degree of damping is present (see Problem 17-2). Therefore, as time goes on, the motion assumes the character of the second term, the *steady-state* part. On this basis, we can neglect the transient part of (28) and assert that for large t the general solution of (25) is essentially equal to the particular solution (27). The frequency of this forced vibration equals the impressed frequency $\omega/2\pi$, and its amplitude is the coefficient

$$\frac{F_0}{\sqrt{(k - \omega^2 M)^2 + \omega^2 c^2}}. \tag{29}$$

This expression for the amplitude holds some interesting secrets, for it depends not only on ω and F_0 but also on k, c, and M. As an example, we note that if c is very small and ω is close to $\sqrt{k/M}$ (so that $k - \omega^2 M$ is very small), which means that the motion is lightly damped and the impressed frequency $\omega/2\pi$ is close to the natural frequency

$$\frac{1}{2\pi} \sqrt{\frac{k}{M} - \frac{c^2}{4M^2}},$$

then the amplitude is very large. This phenomenon is known as *resonance*. A classic example is provided by the forced vibration of a bridge under the impact of the feet of marching columns of men whose pace corresponds closely to the natural frequency of the bridge.

Finally, we mention briefly certain links between the mechanical problem treated above and the electrical problem discussed in Section 13. It was shown in that section that if a periodic electromotive force $E = E_0 \cos \omega t$ acts in a simple circuit containing a resistor, an inductor, and a capacitor, then the charge Q on the capacitor is governed by the differential equation

$$L\frac{d^2 Q}{dt^2} + R\frac{dQ}{dt} + \frac{1}{C}Q = E_0 \cos \omega t. \tag{30}$$

This equation is strikingly similar to (25). In particular, the following correspondences suggest themselves:

$$\text{mass } M \leftrightarrow \text{inductance } L;$$
$$\text{viscosity } c \leftrightarrow \text{resistance } R;$$
$$\text{stiffness of spring } k \leftrightarrow \text{reciprocal of capacitance } \frac{1}{C};$$
$$\text{displacement } x \leftrightarrow \text{charge } Q \text{ on capacitor.}$$

This analogy between the mechanical and electrical systems renders identical the mathematics of the two systems, and enables us to carry over at once all mathematical conclusions from the first to the second. In the given electric circuit we therefore have a critical resistance below which the free behavior of the circuit will be vibratory with a certain natural frequency, a forced steady-state vibration of the charge Q, and resonance phenomena that appear when the circumstances are favorable.

PROBLEMS

1. Consider the forced vibration (27) in the underdamped case, and find the impressed frequency for which the amplitude (29) attains a maximum. Will such an impressed frequency necessarily exist? This value of the impressed frequency (when it exists) is called the *resonance frequency*. Show that it is always less than the natural frequency.

2. Consider the underdamped free vibration described by formula (20). Show that x assumes maximum values for $t = 0, T, 2T, \ldots$, where T is the "period" as given in formula (21). If x_1 and x_2 are any two successive maximum values of x, show that $x_1/x_2 = e^{bT}$. The logarithm of this quantity, bT, is known as the *logarithmic decrement* of the vibration.

3. A spherical buoy of radius r floats half-submerged in water. If it is depressed slightly, a restoring force equal to the weight of the displaced water presses it upward; and if it is then released, it will bob up and down. Find the period of oscillation if the friction of the water is neglected.

4. A cylindrical buoy 2 feet in diameter floats with its axis vertical in fresh water of density 62.4 lb/ft³. When depressed slightly and released, its period of oscillation is observed to be 1.9 seconds. What is the weight of the buoy?

5. Suppose that a straight tunnel is drilled through the earth between any two points on the surface. If tracks are laid, then—neglecting friction—a train placed in the tunnel at one end will roll through the earth under its own weight, stop at the other end, and return. Show that the time required for a complete round trip is the same for all such tunnels, and estimate its value. If the tunnel is $2L$ miles long, what is the greatest speed attained by the train?

6. The cart in Fig. 19 weighs 128 pounds and is attached to the wall by a spring with spring constant $k = 64$ lb/ft. The cart is pulled 6 inches in the direction away from the wall and released with no initial velocity. Simultaneously a periodic external force $F_e = f(t) = 32 \sin 4t$ is applied to the cart. Assuming that there is no air resistance, find the position $x = x(t)$ of the cart at time t. Note particularly that $|x(t)|$ has arbitrarily large values as $t \to \infty$, a phenomenon known as *pure resonance* and caused by the fact that the forcing function has the same period as the free vibrations of the unforced system.

7. (This problem is intended only for students who are not intimidated by calculations with complex numbers.) The correspondence between equations (25) and (30) makes it easy to write down the steady-state solution of (30) by merely changing the notation in (27):

$$Q = \frac{E_0}{\sqrt{(1/C - \omega^2 L)^2 + \omega^2 R^2}} \cos(\omega t - \phi), \tag{*}$$

where $\tan \phi = \omega R/(1/C - \omega^2 L)$. In electrical engineering it is customary to think of $E_0 \cos \omega t$ in (30) as the real part of $E_0 e^{i\omega t}$, and instead of (30) we would then consider the differential equation

$$L\frac{d^2 Q}{dt^2} + R\frac{dQ}{dt} + \frac{1}{C}Q = E_0 e^{i\omega t}.$$

Find a particular solution of this equation by the method of undetermined coefficients, and at the end of the calculation take the real part of this solution and thereby obtain the solution (*) of the differential equation (30).[8]

---

[8] The use of complex numbers in the mathematics of electric circuit problems was pioneered by the mathematician, inventor and electrical engineer Charles Proteus Steinmetz (1865–1923). As a young man in Germany, his student socialist activities got him into trouble with Bismarck's police, and he hastily emigrated to America in 1889. He was employed by the General Electric Company in its earliest period, and he quickly became the scientific brains of the Company and probably the greatest of all electrical engineers. When he came to GE there was no way to mass-produce electric motors or generators, and no economically viable way to transmit electric power more than 3 miles. Steinmetz solved these problems by using mathematics and the power of his own mind, and thereby improved human life forever in ways too numerous to count.

He was a dwarf who was crippled by a congenital deformity and lived with pain, but he was universally admired for his scientific genius and loved for his warm humanity and puckish sense of humor. The following little-known but unforgettable anecdote about him was published in the Letters section of *Life* magazine (May 14, 1965):

Sirs: In your article on Steinmetz (April 23) you mentioned a consultation with Henry Ford. My father, Burt Scott, who was an employee of Henry Ford for many years, related to me the story behind that meeting. Technical troubles developed with a huge new generator at Ford's River Rouge plant. His electrical engineers were unable to locate the difficulty so Ford solicited the aid of Steinmetz. When "the little giant" arrived at the plant, he rejected all assistance, asking only for a notebook, pencil and cot. For two straight days and nights he listened to the generator and made countless computations. Then he asked for a ladder, a measuring tape and a piece of chalk. He laboriously ascended the ladder, made careful measurements, and put a chalk mark on the side of the generator. He descended and told his skeptical audience to remove a plate from the side of the generator and take out 16 windings from the field coil at that location. The corrections were made and the generator then functioned perfectly. Subsequently Ford received a bill for $10,000 signed by Steinmetz for G.E. Ford returned the bill acknowledging the good job done by Steinmetz but respectfully requesting an itemized statement. Steinmetz replied as follows: Making chalk mark on generator $1. Knowing where to make mark $9,999. Total due $10,000.

21 NEWTON'S LAW OF GRAVITATION AND THE MOTION OF THE PLANETS

The inverse square law of attraction underlies so many natural phenomena—the orbits of the planets around the sun, the motion of the moon and artificial satellites about the earth, the paths described by charged particles in atomic physics, etc.—that every person educated in science ought to know something about its consequences. Our purpose in this section is to deduce Kepler's laws of planetary motion from Newton's law of universal gravitation, and to this end we discuss the motion of a small particle of mass m (a planet) under the attraction of a fixed large particle of mass M (the sun).

For problems involving a moving particle in which the force acting on it is always directed along the line from the particle to a fixed point, it is usually simplest to resolve the velocity, acceleration, and force into components along and perpendicular to this line. We therefore place the fixed particle M at the origin of a polar coordinate system (Fig. 23) and express the radius vector from the origin to the moving particle m in the form

$$\mathbf{r} = r\mathbf{u}_r, \tag{1}$$

where $\mathbf{u}_r$ is the unit vector in the direction of $\mathbf{r}$.[9] It is clear that

$$\mathbf{u}_r = \mathbf{i}\cos\theta + \mathbf{j}\sin\theta, \tag{2}$$

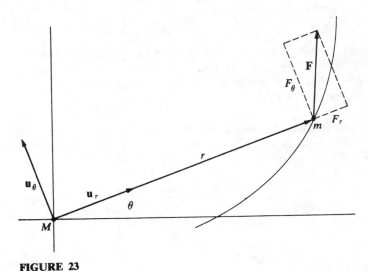

FIGURE 23

---

[9] We here adopt the usual convention of signifying vectors by boldface type.

and also that the corresponding unit vector $\mathbf{u}_\theta$, perpendicular to $\mathbf{u}_r$ in the direction of increasing θ, is given by

$$\mathbf{u}_\theta = -\mathbf{i} \sin \theta + \mathbf{j} \cos \theta. \tag{3}$$

The simple relations

$$\frac{d\mathbf{u}_r}{d\theta} = \mathbf{u}_\theta \quad \text{and} \quad \frac{d\mathbf{u}_\theta}{d\theta} = -\mathbf{u}_r,$$

obtained by differentiating (2) and (3), are essential for computing the velocity and acceleration vectors $\mathbf{v}$ and $\mathbf{a}$. Direct calculation from (1) now yields

$$\mathbf{v} = \frac{d\mathbf{r}}{dt} = r\frac{d\mathbf{u}_r}{dt} + \mathbf{u}_r\frac{dr}{dt} = r\frac{d\mathbf{u}_r}{d\theta}\frac{d\theta}{dt} + \mathbf{u}_r\frac{dr}{dt} = r\frac{d\theta}{dt}\mathbf{u}_\theta + \frac{dr}{dt}\mathbf{u}_r \tag{4}$$

and

$$\mathbf{a} = \frac{d\mathbf{v}}{dt} = \left(r\frac{d^2\theta}{dt^2} + 2\frac{dr}{dt}\frac{d\theta}{dt}\right)\mathbf{u}_\theta + \left[\frac{d^2r}{dt^2} - r\left(\frac{d\theta}{dt}\right)^2\right]\mathbf{u}_r. \tag{5}$$

If the force $\mathbf{F}$ acting on m is written in the form

$$\mathbf{F} = F_\theta\mathbf{u}_\theta + F_r\mathbf{u}_r, \tag{6}$$

then from (5) and (6) and Newton's second law of motion $m\mathbf{a} = \mathbf{F}$, we get

$$m\left(r\frac{d^2\theta}{dt^2} + 2\frac{dr}{dt}\frac{d\theta}{dt}\right) = F_\theta \quad \text{and} \quad m\left[\frac{d^2r}{dt^2} - r\left(\frac{d\theta}{dt}\right)^2\right] = F_r. \tag{7}$$

These differential equations govern the motion of the particle m, and are valid regardless of the nature of the force. Our next task is to extract information from them by making suitable assumptions about the direction and magnitude of $\mathbf{F}$.

Central forces and Kepler's Second Law. $\mathbf{F}$ is called a *central force* if it has no component perpendicular to $\mathbf{r}$, that is, if $F_\theta = 0$. Under this assumption the first of equations (7) becomes

$$r\frac{d^2\theta}{dt^2} + 2\frac{dr}{dt}\frac{d\theta}{dt} = 0.$$

On multiplying through by r, we obtain

$$r^2\frac{d^2\theta}{dt^2} + 2r\frac{dr}{dt}\frac{d\theta}{dt} = 0$$

or

$$\frac{d}{dt}\left(r^2\frac{d\theta}{dt}\right) = 0,$$

so

$$r^2 \frac{d\theta}{dt} = h \tag{8}$$

for some constant h. We shall assume that h is positive, which evidently means that m is moving in a counterclockwise direction. If $A = A(t)$ is the area swept out by $\mathbf{r}$ from some fixed position of reference, so that $dA = r^2 \, d\theta/2$, then (8) implies that

$$dA = \frac{1}{2}\left(r^2 \frac{d\theta}{dt}\right) dt = \frac{1}{2} h \, dt. \tag{9}$$

On integrating (9) from t_1 to t_2, we get

$$A(t_2) - A(t_1) = \frac{1}{2} h(t_2 - t_1). \tag{10}$$

This yields *Kepler's second law*: the radius vector $\mathbf{r}$ from the sun to a planet sweeps out equal areas in equal intervals of time.[10]

Central gravitational forces and Kepler's First Law. We now specialize even further, and assume that $\mathbf{F}$ is a central attractive force whose magnitude—according to Newton's law of gravitation—is directly proportional to the product of the two masses and inversely proportional to the square of the distance between them:

$$F_r = -G\frac{Mm}{r^2}. \tag{11}$$

The letter G represents the *gravitational constant*, which is one of the universal constants of nature. If we write (11) in the slightly simpler form

$$F_r = -\frac{km}{r^2},$$

where $k = GM$, then the second of equations (7) becomes

$$\frac{d^2r}{dt^2} - r\left(\frac{d\theta}{dt}\right)^2 = -\frac{k}{r^2}. \tag{12}$$

The next step in this line of thought is difficult to motivate, because it involves considerable technical ingenuity, but we will try. Our purpose

---

[10] When the Danish astronomer Tycho Brahe died in 1601, his assistant Johannes Kepler (1571–1630) inherited great masses of raw data on the positions of the planets at various times. Kepler worked incessantly on this material for 20 years, and at last succeeded in distilling from it his three beautifully simple laws of planetary motion—which were the climax of thousands of years of purely observational astronomy.

is to use the differential equation (12) to obtain the equation of the orbit in the polar form $r = f(\theta)$, so we want to eliminate t from (12) and consider θ as the independent variable. Also, we want r to be the dependent variable, but if (8) is used to put (12) in the form

$$\frac{d^2r}{dt^2} - \frac{h^2}{r^3} = -\frac{k}{r^2},\tag{13}$$

then the presence of powers of $1/r$ suggests that it might be temporarily convenient to introduce a new dependent variable $z = 1/r$.

To accomplish these various aims, we must first express d^2r/dt^2 in terms of $d^2z/d\theta^2$, by calculating

$$\frac{dr}{dt} = \frac{d}{dt}\left(\frac{1}{z}\right) = -\frac{1}{z^2}\frac{dz}{dt} = -\frac{1}{z^2}\frac{dz}{d\theta}\frac{d\theta}{dt} = -\frac{1}{z^2}\frac{dz}{d\theta}\frac{h}{r^2} = -h\frac{dz}{d\theta}$$

and

$$\frac{d^2r}{dt^2} = -h\frac{d}{dt}\left(\frac{dz}{d\theta}\right) = -h\frac{d}{d\theta}\left(\frac{dz}{d\theta}\right)\frac{d\theta}{dt} = -h\frac{d^2z}{d\theta^2}\frac{h}{r^2} = -h^2z^2\frac{d^2z}{d\theta^2}.$$

When the latter expression is inserted in (13) and $1/r$ is replaced by z, we get

$$-h^2z^2\frac{d^2z}{d\theta^2} - h^2z^3 = -kz^2$$

or

$$\frac{d^2z}{d\theta^2} + z = \frac{k}{h^2}.$$

The general solution of this equation can be written down at once:

$$z = A \sin \theta + B \cos \theta + \frac{k}{h^2}.\tag{14}$$

For the sake of simplicity, we shift the direction of the polar axis in such a way that r is minimal (that is, m is closest to the origin) when $\theta = 0$. This means that z is to be maximal in this direction, so

$$\frac{dz}{d\theta} = 0 \quad \text{and} \quad \frac{d^2z}{d\theta^2} < 0$$

when $\theta = 0$. These conditions imply that $A = 0$ and $B > 0$. If we now replace z by $1/r$, then (14) can be written

$$r = \frac{1}{k/h^2 + B \cos \theta} = \frac{h^2/k}{1 + (Bh^2/k) \cos \theta},$$

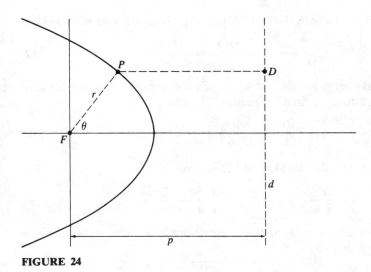

FIGURE 24

and if we put $e = Bh^2/k$, then our equation for the orbit becomes

$$r = \frac{h^2/k}{1 + e \cos \theta},\qquad(15)$$

where e is a positive constant.

At this point we recall (Fig. 24) that the locus defined by $PF/PD = e$ is the conic section with focus F, directrix d, and eccentricity e. When this condition is expressed in terms of r and θ, it is easy to see that

$$r = \frac{pe}{1 + e \cos \theta}$$

is the polar equation of our conic section, which is an ellipse, a parabola, or a hyperbola according as $e < 1$, $e = 1$, or $e > 1$. These remarks show that the orbit (15) is a conic section with eccentricity $e = Bh^2/k$; and since the planets remain in the solar system and do not move infinitely far away from the sun, the ellipse is the only possibility. This yields *Kepler's first law*: the orbit of each planet is an ellipse with the sun at one focus.

The physical meaning of the eccentricity. It follows from equation (4) that the kinetic energy of m is

$$\frac{1}{2}mv^2 = \frac{1}{2}m\left[r^2\left(\frac{d\theta}{dt}\right)^2 + \left(\frac{dr}{dt}\right)^2\right].\qquad(16)$$

The potential energy of the system is the negative of the work required to move m to infinity (where the potential energy is zero), and is therefore

$$-\int_r^\infty \frac{km}{r^2}\, dr = \frac{km}{r}\Big|_r^\infty = -\frac{km}{r}. \tag{17}$$

If E is the total energy of the system, which is constant by the principle of conservation of energy, then (16) and (17) yield

$$\frac{1}{2}m\left[r^2\left(\frac{d\theta}{dt}\right)^2 + \left(\frac{dr}{dt}\right)^2\right] - \frac{km}{r} = E. \tag{18}$$

At the instant when $\theta = 0$, (15) and (18) give

$$r = \frac{h^2/k}{1 + e} \quad \text{and} \quad \frac{mr^2}{2}\frac{h^2}{r^4} - \frac{km}{r} = E.$$

It is easy to eliminate r from these equations; and when the result is solved for e, we find that

$$e = \sqrt{1 + E\left(\frac{2h^2}{mk^2}\right)}.$$

This enables us to write equation (15) for the orbit in the form

$$r = \frac{h^2/k}{1 + \sqrt{1 + E(2h^2/mk^2)}\,\cos\theta}. \tag{19}$$

It is evident from (19) that the orbit is an ellipse, a parabola, or a hyperbola according as $E < 0$, $E = 0$, or $E > 0$. It is therefore clear that the nature of the orbit of m is completely determined by its total energy E. Thus the planets in the solar system have negative energies and move in ellipses, and bodies passing through the solar system at high speeds have positive energies and travel along hyperbolic paths. It is interesting to realize that if a planet like the earth could be given a push from behind, sufficiently strong to speed it up and lift its total energy above zero, it would enter into a hyperbolic orbit and leave the solar system permanently.

The periods of revolution of the planets and Kepler's Third Law. We now restrict our attention to the case in which m has an elliptic orbit (Fig. 25) whose polar and rectangular equations are (15) and

$$\frac{x^2}{a^2} + \frac{y^2}{b^2} = 1.$$

It is well known from elementary analytic geometry that $e = c/a$ and $c^2 = a^2 - b^2$, so $e^2 = (a^2 - b^2)/a^2$ and

$$b^2 = a^2(1 - e^2). \tag{20}$$

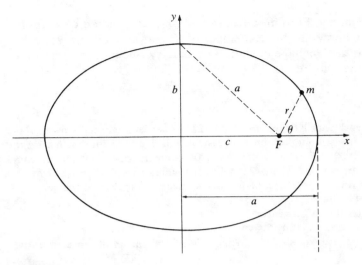

FIGURE 25

In astronomy the semimajor axis of the orbit is called the *mean distance,* because it is one-half the sum of the least and greatest values of r, so (15) and (20) give

$$a = \frac{1}{2}\left(\frac{h^2/k}{1+e} - \frac{h^2/k}{1-e}\right) = \frac{h^2}{k(1-e^2)} = \frac{h^2a^2}{kb^2},$$

and we have

$$b^2 = \frac{h^2a}{k}. \tag{21}$$

If T is the period of m (that is, the time required for one complete revolution in its orbit), then since the area of the ellipse is πab it follows from (10) that $\pi ab = hT/2$. In view of (21), this yields

$$T^2 = \frac{4\pi^2a^2b^2}{h^2} = \left(\frac{4\pi^2}{k}\right)a^3. \tag{22}$$

In the present idealized treatment, the constant $k = GM$ depends on the central mass M but not on m, so (22) holds for all the planets in our solar system and we have *Keplers' third law*: the squares of the periods of revolution of the planets are proportional to the cubes of their mean distances.

The ideas of this section are of course due primarily to Newton (Appendix B). However, the arguments given here are quite different from those that were used in print by Newton himself, for he made no explicit use of the methods of calculus in any of his published works on physics or astronomy. For him calculus was a private method of scientific investigation unknown to his contemporaries, and he had to rewrite his discoveries into the language of classical geometry whenever he wished to communicate them to others.

PROBLEMS

1. In practical work with Kepler's third law (22), it is customary to measure T in years and a in astronomical units (1 astronomical unit = the earth's mean distance $\cong 93,000,000$ miles $\cong 150,000,000$ kilometers). With these convenient units of measurement, (22) takes the simpler form $T^2 = a^3$. What is the period of revolution T of a planet whose mean distance from the sun is
   (a) twice that of the earth?
   (b) three times that of the earth?
   (c) twenty-five times that of the earth?

2. (a) Mercury's "year" is 88 days. What is Mercury's mean distance from the sun?
   (b) The mean distance of the planet Saturn is 9.54 astronomical units. What is Saturn's period of revolution about the sun?

3. Kepler's first two laws, in the form of equations (8) and (15), imply that m is attracted toward the origin with a force whose magnitude is inversely proportional to the square of r. This was Newton's fundamental discovery, for it caused him to propound his law of gravitation and investigate its consequences. Prove this by assuming (8) and (15) and verifying the following statements:

   (a) $F_\theta = 0$;

   (c) $\dfrac{d^2 r}{dt^2} = \dfrac{ke \cos \theta}{r^2}$;

   (b) $\dfrac{dr}{dt} = \dfrac{ke}{h} \sin \theta$;

   (d) $F_r = -\dfrac{mk}{r^2} = -G\dfrac{Mm}{r^2}$.

4. Show that the speed v of a planet at any point of its orbit is given by

$$v^2 = k\left(\frac{2}{r} - \frac{1}{a}\right).$$

5. Suppose that the earth explodes into fragments which fly off at the same speed in different directions into orbits of their own. Use Kepler's third law and the result of Problem 4 to show that all fragments that do not fall into the sun or escape from the solar system will reunite later at the same point where they began to diverge.

22 HIGHER ORDER LINEAR EQUATIONS. COUPLED HARMONIC OSCILLATORS

Even though the main topic of this chapter is second order linear equations, there are several aspects of higher order linear equations that make it worthwhile to discuss them briefly.

Most of the ideas and methods described in Sections 14 to 19 are easily extended to nth order linear equations with constant coefficients,

$$y^{(n)} + a_1 y^{(n-1)} + \cdots + a_{n-1} y' + a_n y = f(x), \tag{1}$$

where $f(x)$ is assumed to be continuous on an interval $[a,b]$. The basic fact to keep in mind is that the general solution of (1) has the form we expect,

$$y(x) = y_g(x) + y_p(x),$$

where $y_p(x)$ is any particular solution of (1) and $y_g(x)$ is the general solution of the reduced homogeneous equation

$$y^{(n)} + a_1 y^{(n-1)} + \cdots + a_{n-1} y' + a_n y = 0. \tag{2}$$

The proof is exactly the same as the proof for the case $n = 2$, and will not be repeated.

We begin by considering the problem of finding the general solution of the homogeneous equation (2). Our experience with the case $n = 2$ tells us that this equation probably has solutions of the form $y = e^{rx}$ for suitable values of the constant r. By substituting $y = e^{rx}$ and its derivatives into (2) and dividing out the nonzero factor e^{rx}, we obtain the *auxiliary equation*

$$r^n + a_1 r^{n-1} + \cdots + a_{n-1} r + a_n = 0. \tag{3}$$

The polynomial on the left side of (3) is called the *auxiliary polynomial*; in principle it can always be factored completely into a product of n linear factors, and equation (3) can then be written in the factored form

$$(r - r_1)(r - r_2) \cdots (r - r_n) = 0.$$

The constants $r_1, r_2, \ldots, r_n$ are the roots of the auxiliary equation (3). If these roots are distinct from one another, then we have n distinct solutions

$$e^{r_1 x}, e^{r_2 x}, \ldots, e^{r_n x} \tag{4}$$

of the homogeneous equation (2). Just as in the case $n = 2$, the linear combination

$$y(x) = c_1 e^{r_1 x} + c_2 e^{r_2 x} + \cdots + c_n e^{r_n x} \tag{5}$$

is also a solution for every choice of the coefficients $c_1, c_2, \ldots, c_n$.

Since (5) contains n arbitrary constants, we have reasonable grounds for hoping that it is the general solution of the nth order equation (2). To elevate this hope into a certainty, e must appeal to a small body of theory that we now sketch very briefly.

When the theorems of Sections 14 and 15 are extended in the natural way, it can be proved that (5) is the general solution of (2) if the solutions (4) are linearly independent.[11] There are several ways of establishing the fact that the solutions (4) are linearly independent whenever the roots $r_1, r_2, \ldots, r_n$ are distinct, but we omit the details. It therefore follows that (5) actually is the general solution of (2) in this case.

Repeated real roots. If the real roots of (3) are not all distinct, then the solutions (4) are linearly dependent and (5) is not the general solution. For example, if $r_1 = r_2$ then the part of (5) consisting of $c_1 e^{r_1 x} + c_2 e^{r_2 x}$ becomes $(c_1 + c_2)e^{r_1 x}$, and the two constants c_1 and c_2 become one constant $c_1 + c_2$. To see what to do when this happens, we recall that in the special case of the second order equation, where we had only the two roots r_1 and r_2, we found that when $r_1 = r_2$ the solution $c_1 e^{r_1 x} + c_2 e^{r_2 x}$ had to be replaced by $c_1 e^{r_1 x} + c_2 x e^{r_1 x} = (c_1 + c_2 x)e^{r_1 x}$. It can be verified by substitution that if $r_1 = r_2$ for the nth order equation (2), then the first two terms of (5) must be replaced by this same expression.

More generally, if $r_1 = r_2 = \cdots = r_k$ is a real root of multiplicity k (that is, a k-fold repeated root) of the auxiliary equation (3), then the first k terms in the solution (5) must be replaced by

$$(c_1 + c_2 x + c_3 x^2 + \cdots + c_k x^{k-1})e^{r_1 x}.$$

A similar family of solutions is needed for each multiple real root, giving a correspondingly modified form of (5). In the next section we will show how to obtain these expressions by operator methods.

Complex roots. Some of the roots of the auxiliary equation (3) may be complex numbers. Since the coefficients of (3) are real, all complex roots occur in conjugate complex pairs $a + ib$ and $a - ib$. As in the case $n = 2$, the part of the solution (5) corresponding to two such roots can be written in the alternative real form

$$e^{ax}(A \cos bx + B \sin bx).$$

If $a + ib$ and $a - ib$ are roots of multiplicity k, then we must take

$$e^{ax}[(A_1 + A_2 x + \cdots + A_k x^{k-1}) \cos bx$$
$$+ (B_1 + B_2 x + \cdots + B_k x^{k-1}) \sin bx]$$

as part of the general solution.

---

[11] This requires establishing the same connections as before among (1) satisfying n initial conditions, (2) the nonvanishing of the Wronskian, (3) Abel's formula, and (4) linear independence. A set of n functions $y_1(x), y_2(x), \ldots, y_n(x)$ is said to be *linearly dependent* if one of them can be expressed as a linear combination of the others, and *linearly independent* if this is not possible. In specific cases this is usually easy to decide by inspection. Equivalently, linear dependence means that there exists a relation of the form $c_1 y_1(x) + c_2 y_2(x) + \cdots + c_n y_n(x) = 0$ in which at least one of the c's is not zero, and linear independence means that any such relation implies that all the c's must be zero.

Example 1. The differential equation

$$y^{(4)} - 5y'' + 4y = 0$$

has auxiliary equation

$$r^4 - 5r^2 + 4 = (r^2 - 1)(r^2 - 4) = (r - 1)(r + 1)(r - 2)(r + 2) = 0.$$

Its general solution is therefore

$$y = c_1 e^x + c_2 e^{-x} + c_3 e^{2x} + c_4 e^{-2x}.$$

Example 2. The equation

$$y^{(4)} - 8y'' + 16y = 0$$

has auxiliary equation

$$r^4 - 8r^2 + 16 = (r^2 - 4)^2 = (r - 2)^2(r + 2)^2 = 0,$$

so the general solution is

$$y = (c_1 + c_2 x)e^{2x} + (c_3 + c_4 x)e^{-2x}.$$

Example 3. The equation

$$y^{(4)} - 2y''' + 2y'' - 2y' + y = 0$$

has auxiliary equation

$$r^4 - 2r^3 + 2r^2 - 2r + 1 = 0,$$

or, after factoring,[12]

$$(r - 1)^2(r^2 + 1) = 0.$$

The general solution is therefore

$$y = (c_1 + c_2 x)e^x + c_3 \cos x + c_4 \sin x.$$

Example 4. Coupled harmonic oscillators. Linear equations of order $n > 2$ arise most often in physics by eliminating variables from simultaneous systems of second order equations. We can see an example of this by linking together two simple harmonic oscillators of the kind discussed at the beginning of Section 20. Accordingly, let two carts of masses m_1 and m_2 be attached to the left and right walls in Fig. 26 by springs with spring constants k_1 and k_2. If there is no damping and these carts are left unconnected, then when disturbed each moves with its own simple harmonic motion, that is, we have two independent harmonic oscillators. We obtain *coupled harmonic oscillators* if we now connect the carts to each

---

[12] To factor the auxiliary equation, notice that $r = 1$ is a root that can be found by inspection, so $r - 1$ is a factor of the auxiliary polynomial and the other factor can be found by long division.

other by a spring with spring constant k_3, as indicated in the figure. By applying Newton's second law of motion, it can be shown (Problem 16) that the displacements x_1 and x_2 of the carts satisfy the following simultaneous system of second order linear equations:

$$m_1 \frac{d^2 x_1}{dt^2} = -k_1 x_1 + k_3(x_2 - x_1)$$

$$= -(k_1 + k_3)x_1 + k_3 x_2,$$

$$m_2 \frac{d^2 x_2}{dt^2} = -k_3(x_2 - x_1) - k_2 x_2 \qquad (6)$$

$$= k_3 x_1 - (k_2 + k_3)x_2.$$

We can now obtain a single fourth order equation for x_1 by solving the first equation for x_2 and substituting in the second equation (Problem 17).

We have not yet addressed the problem of finding a particular solution for the complete equation (1). In this context it suffices to remark that the method of undetermined coefficients discussed in Section 18 continues to apply, with obvious minor changes, for functions $f(x)$ of the types considered in that section. In the next section we shall examine a totally different approach to the problem of finding particular solutions.

Example 5. Find a particular solution of the differential equation $y''' + 2y'' - y' = 3x^2 - 2x + 1$.

Our experience in Section 18 suggests that we take a trial solution of the form

$$y = x(a_0 + a_1 x + a_2 x^2)$$

$$= a_0 x + a_1 x^2 + a_2 x^3.$$

Since $y' = a_0 + 2a_1 x + 3a_2 x^2$, $y'' = 2a_1 + 6a_2 x$, and $y''' = 6a_2$, substitution in the given equation yields

$$6a_2 + 2(2a_1 + 6a_2 x) - (a_0 + 2a_1 x + 3a_2 x^2) = 3x^2 - 2x + 1$$

or, after collecting coefficients of like powers of x,

$$-3a_2 x^2 + (-2a_1 + 12a_2)x + (-a_0 + 4a_1 + 6a_2) = 3x^2 - 2x + 1.$$

Thus,

$$-3a_2 = 3,$$

$$-2a_1 + 12a_2 = -2,$$

$$-a_0 + 4a_1 + 6a_2 = 1,$$

so $a_2 = -1$, $a_1 = -5$, and $a_0 = -27$. We therefore have a particular solution $y = -27x - 5x^2 - x^3$.

PROBLEMS

Find the general solution of each of the following equations.

1. $y''' - 3y'' + 2y' = 0.$
2. $y''' - 3y'' + 4y' - 2y = 0.$
3. $y''' - y = 0.$
4. $y''' + y = 0.$
5. $y''' + 3y'' + 3y' + y = 0.$
6. $y^{(4)} + 4y''' + 6y'' + 4y' + y = 0.$
7. $y^{(4)} - y = 0.$
8. $y^{(4)} + 5y'' + 4y = 0.$
9. $y^{(4)} - 2a^2y'' + a^4y = 0.$
10. $y^{(4)} + 2a^2y'' + a^4y = 0.$
11. $y^{(4)} + 2y''' + 2y'' + 2y' + y = 0.$
12. $y^{(4)} + 2y''' - 2y'' - 6y' + 5y = 0.$
13. $y''' - 6y'' + 11y' - 6y = 0.$
14. $y^{(4)} + y''' - 3y'' - 5y' - 2y = 0.$
15. $y^{(5)} - 6y^{(4)} - 8y''' + 48y'' + 16y' - 96y = 0.$
16. Derive equations (6) for the coupled harmonic oscillators by using the configuration shown in Fig. 26, where both carts are displaced to the right from their equilibrium positions and $x_2 > x_1$, so that the spring on the right is compressed and the other two are stretched.
17. In Example 4, find the fourth order differential equation for x_1 by eliminating x_2 as suggested.
18. In the preceding problem, solve the fourth order equation for x_1 if the masses are equal and the spring constants are equal, so that $m_1 = m_2 = m$ and $k_1 = k_2 = k_3 = k$. In this special case, show directly (that is, without using the symmetry of the situation) that x_2 satisfies the same differential equation as x_1. The two frequencies associated with these coupled harmonic oscillators are called the *normal frequencies* of the system. What are they?
19. Find the general solution of $y^{(4)} = 0.$ Of $y^{(4)} = \sin x + 24.$
20. Find the general solution of $y''' - 3y'' + 2y' = 10 + 42e^{3x}.$

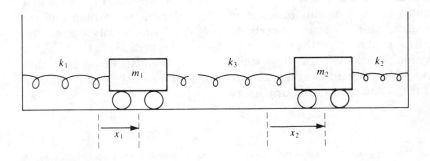

FIGURE 26

21. Find the solution of $y''' - y' = 1$ that satisfies the initial conditions $y(0) = y'(0) = y''(0) = 4$.

22. Show that the change of independent variable $x = e^z$ transforms the third order Euler equidimensional equation

$$x^3 y''' + a_1 x^2 y'' + a_2 xy' + a_3 y = 0$$

into a third order linear equation with constant coefficients. (This transformation also works for the nth order Euler equation.) Solve the following equations by this method:

(a) $x^3 y''' + 3x^2 y'' = 0$;   (c) $x^3 y''' + 2x^2 y'' + xy' - y = 0$.

(b) $x^3 y''' + x^2 y'' - 2xy' + 2y = 0$;

23. In determining the drag on a small sphere moving at a constant speed through a viscous fluid, it is necessary to solve the differential equation

$$x^3 y^{(4)} + 8x^2 y''' + 8xy'' - 8y' = 0.$$

Notice that this is an Euler equation for y' and use the method of Problem 22 to show that the general solution is

$$y = c_1 x^2 + c_2 x^{-1} + c_3 x^{-3} + c_4.$$

These ideas are part of the mathematical background used by Robert A. Millikan in his famous oil-drop experiment of 1909 for measuring the charge on an electron, for which he won the 1923 Nobel Prize.[13]

23 OPERATOR METHODS FOR FINDING PARTICULAR SOLUTIONS

At the end of Section 22 we referred to the problem of finding particular solutions for nonhomogeneous equations of the form

$$\frac{d^n y}{dx^n} + a_1 \frac{d^{n-1} y}{dx^{n-1}} + \cdots + a_{n-1} \frac{dy}{dx} + a_n y = f(x). \tag{1}$$

In this section we give a very brief sketch of the use of differential operators for solving this problem in more efficient ways than any we have seen before. These "operational methods" are mainly due to the English applied mathematician Oliver Heaviside (1850–1925). Heaviside's methods seemed so strange to the scientists of his time that he was widely regarded as a crackpot, which unfortunately is a common fate for thinkers of unusual originality.

---

[13] For a clear explanation of this exceedingly ingenious experiment, with a good drawing of the apparatus, see pp. 50–51 of the book by Linus Pauling mentioned in Section 4 [Note 12].

Let us represent derivatives by powers of D, so that

$$Dy = \frac{dy}{dx}, \qquad D^2 y = \frac{d^2 y}{dx^2}, \qquad \cdots, \qquad D^n y = \frac{d^n y}{dx^n}.$$

Then (1) can be written as

$$D^n y + a_1 D^{n-1} y + \cdots + a_{n-1} Dy + a_n y = f(x), \qquad (2)$$

or as

$$(D^n + a_1 D^{n-1} + \cdots + a_{n-1} D + a_n) y = f(x),$$

or as

$$p(D) y = f(x), \qquad (3)$$

where the differential operator $p(D)$ is simply the auxiliary polynomial $p(r)$ with r replaced by D. The successive application of two or more such operators can be made by first multiplying the operators together by the usual rules of algebra and then applying the product operator. For example, we know that $p(D)$ can be formally factored into

$$p(D) = (D - r_1)(D - r_2) \cdots (D - r_n), \qquad (4)$$

where $r_1, r_2, \ldots, r_n$ are the roots of the auxiliary equation; and these factors can then be applied successively in any order to yield the same result as a single application of $p(D)$. As an illustration of this idea, we point out that if the auxiliary equation is of the second decree and therefore has only two roots r_1 and r_2, then formally we have

$$(D - r_1)(D - r_2) = D^2 - (r_1 + r_2)D + r_1 r_2; \qquad (5)$$

and since

$$(D - r_2)y = \left(\frac{d}{dx} - r_2\right) y = \frac{dy}{dx} - r_2 y,$$

we can verify (5) by writing

$$\begin{aligned}
(D - r_1)(D - r_2)y &= \left(\frac{d}{dx} - r_1\right)\left(\frac{dy}{dx} - r_2 y\right) \\
&= \frac{d}{dx}\left(\frac{dy}{dx} - r_2 y\right) - r_1\left(\frac{dy}{dx} - r_2 y\right) \\
&= \frac{d^2 y}{dx^2} - (r_1 + r_2)\frac{dy}{dx} + r_1 r_2 y \\
&= D^2 y - (r_1 + r_2)Dy + r_1 r_2 y \\
&= [D^2 - (r_1 + r_2)D + r_1 r_2]y,
\end{aligned}$$

for this is the meaning of (5).

We have no difficulty with the meaning of the expression $p(D)y$ on the left of (3): it has the same meaning as the left side of (2) or (1). Our purpose now is to learn how to treat $p(D)$ as a separate entity, and in doing this to develop the methods for solving (3) that are the subject of this section. Without beating around the bush, we wish to "solve formally" for y in (3), obtaining

$$y = \frac{1}{p(D)} f(x). \tag{6}$$

Here $1/p(D)$ represents an operation to be performed on $f(x)$ to yield y. The question is, what is the nature of this operation, and how can we carry it out? In order to begin to understand these matters, we consider the simple equation $Dy = f(x)$, which gives

$$y = \frac{1}{D} f(x).$$

But $Dy = f(x)$, or equivalently $dy/dx = f(x)$, is easily solved by writing

$$y = \int f(x)\, dx,$$

so it is natural to make the definition

$$\frac{1}{D} f(x) = \int f(x)\, dx. \tag{7}$$

This tells us that the operator $1/D$ applied to a function means integrate the function. Similarly, the operator $1/D^2$ applied to a function means integrate the function twice in succession, and so on. Operators like $1/D$ and $1/D^2$ are called *inverse operators*. We continue this line of investigation and examine other inverse operators. Consider

$$(D - r)y = f(x), \tag{8}$$

where r is a constant. Formally, we have

$$y = \frac{1}{D - r} f(x).$$

But (8) is the simple first order linear equation

$$\frac{dy}{dx} - ry = f(x),$$

whose solution by Section 10 is

$$y = e^{rx} \int e^{-rx} f(x)\, dx.$$

(We suppress constants of integration because we are only seeking particular solutions.) It is therefore natural to make the definition

$$\frac{1}{D-r}f(x) = e^{rx}\int e^{-rx}f(x)\,dx. \tag{9}$$

Notice that this reduces to (7) when $r = 0$. We are now ready to begin carrying out the problem-solving procedures that arise from (6).

METHOD 1: SUCCESSIVE INTEGRATIONS. By using the factorization (4), we can write formula (6) as

$$y = \frac{1}{p(D)}f(x) = \frac{1}{(D-r_1)(D-r_2)\cdots(D-r_n)}f(x)$$

$$= \frac{1}{D-r_1}\frac{1}{D-r_2}\cdots\frac{1}{D-r_n}f(x).$$

Here we may apply the n inverse operators in any convenient order, and by (9) we know that the complete process requires n successive integrations. That the resulting function $y = y(x)$ is a particular solution of (3) is easily seen; for by applying to y the factors of $p(D)$ in suitable order, we undo the successive integrations and arrive back at $f(x)$.

Example 1. Find a particular solution of $y'' - 3y' + 2y = xe^x$.

Solution. We have $(D^2 - 3D + 2)y = xe^x$, so

$$(D-1)(D-2)y = xe^x \qquad \text{and} \qquad y = \frac{1}{D-1}\frac{1}{D-2}xe^x.$$

By (9) and an integration by parts, we obtain

$$\frac{1}{D-2}xe^x = e^{2x}\int e^{-2x}xe^x\,dx = -(1+x)e^x,$$

so

$$y = \frac{1}{D-1}[-(1+x)e^x] = -e^x\int e^{-x}(1+x)e^x\,dx = -\tfrac{1}{2}(1+x)^2e^x.$$

Example 2. Find a particular solution of $y'' - y = e^{-x}$.

Solution. We have $(D^2 - 1)y = e^{-x}$, so

$$(D-1)(D+1)y = e^{-x}, \qquad y = \frac{1}{D-1}\frac{1}{D+1}e^{-x},$$

$$\frac{1}{D+1}e^{-x} = e^{-x}\int e^x e^{-x}\,dx = xe^{-x},$$

$$y = \frac{1}{D-1}xe^{-x} = e^x\int e^{-x}xe^{-x}\,dx = (-\tfrac{1}{2}x - \tfrac{1}{4})e^{-x}.$$

METHOD 2: PARTIAL FRACTIONS DECOMPOSITIONS OF OPER-ATORS. The successive integrations of method 1 are likely to become complicated and time-consuming to carry out. The formula

$$y = \frac{1}{p(D)} f(x) = \frac{1}{(D - r_1)(D - r_2) \cdots (D - r_n)} f(x)$$

suggests a way to avoid this work, for it suggests the possibility of decomposing the operator on the right into partial fractions. If the factors of $p(D)$ are distinct, we can write

$$y = \frac{1}{p(D)} f(x) = \left[\frac{A_1}{D - r_1} + \frac{A_2}{D - r_2} + \cdots + \frac{A_n}{D - r_n} \right] f(x)$$

for suitable constants $A_1, A_2, \ldots, A_n$, and each term on the right can be found by using (9). The operator in brackets here is sometimes called the *Heaviside expansion* of the inverse operator $1/p(D)$.

Example 3. Solve the problem in Example 1 by this method.

Solution. We have

$$y = \frac{1}{(D - 1)(D - 2)} xe^x = \left[\frac{1}{D - 2} - \frac{1}{D - 1} \right] xe^x$$

$$= \frac{1}{D - 2} xe^x - \frac{1}{D - 1} xe^x$$

$$= e^{2x} \int e^{-2x} xe^x \, dx - e^x \int e^{-x} xe^x \, dx$$

$$= -1(1 + x)e^x - \tfrac{1}{2}x^2 e^x = -(1 + x + \tfrac{1}{2}x^2)e^x.$$

The student will notice that this solution is not quite the same as the solution found in Example 1. However, it is easy to see that they differ only by a solution of the reduced homogeneous equation, so all is well.

Example 4. Solve the problem in Example 2 by this method.

Solution. We have

$$y = \frac{1}{(D - 1)(D + 1)} e^{-x} = \frac{1}{2} \left[\frac{1}{D - 1} - \frac{1}{D + 1} \right] e^{-x}$$

$$= \tfrac{1}{2} e^x \int e^{-x} e^{-x} \, dx - \tfrac{1}{2} e^{-x} \int e^x e^{-x} \, dx$$

$$= -\tfrac{1}{4} e^{-x} - \tfrac{1}{2} xe^{-x}.$$

If some of the factors of $p(D)$ are repeated, then we know that the form of the partial fractions decomposition is different. For example, if $D - r_1$ is a k-fold repeated factor, then the decomposition contains the terms

$$\frac{A_1}{D - r_1} + \frac{A_2}{(D - r_1)^2} + \cdots + \frac{A_k}{(D - r_1)^k}.$$

These operators can be applied to $f(x)$ in order from left to right, each requiring an integration based on the result of the preceding step, as in method 1.

METHOD 3: SERIES EXPANSIONS OF OPERATORS. For problems in which $f(x)$ is a polynomial, it is often useful to expand the inverse operator $1/p(D)$ in a power series in D, so that

$$y = \frac{1}{p(D)} f(x) = (1 + b_1 D + b_2 D^2 + \cdots) f(x).$$

The reason for this is that high derivatives of polynomials disappear, because $D^k x^n = 0$ if $k > n$.

Example 5. Find a particular solution of $y''' - 2y'' + y = x^4 + 2x + 5$.

Solution. We have $(D^3 - 2D^2 + 1)y = x^4 + 2x + 5$, so

$$y = \frac{1}{1 - 2D^2 + D^3} (x^4 + 2x + 5).$$

By ordinary long division we find that

$$\frac{1}{1 - 2D^2 + D^3} = 1 + 2D^2 - D^3 + 4D^4 - 4D^5 + \cdots,$$

so

$$y = (1 + 2D^2 - D^3 + 4D^4 - 4D^5 + \cdots)(x^4 + 2x + 5)$$
$$= (x^4 + 2x + 5) + 2(12x^2) - (24x) + 4(24)$$
$$= x^4 + 24x^2 - 22x + 101.$$

In order to make the fullest use of this method, it is desirable to keep in mind the following series expansions from elementary algebra:

$$\frac{1}{1 - r} = 1 + r + r^2 + r^3 + \cdots \quad \text{and} \quad \frac{1}{1 + r} = 1 - r + r^2 - r^3 + \cdots.$$

In this context we are only interested in these formulas as "formal" series expansions, and have no need to concern ourselves with their convergence behavior.

Example 6. Find a particular solution of $y''' + y'' + y' + y = x^5 - 2x^2 + x$.

Solution. We have $(D^3 + D^2 + D + 1)y = x^5 - 2x^2 + x$, so

$$y = \frac{1}{1 + D + D^2 + D^3}(x^5 - 2x^2 + x)$$

$$= \frac{1}{1 - D^4}(1 - D)(x^5 - 2x^2 + x)$$

$$= \frac{1}{1 - D^4}[(x^5 - 2x^2 + x) - (5x^4 - 4x + 1)]$$

$$= (1 + D^4 + D^8 + \cdots)[x^5 - 5x^4 - 2x^2 + 5x - 1]$$

$$= (x^5 - 5x^4 - 2x^2 + 5x - 1) + (120x - 120)$$

$$= x^5 - 5x^4 - 2x^2 + 125x - 121.$$

The remarkable thing about the procedures illustrated in these examples is that they actually work!

METHOD 4: THE EXPONENTIAL SHIFT RULE. As we know, exponential functions behave in a special way under differentiation. This fact enables us to simplify our work whenever $f(x)$ contains a factor of the form e^{kx}. Thus, if $f(x) = e^{kx}g(x)$, we begin by noticing that

$$(D - r)f(x) = (D - r)e^{kx}g(x)$$

$$= e^{kx}Dg(x) + ke^{kx}g(x) - re^{kx}g(x)$$

$$= e^{kx}(D + k - r)g(x).$$

By applying this formula to the successive factors $D - r_1$, $D - r_2, \ldots, D - r_n$, we see that for the polynomial operator $p(D)$,

$$p(D)e^{kx}g(x) = e^{kx}p(D + k)g(x). \tag{10}$$

This says that we can move the factor e^{kx} to the left of the operator $p(D)$ if we replace D by $D + k$ in the operator.

The same property is valid for the inverse operator $1/p(D)$, that is,

$$\frac{1}{p(D)}e^{kx}g(x) = e^{kx}\frac{1}{p(D + k)}g(x). \tag{11}$$

To see this, we simply apply $p(D)$ to the right side and use (10):

$$p(D)e^{kx}\frac{1}{p(D - k)}g(x) = e^{kx}p(D + k)\frac{1}{p(D + k)}g(x) = e^{kx}g(x).$$

Properties (10) and (11) are called the *exponential shift rule*. They are useful in moving exponential functions out of the way of operators.

Example 7. Solve the problem in Example 1 by this method.

Solution. We have $(D^2 - 3D + 2)y = xe^x$, so

$$y = \frac{1}{D^2 - 3D + 2}xe^x = e^x\frac{1}{(D + 1)^2 - 3(D + 1) + 2}x$$

$$= e^x\frac{1}{D^2 - D}x = -e^x\frac{1}{D}\frac{1}{1 - D}x$$

$$= -e^x\left(\frac{1}{D} + 1 + D + D^2 + \cdots\right)x$$

$$= -e^x(\tfrac{1}{2}x^2 + x + 1),$$

as we have already seen in Examples 1 and 3.

Interested readers will find additional material on the methods of this section in the "Historical Introduction" to H. S. Carslaw and J. C. Jaeger, *Operational Methods In Applied Mathematics*, Dover, New York, 1963; and in E. Stephens, *The Elementary Theory of Operational Mathematics*, McGraw-Hill, New York, 1937.

PROBLEMS

1. Find a particular solution of $y'' - 4y = e^{2x}$ by using each of Methods 1 and 2.
2. Find a particular solution of $y'' - y = x^2e^{2x}$ by using each of Methods 1, 2, and 4.

In Problems 3 to 6, find a particular solution by using Method 1.

3. $y'' + 4y' + 4y = 10x^3e^{-2x}$.
4. $y'' - 2y' + y = e^x$.
5. $y'' - y = e^{-x}$.
6. $y'' - 2y' - 3y = 6e^{5x}$.

In Problems 7 to 15, find a particular solution by using Method 3.

7. $y'' - y' + y = x^3 - 3x^2 + 1$.
8. $y''' - 2y' + y = 2x^3 - 3x^2 + 4x + 5$.
9. $4y'' + y = x^4$.
10. $y^{(5)} - y''' = x^2$.
11. $y^{(6)} - y = x^{10}$.
12. $y'' + y' - y = 3x - x^4$.
13. $y'' + y = x^4$.
14. $y''' - y'' = 12x - 2$.
15. $y''' + y'' = 9x^2 - 2x + 1$.

In Problems 16 to 18, find a particular solution by using Method 4.

16. $y'' - 4y' + 3y = x^3 e^{2x}$.
17. $y'' - 7y' + 12y = e^{2x}(x^3 - 5x^2)$.
18. $y'' + 2y' + y = 2x^2 e^{-2x} + 3e^{2x}$.

In Problems 19 to 24, find a particular solution by any method.

19. $y''' - 8y = 16x^2$.
20. $y^{(4)} - y = 1 - x^3$.
21. $y''' - \frac{1}{4}y' = x$.
22. $y^{(4)} = x^{-3}$.
23. $y''' - y'' + y' = x + 1$.
24. $y''' + 2y'' = x$.
25. Use the exponential shift rule to find the general solution of each of the following equations:
  (a) $(D - 2)^3 y = e^{2x}$ [hint: multiply by e^{-2x} and use (10)];
  (b) $(D + 1)^3 y = 12e^{-x}$;
  (c) $(D - 2)^2 y = e^{2x} \sin x$.
26. Consider the nth order homogeneous equation $p(D)y = 0$.
  (a) If a polynomial $q(r)$ is a factor of the auxiliary polynomial $p(r)$, show that any solution of the differential equation $q(D)y = 0$ is also a solution of $p(D)y = 0$.
  (b) If r_1 is a root of multiplicity k of the auxiliary equation $p(r) = 0$, show that any solution of $(D - r_1)^k y = 0$ is also a solution of $p(D)y = 0$.
  (c) Use the exponential shift rule to show that $(D - r_1)^k y = 0$ has

$$y = (c_1 + c_2 x + c_3 x^2 + \cdots + c_k x^{k-1})e^{r_1 x}$$

  as its general solution. *Hint:* $(D - r_1)^k y = 0$ is equivalent to $e^{r_1 x} D^k (e^{-r_1 x} y) = 0$.

APPENDIX A. EULER

Leonhard Euler (1707–1783) was Switzerland's foremost scientist and one of the three greatest mathematicians of modern times (the other two being Gauss and Riemann).

He was perhaps the most prolific author of all time in any field. From 1727 to 1783 his writings poured out in a seemingly endless flood, constantly adding knowledge to every known branch of pure and applied mathematics, and also to many that were not known until he created them. He averaged about 800 printed pages a year throughout his long life, and yet he almost always had something worthwhile to say and never seems long-winded. The publication of his complete works was started in 1911, and the end is not in sight. This edition was planned to include 887 titles in 72 volumes, but since that time extensive new deposits of previously unknown manuscripts have been unearthed, and it is now estimated that more than 100 large volumes will be required for completion of the project. Euler evidently wrote mathematics with the ease and fluency of a skilled speaker discoursing on subjects with which he is intimately familiar. His writings are models of relaxed clarity. He never condensed, and he reveled in the rich abundance of his ideas and the vast scope of his interests. The French physicist Arago, in speaking of Euler's incomparable mathematical facility, remarked that "He calculated without apparent effort, as men breathe, or as eagles sustain themselves in the wind." He suffered total blindness during the last 17 years of his life, but with the aid of his powerful memory and fertile imagination, and with helpers to write his books and scientific papers from dictation, he actually increased his already prodigious output of work.

Euler was a native of Basel and a student of John Bernoulli at the University, but he soon outstripped his teacher. His working life was spent as a member of the Academies of Science at Berlin and St. Petersburg, and most of his papers were published in the journals of these organizations. His business was mathematical research, and he knew his business. He was also a man of broad culture, well versed in the classical languages and literatures (he knew the *Aeneid* by heart), many modern languages, physiology, medicine, botany, geography, and the entire body of physical science as it was known in his time. However, he had little talent for metaphysics or disputation, and came out second best in many good-natured verbal encounters with Voltaire at the court of Frederick the Great. His personal life was as placid and uneventful as is possible for a man with 13 children.

Though he was not himself a teacher, Euler has had a deeper influence on the teaching of mathematics than any other man. This came about chiefly through his three great treatises: *Introductio in Analysin Infinitorum* (1748); *Institutiones Calculi Differentialis* (1755); and *Institutiones Calculi Integralis* (1768–1794). There is considerable truth in the old saying that all elementary and advanced calculus textbooks since 1748 are essentially copies of Euler or copies of copies of Euler.[14] These works summed up and codified the discoveries of his predecessors, and are full of Euler's own ideas. He extended and perfected plane and solid analytic geometry, introduced the analytic approach to trigonometry, and was responsible for the modern treatment of the functions $\log x$ and e^x. He created a consistent theory of logarithms of negative and imaginary numbers, and discovered that $\log x$ has an infinite number of values. It

---

[14] See C. B. Boyer, "The Foremost Textbook of Modern Times," *Am. Math. Monthly*, Vol. 58, pp. 223–226, 1951.

was through his work that the symbols e, π, and i $(=\sqrt{-1})$ became common currency for all mathematicians, and it was he who linked them together in the astonishing relation $e^{\pi i} = -1$. This is merely a special case (put $\theta = \pi$) of his famous formula $e^{i\theta} = \cos \theta + i \sin \theta$, which connects the exponential and trigonometric functions and is absolutely indispensable in higher analysis.[15] Among his other contributions to standard mathematical notation were $\sin x$, $\cos x$, the use of $f(x)$ for an unspecified function, and the use of Σ for summation.[16] Good notations are important, but the ideas behind them are what really count, and in this respect Euler's fertility was almost beyond belief. He preferred concrete special problems to the general theories in vogue today, and his unique insight into the connections between apparently unrelated formulas blazed many trails into new areas of mathematics which he left for his successors to cultivate.

He was the first and greatest master of infinite series, infinite products, and continued fractions, and his works are crammed with striking discoveries in these fields. James Bernoulli (John's older brother) found the sums of several infinite series, but he was not able to find the sum of the reciprocals of the squares, $1 + \frac{1}{4} + \frac{1}{9} + \frac{1}{16} + \cdots$. He wrote, "If someone should succeed in finding this sum, and will tell me about it, I shall be much obliged to him." In 1736, long after James's death, Euler made the wonderful discovery that

$$1 + \frac{1}{4} + \frac{1}{9} + \frac{1}{16} + \cdots = \frac{\pi^2}{6}.$$

He also found the sums of the reciprocals of the fourth and sixth powers,

$$1 + \frac{1}{2^4} + \frac{1}{3^4} + \cdots = 1 + \frac{1}{16} + \frac{1}{81} + \cdots = \frac{\pi^4}{90}$$

and

$$1 + \frac{1}{2^6} + \frac{1}{3^6} + \cdots = 1 + \frac{1}{64} + \frac{1}{729} + \cdots = \frac{\pi^6}{945}.$$

---

[15] An even more astonishing consequence of his formula is the fact that an imaginary power of an imaginary number can be real, in particular $i^i = e^{-\pi/2}$; for if we put $\theta = \pi/2$, we obtain $e^{\pi i/2} = i$, so

$$i^i = (e^{\pi i/2})^i = e^{\pi i^2/2} = e^{-\pi/2}.$$

Euler further showed that i^i has infinitely many values, of which this calculation produces only one.

[16] See F. Cajori, *A History of Mathematical Notations*, Open Court, Chicago, 1929.

When John heard about these feats, he wrote, "If only my brother were alive now."[17] Few would believe that these formulas are related—as they are—to Wallis's infinite product (1656),

$$\frac{\pi}{2} = \frac{2}{1} \cdot \frac{2}{3} \cdot \frac{4}{3} \cdot \frac{4}{5} \cdot \frac{6}{5} \cdot \frac{6}{7} \cdots.$$

Euler was the first to explain this in a satisfactory way, in terms of his infinite product expansion of the sine,

$$\frac{\sin x}{x} = \left(1 - \frac{x^2}{\pi^2}\right)\left(1 - \frac{x^2}{4\pi^2}\right)\left(1 - \frac{x^2}{9\pi^2}\right)\cdots.$$

Wallis's product is also related to Brouncker's remarkable continued fraction,

$$\frac{\pi}{4} = \cfrac{1}{1 + \cfrac{1^2}{2 + \cfrac{3^2}{2 + \cfrac{5^2}{2 + \cfrac{7^2}{2 + \cdots}}}}},$$

which became understandable only in the context of Euler's extensive researches in this field.

His work in all departments of analysis strongly influenced the further development of this subject through the next two centuries. He contributed many important ideas to differential equations, including substantial parts of the theory of second order linear equations and the method of solution by power series. He gave the first systematic discussion of the calculus of variations, which he founded on his basic differential equation for a minimizing curve. He introduced the number now known as *Euler's constant,*

$$\gamma = \lim_{n \to \infty} \left(1 + \frac{1}{2} + \frac{1}{3} + \cdots + \frac{1}{n} - \log n\right) = 0.5772\ldots,$$

which is the most important special number in mathematics after π and e. He discovered the integral defining the gamma function,

$$\Gamma(x) = \int_0^\infty t^{x-1} e^{-t}\, dt,$$

---

[17] The world is still waiting—more than 200 years later—for someone to discover the sum of the reciprocals of the cubes.

which is often the first of the so-called higher transcendental functions that students meet beyond the level of calculus, and he developed many of its applications and special properties. He also worked with Fourier series, encountered the Bessel functions in his study of the vibrations of a stretched circular membrane, and applied Laplace transforms to solve differential equations—all before Fourier, Bessel, and Laplace were born. Even though Euler died about 200 years ago, he lives everywhere in analysis.

E. T. Bell, the well-known historian of mathematics, observed that "One of the most remarkable features of Euler's universal genius was its equal strength in both of the main currents of mathematics, the continuous and the discrete." In the realm of the discrete, he was one of the originators of modern number theory and made many far-reaching contributions to this subject throughout his life. In addition, the origins of topology—one of the dominant forces in modern mathematics—lie in his solution of the Königsberg bridge problem and his formula $V - E + F = 2$ connecting the numbers of vertices, edges, and faces of a simple polyhedron. In the following paragraphs, we briefly describe some of his activities in these fields.

In number theory, Euler drew much of his inspiration from the challenging marginal notes left by Fermat in his copy of the works of Diophantus. He gave the first published proofs of both Fermat's theorem and Fermat's two squares theorem. He later generalized the first of these classic results by introducing the Euler ϕ function; his proof of the second cost him 7 years of intermittent effort. In addition, he proved that every positive integer is a sum of four squares and investigated the law of quadratic reciprocity.

Some of his most interesting work was connected with the sequence of prime numbers, that is, with those integers $p > 1$ whose only positive divisors are 1 and p. His use of the divergence of the harmonic series $1 + \frac{1}{2} + \frac{1}{3} + \cdots$ to prove Euclid's theorem that there are infinitely many primes is so simple and ingenious that we venture to give it here. Suppose that there are only N primes, say $p_1, p_2, \ldots, p_N$. Then each integer $n > 1$ is uniquely expressible in the form $n = p_1^{a_1} p_2^{a_2} \cdots p_N^{a_N}$. If a is the largest of these exponents, then it is easy to see that

$$1 + \frac{1}{2} + \frac{1}{3} + \cdots + \frac{1}{n} \leq \left(1 + \frac{1}{p_1} + \frac{1}{p_1^2} + \cdots + \frac{1}{p_1^a}\right)$$

$$\times \left(1 + \frac{1}{p_2} + \frac{1}{p_2^2} + \cdots + \frac{1}{p_2^a}\right) \cdots \left(1 + \frac{1}{p_N} + \frac{1}{p_N^2} + \cdots + \frac{1}{p_N^a}\right),$$

by multiplying out the factors on the right. But the simple formula $1 + x + x^2 + \cdots = 1/(1 - x)$, which is valid for $|x| < 1$, shows that the factors in the above product are less than the numbers

$$\frac{1}{1 - 1/p_1}, \frac{1}{1 - 1/p_2}, \ldots, \frac{1}{1 - 1/p_N},$$

so

$$1 + \frac{1}{2} + \frac{1}{3} + \cdots + \frac{1}{n} < \frac{p_1}{p_1 - 1}\frac{p_2}{p_2 - 1} \cdots \frac{p_N}{p_N - 1}$$

for every *n*. This contradicts the divergence of the harmonic series and shows that there cannot exist only a finite number of primes. He also proved that the series

$$\frac{1}{2} + \frac{1}{3} + \frac{1}{5} + \frac{1}{7} + \frac{1}{11} + \frac{1}{13} + \frac{1}{17} + \cdots$$

of the reciprocals of the primes diverges, and discovered the following wonderful identity: if $s > 1$, then

$$\sum_{n=1}^{\infty} \frac{1}{n^s} = \prod_{p} \frac{1}{1 - 1/p^s},$$

where the expression on the right denotes the product of the numbers $(1 - p^{-s})^{-1}$ for all primes *p*. We shall return to this identity later, in our note on Riemann in Appendix E in Chapter 5.

He also initiated the theory of partitions, a little-known branch of number theory that turned out much later to have applications in statistical mechanics and the kinetic theory of gases. A typical problem of this subject is to determine the number $p(n)$ of ways in which a given positive integer *n* can be expressed as a sum of positive integers, and if possible to discover some properties of this function. For example, 4 can be partitioned into $4 = 3 + 1 = 2 + 2 = 2 + 1 + 1 = 1 + 1 + 1 + 1$, so $p(4) = 5$, and similarly $p(5) = 7$ and $p(6) = 11$. It is clear that $p(n)$ increases very rapidly with *n*, so rapidly, in fact, that[18]

$$p(200) = 3,972,999,029,388.$$

Euler began his investigations by noticing (only geniuses notice such things) that $p(n)$ is the coefficient of x^n when the function

---

[18] This evaluation required a month's work by a skilled computer in 1918. His motive was to check an approximate formula for $p(n)$, namely

$$p(n) \cong \frac{1}{4n\sqrt{3}} e^{\pi\sqrt{2n/3}}$$

(the error was extremely small).

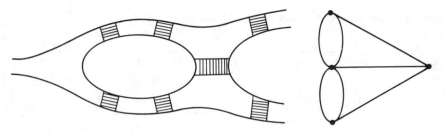

FIGURE 27
The Königsberg bridges.

$[(1 - x)(1 - x^2)(1 - x^3) \cdots]^{-1}$ is expanded in a power series:

$$\frac{1}{(1 - x)(1 - x^2)(1 - x^3) \cdots} = 1 + p(1)x + p(2)x^2 + p(3)x^3 + \cdots .$$

By building on this foundation, he derived many other remarkable identities related to a variety of problems about partitions.[19]

The Königsberg bridge problem originated as a pastime of Sunday strollers in the town of Königsberg (now Kaliningrad) in what was formerly East Prussia. There were seven bridges across the river that flows through the town (see Fig. 27). The residents used to enjoy walking from one bank to the islands and then to the other bank and back again, and the conviction was widely held that it is impossible to do this by crossing all seven bridges without crossing any bridge more than once. Euler analyzed the problem by examining the schematic diagram given on the right in the figure, in which the land areas are represented by points and the bridges by lines connecting these points. The points are called vertices, and a vertex is said to be odd or even according as the number of lines leading to it is odd or even. In modern terminology, the entire configuration is called a *graph,* and a path through the graph that traverses every line but no line more than once is called an *Euler path.* An Euler path need not end at the vertex where it began, but if it does, it is called an *Euler circuit.* By the use of combinatorial reasoning, Euler arrived at the following theorems about any such graph: (1) there are an

---

[19] See Chapter XIX of G. H. Hardy and E. M. Wright, *An Introduction to the Theory of Numbers,* Oxford University Press, 1938; or Chapters 12–14 of G. E. Andrews, *Number Theory,* W. B. Saunders, San Francisco, 1971. These treatments are "elementary" in the technical sense that they do not use the high-powered machinery of advanced analysis, but nevertheless they are far from simple. For students who wish to experience some of Euler's most interesting work in number theory at first hand, and in a context not requiring much previous knowledge, we recommend Chapter VI of G. Polya's fine book, *Induction and Analogy in Mathematics,* Princeton University Press, 1954.

even number of odd vertices; (2) if there are no odd vertices, there is an Euler circuit starting at any point; (3) if there are two odd vertices, there is no Euler circuit, but there is an Euler path starting at one odd vertex and ending at the other; (4) if there are more than two odd vertices, there are no Euler paths.[20] The graph of the Königsberg bridges has four odd vertices, and therefore, by the last theorem, has no Euler paths.[21] The branch of mathematics that has developed from these ideas is known as *graph theory*; it has applications to chemical bonding, economics, psychosociology, the properties of networks of roads and railroads, and other subjects.

A polyhedron is a solid whose surface consists of a number of polygonal faces, and a regular polyhedron has faces that are regular polygons. As we know, there exists a regular polygon with n sides for each positive integer $n = 3, 4, 5, \ldots$, and they even have special names—equilateral triangle, square, regular pentagon, etc. However, it is a curious fact—and has been known since the time of the ancient Greeks—that there are only five regular polyhedra, those shown in Fig. 28, with names given in the table below.

The Greeks studied these figures assiduously, but it remained for

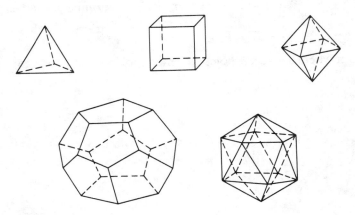

FIGURE 28
Regular polyhedra.

---

[20] Euler's original paper of 1736 is interesting to read and easy to understand; it can be found on pp. 573–580 of J. R. Newman (ed), *The World of Mathematics*, Simon and Schuster, New York, 1956.

[21] It is easy to see—without appealing to any theorems— that this graph contains no Euler circuit, for if there were such a circuit, it would have to enter each vertex as many times as it leaves it, and therefore every vertex would be even. Similar reasoning shows also that if there were an Euler path that is not a circuit, there would be two odd vertices.

Euler to discover the simplest of their common properties: If V, E, and F denote the numbers of vertices, edges, and faces of any one of them, then in every case we have

$$V - E + F = 2.$$

This fact is known as *Euler's formula for polyhedra*, and it is easy to verify from the data summarized in the following table.

	V	E	F
Tetrahedron	4	6	4
Cube	8	12	6
Octahedron	6	12	8
Dodecahedron	20	30	12
Icosahedron	12	30	20

This formula is also valid for any irregular polyhedron as long as it is *simple*—which means that it has no "holes" in it, so that its surface can be deformed continuously into the surface of a sphere. Figure 29 shows two simple irregular polyhedra for which $V - E + F = 6 - 10 + 6 = 2$ and $V - E + F = 6 - 9 + 5 = 2$. However, Euler's formula must be extended to

$$V - E + F = 2 - 2p$$

in the case of a polyhedron with p holes (a simple polyhedron is one for

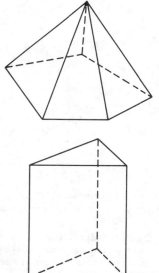

FIGURE 29

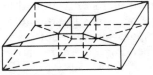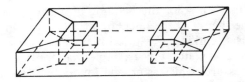

FIGURE 30

which $p = 0$). Figure 30 illustrates the cases $p = 1$ and $p = 2$; here we have $V - E + F = 16 - 32 + 16 = 0$ when $p = 1$, and $V - E + F = 24 - 44 + 18 = -2$ when $p = 2$. The significance of these ideas can best be understood by imagining a polyhedron to be a hollow figure with a surface made of thin rubber, and inflating it until it becomes smooth. We no longer have flat faces and straight edges, but instead a map on the surface consisting of curved regions, their boundaries, and points where boundaries meet. The number $V - E + F$ has the same value for all maps on our surface, and is called the *Euler characteristic* of this surface. The number p is called the *genus* of the surface. These two numbers, and the relation between them given by the equation $V - E + F = 2 - 2p$, are evidently unchanged when the surface is continuously deformed by stretching or bending. Intrinsic geometric properties of this kind—which have little connection with the type of geometry concerned with lengths, angles, and areas—are called *topological*. The serious study of such topological properties has greatly increased during the past century, and has furnished valuable insights to many branches of mathematics and science.[22]

The distinction between pure and applied mathematics did not exist in Euler's day, and for him the entire physical universe was a convenient object whose diverse phenomena offered scope for his methods of analysis. The foundations of classical mechanics had been laid down by Newton, but Euler was the principal architect. In his treatise of 1736 he was the first to explicitly introduce the concept of a mass-point or particle, and he was also the first to study the acceleration of a particle moving along any curve and to use the notion of a vector in connection with velocity and acceleration. His continued successes in mathematical physics were so numerous, and his influence was so pervasive, that most of his discoveries are not credited to him at all and are taken for granted by physicists as part of the natural order of things. However, we do have Euler's equations of motion for the rotation of a rigid body, Euler's hydrodynamic equation for the flow of an ideal incompressible fluid,

---

[22] Proofs of Euler's formula and its extension are given on pp. 236–240 and 256–259 of R. Courant and H. Robbins, *What Is Mathematics?*, Oxford University Press, 1941. See also G. Polya, *op. cit.*, pp. 35–43.

Euler's law for the bending of elastic beams, and Euler's critical load in the theory of the buckling of columns. On several occasions the thread of his scientific thought led him to ideas his contemporaries were not ready to assimilate. For example, he foresaw the phenomenon of radiation pressure, which is crucial for the modern theory of the stability of stars, more than century before Maxwell rediscovered it in his own work on electromagnetism.

Euler was the Shakespeare of mathematics—universal, richly detailed, and inexhaustible.[23]

APPENDIX B.   NEWTON

Most people are acquainted in some degree with the name and reputation of Isaac Newton (1642–1727), for his universal fame as the discoverer of the law of gravitation has continued undiminished over the two and a half centuries since his death. It is less well known, however, that in the immense sweep of his vast achievements he virtually created modern physical science, and in consequence has had a deeper influence on the direction of civilized life than the rise and fall of nations. Those in a position to judge have been unanimous in considering him one of the very few supreme intellects that the human race has produced.

Newton was born to a farm family in the village of Woolsthorpe in northern England. Little is known of his early years, and his undergraduate life at Cambridge seems to have been outwardly undistinguished. In 1665 an outbreak of the plague caused the universities to close, and Newton returned to his home in the country, where he remained until 1667. There, in 2 years of rustic solitude—from age 22 to 24—his creative genius burst forth in a flood of discoveries unmatched in the history of human thought: the binomial series for negative and fractional exponents; differential and integral calculus; universal gravitation as the key to the mechanism of the solar system; and the resolution of sunlight into the visual spectrum by means of a prism, with its implications for understanding the colors of the rainbow and the nature of light in general. In his old age he reminisced as follows about this miraculous period of his youth: "In those days I was in the prime of my age for

---

[23] For further information, see C. Truesdell, "Leonhard Euler, Supreme Geometer (1707–1783)," in *Studies in Eighteenth-Century Culture,* Case Western Reserve University Press, 1972. Also, the November 1983 issue of *Mathematics Magazine* is wholly devoted to Euler and his work.

invention and minded Mathematicks and Philosophy [i.e., science] more than at any time since."[24]

Newton was always an inward and secretive man, and for the most part kept his monumental discoveries to himself. He had no itch to publish, and most of his great works had to be dragged out of him by the cajolery and persistence of his friends. Nevertheless, his unique ability was so evident to his teacher, Isaac Barrow, that in 1669 Barrow resigned his professorship in favor of his pupil (an unheard-of event in academic life), and Newton settled down at Cambridge for the next 27 years. His mathematical discoveries were never really published in connected form; they became known in a limited way almost by accident, through conversations and replies to questions put to him in correspondence. He seems to have regarded his mathematics mainly as a fruitful tool for the study of scientific problems, and of comparatively little interest in itself. Meanwhile, Leibniz in Germany had also invented calculus independently; and by his active correspondence with the Bernoullis and the later work of Euler, leadership in the new analysis passed to the Continent, where it remained for 200 years.[25]

Not much is known about Newton's life at Cambridge in the early years of his professorship, but it is certain that optics and the construction of telescopes were among his main interests. He experimented with many techniques for grinding lenses (using tools which he made himself), and about 1670 built the first reflecting telescope, the earliest ancestor of the great instruments in use today at Mount Palomar and throughout the world. The pertinence and simplicity of his prismatic analysis of sunlight have always marked this early work as one of the timeless classics of experimental science. But this was only the beginning, for he went further and further in penetrating the mysteries of light, and all his efforts in this direction continued to display experimental genius of the highest order. He published some of his discoveries, but they were greeted with such contentious stupidity by the leading scientists of the day that he retired back into his shell with a strengthened resolve to work thereafter

---

[24] The full text of this autobiographical statement (probably written sometime in the period 1714–1720) is given on pp. 291–292 of I. Bernard Cohen, *Introduction to Newton's 'Principia,'* Harvard University Press, 1971. The present writer owns a photograph of the original document.

[25] It is interesting to read Newton's correspondence with Leibniz (via Oldenburg) in 1676 and 1677 (see *The Correspondence of Isaac Newton*, Cambridge University Press, 1959–1976, 6 volumes so far). In Items 165, 172, 188, and 209, Newton discusses his binomial series but conceals in anagrams his ideas about calculus and differential equations, while Leibniz freely reveals his own version of calculus. Item 190 is also of considerable interest, for in it Newton records what is probably the earliest statement and proof of the Fundamental Theorem of Calculus.

for his own satisfaction alone. Twenty years later he unburdened himself to Leibniz in the following words: "As for the phenomena of colours . . . I conceive myself to have discovered the surest explanation, but I refrain from publishing books for fear that disputes and controversies may be raised against me by ignoramuses."[26]

In the late 1670s Newton lapsed into one of his periodic fits of distaste for science, and directed his energies into other channels. As yet he had published nothing about dynamics or gravity, and the many discoveries he had already made in these areas lay unheeded in his desk. At last, however, under the skillful prodding of the astronomer Edmund Halley (of Halley's Comet), he turned his mind once again to these problems and began to write his greatest work, the *Principia*.[27]

It all seems to have started in 1684 with three men in deep conversation in a London inn—Halley, and his friends Christopher Wren and Robert Hooke. By thinking about Kepler's third law of planetary motion, Halley had come to the conclusion that the attractive gravitational force holding the planets in their orbits was probably inversely proportional to the square of the distance from the sun.[28] However, he was unable to do anything more with the idea than formulate it as a conjecture. As he later wrote (in 1686):

> I met with Sir Christopher Wren and Mr. Hooke, and falling in discourse about it, Mr. Hooke affirmed that upon that principle all the Laws of the celestiall motions were to be demonstrated, and that he himself had done it. I declared the ill success of my attempts; and Sir Christopher, to encourage the Inquiry, said that he would give Mr. Hooke or me two months' time to bring him a convincing demonstration therof, and besides the honour, he of us that did it, should have from him a present of a book of 40 shillings. Mr. Hooke then said that he had it, but that he would conceale it for some time, that others triing and failing, might know how to value it, when he should make it publick; however, I remember Sir

---

[26] *Correspondence*, Item 427.

[27] The full title is *Philosophiae Naturalis Principia Mathematica* (*Mathematical Principles of Natural Philosophy*).

[28] At that time this was quite easy to prove under the simplifying assumption—which contradicts Kepler's other two laws—that each planet moves with constant speed v in a circular orbit of radius r. [Proof: In 1673 Huygens had shown, in effect, that the acceleration a of such a planet is given by $a = v^2/r$. If T is the periodic time, then

$$a = \frac{(2\pi r/T)^2}{r} = \frac{4\pi^2}{r^2} \cdot \frac{r^3}{T^2}.$$

By Kepler's third law, T^2 is proportional to r^3, so r^3/T^2 is constant, and a is therefore inversely proportional to r^2. If we now suppose that the attractive force F is proportional to the acceleration, then it follows that F is also inversely proportional to r^2.]

Christopher was little satisfied that he could do it, and tho Mr. Hooke then promised to show it him, I do not yet find that in that particular he has been as good as his word.[29]

It seems clear that Halley and Wren considered Hooke's assertions to be merely empty boasts. A few months later Halley found an opportunity to visit Newton in Cambridge, and put the question to him: "What would be the curve described by the planets on the supposition that gravity diminishes as the square of the distance?" Newton answered immediately, "An ellipse." Struck with joy and amazement, Halley asked him how he knew that. "Why," said Newton, "I have calculated it." Not guessed, or surmised, or conjectured, but *calculated*. Halley wanted to see the calculations at once, but Newton was unable to find the papers. It is interesting to speculate on Halley's emotions when he realized that the age-old problem of how the solar system works had at last been solved—but that the solver hadn't bothered to tell anybody and had even lost his notes. Newton promised to write out the theorems and proofs again and send them to Halley, which he did. In the course of fulfilling his promise he rekindled his own interest in the subject, and went on, and greatly broadened the scope of his researches.[30]

In his scientific efforts Newton somewhat resembled a live volcano, with long periods of quiescence punctuated from time to time by massive eruptions of almost superhuman activity. The *Principia* was written in 18 incredible months of total concentration, and when it was published in 1687 it was immediately recognized as one of the supreme achievements of the human mind. It is still universally considered to be the greatest contribution to science ever made by one man. In it he laid down the basic principles of theoretical mechanics and fluid dynamics; gave the first mathematical treatment of wave motion; deduced Kepler's laws from the inverse square law of gravitation, and explained the orbits of comets; calculated the masses of the earth, the sun, and the planets with satellites; accounted for the flattened shape of the earth, and used this to explain the precession of the equinoxes; and founded the theory of tides. These are only a few of the splendors of this prodigious work.[31] The *Principia* has always been a difficult book to read, for the style has an inhuman quality of icy remoteness, which perhaps is appropriate to the

---

[29] *Correspondence*, Item 289.

[30] For additional details and the sources of our information about these events, see Cohen, *op. cit.*, pp. 47–54.

[31] A valuable outline of the contents of the *Principia* is given in Chapter VI of W. W. Rouse Ball, *An Essay on Newton's Principia* (first published in 1893; reprinted in 1972 by Johnson Reprint Corp, New York).

grandeur of the theme. Also, the densely packed mathematics consists almost entirely of classical geometry, which was little cultivated then and is less so now.[32] In his dynamics and celestial mechanics, Newton achieved the victory for which Copernicus, Kepler, and Galileo had prepared the way. This victory was so complete that the work of the greatest scientists in these fields over the next two centuries amounted to little more than footnotes to his colossal synthesis. It is also worth remembering in this context that the science of spectroscopy, which more than any other has been responsible for extending astronomical knowledge beyond the solar system to the universe at large, had its origin in Newton's spectral analysis of sunlight.

After the mighty surge of genius that went into the creation of the *Principia,* Newton again turned away from science. However, in a famous letter to Bentley in 1692, he offered the first solid speculations on how the universe of stars might have developed out of a primordial featureless cloud of cosmic dust:

> It seems to me, that if the matter of our Sun and Planets and all the matter in the Universe was evenly scattered throughout all the heavens, and every particle has an innate gravity towards all the rest ... some of it would convene into one mass and some into another, so as to make an infinite number of great masses scattered at great distances from one to another throughout all that infinite space. And thus might the Sun and Fixt stars be formed, supposing the matter were of a lucid nature.[33]

This was the beginning of scientific cosmology, and later led, through the ideas of Thomas Wright, Kant, Herschel, and their successors, to the elaborate and convincing theory of the nature and origin of the universe provided by late twentieth century astronomy.

In 1693 Newton suffered a severe mental illness accompanied by delusions, deep melancholy, and fears of persecution. He complained that he could not sleep, and said that he lacked his "former consistency of mind." He lashed out with wild accusations in shocking letters to his friends Samuel Pepys and John Locke. Pepys was informed that their friendship was over and that Newton would see him no more; Locke was charged with trying to entangle him with women and with being a

---

[32] The nineteenth century British philosopher Whewell has a vivid remark about this: "Nobody since Newton has been able to use geometrical methods to the same extent for the like purposes; and as we read the *Principia* we feel as when we are in an ancient armoury where the weapons are of gigantic size; and as we look at them we marvel what manner of man he was who could use as a weapon what we can scarcely lift as a burden."

[33] *Correspondence,* Item 398.

"Hobbist" (a follower of Hobbes, i.e., an atheist and materialist).[34] Both men feared for Newton's sanity. They responded with careful concern and wise humanity, and the crisis passed.

In 1696 Newton left Cambridge for London to become Warden (and soon Master) of the Mint, and during the remainder of his long life he entered a little into society and even began to enjoy his unique position at the pinnacle of scientific fame. These changes in his interests and surroundings did not reflect any decrease in his unrivaled intellectual powers. For example, late one afternoon, at the end of a hard day at the Mint, he learned of a now-famous problem that the Swiss scientist John Bernoulli had posed as a challenge "to the most acute mathematicians of the entire world." The problem can be stated as follows: Suppose two nails are driven at random into a wall, and let the upper nail be connected to the lower by a wire in the shape of a smooth curve. What is the shape of the wire down which a bead will slide (without friction) under the influence of gravity so as to pass from the upper nail to the lower in the least possible time? This is Bernoulli's *brachistochrone* ("shortest time") *problem*. Newton recognized it at once as a challenge to himself from the Continental mathematicians; and in spite of being out of the habit of scientific thought, he summoned his resources and solved it that evening before going to bed. His solution was published anonymously, and when Bernoulli saw it, he wryly remarked, "I recognize the lion by his claw."

Of much greater significance for science was the publication of his *Opticks* in 1704. In this book he drew together and extended his early work on light and color. As an appendix he added his famous Queries, or speculations on areas of science that lay beyond his grasp in the future. In part the Queries relate to his lifelong preoccupation with chemistry (or alchemy, as it was then called). He formed many tentative but exceedingly careful conclusions—always founded on experiment—about the probable nature of matter; and though the testing of his speculations about atoms (and even nuclei) had to await the refined experimental work of the late nineteenth and early twentieth centuries, he has been proven absolutely correct in the main outlines of his ideas.[35] So, in this field of science too, in the prodigious reach and accuracy of his scientific imagination, he passed far beyond not only his contemporaries but also many generations of his successors. In addition, we quote two astonishing remarks from Queries 1 and 30, respectively: "Do Not Bodies act upon Light at a distance, and by their action bend its Rays?" and "Are not

---

[34] *Correspondence*, Items 420, 421, and 426.

[35] See S. I. Vavilov, "Newton and the Atomic Theory," in *Newton Tercentenary Celebrations*, Cambridge University Press, 1947.

gross Bodies and Light convertible into one another?" It seems as clear as words can be that Newton is here conjecturing the gravitational bending of light and the equivalence of mass and energy, which are prime consequences of the theory of relativity. The former phenomenon was first observed during the total solar eclipse of May 1919, and the latter is now known to underlie the energy generated by the sun and the stars. On other occasions as well he seems to have known, in some mysterious intuitive way, far more than he was ever willing or able to justify, as in this cryptic sentence in a letter to a friend: "It's plain to me by the fountain I draw it from, though I will not undertake to prove it to others."[36] Whatever the nature of this "fountain" may have been, it undoubtedly depended on his extraordinary powers of concentration. When asked how he made his discoveries, he said, "I keep the subject constantly before me and wait till the first dawnings open little by little into the full light." This sounds simple enough, but everyone with experience in science or mathematics knows how very difficult it is to hold a problem continuously in mind for more than a few seconds or a few minutes. One's attention flags; the problem repeatedly slips away and repeatedly has to be dragged back by an effort of will. From the accounts of witnesses, Newton seems to have been capable of almost effortless sustained concentration on his problems for hours and days and weeks, with even the need for occasional food and sleep scarcely interrupting the steady squeezing grip of his mind.

In 1695 Newton received a letter from his Oxford mathematical friend John Wallis, containing news that cast a cloud over the rest of life. Writing about Newton's early mathematical discoveries, Wallis warned him that in Holland "your Notions" are known as "Leibniz's *Calculus Differentialis*," and he urged Newton to take steps to protect his reputation.[37] At that time the relations between Newton and Leibniz were still cordial and mutually respectful. However, Wallis's letters soon curdled the atmosphere, and initiated the most prolonged, bitter, and damaging of all scientific quarrels: the famous (or infamous) Newton–Leibniz priority controversy over the invention of calculus.

It is now well established that each man developed his own form of calculus independently of the other, that Newton was first by 8 or 10 years but did not publish his ideas, and that Leibniz's papers of 1684 and 1686 were the earliest publications on the subject. However, what are now perceived as simple facts were not nearly so clear at the time. There were ominous minor rumblings for years after Wallis's letters, as the

---

[36] *Correspondence*, Item 193.

[37] *Correspondence*, Items 498 and 503.

storm gathered:

> What began as mild innuendoes rapidly escalated into blunt charges of plagiarism on both sides. Egged on by followers anxious to win a reputation under his auspices, Newton allowed himself to be drawn into the centre of the fray; and, once his temper was aroused by accusations of dishonesty, his anger was beyond constraint. Leibniz's conduct of the controversy was not pleasant, and yet it paled beside that of Newton. Although he never appeared in public, Newton wrote most of the pieces that appeared in his defense, publishing them under the names of his young men, who never demurred. As president of the Royal Society, he appointed an "impartial" committee to investigate the issue, secretly wrote the report officially published by the society [in 1712], and reviewed it anonymously in the *Philosophical Transactions*. Even Leibniz's death could not allay Newton's wrath, and he continued to pursue the enemy beyond the grave. The battle with Leibniz, the irrepressible need to efface the charge of dishonesty, dominated the final 25 years of Newton's life. Almost any paper on any subject from those years is apt to be interrupted by a furious paragraph against the German philosopher, as he honed the instruments of his fury ever more keenly.[38]

All this was bad enough, but the disastrous effect of the controversy on British science and mathematics was much more serious. It became a matter of patriotic loyalty for the British to use Newton's geometrical methods and clumsy calculus notations, and to look down their noses at the upstart work being done on the Continent. However, Leibniz's analytical methods proved to be far more fruitful and effective, and it was his followers who were the moving spirits in the richest period of development in mathematical history. What has been called "the Great Sulk" continued; for the British, the work of the Bernoullis, Euler, Lagrange, Laplace, Gauss, and Riemann remained a closed book; and British mathematics sank into a coma of impotence and irrelevancy that lasted through most of the eighteenth and nineteenth centuries.

Newton has often been thought of and described as the ultimate rationalist, the embodiment of the Age of Reason. His conventional image is that of a worthy but dull absent-minded professor in a foolish powdered wig. But nothing could be further from the truth. This is not the place to discuss or attempt to analyze his psychotic flaming rages; or his monstrous vengeful hatreds that were unquenched by the death of his enemies and continued at full strength to the end of his own life; or the 58 sins he listed in the private confession he wrote in 1662; or his secretiveness and shrinking insecurity; or his peculiar relations with

---

[38] Richard S. Westfall, in the *Encyclopaedia Britannica*.

women, especially with his mother, who he thought had abandoned him at the age of 3. And what are we to make of the bushels of unpublished manuscripts (millions of words and thousands of hours of thought!) that reflect his secret lifelong studies of ancient chronology, early Christian doctrine, and the prophecies of Daniel and St. John? Newton's desire to know had little in common with the smug rationalism of the eighteenth century; on the contrary, it was a form of desperate self-preservation against the dark forces that he felt pressing in around him. As an original thinker in science and mathematics he was a stupendous genius whose impact on the world can be seen by everyone; but as a man he was so strange in every way that normal people can scarcely begin to understand him.[39] It is perhaps most accurate to think of him in medieval terms—as a consecrated, solitary, intuitive mystic for whom science and mathematics were means of reading the riddle of the universe.

---

[39] The best effort is Frank E. Manuel's excellent book, *A Portrait of Isaac Newton,* Harvard University Press, 1968.

CHAPTER
4

QUALITATIVE
PROPERTIES OF
SOLUTIONS

24 OSCILLATIONS AND THE STURM SEPARATION THEOREM

It is natural to feel that a differential equation should be solved, and one of the main aims of our work in Chapter 3 was to develop ways of finding explicit solutions of the second order linear equation

$$y'' + P(x)y' + Q(x)y = 0. \tag{1}$$

Unfortunately, however—as we have tried to emphasize—it is rarely possible to solve this equation in terms of familiar elementary functions. This situation leads us to seek wider vistas by formulating the problem at a higher level, and to recognize that our real goal is to understand the nature and properties of the solutions of (1). If this goal can be attained by means of elementary formulas for these solutions, well and good. If not, then we try to open up other paths to the same destination. In this brief chapter we turn our attention to the problem of learning what we can about the essential characteristics of the solutions of (1) by direct analysis of the equation itself, in the absence of formal expressions for these solutions. It is surprising how much interesting and useful information can be gained in this way.

As an illustration of the idea that many properties of the solutions of a differential equation can be discovered by studying the equation itself, without solving it in any traditional sense, we discuss the familiar equation

$$y'' + y = 0. \tag{2}$$

We know perfectly well that $y_1(x) = \sin x$ and $y_2(x) = \cos x$ are two linearly independent solutions of (2); that they are fully determined by the initial conditions $y_1(0) = 0$, $y_1'(0) = 1$ and $y_2(0) = 1$, $y_2'(0) = 0$; and that the general solution is $y(x) = c_1 y_1(x) + c_2 y_2(x)$. Normally we regard (2) as completely solved by these observations, for the functions $\sin x$ and $\cos x$ are old friends and we know a great deal about them. However, our knowledge of $\sin x$ and $\cos x$ can be thought of as an accident of history; and for the sake of emphasizing our present point of view, we now pretend total ignorance of these familiar functions. Our purpose is to see how their properties can be squeezed out of (2) and the initial conditions they satisfy. The only tools we shall use are qualitative arguments and the general principles described in Sections 14 and 15.

Accordingly, let $y = s(x)$ be defined as the solution of (2) determined by the initial conditions $s(0) = 0$ and $s'(0) = 1$. If we try to sketch the graph of $s(x)$ by letting x increase from 0, the initial conditions tell us to start the curve at the origin and let it rise with slope beginning at 1 (Fig. 31). From the equation itself we have $s''(x) = -s(x)$, so when the curve is above the x-axis, $s''(x)$ is a negative number that increases in magnitude as the curve rises. Since $s''(x)$ is the rate of change of the slope $s'(x)$, this slope decreases at an increasing rate as the curve lifts, and it must reach 0 at some point $x = m$. As x continues to increase, the curve falls toward the x-axis, $s'(x)$ decreases at a decreasing rate, and the curve crosses the x-axis at a point we can *define* to be π. Since $s''(x)$ depends only on $s(x)$, we see that the graph between $x = 0$ and $x = \pi$ is symmetric about the line $x = m$, so $m = \pi/2$ and $s'(\pi) = -1$. A similar argument shows that the next portion of the curve is an inverted replica of the first arch, and so on indefinitely.

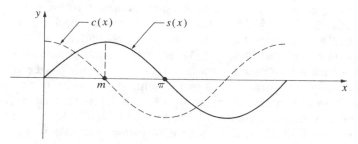

FIGURE 31

In order to make further progress, it is convenient at this stage to introduce $y = c(x)$ as the solution of (2) determined by the initial conditions $c(0) = 1$ and $c'(0) = 0$. These conditions tell us (Fig. 31) that the graph of $c(x)$ starts at the point $(0,1)$ and moves to the right with slope beginning at 0. since by equation (2) we know that $c''(x) = -c(x)$, the same reasoning as before shows that the curve bends down and crosses the x-axis. It is natural to conjecture that the height of the first arch of $s(x)$ is 1, that the first zero of $c(x)$ is $\pi/2$, etc.; but to establish these guesses as facts, we begin by showing that

$$s'(x) = c(x) \quad \text{and} \quad c'(x) = -s(x). \tag{3}$$

To prove the first statement, we start by observing that (2) yields $y''' + y' = 0$ or $(y')'' + y' = 0$, so the derivative of any solution of (2) is again a solution (see Problem 17-4). Thus $s'(x)$ and $c(x)$ are both solutions of (2), and by Theorem 14-A it suffices to show that they have the same values and the same derivatives at $x = 0$. This follows at once from $s'(0) = 1$, $c(0) = 1$ and $s''(0) = -s(0) = 0$, $c'(0) = 0$. The second formula in (3) is an immediate consequence of the first, for $c'(x) = s''(x) = -s(x)$. We now use (3) to prove

$$s(x)^2 + c(x)^2 = 1. \tag{4}$$

Since the derivative of the left side of (4) is

$$2s(x)c(x) - 2c(x)s(x),$$

which is 0, we see that $s(x)^2 + c(x)^2$ equals a constant, and this constant must be 1 because $s(0)^2 + c(0)^2 = 1$. It follows at once from (4) that the height of the first arch of $s(x)$ is 1 and that the first zero of $c(x)$ is $\pi/2$. This result also enables us to show that $s(x)$ and $c(x)$ are linearly independent, for their Wronskian is

$$W[s(x),c(x)] = s(x)c'(x) - c(x)s'(x)$$
$$= -s(x)^2 - c(x)^2 = -1.$$

In much the same way, we can continue and establish the following additional facts:

$$s(x + a) = s(x)c(a) + c(x)s(a); \tag{5}$$

$$c(x + a) = c(x)c(a) - s(x)s(a); \tag{6}$$

$$s(2x) = 2s(x)c(x); \tag{7}$$

$$c(2x) = c(x)^2 - s(x)^2; \tag{8}$$

$$s(x + 2\pi) = s(x); \tag{9}$$

$$c(x + 2\pi) = c(x). \tag{10}$$

The proofs are not difficult, and we leave them to the reader (see Problem 1). Among other things, it is easy to see from the above results that the positive zeros of $s(x)$ and $c(x)$ are, respectively, π, 2π, 3π, ... and $\pi/2$, $\pi/2 + \pi$, $\pi/2 + 2\pi$, ...

There are two main points to be made about the above discussion. First, we have extracted almost every significant property of the functions $\sin x$ and $\cos x$ from equation (2) *by the methods of differential equations alone*, without using any prior knowledge of trigonometry. Second, the tools we did use consisted chiefly of convexity arguments (involving the sign and magnitude of the second derivative) and the basic properties of linear equations set forth in Sections 14 and 15.

It goes without saying that most of the above properties of $\sin x$ and $\cos x$ are peculiar to these functions alone. Nevertheless, the central feature of their behavior—the fact that they oscillate in such a manner that their zeros are distinct and occur alternately—can be generalized far beyond these particular functions. The following result in this direction is called the *Sturm separation theorem.*[1]

Theorem A. *If $y_1(x)$ and $y_2(x)$ are two linearly independent solutions of*

$$y'' + P(x)y' + Q(x)y = 0,$$

then the zeros of these functions are distinct and occur alternately—in the sense that $y_1(x)$ vanishes exactly once between any two successive zeros of $y_2(x)$, and conversely.

Proof. The argument rests primarily on the fact (see the lemmas in Section 15) that since y_1 and y_2 are linearly independent, their Wronskian

$$W(y_1,y_2) = y_1(x)y_2'(x) - y_2(x)y_1'(x)$$

does not vanish, and therefore—since it is continuous—must have constant sign. First, it is easy to see that y_1 and y_2 cannot have a common zero; for if they do, then the Wronskian will vanish at that point, which is impossible. We now assume that x_1 and x_2 are successive zeros of y_2 and show that y_1 vanishes between these points. The Wronskian clearly reduces to $y_1(x)y_2'(x)$ at x_1 and x_2, so both factors $y_1(x)$ and $y_2'(x)$ are $\neq 0$ at each of these points. Furthermore, $y_2'(x_1)$ and $y_2'(x_2)$ must have opposite signs, because if y_2 is increasing at x_1 it must be decreasing at x_2, and vice versa. Since the

---

[1] Jacques Charles François Sturm (1803–1855) was a Swiss mathematician who spent most of his life in Paris. For a time he was tutor to the de Broglie family, and after holding several other positions he at last succeeded Poisson in the Chair of Mechanics at the Sorbonne. His main work was done in what is now called the Sturm–Liouville theory of differential equations, which has been of steadily increasing importance ever since in both pure mathematics and mathematical physics.

Wronskian has constant sign, $y_1(x_1)$ and $y_1(x_2)$ must also have opposite signs, and therefore, by continuity, $y_1(x)$ must vanish at some point between x_1 and x_2. Note that y_1 cannot vanish more than once between x_1 and x_2; for if it does, then the same argument shows that y_2 must vanish between these zeros of y_1, which contradicts the original assumption that x_1 and x_2 are successive zeros of y_2.

The convexity arguments given above in connection with the equation $y'' + y = 0$ make it clear that in discussing the oscillation of solutions it is convenient to deal with equations in which the first derivative term is missing. We now show that any equation of the form

$$y'' + P(x)y' + Q(x)y = 0 \tag{11}$$

can be written as

$$u'' + q(x)u = 0 \tag{12}$$

by a simple change of the dependent variable. It is customary to refer to (11) as the *standard form*, and to (12) as the *normal form*, of a homogeneous second order linear equation. To write (11) in normal form, we put $y(x) = u(x)v(x)$, so that $y' = uv' + u'v$ and $y'' = uv'' + 2u'v' + u''v$. When these expressions are substituted in (11), we obtain

$$vu'' + (2v' + Pv)u' + (v'' + Pv' + Qv)u = 0. \tag{13}$$

On setting the coefficient of u' equal to zero and solving, we find that

$$v = e^{-\frac{1}{2}\int P\,dx} \tag{14}$$

reduces (13) to the normal form (12) with

$$q(x) = Q(x) - \frac{1}{4}P(x)^2 - \frac{1}{2}P'(x). \tag{15}$$

Since $v(x)$ as given by (14) never vanishes, the above transformation of (11) into (12) has no effect whatever on the zeros of solutions, and therefore leaves unaltered the oscillation phenomena which are the objects of our present interest.

We next show that if $q(x)$ in (12) is a negative function, then the solutions of this equation do not oscillate at all.

Theorem B. *If $q(x) < 0$, and if $u(x)$ is a nontrivial solution of $u'' + q(x)u = 0$, then $u(x)$ has at most one zero.*

Proof. Let x_0 be a zero of $u(x)$, so that $u(x_0) = 0$. Since $u(x)$ is nontrivial (i.e., is not identically zero), Theorem 14-A implies that $u'(x_0) \neq 0$. For the sake of concreteness, we now assume that $u'(x_0) > 0$, so that $u(x)$ is positive over some interval to the right of x_0. Since $q(x) < 0$, $u''(x) = -q(x)u(x)$ is a positive function on the same interval. This implies that the slope $u'(x)$ is an increasing function, so $u(x)$ cannot have a zero to the right of x_0, and in the same way it has none to the left of x_0. A similar argument holds when $u'(x_0) < 0$, so $u(x)$ has either no zeros at all or only one, and the proof is complete.

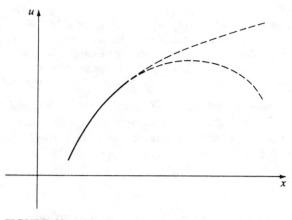

FIGURE 32

Since our interest is in the oscillation of solutions, this result leads us to confine our study of (12) to the special case in which $q(x)$ is a positive function.

Even in this case, however, it is not necessarily true that solutions will oscillate. To get an idea of what is involved, let $u(x)$ be a nontrivial solution of (12) with $q(x) > 0$. If we consider a portion of the graph above the x-axis (Fig. 32), then $u''(x) = -q(x)u(x)$ is negative, so the graph is concave down and the slope $u'(x)$ is decreasing. If this slope ever becomes negative, then the curve plainly crosses the x-axis somewhere to the right and we get a zero for $u(x)$. We know that this happens when $q(x)$ is constant. The alternative is that although $u'(x)$ decreases, it never reaches zero and the curve continues to rise, as in the upper part of Fig. 32. It is reasonably clear from these remarks that $u(x)$ will have zeros as x increases whenever $q(x)$ does not decrease too rapidly. This leads us to the next theorem.

Theorem C. *Let $u(x)$ be any nontrivial solution of $u'' + q(x)u = 0$, where $q(x) > 0$ for all $x > 0$. If*

$$\int_1^\infty q(x)\, dx = \infty, \qquad (16)$$

then $u(x)$ has infinitely many zeros on the positive x-axis.

Proof. Assume the contrary, namely, that $u(x)$ vanishes at most a finite number of times for $0 < x < \infty$, so that a point $x_0 > 1$ exists with the property that $u(x) \neq 0$ for all $x \geq x_0$. We may clearly suppose, without any loss of generality, that $u(x) > 0$ for all $x \geq x_0$, since $u(x)$ can be replaced by its negative if necessary. Our purpose is to contradict the assumption by showing that $u'(x)$ is negative somewhere to the right of x_0—for, by the above remarks, this will imply that $u(x)$ has a zero to the right of x_0. If we put

$$v(x) = -\frac{u'(x)}{u(x)}$$

for $x \geq x_0$, then a simple calculation shows that

$$v'(x) = q(x) + v(x)^2;$$

and on integrating this from x_0 to x, where $x > x_0$, we get

$$v(x) - v(x_0) = \int_{x_0}^{x} q(x)\, dx + \int_{x_0}^{x} v(x)^2\, dx.$$

We now use (16) to conclude that $v(x)$ is positive if x is taken large enough. This shows that $u(x)$ and $u'(x)$ have opposite signs if x is sufficiently large, so $u'(x)$ is negative and the proof is complete.

PROBLEMS

1. Prove formulas (5) to (10) by arguments consistent with the spirit of the preceding discussion.
2. Show that the zeros of the functions $a \sin x + b \cos x$ and $c \sin x + d \cos x$ are distinct and occur alternately whenever $ad - bc \neq 0$.
3. Find the normal form of Bessel's equation

$$x^2 y'' + xy' + (x^2 - p^2)y = 0,$$

and use it to show that every nontrivial solution has infinitely many positive zeros.
4. The hypothesis of Theorem C is false for the Euler equation $y'' + (k/x^2)y = 0$, but the conclusion is sometimes true and sometimes false, depending on the magnitude of the positive constant k. Show that every nontrivial solution has an infinite number of positive zeros if $k > 1/4$, and only a finite number if $k \leq 1/4$.

25 THE STURM COMPARISON THEOREM

In this section we continue our study of the oscillation behavior of nontrivial solutions of the differential equation

$$y'' + q(x)y = 0, \tag{1}$$

where $q(x)$ is a positive function. We begin with a theorem that rules out the possibility of infinitely many oscillations on closed intervals.

Theorem A. *Let $y(x)$ be a nontrivial solution of equation (1) on a closed interval $[a,b]$. Then $y(x)$ has at most a finite number of zeros in this interval.*

Proof. We assume the contrary, namely, that $y(x)$ has an infinite number of zeros in $[a,b]$. It follows from this that there exist in $[a,b]$ a point x_0 and a sequence of zeros $x_n \neq x_0$ such that $x_n \to x_0$.[2] Since $y(x)$ is continuous and differentiable at x_0, we have

$$y(x_0) = \lim_{x_n \to x_0} y(x_n) = 0$$

and

$$y'(x_0) = \lim_{x_n \to x_0} \frac{y(x_n) - y(x_0)}{x_n - x_0} = 0.$$

By Theorem 14-A, these statements imply that $y(x)$ is the trivial solution of (1), and this contradiction completes the proof.

We now recall that the Sturm separation theorem tells us that the zeros of any two (nontrivial) solutions of (1) either coincide or occur alternately, depending on whether these solutions are linearly dependent or independent. Thus, all solutions of (1) oscillate with essentially the same rapidity, in the sense that on a given interval the number of zeros of any solution cannot differ by more than one from the number of zeros of any other solution. On the other hand, it is clear that solutions of

$$y'' + 4y = 0 \tag{2}$$

oscillate more rapidly—that is, have more zeros—than solutions of

$$y'' + y = 0; \tag{3}$$

for the zeros of a solution of (2) such as $y = \sin 2x$ are only half as far apart as the zeros of a solution $y = \sin x$ of (3). The following result, which is known as the *Sturm comparison theorem*, shows that this behavior is typical in the sense that the solutions of (1) oscillate more rapidly when $q(x)$ is increased.

Theorem B. *Let $y(x)$ and $z(x)$ be nontrivial solutions of*

$$y'' + q(x)y = 0$$

and

$$z'' + r(x)z = 0,$$

where $q(x)$ and $r(x)$ are positive functions such that $q(x) > r(x)$. Then $y(x)$ vanishes at least once between any two successive zeros of $z(x)$.

Proof. Let x_1 and x_2 be successive zeros of $z(x)$, so that $z(x_1) = z(x_2) = 0$ and $z(x)$ does not vanish on the open interval (x_1, x_2). We assume that $y(x)$

---

[2] In this inference we use the *Bolzano–Weierstrass theorem* of advanced calculus, which expresses one of the basic topological properties of the real number system.

does not vanish on (x_1, x_2), and prove the theorem by deducing a contradiction. It is clear that no loss of generality is involved in supposing that both $y(x)$ and $z(x)$ are positive on (x_1, x_2), for either function can be replaced by its negative if necessary. If we emphasize that the Wronskian

$$W(y, z) = y(x)z'(x) - z(x)y'(x)$$

is a function of x by writing it $W(x)$, then

$$\frac{dW(x)}{dx} = yz'' - zy''$$

$$= y(-rz) - z(-qy)$$

$$= (q - r)yz > 0$$

on (x_1, x_2). We now integrate both sides of this inequality from x_1 to x_2 and obtain

$$W(x_2) - W(x_1) > 0 \quad \text{or} \quad W(x_2) > W(x_1).$$

However, the Wronskian reduces to $y(x)z'(x)$ at x_1 and x_2, so

$$W(x_1) \geq 0 \quad \text{and} \quad W(x_2) \leq 0,$$

which is the desired contradiction.

It follows from this theorem that if we have $q(x) > k^2 > 0$ in equation (1), then any solution must vanish between any two successive zeros of a solution $y(x) = \sin k(x - x_0)$ of the equation $y'' + k^2 y = 0$, and therefore must vanish in any interval of length π/k. For example, if we consider Bessel's equation

$$x^2 y'' + xy' + (x^2 - p^2)y = 0$$

in normal form

$$u'' + \left(1 + \frac{1 - 4p^2}{4x^2}\right)u = 0,$$

and compare this with $u'' + u = 0$, then we at once have the next theorem.

Theorem C. *Let $y_p(x)$ be a nontrivial solution of Bessel's equation on the positive x-axis. If $0 \leq p < 1/2$, then every interval of length π contains at least one zero of $y_p(x)$; if $p = 1/2$, then the distance between successive zeros of $y_p(x)$ is exactly π; and if $p > 1/2$, then every interval of length π contains at most one zero of $y_p(x)$.*

Bessel's equation is of considerable importance in mathematical physics. The oscillation properties of its solutions expressed in Theorem C, and also in Problem 24-3 and Problem 1 below, are clearly of fundamental significance for understanding the nature of these solutions. In Chapter 8 we shall devote a good deal of effort to finding explicit solutions for Bessel's equation in terms of power series. However, these series solutions are awkward tools to try to use in studying oscillation properties, and it is a great convenience to be able to turn to qualitative reasoning of the kind discussed in this chapter.

PROBLEMS

1. Let x_1 and x_2 be successive positive zeros of a nontrivial solution $y_p(x)$ of Bessel's equation.
   (a) If $0 \le p < 1/2$, show that $x_2 - x_1$ is less than π and approaches π as $x_1 \to \infty$.
   (b) If $p > 1/2$, show that $x_2 - x_1$ is greater than π and approaches π as $x_1 \to \infty$.
2. If $y(x)$ is a nontrivial solution of $y'' + q(x)y = 0$, show that $y(x)$ has an infinite number of positive zeros if $q(x) > k/x^2$ for some $k > 1/4$, and only a finite number if $q(x) < 1/4x^2$.
3. Every nontrivial solution of $y'' + (\sin^2 x + 1)y = 0$ has an infinite number of positive zeros. Formulate and prove a theorem that includes this statement as a special case.

CHAPTER

5

POWER
SERIES
SOLUTIONS
AND SPECIAL
FUNCTIONS

26 INTRODUCTION. A REVIEW OF POWER SERIES

Most of the specific functions encountered in elementary analysis belong to a class known as the *elementary functions*. In order to describe this class, we begin by recalling that an *algebraic function* is a polynomial, a rational function, or more generally any function $y = f(x)$ that satisfies an equation of the form

$$P_n(x)y^n + P_{n-1}(x)y^{n-1} + \cdots + P_1(x)y + P_0(x) = 0,$$

where each $P_i(x)$ is a polynomial. The elementary functions consist of the algebraic functions; the elementary *transcendental* (or nonalgebraic) functions occurring in calculus—i.e., the trigonometric, inverse trigonometric, exponential, and logarithmic functions; and all others that can be constructed from these by adding, subtracting, multiplying, dividing, or forming a function of a function. Thus,

$$y = \tan \left[\frac{xe^{1/x} + \tan^{-1}(1 + x^2)}{\sin x \cos 2x - \sqrt{\log x}} \right]^{1/3}$$

is an elementary function.

Beyond the elementary functions lie the *higher transcendental functions,* or, as they are often called, the *special functions.* Since the beginning of the eighteenth century, many hundreds of special functions have been considered sufficiently interesting or important to merit some degree of study. Most of these are almost completely forgotten, but some, such as the gamma function, the Riemann zeta function, the elliptic functions, and those that continue to be useful in mathematical physics, have generated extensive theories. And among these, a few are so rich in meaning and influence that the mere history of any one of them would fill a large book.[1]

The field of special functions was cultivated with enthusiastic devotion by many of the greatest mathematicians of the eighteenth and nineteenth centuries—by Euler, Gauss, Abel, Jacobi, Weierstrass, Riemann, Hermite, and Poincaré, among others. But tastes change with the times, and today most mathematicians prefer to study large classes of functions (continuous functions, integrable functions, etc.) instead of outstanding individuals. Nevertheless, there are still many who favor biography over sociology, and a balanced treatment of analysis cannot neglect either view.

Special functions vary rather widely with respect to their origin, nature, and applications. However, one large group with a considerable degree of unity consists of those that arise as solutions of second order linear differential equations. Many of these find applications in connection with the partial differential equations of mathematical physics. They are also important, through the theory of orthogonal expansions, as the main historical source of linear analysis, which has played a central role in shaping much of modern pure mathematics.

Let us try to understand in a general way how these functions arise. It will be recalled that if we wish to solve the simple equation

$$y'' + y = 0, \tag{1}$$

then the familiar functions $y = \sin x$ and $y = \cos x$ are already available for this purpose from elementary calculus. The situation with respect to the equation

$$xy'' + y' + xy = 0 \tag{2}$$

is quite different, for this equation cannot be solved in terms of elementary functions. As a matter of fact, there is no known type of

---

[1] The reader who wishes to form an impression of the extent of this part of analysis would do well to look through the three volumes of *Higher Transcendental Functions,* A. Erdélyi (ed.), McGraw-Hill, New York, 1953–1955.

second order linear equation—apart from those with constant coefficients, and equations reducible to these by changes of the independent variable—which can be solved in terms of elementary functions. In Chapter 4 we found that certain general properties of the solutions of such an equation can often be established without solving the equation at all. But if a particular equation of this kind seems important enough to demand some sort of explicit solution, what can we do? The approach we develop in this chapter is to solve it in terms of power series and to use these series to define new special functions. We then investigate the properties of these functions by means of their series expansions. If we succeed in learning enough about them, then they attain the status of "familiar functions" and can be used as tools for studying the problem that gave rise to the original differential equation. Needless to say, this program is easier to describe than to carry out, and is worthwhile only in the case of functions with a variety of significant applications.

It is clear from the above remarks that we will be using power series extensively throughout this chapter. We take it for granted that most readers are reasonably well acquainted with these series from an earlier course in calculus. Nevertheless, for the benefit of those whose familiarity with this topic may have faded slightly, we present a brief review of the main facts.

A. An infinite series of the form

$$\sum_{n=0}^{\infty} a_n x^n = a_0 + a_1 x + a_2 x^2 + \cdots \tag{3}$$

is called a *power series in x*. The series

$$\sum_{n=0}^{\infty} a_n (x - x_0)^n = a_0 + a_1(x - x_0) + a_2(x - x_0)^2 + \cdots \tag{4}$$

is a *power series in $x - x_0$*, and is somewhat more general than (3). However, (4) can always be reduced to (3) by replacing $x - x_0$ by x—which is merely a translation of the coordinate system—so for the most part we shall confine our discussion to power series of the form (3).

B. The series (3) is said to *converge* at a point x if the limit

$$\lim_{m \to \infty} \sum_{n=0}^{m} a_n x^n$$

exists, and in this case the *sum* of the series is the value of this limit. It is obvious that (3) always converges at the point $x = 0$. With respect to the arrangement of their points of convergence, all power series in x fall into one or another or three major categories. These are typified by the following examples:

$$\sum_{n=0}^{\infty} n!x^n = 1 + x + 2!x^2 + 3!x^3 + \cdots; \tag{5}$$

$$\sum_{n=0}^{\infty} \frac{x^n}{n!} = 1 + x + \frac{x^2}{2!} + \frac{x^3}{3!} + \cdots; \tag{6}$$

$$\sum_{n=0}^{\infty} x^n = 1 + x + x^2 + x^3 + \cdots. \tag{7}$$

The first of these series diverges (i.e., fails to converge) for all $x \neq 0$; the second converges for all x; and the third converges for $|x| < 1$ and diverges for $|x| > 1$. Some power series in x behave like (5), and converge only for $x = 0$. These are of no interest to us. Some, like (6), converge for all x. These are the easiest to work with. All others are roughly similar to (7). This means that to each series of this kind there corresponds a positive real number R, called the *radius of convergence*, with the property that the series converges if $|x| < R$ and diverges if $|x| > R$ [$R = 1$ in the case of (7)].

It is customary to put R equal to 0 when the series converges only for $x = 0$, and equal to ∞ when it converges for all x. This convention allows us to cover all possibilities in a single statement: each power series in x has a radius of convergence R, where $0 \leq R \leq \infty$, with the property that the series converges if $|x| < R$ and diverges if $|x| > R$. It should be noted that if $R = 0$ then no x satisfies $|x| < R$, and if $R = \infty$ then no x satisfies $|x| > R$.

In many important cases the value of R can be found as follows. Let

$$\sum_{n=0}^{\infty} u_n = u_0 + u_1 + u_2 + \cdots$$

be a series of nonzero constants. We recall from elementary calculus that if the limit

$$\lim_{n \to \infty} \left| \frac{u_{n+1}}{u_n} \right| = L$$

exists, then the *ratio test* asserts that the series converges if $L < 1$ and diverges if $L > 1$. In the case of our power series (3), this tells us that if each $a_n \neq 0$, and if for a fixed point $x \neq 0$ we have

$$\lim_{n \to \infty} \left| \frac{a_{n+1}x^{n+1}}{a_n x^n} \right| = \lim_{n \to \infty} \left| \frac{a_{n+1}}{a_n} \right| |x| = L,$$

then (3) converges if $L < 1$ and diverges if $L > 1$. These considerations yield the formula

$$R = \lim_{n \to \infty} \left| \frac{a_n}{a_{n+1}} \right|$$

if this limit exists (we put $R = \infty$ if $|a_n/a_{n+1}| \to \infty$). Regardless of whether this formula can be used or not, it is known that R always exists; and if R is finite and nonzero, then it determines an *interval of convergence* $-R < x < R$ such that inside the interval the series converges and outside the interval it diverges. A power series may or may not converge at either endpoint of its interval of convergence.

C. Suppose that (3) converges for $|x| < R$ with $R > 0$, and denote its sum by $f(x)$:

$$f(x) = \sum_{n=0}^{\infty} a_n x^n = a_0 + a_1 x + a_2 x^2 + \cdots. \tag{8}$$

Then $f(x)$ is automatically continuous and has derivatives of all orders for $|x| < R$. Also, the series can be differentiated termwise in the sense that

$$f'(x) = \sum_{n=1}^{\infty} n a_n x^{n-1} = a_1 + 2a_2 x + 3a_3 x^2 + \cdots,$$

$$f''(x) = \sum_{n=2}^{\infty} n(n-1) a_n x^{n-2} = 2a_2 + 3 \cdot 2a_3 x + \cdots,$$

and so on, and each of the resulting series converges for $|x| < R$. These successive differentiated series yield the following basic formula linking the a_n to $f(x)$ and its derivatives:

$$a_n = \frac{f^{(n)}(0)}{n!}. \tag{9}$$

Furthermore, it is often useful to know that the series (8) can be integrated termwise provided the limits of integration lie inside the interval of convergence.

If we have a second power series in x that converges to a function $g(x)$ for $|x| < R$, so that

$$g(x) = \sum_{n=0}^{\infty} b_n x^n = b_0 + b_1 x + b_2 x^2 + \cdots, \tag{10}$$

then (8) and (10) can be added or subtracted termwise:

$$f(x) \pm g(x) = \sum_{n=0}^{\infty} (a_n \pm b_n) x^n = (a_0 \pm b_0) + (a_1 \pm b_1) x + \cdots.$$

They can also be multiplied as if they were polynomials, in the sense that

$$f(x)g(x) = \sum_{n=0}^{\infty} c_n x^n$$

where $c_n = a_0 b_n + a_1 b_{n-1} + \cdots + a_n b_0.$[2] If it happens that both series converge to the same function, so that $f(x) = g(x)$ for $|x| < R$, then formula (9) implies that they must have the same coefficients: $a_0 = b_0, a_1 = b_1, \ldots$. In particular, if $f(x) = 0$ for $|x| < R$, then $a_0 = 0$, $a_1 = 0, \ldots$.

D. Let $f(x)$ be a continuous function that has derivatives of all orders for $|x| < R$ with $R > 0$. Can $f(x)$ be represented by a power series? If we use (9) to *define* the a_n, then it is natural to hope that the expansion

$$f(x) = \sum_{n=0}^{\infty} \frac{f^{(n)}(0)}{n!} x^n = f(0) + f'(0)x + \frac{f''(0)}{2!} x^2 + \cdots \qquad (11)$$

will hold throughout the interval. This is often true, but unfortunately it is sometimes false. One way of investigating the validity of this expansion for a specific point x in the interval is to use *Taylor's formula*:

$$f(x) = \sum_{k=0}^{n} \frac{f^{(k)}(0)}{k!} x^k + R_n(x),$$

where the remainder $R_n(x)$ is given by

$$R_n(x) = \frac{f^{(n+1)}(\bar{x})}{(n+1)!} x^{n+1}$$

for some point $\bar{x}$ between 0 and x. To verify (11), it suffices to show that $R_n(x) \to 0$ as $n \to \infty$. By means of this procedure, it is quite easy to obtain the following familiar expansions, which are valid for all x:

$$e^x = \sum_{n=0}^{\infty} \frac{x^n}{n!} = 1 + x + \frac{x^2}{2!} + \frac{x^3}{3!} + \cdots; \qquad (12)$$

$$\sin x = \sum_{n=0}^{\infty} (-1)^n \frac{x^{2n+1}}{(2n+1)!} = x - \frac{x^3}{3!} + \frac{x^5}{5!} - \cdots; \qquad (13)$$

$$\cos x = \sum_{n=0}^{\infty} (-1)^n \frac{x^{2n}}{(2n)!} = 1 - \frac{x^2}{2!} + \frac{x^4}{4!} - \cdots. \qquad (14)$$

---

[2] It will be useful later to notice that c_n can be written in two equivalent forms:

$$c_n = \sum_{k=0}^{n} a_k b_{n-k} \quad \text{and} \quad c_n = \sum_{k=0}^{n} a_{n-k} b_k.$$

If a specific convergent power series is given to us, how can we recognize the function that is its sum? In general it is impossible to do this, for very few power series have sums that are familiar elementary functions.

E. A function $f(x)$ with the property that a power series expansion of the form

$$f(x) = \sum_{n=0}^{\infty} a_n(x - x_0)^n \tag{15}$$

is valid in some neighborhood of the point x_0 is said to be *analytic* at x_0. In this case the a_n are necessarily given by

$$a_n = \frac{f^{(n)}(x_0)}{n!},$$

and (15) is called the *Taylor series* of $f(x)$ at x_0. Thus, (12), (13), and (14) tell us that e^x, $\sin x$, and $\cos x$ are analytic at $x_0 = 0$, and the given series are the Taylor series of these functions at this point. Most questions about analyticity can be answered by means of the following facts:

1. Polynomials and the functions e^x, $\sin x$, and $\cos x$ are analytic at all points.
2. If $f(x)$ and $g(x)$ are analytic at x_0, then $f(x) + g(x)$, $f(x)g(x)$, and $f(x)/g(x)$ [if $g(x_0) \neq 0$] are also analytic at x_0.
3. If $f(x)$ is analytic at x_0 and $f^{-1}(x)$ is a continuous inverse, then $f^{-1}(x)$ is analytic at $f(x_0)$ if $f'(x_0) \neq 0$.
4. If $g(x)$ is analytic at x_0 and $f(x)$ is analytic at $g(x_0)$, then $f(g(x))$ is analytic at x_0.
5. The sum of a power series is analytic at all points inside the interval of convergence.

Some of these statements are quite easy to prove by elementary methods, but others are not. Generally speaking, the behavior of analytic functions can be fully understood only in the broader context of the theory of functions of a complex variable.

PROBLEMS

1. Use the ratio test to verify that $R = 0$, $R = \infty$, and $R = 1$ for the series (5), (6), and (7).
2. If p is not zero or a positive integer, show that the series

$$\sum_{n=1}^{\infty} \frac{p(p-1)(p-2)\cdots(p-n+1)}{n!} x^n$$

converges for $|x| < 1$ and diverges for $|x| > 1$.

3. Show that $R = \infty$ for the series on the right sides of expansions (13) and (14).
4. Use Taylor's formula to establish the validity of the expansions (12), (13), and (14) for all x. Hint: $a^n/n! \to 0$ for every constant a (why?).
5. It is well known from elementary algebra that

$$1 + x + x^2 + \cdots + x^n = \frac{1 - x^{n+1}}{1 - x} \qquad \text{if } x \neq 1.$$

Use this to show that the expansions

$$\frac{1}{1 - x} = 1 + x + x^2 + x^3 + \cdots$$

and

$$\frac{1}{1 + x} = 1 - x + x^2 - x^3 + \cdots$$

are valid for $|x| < 1$. Apply the latter to show that

$$\log (1 + x) = x - \frac{x^2}{2} + \frac{x^3}{3} - \frac{x^4}{4} + \cdots$$

and

$$\tan^{-1} x = x - \frac{x^3}{3} + \frac{x^5}{5} - \frac{x^7}{7} + \cdots$$

for $|x| < 1$.
6. Use the first expansion given in Problem 5 to find the power series for $1/(1 - x)^2$
   (a) by squaring;
   (b) by differentiating.
7. (a) Show that the series for $\cos x$,

$$y = 1 - \frac{x^2}{1 \cdot 2} + \frac{x^4}{1 \cdot 2 \cdot 3 \cdot 4} - \frac{x^6}{1 \cdot 2 \cdot 3 \cdot 4 \cdot 5 \cdot 6} + \cdots,$$

has the property that $y'' = -y$, and is therefore a solution of equation (1).
   (b) Show that the series

$$y = 1 - \frac{x^2}{2^2} + \frac{x^4}{2^2 \cdot 4^2} - \frac{x^6}{2^2 \cdot 4^2 \cdot 6^2} + \cdots$$

converges for all x, and verify that it is a solution of equation (2). [Observe that this series can be obtained from the one in (a) by replacing each odd factor in the denominators by the next greater even number. The sum of this series is a useful special function denoted by $J_0(x)$ and called the *Bessel function of order* 0; it will be studied in detail in Chapter 8.]

27 SERIES SOLUTIONS OF FIRST ORDER EQUATIONS

We have repeatedly emphasized that many interesting and important differential equations cannot be solved by any of the methods discussed in earlier chapters, and also that solutions for equations of this kind can often be found in terms of power series. Our purpose in this section is to explain the procedure by showing how it works in the case of first order equations that are easy to solve by elementary methods.

As our first example, we consider the equation

$$y' = y. \tag{1}$$

We assume that this equation has a power series solution of the form

$$y = a_0 + a_1 x + a_2 x^2 + \cdots + a_n x^n + \cdots \tag{2}$$

that converges for $|x| < R$ with $R > 0$; that is, we assume that (1) has a solution that is analytic at the origin. A power series can be differentiated term by term in its interval of convergence, so

$$y' = a_1 + 2a_2 x + 3a_3 x^2 + \cdots + (n + 1)a_{n+1} x^n + \cdots. \tag{3}$$

Since $y' = y$, the series (2) and (3) must have the same coefficients:

$$a_1 = a_0, \qquad 2a_2 = a_1, \qquad 3a_3 = a_2, \ldots, (n + 1)a_{n+1} = a_n, \ldots.$$

These equations enable us to express each a_n in terms of a_0:

$$a_1 = a_0, \qquad a_2 = \frac{a_1}{2} = \frac{a_0}{2}, \qquad a_3 = \frac{a_2}{3} = \frac{a_0}{2 \cdot 3}, \ldots, a_n = \frac{a_0}{n!}, \ldots.$$

When these coefficients are inserted in (2), we obtain our power series solution

$$y = a_0 \left(1 + x + \frac{x^2}{2!} + \frac{x^3}{3!} + \cdots + \frac{x^n}{n!} + \cdots \right), \tag{4}$$

where no condition is imposed on a_0. It is essential to understand that so far this solution is only tentative, because we have no guarantee that (1) actually has a power series solution of the form (2). The above argument shows only that *if* (1) has such a solution, *then* that solution must be (4). However, it follows at once from the ratio test that the series in (4) converges for all x, so the term-by-term differentiation is valid and (4) really is a solution of (1). In this case we can easily recognize the series in (4) as the power series expansion of e^x, so (4) can be written as

$$y = a_0 e^x.$$

Needless to say, we can get this solution directly from (1) by separating variables and integrating. Nevertheless, it is important to realize that (4) would still be a perfectly respectable solution even if (1) were unsolvable by elementary methods and the series in (4) could not be recognized as the expansion of a familiar function.

This example suggests a useful method for obtaining the power series expansion of a given function: find the differential equation satisfied by the function, and then solve this equation by power series.

As an illustration of this idea we consider the function

$$y = (1 + x)^p, \tag{5}$$

where p is an arbitrary constant. It is easy to see that (5) is the indicated particular solution of the following differential equation:

$$(1 + x)y' = py, \qquad y(0) = 1. \tag{6}$$

As before, we assume that (6) has a power series solution

$$y = a_0 + a_1x + a_2x^2 + \cdots + a_nx^n + \cdots \tag{7}$$

with positive radius of convergence. It follows from this that

$$
\begin{aligned}
y' &= a_1 + 2a_2x + 3a_3x^2 + \cdots + (n + 1)a_{n+1}x^n + \cdots, \\
xy' &= \qquad\quad a_1x + 2a_2x^2 + \cdots + \qquad na_nx^n + \cdots, \\
py &= pa_0 + pa_1x + pa_2x^2 + \cdots + \qquad pa_nx^n + \cdots.
\end{aligned}
$$

By equation (6), the sum of the first two series must equal the third, so equating the coefficients of successive powers of x gives

$$a_1 = pa_0, \qquad 2a_2 + a_1 = pa_1, \qquad 3a_3 + 2a_2 = pa_2, \ldots,$$

$$(n + 1)a_{n+1} + na_n = pa_n, \ldots.$$

The initial condition in (6) implies that $a_0 = 1$, so

$$a_1 = p, \qquad a_2 = \frac{a_1(p - 1)}{2} = \frac{p(p - 1)}{2},$$

$$a_3 = \frac{a_2(p - 2)}{3} = \frac{p(p - 1)(p - 2)}{2\cdot3}, \ldots,$$

$$a_n = \frac{p(p - 1)(p - 2) \cdots (p - n + 1)}{n!}, \ldots.$$

With these coefficients, (7) becomes

$$y = 1 + px + \frac{p(p - 1)}{2!}x^2 + \frac{p(p - 1)(p - 2)}{3!}x^3 + \cdots$$

$$+ \frac{p(p - 1)(p - 2) \cdots (p - n + 1)}{n!}x^n + \cdots. \tag{8}$$

To conclude that (8) actually is the desired solution, it suffices to observe that this series converges for $|x| < 1$ (see Problem 26-2). On comparing the two solutions (5) and (8), and using the fact that (6) has only one solution, we have

$$(1 + x)^p = 1 + px + \frac{p(p - 1)}{2!}x^2 + \cdots$$

$$+ \frac{p(p - 1) \cdots (p - n + 1)}{n!}x^n + \cdots \quad (9)$$

for $|x| < 1$. This expansion is called the *binomial series,* and generalizes the binomial theorem to the case of an arbitrary exponent.[3]

PROBLEMS

1. Consider the following differential equations:
   (a) $y' = 2xy$;
   (b) $y' + y = 1$.
   In each case, find a power series solution of the form $\sum a_n x^n$, try to recognize the resulting series as the expansion of a familiar function, and verify your conclusion by solving the equation directly.
2. Consider the following differential equations:
   (a) $xy' = y$;
   (b) $x^2 y' = y$.
   In each case, find a power series solution of the form $\sum a_n x^n$, solve the equation directly, and explain any discrepancies that arise.
3. Express $\sin^{-1} x$ in the form of a power series $\sum a_n x^n$ by solving $y' = (1 - x^2)^{-1/2}$ in two ways. (*Hint:* Remember the binomial series.) Use this result to obtain the formula

   $$\frac{\pi}{6} = \frac{1}{2} + \frac{1}{2} \cdot \frac{1}{3 \cdot 2^3} + \frac{1 \cdot 3}{2 \cdot 4} \cdot \frac{1}{5 \cdot 2^5} + \frac{1 \cdot 3 \cdot 5}{2 \cdot 4 \cdot 6} \cdot \frac{1}{7 \cdot 2^7} + \cdots.$$

4. The differential equations considered in the text and preceding problems are all linear. The equation

   $$y' = 1 + y^2 \quad (*)$$

---

[3] As the reader will recall from elementary algebra, the binomial theorem states that if n is a positive integer, then

$$(1 + x)^n = 1 + nx + \frac{n(n - 1)}{2!}x^2 + \cdots + \frac{n(n - 1) \cdots (n - k + 1)}{k!}x^k + \cdots + x^n.$$

More concisely,

$$(1 + x)^n = \sum_{k=0}^{n} \binom{n}{k}x^k,$$

where the *binomial coefficient* $\binom{n}{k}$ is defined by

$$\binom{n}{k} = \frac{n!}{k!(n - k)!} = \frac{n(n - 1) \cdots (n - k + 1)}{k!}.$$

is nonlinear, and it is easy to see directly that $y = \tan x$ is the particular solution for which $y(0) = 0$. Show that

$$\tan x = x + \frac{1}{3}x^3 + \frac{2}{15}x^5 + \cdots$$

by assuming a solution for equation (*) in the form of a power series $\sum a_n x^n$ and finding the a_n in two ways:

(a) by the method of the examples in the text (note particularly how the nonlinearity of the equation complicates the formulas);
(b) by differentiating equation (*) repeatedly to obtain

$$y'' = 2yy', \qquad y''' = 2yy'' + 2(y')^2, \ldots,$$

and using the formula $a_n = f^{(n)}(0)/n!$.

5. Solve the equation

$$y' = x - y, \qquad y(0) = 0$$

by each of the methods suggested in Problem 4. What familiar function does the resulting series represent? Verify your conclusion by solving the equation directly as a first order linear equation.

28 SECOND ORDER LINEAR EQUATIONS. ORDINARY POINTS

We now turn our attention to the general homogeneous second order linear equation

$$y'' + P(x)y' + Q(x)y = 0. \tag{1}$$

As we know, it is occasionally possible to solve such an equation in terms of familiar elementary functions. This is true, for instance, when $P(x)$ and $Q(x)$ are constants, and in a few other cases as well. For the most part, however, the equations of this type having the greatest significance in both pure and applied mathematics are beyond the reach of elementary methods, and can only be solved by means of power series.

The central fact about equation (1) is that the behavior of its solutions near a point x_0 depends on the behavior of its coefficient functions $P(x)$ and $Q(x)$ near this point. In this section we confine ourselves to the case in which $P(x)$ and $Q(x)$ are "well behaved" in the sense of being analytic at x_0, which means that each has a power series expansion valid in some neighborhood of this point. In this case x_0 is called an *ordinary point* of equation (1), and it turns out that every solution of the equation is also analytic at this point. In other words, the analyticity of the coefficients of (1) at a certain point implies that its solutions are also analytic there. Any point that is not an ordinary point of (1) is called a *singular point*.

We shall prove the statement made in the above paragraph, but first we consider some illustrative examples.

In the case of the familiar equation

$$y'' + y = 0, \tag{2}$$

the coefficient functions are $P(x) = 0$ and $Q(x) = 1$. These functions are analytic at all points, so we seek a solution of the form

$$y = a_0 + a_1 x + a_2 x^2 + \cdots + a_n x^n + \cdots. \tag{3}$$

Differentiating (3) yields

$$y' = a_1 + 2a_2 x + 3a_3 x^2 + \cdots + (n + 1)a_{n+1} x^n + \cdots \tag{4}$$

and

$$y'' = 2a_2 + 2 \cdot 3a_3 x + 3 \cdot 4a_4 x^2 + \cdots + (n + 1)(n + 2)a_{n+2} x^n + \cdots. \tag{5}$$

If we substitute (5) and (3) into (2) and add the two series term by term, we get

$$(2a_2 + a_0) + (2 \cdot 3a_3 + a_1)x + (3 \cdot 4a_4 + a_2)x^2 + (4 \cdot 5a_5 + a_3)x^3$$
$$+ \cdots + [(n + 1)(n + 2)a_{n+2} + a_n]x^n + \cdots = 0;$$

and equating to zero the coefficients of successive powers of x gives

$$2a_2 + a_0 = 0, \qquad 2 \cdot 3a_3 + a_1 = 0, \qquad 3 \cdot 4a_4 + a_2 = 0,$$
$$4 \cdot 5a_5 + a_3 = 0, \ldots, (n + 1)(n + 2)a_{n+2} + a_n = 0, \ldots.$$

By means of these equations we can express a_n in terms of a_0 or a_1 according as n is even or odd:

$$a_2 = -\frac{a_0}{2}, \qquad a_3 = -\frac{a_1}{2 \cdot 3}, \qquad a_4 = -\frac{a_2}{3 \cdot 4} = \frac{a_0}{2 \cdot 3 \cdot 4},$$

$$a_5 = -\frac{a_3}{4 \cdot 5} = \frac{a_1}{2 \cdot 3 \cdot 4 \cdot 5}, \ldots.$$

With these coefficients, (3) becomes

$$y = a_0 + a_1 x - \frac{a_0}{2}x^2 - \frac{a_1}{2 \cdot 3}x^3 + \frac{a_0}{2 \cdot 3 \cdot 4}x^4 + \frac{a_1}{2 \cdot 3 \cdot 4 \cdot 5}x^5 - \cdots$$

$$= a_0\left(1 - \frac{x^2}{2!} + \frac{x^4}{4!} - \cdots\right) + a_1\left(x - \frac{x^3}{3!} + \frac{x^5}{5!} - \cdots\right). \tag{6}$$

Let $y_1(x)$ and $y_2(x)$ denote the two series in parentheses. We have shown formally that (6) satisfies (2) for any two constants a_0 and a_1. In particular, by choosing $a_0 = 1$ and $a_1 = 0$ we see that y_1 satisfies this equation, and the choice $a_0 = 0$ and $a_1 = 1$ shows that y_2 also satisfies the equation. Just as in the examples of the previous section, the only remaining issue concerns the convergence of the two series defining y_1 and y_2. But the ratio test shows at once that each of these series—and therefore the series (6)—converges for all x (see Problem 26-3). It follows that all the operations performed on (3) are legitimate, so (6) is a valid solution of (2) as opposed to a merely formal solution. Furthermore, y_1 and y_2 are linearly independent since it is obvious that neither series is a constant multiple of the other. We therefore see that (6) is the general solution of (2), and that any particular solution is obtained by specifying the values of $y(0) = a_0$ and $y'(0) = a_1$.

In the above example the two series in parentheses are easily recognizable as the expansions of $\cos x$ and $\sin x$, so (6) can be written in the form

$$y = a_0 \cos x + a_1 \sin x.$$

Naturally, this conclusion could have been foreseen in the beginning, since (2) is a very simple equation whose solutions are perfectly familiar to us. ·However, this result should be regarded as only a lucky accident, for most series solutions found in this way are quite impossible to identify and represent previously unknown functions.

As an illustration of this remark, we use the same procedure to solve *Legendre's equation*

$$(1 - x^2)y'' - 2xy' + p(p + 1)y = 0, \qquad (7)$$

where p is a constant. It is clear that the coefficient functions

$$P(x) = \frac{-2x}{1 - x^2} \quad \text{and} \quad Q(x) = \frac{p(p + 1)}{1 - x^2} \qquad (8)$$

are analytic at the origin. The origin is therefore an ordinary point, and we expect a solution of the form $y = \sum a_n x^n$. Since $y' = \sum (n + 1)a_{n+1}x^n$, we get the following expansions for the individual terms on the left side of equation (7):

$$y'' = \sum (n + 1)(n + 2)a_{n+2}x^n,$$

$$-x^2y'' = \sum - (n - 1)na_n x^n,$$

$$-2xy' = \sum - 2na_n x^n,$$

and

$$p(p + 1)y = \sum p(p + 1)a_n x^n.$$

By equation (7), the sum of these series is required to be zero, so the coefficient of x^n must be zero for every n:

$$(n + 1)(n + 2)a_{n+2} - (n - 1)na_n - 2na_n + p(p + 1)a_n = 0.$$

With a little manipulation, this becomes

$$a_{n+2} = -\frac{(p - n)(p + n + 1)}{(n + 1)(n + 2)} a_n. \tag{9}$$

Just as in the previous example, this *recursion formula* enables us to express a_n in terms of a_0 or a_1 according as n is even or odd:

$$a_2 = -\frac{p(p + 1)}{1 \cdot 2} a_0,$$

$$a_3 = -\frac{(p - 1)(p + 2)}{2 \cdot 3} a_1,$$

$$a_4 = -\frac{(p - 2)(p + 3)}{3 \cdot 4} a_2 = \frac{p(p - 2)(p + 1)(p + 3)}{4!} a_0,$$

$$a_5 = -\frac{(p - 3)(p + 4)}{4 \cdot 5} a_3 = \frac{(p - 1)(p - 3)(p + 2)(p + 4)}{5!} a_1,$$

$$a_6 = -\frac{(p - 4)(p + 5)}{5 \cdot 6} a_4$$

$$= -\frac{p(p - 2)(p - 4)(p + 1)(p + 3)(p + 5)}{6!} a_0,$$

$$a_7 = -\frac{(p - 5)(p + 6)}{6 \cdot 7} a_5$$

$$= -\frac{(p - 1)(p - 3)(p - 5)(p + 2)(p + 4)(p + 6)}{7!} a_1,$$

and so on. By inserting these coefficients into the assumed solution $y = \Sigma a_n x^n$, we obtain

$$y = a_0 \left[1 - \frac{p(p + 1)}{2!} x^2 + \frac{p(p - 2)(p + 1)(p + 3)}{4!} x^4 \right.$$

$$\left. - \frac{p(p - 2)(p - 4)(p + 1)(p + 3)(p + 5)}{6!} x^6 + \cdots \right]$$

$$+ a_1 \left[x - \frac{(p - 1)(p + 2)}{3!} x^3 + \frac{(p - 1)(p - 3)(p + 2)(p + 4)}{5!} x^5 \right.$$

$$\left. - \frac{(p - 1)(p - 3)(p - 5)(p + 2)(p + 4)(p + 6)}{7!} x^7 + \cdots \right]$$

$$\tag{10}$$

as our formal solution of (7).

When p is not an integer, each series in brackets has radius of convergence $R = 1$. This is most easily seen by using the recursion formula (9): for the first series, this formula (with n replaced by $2n$) yields

$$\left| \frac{a_{2n+2}x^{2n+2}}{a_{2n}x^{2n}} \right| = \left| -\frac{(p - 2n)(p + 2n + 1)}{(2n + 1)(2n + 2)} \right| |x|^2 \to |x|^2$$

as $n \to \infty$, and similarly for the second series. As before, the fact that each series has positive radius of convergence justifies the operations we have performed and shows that (10) is a valid solution of (7) for every choice of the constants a_0 and a_1. Each bracketed series is a particular solution; and since it is clear that the functions defined by these series are linearly independent, (10) is the general solution of (7) on the interval $|x| < 1$.

The functions defined by (10) are called *Legendre functions*, and in general they are not elementary. However, when p is a nonnegative integer, one of the series terminates and is thus a polynomial—the first series if p is even and the second series if p is odd—while the other does not and remains an infinite series. This observation leads to the particular solutions of (7) known as *Legendre polynomials*, whose properties and applications we discuss in Chapter 8.

We now apply the method of these examples to establish the following general theorem about the nature of solutions near ordinary points.

Theorem A. *Let x_0 be an ordinary point of the differential equation*

$$y'' + P(x)y' + Q(x)y = 0, \tag{11}$$

and let a_0 and a_1 be arbitrary constants. Then there exists a unique function $y(x)$ that is analytic at x_0, is a solution of equation (11) in a certain neighborhood of this point, and satisfies the initial conditions $y(x_0) = a_0$ and $y'(x_0) = a_1$. Furthermore, if the power series expansions of $P(x)$ and $Q(x)$ are valid on an interval $|x - x_0| < R$, $R > 0$, then the power series expansion of this solution is also valid on the same interval.

Proof. For the sake of convenience, we restrict our argument to the case in which $x_0 = 0$. This permits us to work with power series in x rather than $x - x_0$, and involves no real loss of generality. With this slight simplification, the hypothesis of the theorem is that $P(x)$ and $Q(x)$ are analytic at the origin and therefore have power series expansions

$$P(x) = \sum_{n=0}^{\infty} p_n x^n = p_0 + p_1 x + p_2 x^2 + \cdots \tag{12}$$

and

$$Q(x) = \sum_{n=0}^{\infty} q_n x^n = q_0 + q_1 x + q_2 x^2 + \cdots \tag{13}$$

that converge on an interval $|x| < R$ for some $R > 0$. Keeping in mind the specified initial conditions, we try to find a solution for (11) in the form of a power series

$$y = \sum_{n=0}^{\infty} a_n x^n = a_0 + a_1 x + a_2 x^2 + \cdots \tag{14}$$

with radius of convergence at least R. Differentiation of (14) yields

$$y' = \sum_{n=0}^{\infty} (n + 1)a_{n+1}x^n = a_1 + 2a_2 x + 3a_3 x^2 + \cdots \tag{15}$$

and

$$y'' = \sum_{n=0}^{\infty} (n + 1)(n + 2)a_{n+2}x^n$$
$$= 2a_2 + 2 \cdot 3a_3 x + 3 \cdot 4a_4 x^2 + \cdots. \tag{16}$$

It now follows from the rule for multiplying power series that

$$P(x)y' = \left(\sum_{n=0}^{\infty} p_n x^n \right) \left[\sum_{n=0}^{\infty} (n + 1)a_{n+1}x^n \right]$$
$$= \sum_{n=0}^{\infty} \left[\sum_{k=0}^{n} p_{n-k}(k + 1)a_{k+1} \right] x^n \tag{17}$$

and

$$Q(x)y = \left(\sum_{n=0}^{\infty} q_n x^n \right) \left(\sum_{n=0}^{\infty} a_n x^n \right)$$
$$= \sum_{n=0}^{\infty} \left(\sum_{k=0}^{n} q_{n-k}a_k \right) x^n. \tag{18}$$

On substituting (16), (17), and (18) into (11) and adding the series term by term, we obtain

$$\sum_{n=0}^{\infty} \left[(n + 1)(n + 2)a_{n+2} + \sum_{k=0}^{n} p_{n-k}(k + 1)a_{k+1} + \sum_{k=0}^{n} q_{n-k}a_k \right] x^n = 0,$$

so we have the following recursion formula for the a_n:

$$(n + 1)(n + 2)a_{n+2} = - \sum_{k=0}^{n} [(k + 1)p_{n-k}a_{k+1} + q_{n-k}a_k]. \tag{19}$$

For $n = 0, 1, 2, \ldots$ this formula becomes

$$2a_2 = -(p_0 a_1 + q_0 a_0),$$
$$2 \cdot 3a_3 = -(p_1 a_1 + 2p_0 a_2 + q_1 a_0 + q_0 a_1),$$
$$3 \cdot 4a_4 = -(p_2 a_1 + 2p_1 a_2 + 3p_0 a_3 + q_2 a_0 + q_1 a_1 + q_0 a_2),$$
$$\cdots \cdots$$

These formulas determine $a_2, a_3, \ldots$ in terms of a_0 and a_1, so the resulting series (14), which formally satisfies (11) and the given initial conditions, is uniquely determined by these requirements.

Suppose now that we can prove that the series (14), with its coefficients defined by formula (19), actually converges for $|x| < R$. Then by the general theory of power series it will follow that the formal operations by which (14) was made to satisfy (11)—termwise differentiation, multiplication, and term-by-term addition—are justified, and the proof will be complete. This argument is not easy. We give the details in Appendix A, where they can be omitted conveniently by any reader who wishes to do so.

A few final remarks are in order. In our examples we encountered only what are known as *two-term recursion formulas* for the coefficients of the unknown series solutions. The simplicity of these formulas makes it fairly easy to determine the general terms of the resulting series and to obtain precise information about their radii of convergence. However, it is apparent from formula (19) that this simplicity is not to be expected in general. In most cases the best we can do is to find the radii of convergence of the series expansions of $P(x)$ and $Q(x)$ and to conclude from the theorem that the radius for the series solution must be at least as large as the smaller of these numbers. Thus, for Legendre's equation it is clear from (8) and the familiar expansion

$$\frac{1}{1 - x^2} = 1 + x^2 + x^4 + \cdots, \qquad R = 1,$$

that $R = 1$ for both $P(x)$ and $Q(x)$. We therefore know at once, without further calculation, that any solution of the form $y = \sum a_n x^n$ must be valid at least on the interval $|x| < 1$.

PROBLEMS

1. Find the general solution of $(1 + x^2)y'' + 2xy' - 2y = 0$ in terms of power series in x. Can you express this solution by means of elementary functions?

2. Consider the equation $y'' + xy' + y = 0$.
   (a) Find its general solution $y = \sum a_n x^n$ in the form $y = a_0 y_1(x) + a_1 y_2(x)$, where $y_1(x)$ and $y_2(x)$ are power series.
   (b) Use the ratio test to verify that the two series $y_1(x)$ and $y_2(x)$ converge for all x, as Theorem A asserts.
   (c) Show that $y_1(x)$ is the series expansion of $e^{-x^2/2}$, use this fact to find a second independent solution by the method of Section 16, and convince yourself that this second solution is the function $y_2(x)$ found in (a).

3. Verify that the equation $y'' + y' - xy = 0$ has a three-term recursion formula, and find its series solutions $y_1(x)$ and $y_2(x)$ such that
   (a) $y_1(0) = 1$, $y_1'(0) = 0$;
   (b) $y_2(0) = 0$, $y_2'(0) = 1$.
   Theorem A guarantees that both series converge for all x. Notice how difficult this would be to prove by working with the series themselves.

4. The equation $y'' + (p + \frac{1}{2} - \frac{1}{4}x^2)y = 0$, where p is a constant, certainly has a series solution of the form $y = \sum a_n x^n$.

(a) Show that the coefficients a_n are related by the three-term recursion formula

$$(n + 1)(n + 2)a_{n+2} + \left(p + \frac{1}{2}\right)a_n - \frac{1}{4}a_{n-2} = 0.$$

(b) If the dependent variable is changed from y to w by means of $y = we^{-x^2/4}$, show that the equation is transformed into $w'' - xw' + pw = 0$.

(c) Verify that the equation in (b) has a two-term recursion formula and find its general solution.

5. Solutions of *Airy's equation* $y'' + xy = 0$ are called *Airy functions*, and have applications to the theory of diffraction.[4]

(a) Apply the theorems of Section 24 to verify that every nontrivial Airy function has infinitely many positive zeros and at most one negative zero.

(b) Find the Airy functions in the form of power series, and verify directly that these series converge for all x.

(c) Use the results of (b) to write down the general solution of $y'' - xy = 0$ without calculation.

6. *Chebyshev's equation* is

$$(1 - x^2)y'' - xy' + p^2 y = 0,$$

where p is a constant.

(a) Find two linearly independent series solutions valid for $|x| < 1$.

(b) Show that if $p = n$ where n is an integer ≥ 0, then there is a polynomial solution of degree n. When these are multiplied by suitable constants, they are called the *Chebyshev polynomials*. We shall return to this topic in the problems of Section 31 and in Appendix D.

7. *Hermite's equation* is

$$y'' - 2xy' + 2py = 0,$$

where p is a constant.

(a) Show that its general solution is $y(x) = a_0 y_1(x) + a_1 y_2(x)$, where

$$y_1(x) = 1 - \frac{2p}{2!}x^2 + \frac{2^2 p(p - 2)}{4!}x^4 - \frac{2^3 p(p - 2)(p - 4)}{6!}x^6 + \cdots$$

---

[4] Sir George Biddell Airy (1801–1892), Astronomer Royal of England for many years, was a hard-working, systematic plodder whose sense of decorum almost deprived John Couch Adams of credit for discovering the planet Neptune. As a boy Airy was notorious for his skill in designing peashooters; but in spite of this promising start and some early work in the theory of light—in connection with which he was the first to draw attention to the defect of vision known as astigmatism—he developed into the excessively practical type of scientist who is obsessed by elaborate numerical computations and has little use for general scientific ideas.

and

$$y_2(x) = x - \frac{2(p-1)}{3!}x^3 + \frac{2^2(p-1)(p-3)}{5!}x^5$$

$$- \frac{2^3(p-1)(p-3)(p-5)}{7!}x^7 + \cdots.$$

By Theorem A, both series converge for all x. Verify this directly.

(b) If p is a nonnegative integer, then one of these series terminates and is thus a polynomial—$y_1(x)$ if p is even, and $y_2(x)$ if p is odd—while the other remains an infinite series. Verify that for $p = 0, 1, 2, 3, 4, 5$, these polynomials are $1, x, 1 - 2x^2, x - \frac{2}{3}x^3, 1 - 4x^2 + \frac{4}{3}x^4, x - \frac{4}{3}x^3 + \frac{4}{15}x^5$.

(c) It is clear that the only polynomial solutions of Hermite's equation are constant multiples of the polynomials described in (b). Those constant multiples with the property that the terms containing the highest powers of x are of the form $2^n x^n$ are denoted by $H_n(x)$ and called the *Hermite polynomials*. Verify that $H_0(x) = 1$, $H_1(x) = 2x$, $H_2(x) = 4x^2 - 2$, $H_3(x) = 8x^3 - 12x$, $H_4(x) = 16x^4 - 48x^2 + 12$, and $H_5(x) = 32x^5 - 160x^3 + 120x$.

(d) Verify that the polynomials listed in (c) are given by the general formula

$$H_n(x) = (-1)^n e^{x^2} \frac{d^n}{dx^n} e^{-x^2}.$$

In Appendix B we show how the formula in (d) can be deduced from the series in (a), we prove several of the most useful properties of the Hermite polynomials, and we show briefly how these polynominals arise in a fundamental problem of quantum mechanics.

29 REGULAR SINGULAR POINTS

We recall that a point x_0 is a *singular point* of the differential equation

$$y'' + P(x)y' + Q(x)y = 0 \tag{1}$$

if one or the other (or both) of the coefficient functions $P(x)$ and $Q(x)$ fails to be analytic at x_0. In this case the theorem and methods of the previous section do not apply, and new ideas are necessary if we wish to study the solutions of (1) near x_0. This is a matter of considerable practical importance; for many differential equations that arise in physical problems have singular points, and the choice of physically appropriate solutions is often determined by their behavior near these points. Thus, while we might want to avoid the singular points of a differential equation, it is precisely these points that usually demand particular attention. As a simple example, the origin is clearly a singular point of

$$y'' + \frac{2}{x}y' - \frac{2}{x^2}y = 0.$$

It is easy to verify that $y_1 = x$ and $y_2 = x^{-2}$ are independent solutions for $x > 0$, so $y = c_1 x + c_2 x^{-2}$ is the general solution on this interval. If we happen to be interested only in solutions that are bounded near the origin, then it is evident from this general solution that these are obtained by putting $c_2 = 0$.

In general, there is very little that can be said about the solutions of (1) near the singular point x_0. Fortunately, however, in most of the applications the singular points are rather "weak," in the sense that the coefficient functions are only mildly nonanalytic, and simple modifications of our previous methods yield satisfactory solutions. These are the regular singular points, which are defined as follows. A singular point x_0 of equation (1) is said to be *regular* if the functions $(x - x_0)P(x)$ and $(x - x_0)^2 Q(x)$ are analytic, and *irregular* otherwise.[5] Roughly speaking, this means that the singularity in $P(x)$ cannot be worse than $1/(x - x_0)$, and that in $Q(x)$ cannot be worse than $1/(x - x_0)^2$.

If we consider Legendre's equation 28-(7) in the form

$$y'' - \frac{2x}{1 - x^2} y' + \frac{p(p + 1)}{1 - x^2} y = 0,$$

it is clear that $x = 1$ and $x = -1$ are singular points. The first is regular because

$$(x - 1)P(x) = \frac{2x}{x + 1} \qquad \text{and} \qquad (x - 1)^2 Q(x) = -\frac{(x - 1)p(p + 1)}{x + 1}$$

are analytic at $x = 1$, and the second is also regular for similar reasons. As another example, we mention *Bessel's equation of order p*, where p is a nonnegative constant:

$$x^2 y'' + xy' + (x^2 - p^2)y = 0. \tag{2}$$

If this is written in the form

$$y'' + \frac{1}{x} y' + \frac{x^2 - p^2}{x^2} y = 0,$$

it is apparent that the origin is a regular singular point because

$$xP(x) = 1 \qquad \text{and} \qquad x^2 Q(x) = x^2 - p^2$$

are analytic at $x = 0$. In the remainder of this chapter we will often use

---

[5] This terminology follows a time-honored tradition in mathematics, according to which situations that elude simple analysis are dismissed by such pejorative terms as "improper," "inadmissible," "degenerate," "irregular," and so on.

Bessel's equation as an illustrative example, and in Chapter 8 its solutions and their applications will be examined in considerable detail.

Now let us try to understand the reasons behind the definition of a regular singular point. To simplify matters, we may assume that the singular point x_0 is located at the origin; for if it is not, then we can always move it to the origin by changing the independent variable from x to $x - x_0$. Our starting point is the fact that the general form of a function analytic at $x = 0$ is $a_0 + a_1x + a_2x^2 + \cdots$. As a consequence, the origin will certainly be a singular point of (1) if

$$P(x) = \cdots + \frac{b_{-2}}{x^2} + \frac{b_{-1}}{x} + b_0 + b_1x + b_2x^2 + \cdots$$

and

$$Q(x) = \cdots + \frac{c_{-2}}{x^2} + \frac{c_{-1}}{x} + c_0 + c_1x + c_2x^2 + \cdots,$$

and at least one of the coefficients with negative subscripts is nonzero. The type of solution we are aiming at for (1), for reasons that will appear below, is a "quasi power series" of the form

$$\begin{aligned} y &= x^m(a_0 + a_1x + a_2x^2 + \cdots) \\ &= a_0x^m + a_1x^{m+1} + a_2x^{m+2} + \cdots, \end{aligned} \tag{3}$$

where the exponent m may be a negative integer, a fraction, or even an irrational real number. We will see in Problems 6 and 7 that two independent solutions of this kind are possible only if the above expressions for $P(x)$ and $Q(x)$ do not contain, respectively, more than the first term or more than the first two terms to the left of the constant terms b_0 and c_0. An equivalent statement is that $xP(x)$ and $x^2Q(x)$ must be analytic at the origin; and according to the definition, this is precisely what is meant by saying that the singular point $x = 0$ is regular.

The next question we attempt to answer is: where do we get the idea that series of the form (3) might be suitable solutions for equation (1) near the regular singular point $x = 0$? At this stage, the only second order linear equation we can solve completely near a singular point is the Euler equation discussed in Problem 17-5:

$$x^2y'' + pxy' + qy = 0. \tag{4}$$

If this is written in the form

$$y'' + \frac{p}{x}y' + \frac{q}{x^2}y = 0, \tag{5}$$

so that $P(x) = p/x$ and $Q(x) = q/x^2$, then it is clear that the origin is a regular singular point whenever the constants p and q are not both zero. The solutions of this equation provide a very suggestive bridge to the general case, so we briefly recall the details. The key to finding these solutions is the fact that changing the independent variable from x to $z = \log x$ transforms (4) into an equation with constant coefficients. To carry out this process, we assume that $x > 0$ (so that z is a real variable) and write

$$y' = \frac{dy}{dx} = \frac{dy}{dz}\frac{dz}{dx} = \frac{dy}{dz}\frac{1}{x}$$

and

$$y'' = \frac{d^2y}{dx^2} = \frac{d}{dx}\left(\frac{dy}{dx}\right) = \frac{dy}{dz}\left(-\frac{1}{x^2}\right) + \frac{1}{x}\frac{d}{dx}\left(\frac{dy}{dz}\right)$$

$$= -\frac{1}{x^2}\frac{dy}{dz} + \frac{1}{x}\frac{d}{dz}\left(\frac{dy}{dz}\right)\frac{dz}{dx} = \frac{1}{x^2}\frac{d^2y}{dz^2} - \frac{1}{x^2}\frac{dy}{dz}.$$

When these expressions are inserted in (4), the transformed equation is clearly

$$\frac{d^2y}{dz^2} + (p-1)\frac{dy}{dz} + qy = 0, \tag{6}$$

whose auxiliary equation is

$$m^2 + (p-1)m + q = 0. \tag{7}$$

If the roots of (7) are m_1 and m_2, then we know that (6) has the following independent solutions:

$$e^{m_1 z} \quad \text{and} \quad e^{m_2 z} \quad \text{if } m_2 \neq m_1;$$

$$e^{m_1 z} \quad \text{and} \quad ze^{m_1 z} \quad \text{if } m_2 = m_1.$$

Since $e^z = x$, the corresponding pairs of solutions for (4) are

$$\begin{aligned} x^{m_1} \quad &\text{and} \quad x^{m_2} \quad &&\text{if } m_2 \neq m_1; \\ x^{m_1} \quad &\text{and} \quad x^{m_1}\log x \quad &&\text{if } m_2 = m_1. \end{aligned} \tag{8}$$

If we seek solutions valid on the interval $x < 0$, we have only to change the variable to $t = -x$ and solve the resulting equation for $t > 0$.

We have presented this discussion of Euler's equation and its solutions for two reasons. First, we point out that the most general differential equation with a regular singular point at the origin is simply equation (5) with the constant numerators p and q replaced by power series:

$$y'' + \left(\frac{p_0 + p_1 x + p_2 x^2 + \cdots}{x}\right)y' + \left(\frac{q_0 + q_1 x + q_2 x^2 + \cdots}{x^2}\right)y = 0. \tag{9}$$

Second, if the transition from (5) to (9) is accomplished by replacing constants by power series, then it is natural to guess that the corresponding transition from (8) to the solutions of (9) might be accomplished by replacing power functions x^m by series of the form (3). We therefore expect that (9) will have two independent solutions of the form (3), or perhaps one of this form and one of the form

$$y = x^m \log x(a_0 + a_1x + a_2x^2 + \cdots), \tag{10}$$

where we assume that $x > 0$. The next section will show that these are very good guesses.

One final remark is necessary before we leave these generalities. Notice that if $a_0 = 0$ in expressions like (3) and (10), then some positive integral power of x can be factored out of the power series part and combined with x^m. We therefore always assume that $a_0 \neq 0$ in such expressions; and this assumption means only that the highest possible power of x is understood to be factored out before any calculations are performed. Series of the form (3) are called *Frobenius series,* and the procedure described below for finding solutions of this type is known as the *method of Frobenius.*[6] Frobenius series evidently include power series as special cases, whenever m is zero or a positive integer.

To illustrate the above ideas, we consider the equation

$$2x^2y'' + x(2x + 1)y' - y = 0. \tag{11}$$

If this is written in the more revealing form

$$y'' + \frac{1/2 + x}{x}y' + \frac{-1/2}{x^2}y = 0, \tag{12}$$

then we see at once that $xP(x) = \frac{1}{2} + x$ and $x^2Q(x) = -\frac{1}{2}$, so $x = 0$ is a regular singular point. We now introduce our assumed Frobenius series solution

$$
\begin{aligned}
y &= x^m(a_0 + a_1x + a_2x^2 + \cdots) \\
&= a_0x^m + a_1x^{m+1} + a_2x^{m+2} + \cdots,
\end{aligned}
\tag{13}
$$

and its derivatives

$$y' = a_0mx^{m-1} + a_1(m + 1)x^m + a_2(m + 2)x^{m+1} + \cdots$$

---

[6] Ferdinand Georg Frobenius (1849–1917) taught in Berlin and Zurich. He made several valuable contributions to the theory of elliptic functions and differential equations. However, his most influential work was in the field of algebra, where he invented and applied the important concept of group characters and proved a famous theorem about possible extensions of the complex number system.

and

$$y'' = a_0 m(m - 1)x^{m-2} + a_1(m + 1)mx^{m-1}$$
$$+ a_2(m + 2)(m + 1)x^m + \cdots.$$

To find the coefficients in (13), we proceed in essentially the same way as in the case of an ordinary point, with the significant difference that now we must also find the appropriate value (or values) of the exponent m. When the three series above are inserted in (12), and the common factor x^{m-2} is canceled, the result is

$$a_0 m(m - 1) + a_1(m + 1)mx + a_2(m + 2)(m + 1)x^2 + \cdots$$

$$+ \left(\frac{1}{2} + x\right)[a_0 m + a_1(m + 1)x + a_2(m + 2)x^2 + \cdots]$$

$$- \frac{1}{2}(a_0 + a_1 x + a_2 x^2 + \cdots) = 0.$$

By inspection, we combine corresponding powers of x and equate the coefficient of each power of x to zero. This yields the following system of equations:

$$a_0\left[m(m - 1) + \frac{1}{2}m - \frac{1}{2}\right] = 0,$$

$$a_1\left[(m + 1)m + \frac{1}{2}(m + 1) - \frac{1}{2}\right] + a_0 m = 0, \quad (14)$$

$$a_2\left[(m + 2)(m + 1) + \frac{1}{2}(m + 2) - \frac{1}{2}\right] + a_1(m + 1) = 0,$$

$$\cdots.$$

As we explained above, it is understood that $a_0 \neq 0$. It therefore follows from the first of these equations that

$$m(m - 1) + \frac{1}{2}m - \frac{1}{2} = 0. \quad (15)$$

This is called the *indicial equation* of the differential equation (11). Its roots are

$$m_1 = 1 \quad \text{and} \quad m_2 = -\frac{1}{2},$$

and these are the only possible values for the exponent m in (13). For each of these values of m, we now use the remaining equations of (14) to calculate $a_1, a_2, \ldots$ in terms of a_0. For $m_1 = 1$, we obtain

$$a_1 = -\frac{a_0}{2 \cdot 1 + \frac{1}{2} \cdot 2 - \frac{1}{2}} = -\frac{2}{5}a_0,$$

$$a_2 = -\frac{2a_1}{3 \cdot 2 + \frac{1}{2} \cdot 3 - \frac{1}{2}} = -\frac{2}{7}a_1 = \frac{4}{35}a_0,$$

$$\cdots.$$

And for $m_2 = -\frac{1}{2}$, we obtain

$$a_1 = \frac{\frac{1}{2}a_0}{\frac{1}{2}\left(-\frac{1}{2}\right) + \frac{1}{2} \cdot \frac{1}{2} - \frac{1}{2}} = -a_0,$$

$$a_2 = -\frac{\frac{1}{2}a_1}{\frac{3}{2} \cdot \frac{1}{2} + \frac{1}{2} \cdot \frac{3}{2} - \frac{1}{2}} = -\frac{1}{2}a_1 = \frac{1}{2}a_0,$$

$$\cdots.$$

We therefore have the following two Frobenius series solutions, in each of which we have put $a_0 = 1$:

$$y_1 = x\left(1 - \frac{2}{5}x + \frac{4}{35}x^2 + \cdots\right), \tag{16}$$

$$y_2 = x^{-1/2}\left(1 - x + \frac{1}{2}x^2 + \cdots\right). \tag{17}$$

These solutions are clearly independent for $x > 0$, so the general solution of (11) on this interval is

$$y = c_1 x\left(1 - \frac{2}{5}x + \frac{4}{35}x^2 + \cdots\right) + c_2 x^{-1/2}\left(1 - x + \frac{1}{2}x^2 + \cdots\right).$$

The problem of determining the interval of convergence for the two power series in parentheses will be discussed in the next section.

If we look closely at the way in which (15) arises from (12), it is easy to see that the indicial equation of the more general differential equation (9) is

$$m(m - 1) + mp_0 + q_0 = 0. \tag{18}$$

In our example, the indicial equation had two distinct real roots leading to the two independent series solutions (16) and (17). It is natural to expect such a result whenever the indicial equation (18) has distinct real roots m_1 and m_2. This turns out to be true if the difference between m_1 and m_2 is not an integer. If, however, this difference is an integer, then it often (but not always) happens that one of the two expected series solutions does not exist. In this case it is necessary—just as in the case $m_1 = m_2$—to find a second independent solution by other methods. In the next section we investigate these difficulties in greater detail.

PROBLEMS

1. For each of the following differential equations, locate and classify its singular points on the x-axis:
   (a) $x^3(x - 1)y'' - 2(x - 1)y' + 3xy = 0$;
   (b) $x^2(x^2 - 1)^2y'' - x(1 - x)y' + 2y = 0$;
   (c) $x^2y'' + (2 - x)y' = 0$;
   (d) $(3x + 1)xy'' - (x + 1)y' + 2y = 0$.

2. Determine the nature of the point $x = 0$ for each of the following equations:
   (a) $y'' + (\sin x)y = 0$;   (d) $x^3y'' + (\sin x)y = 0$;
   (b) $xy'' + (\sin x)y = 0$;   (e) $x^4y'' + (\sin x)y = 0$.
   (c) $x^2y'' + (\sin x)y = 0$;

3. Find the indicial equation and its roots for each of the following differential equations:
   (a) $x^3y'' + (\cos 2x - 1)y' + 2xy = 0$;
   (b) $4x^2y'' + (2x^4 - 5x)y' + (3x^2 + 2)y = 0$.

4. For each of the following equations, verify that the origin is a regular singular point and calculate two independent Frobenius series solutions:
   (a) $4xy'' + 2y' + y = 0$;   (c) $2xy'' + (x + 1)y' + 3y = 0$;
   (b) $2xy'' + (3 - x)y' - y = 0$;   (d) $2x^2y'' + xy' - (x + 1)y = 0$.

5. When $p = 0$, Bessel's equation (2) becomes

$$x^2y'' + xy' + x^2y = 0.$$

Show that its indicial equation has only one root, and use the method of this section to deduce that

$$y = \sum_{n=0}^{\infty} \frac{(-1)^n}{2^{2n}(n!)^2}x^{2n}$$

is the corresponding Frobenius series solution [see Problem 26-7(b)].

6. Consider the differential equation

$$y'' + \frac{1}{x^2}y' - \frac{1}{x^3}y = 0.$$

   (a) Show that $x = 0$ is an irregular singular point.
   (b) Use the fact that $y_1 = x$ is a solution to find a second independent solution y_2 by the method of Section 16.

(c) Show that the second solution y_2 found in (b) cannot be expressed as a Frobenius series.

7. Consider the differential equation

$$y'' + \frac{p}{x^b} y' + \frac{q}{x^c} y = 0,$$

where p and q are nonzero real numbers and b and c are positive integers. It is clear that $x = 0$ is an irregular singular point if $b > 1$ or $c > 2$.

(a) If $b = 2$ and $c = 3$, show that there is only one possible value of m for which there might exist a Frobenius series solution.

(b) Show similarly that m satisfies a quadratic equation—and hence we can hope for two Frobenius series solutions, corresponding to the roots of this equation—if and only if $b = 1$ and $c \le 2$. Observe that these are exactly the conditions that characterize $x = 0$ as a "weak" or regular singular point as opposed to a "strong" or irregular singular point.

8. The differential equation

$$x^2 y'' + (3x - 1)y' + y = 0$$

has $x = 0$ as an irregular singular point. If (3) is inserted into this equation, show that $m = 0$ and the corresponding Frobenius series "solution" is the power series

$$y = \sum_{n=0}^{\infty} n! x^n,$$

which converges only at $x = 0$. This demonstrates that even when a Frobenius series formally satisfies such an equation, it is not necessarily a valid solution.

30 REGULAR SINGULAR POINTS (CONTINUED)

Our work in the previous section was mainly directed at motivation and technique. We now confront the theoretical side of the problem of solving the general second order linear equation

$$y'' + P(x)y' + Q(x)y = 0 \tag{1}$$

near the regular singular point $x = 0$. The ideas developed above suggest that we attempt a formal calculation of any solutions of (1) that have the Frobenius form

$$y = x^m(a_0 + a_1 x + a_2 x^2 + \cdots), \tag{2}$$

where $a_0 \ne 0$ and m is a number to be determined. Our hope is that any formal solution that arises in this way can be legitimized by a proof and established as a valid solution. The generality of this approach will also serve to illuminate the circumstances under which equation (1) has only one solution of the form (2).[7] For reasons already explained, we confine our attention to the interval $x > 0$. The behavior of solutions on the interval $x < 0$ can be studied by changing the variable to $t = -x$ and solving the resulting equation for $t > 0$.

Our hypothesis is that $xP(x)$ and $x^2Q(x)$ are analytic at $x = 0$, and therefore have power series expansions

$$xP(x) = \sum_{n=0}^{\infty} p_n x^n \quad \text{and} \quad x^2Q(x) = \sum_{n=0}^{\infty} q_n x^n \tag{3}$$

which are valid on an interval $|x| < R$ for some $R > 0$. Just as in the example of the previous section, we must find the possible values of m in (2); and then, for each acceptable m, we must calculate the corresponding coefficients $a_0, a_1, a_2, \ldots$. If we write (2) in the form

$$y = x^m \sum_{n=0}^{\infty} a_n x^n = \sum_{n=0}^{\infty} a_n x^{m+n},$$

then differentiation yields

$$y' = \sum_{n=0}^{\infty} a_n(m + n)x^{m+n-1}$$

and

$$y'' = \sum_{n=0}^{\infty} a_n(m + n)(m - n - 1)x^{m+n-2}$$

$$= x^{m-2} \sum_{n=0}^{\infty} a_n(m + n)(m + n - 1)x^n.$$

The terms $P(x)y'$ and $Q(x)y$ in (1) can now be written as

$$P(x)y' = \frac{1}{x}\left(\sum_{n=0}^{\infty} p_n x^n\right)\left[\sum_{n=0}^{\infty} a_n(m + n)x^{m+n-1}\right]$$

$$= x^{m-2}\left(\sum_{n=0}^{\infty} p_n x^n\right)\left[\sum_{n=0}^{\infty} a_n(m + n)x^n\right]$$

$$= x^{m-2} \sum_{n=0}^{\infty}\left[\sum_{k=0}^{n} p_{n-k}a_k(m + k)\right]x^n$$

$$= x^{m-2} \sum_{n=0}^{\infty}\left[\sum_{k=0}^{n} p_{n-k}a_k(m + k) + p_0 a_n(m + n)\right]x^n$$

---

[7] When we say that (1) has "only one" solution of the form (2), we mean that a second independent solution of this form does not exist.

and

$$Q(x)y = \frac{1}{x^2}\left(\sum_{n=0}^{\infty} q_n x^n\right)\left(\sum_{n=0}^{\infty} a_n x^{m+n}\right)$$

$$= x^{m-2}\left(\sum_{n=0}^{\infty} q_n x^n\right)\left(\sum_{n=0}^{\infty} a_n x^n\right)$$

$$= x^{m-2}\sum_{n=0}^{\infty}\left(\sum_{k=0}^{n} q_{n-k}a_k\right)x^n$$

$$= x^{m-2}\sum_{n=0}^{\infty}\left(\sum_{k=0}^{n-1} q_{n-k}a_k + q_0 a_n\right)x^n.$$

When these expressions for y'', $P(x)y'$, and $Q(x)y$ are inserted in (1) and the common factor x^{m-2} is canceled, then the differential equation becomes

$$\sum_{n=0}^{\infty}\left\{a_n[(m+n)(m+n-1) + (m+n)p_0 + q_0]\right.$$

$$\left. + \sum_{k=0}^{n-1} a_k[(m+k)p_{n-k} + q_{n-k}]\right\}x^n = 0;$$

and equating to zero the coefficient of x^n yields the following recursion formula for the a_n:

$$a_n[(m+n)(m+n-1) + (m+n)p_0 + q_0]$$

$$+ \sum_{k=0}^{n-1} a_k[(m+k)p_{n-k} + q_{n-k}] = 0. \quad (4)$$

On writing this out for the successive values of n, we get

$$a_0[m(m-1) + mp_0 + q_0] = 0,$$

$$a_1[(m+1)m + (m+1)p_0 + q_0] + a_0(mp_1 + q_1) = 0,$$

$$a_2[(m+2)(m+1) + (m+2)p_0 + q_0]$$

$$+ a_0(mp_2 + q_2) + a_1[(m+1)p_1 + q_1] = 0,$$

$$\cdots$$

$$a_n[(m+n)(m+n-1) + (m+n)p_0 + q_0]$$

$$+ a_0(mp_n + q_n) + \cdots + a_{n-1}[(m+n-1)p_1 + q_1] = 0,$$

$$\cdots.$$

If we put $f(m) = m(m-1) + mp_0 + q_0$, then these equations become

$$a_0 f(m) = 0,$$

$$a_1 f(m+1) + a_0(mp_1 + q_1) = 0,$$

$$a_2 f(m + 2) + a_0(mp_2 + q_2) + a_1[(m + 1)p_1 + q_1] = 0,$$

$$\cdots$$

$$a_n f(m + n) + a_0(mp_n + q_n) + \cdots + a_{n-1}[(m + n - 1)p_1 + q_1] = 0,$$

$$\cdots.$$

Since $a_0 \neq 0$, we conclude from the first of these equations that $f(m) = 0$ or, equivalently, that

$$m(m - 1) + mp_0 + q_0 = 0. \tag{5}$$

This is the *indicial equation*, and its roots m_1 and m_2—which are possible values for m in our assumed solution (2)—are called the *exponents* of the differential equation (1) at the regular singular point $x = 0$. The following equations give a_1 in terms of a_0, a_2 in terms of a_0 and a_1, and so on. The a_n are therefore determined in terms of a_0 for each choice of m *unless* $f(m + n) = 0$ for some positive integer n, in which case the process breaks off. Thus, if $m_1 = m_2 + n$ for some integer $n \geq 1$, the choice $m = m_1$ gives a formal solution but in general $m = m_2$ does not—since $f(m_2 + n) = f(m_1) = 0$. If $m_1 = m_2$ we also obtain only one formal solution. In all other cases where m_1 and m_2 are real numbers, this procedure yields two independent formal solutions. It is possible, of course, for m_1 and m_2 to be conjugate complex numbers, but we do not discuss this case because an adequate treatment would lead us too far into complex analysis. The specific difficulty here is that if the m's are allowed to be complex, then the a_n will also be complex, and we do not assume that the reader is familiar with power series having complex coefficients.

These ideas are formulated more precisely in the following theorem.

Theorem A. *Assume that $x = 0$ is a regular singular point of the differential equation* (1) *and that the power series expansions* (3) *of $xP(x)$ and $x^2 Q(x)$ are valid on an interval $|x| < R$ with $R > 0$. Let the indicial equation* (5) *have real roots m_1 and m_2 with $m_2 \leq m_1$. Then equation* (1) *has at least one solution*

$$y_1 = x^{m_1} \sum_{n=0}^{\infty} a_n x^n \quad (a_0 \neq 0) \tag{6}$$

on the interval $0 < x < R$, where the a_n are determined in terms of a_0 by the recursion formula (4) *with m replaced by m_1, and the series $\sum a_n x^n$ converges for $|x| < R$. Furthermore, if $m_1 - m_2$ is not zero or a positive integer, then equation* (1) *has a second independent solution*

$$y_2 = x^{m_2} \sum_{n=0}^{\infty} a_n x^n \quad (a_0 \neq 0) \tag{7}$$

on the same interval, where in this case the a_n are determined in terms of a_0 by formula (4) with m replaced by m_2, and again the series $\sum a_n x^n$ converges for $|x| < R$.

In view of what we have already done, the proof of this theorem can be completed by showing that in each case the series $\sum a_n x^n$ converges on the interval $|x| < R$. Readers who are interested in the details of this argument will find them in Appendix A. We emphasize that in a specific problem it is much simpler to substitute the general Frobenius series (2) directly into the differential equation than to use the recursion formula (4) to calculate the coefficients. This recursion formula finds its main application in the delicate convergence proof given in Appendix A.

Theorem A unfortunately fails to answer the question of how to find a second solution when the difference $m_1 - m_2$ is zero or a positive integer. In order to convey an idea of the possibilities here, we distinguish three cases.

CASE A. If $m_1 = m_2$, there cannot exist a second Frobenius series solution.

The other two cases, in both of which $m_1 - m_2$ is a positive integer, will be easier to grasp if we insert $m = m_2$ in the recursion formula (4) and write it as

$$a_n f(m_2 + n) = -a_0(m_2 p_n + q_n) - \cdots - a_{n-1}[(m_2 + n - 1)p_1 + q_1].$$
$$(8)$$

As we know, the difficulty in calculating the a_n arises because $f(m_2 + n) = 0$ for a certain positive integer n. The next two cases deal with this problem.

CASE B. If the right side of (8) is not zero when $f(m_2 + n) = 0$, then there is no possible way of continuing the calculation of the coefficients and there cannot exist a second Frobenius series solution.

CASE C. If the right side of (8) happens to be zero when $f(m_2 + n) = 0$, then a_n is unrestricted and can be assigned any value whatever. In particular, we can put $a_n = 0$ and continue to compute the coefficients without any further difficulties. Hence in this case there does exist a second Frobenius series solution.

The problems below will demonstrate that each of these three possibilities actually occurs.

The following calculations enable us to discover what form the second solution takes when $m_1 - m_2$ is zero or a positive integer. We begin by defining a positive integer k by $k = m_1 - m_2 + 1$. The indicial equation (5) can be written as

$$(m - m_1)(m - m_2) = m^2 - (m_1 + m_2)m + m_1 m_2 = 0,$$

so equating the coefficients of m yields $p_0 - 1 = -(m_1 + m_2)$ or $m_2 = 1 - p_0 - m_1$, and we have $k = 2m_1 + p_0$. By using the method of Section 16, we can find a second solution y_2 from the known solution $y_1 = x^{m_1}(a_0 + a_1 x + \cdots)$ by writing $y_2 = v y_1$, where

$$v' = \frac{1}{y_1^2} e^{-\int P(x)\,dx}$$

$$= \frac{1}{x^{2m_1}(a_0 + a_1 x + \cdots)^2} e^{-\int ((p_0/x) + p_1 + \cdots)\,dx}$$

$$= \frac{1}{x^{2m_1}(a_0 + a_1 x + \cdots)^2} e^{(-p_0 \log x - p_1 x - \cdots)}$$

$$= \frac{1}{x^k (a_0 + a_1 x + \cdots)^2} e^{(-p_1 x - \cdots)} = \frac{1}{x^k} g(x).$$

The function $g(x)$ defined by the last equality is clearly analytic at $x = 0$, with $g(0) = 1/a_0^2$, so in some interval about the origin we have

$$g(x) = b_0 + b_1 x + b_2 x^2 + \cdots, \qquad b_0 \neq 0. \tag{9}$$

It follows that

$$v' = b_0 x^{-k} + b_1 x^{-k+1} + \cdots + b_{k-1} x^{-1} + b_k + \cdots,$$

so

$$v = \frac{b_0 x^{-k+1}}{-k + 1} + \frac{b_1 x^{-k+2}}{-k + 2} + \cdots + b_{k-1} \log x + b_k x + \cdots$$

and

$$y_2 = y_1 v = y_1 \left(\frac{b_0 x^{-k+1}}{-k + 1} + \cdots + b_{k-1} \log x + b_k x + \cdots \right)$$

$$= b_{k-1} y_1 \log x + x^{m_1}(a_0 + a_1 x + \cdots) \left(\frac{b_0 x^{-k+1}}{-k + 1} + \cdots \right).$$

If we factor x^{-k+1} out of the series last written, use $m_1 - k + 1 = m_2$, and multiply the two remaining power series, then we obtain

$$y_2 = b_{k-1} y_1 \log x + x^{m_2} \sum_{n=0}^{\infty} c_n x^n \tag{10}$$

as our second solution.

Formula (10) has only limited value as a practical tool; but it does yield several grains of information. First, if the exponents m_1 and m_2 are equal, then $k = 1$ and $b_{k-1} = b_0 \neq 0$; so in this case—which is Case A above—the term containing $\log x$ is definitely present in the second solution (10). However, if $m_1 - m_2 = k - 1$ is a positive integer, then sometimes $b_{k-1} \neq 0$ and the logarithmic term is present (Case B), and sometimes $b_{k-1} = 0$ and there is no logarithmic term (Case C). The practical difficulty here is that we cannot readily find b_{k-1} because we have no direct means of calculating the coefficients in (9). In any event, we at least know that in Cases A and B, when $b_{k-1} \neq 0$ and the method of Frobenius is only partly successful, the general form of a second solution is

$$y_2 = y_1 \log x + x^{m_2} \sum_{n=0}^{\infty} c_n x^n, \tag{11}$$

where the c_n are certain unknown constants that can be determined by substituting (11) directly into the differential equation. Notice that this expression is similar to formula 29-(10) but somewhat more complicated.

PROBLEMS

1. The equation

$$x^2 y'' - 3xy' + (4x + 4)y = 0$$

has only one Frobenius series solution. Find it.

2. The equation

$$4x^2 y'' - 8x^2 y' + (4x^2 + 1)y = 0$$

has only one Frobenius series solution. Find the general solution.

3. Find two independent Frobenius series solutions of each of the following equations:
(a) $xy'' + 2y' + xy = 0$; (c) $xy'' - y' + 4x^3 y = 0$.
(b) $x^2 y'' - x^2 y' + (x^2 - 2)y = 0$;

4. Bessel's equation of order $p = 1$ is

$$x^2 y'' + xy' + (x^2 - 1)y = 0.$$

Show that $m_1 - m_2 = 2$ and that the equation has only one Frobenius series solution. Then find it.

5. Bessel's equation of order $p = \dfrac{1}{2}$ is

$$x^2 y'' + xy' + \left(x^2 - \frac{1}{4}\right)y = 0.$$

Show that $m_1 - m_2 = 1$, but that nevertheless the equation has two independent Frobenius series solutions. Then find them.

6. The only Frobenius series solution of Bessel's equation of order $p = 0$ is given in Problem 29-5. By taking this as y_1, and substituting formula (11) into the differential equation, obtain the second independent solution

$$y_2 = y_1 \log x + \sum_{n=1}^{\infty} \frac{(-1)^{n+1}}{2^{2n}(n!)^2} \left(1 + \frac{1}{2} + \cdots + \frac{1}{n}\right) x^{2n}.$$

31 GAUSS'S HYPERGEOMETRIC EQUATION

This famous differential equation is

$$x(1 - x)y'' + [c - (a + b + 1)x]y' - aby = 0, \tag{1}$$

where a, b, and c are constants. The coefficients of (1) may look rather strange, but we shall find that they are perfectly adapted to the use of its solutions in a wide variety of situations. The best way to understand this is to solve the equation for ourselves and see what happens.

We have

$$P(x) = \frac{c - (a + b + 1)x}{x(1 - x)} \quad \text{and} \quad Q(x) = \frac{-ab}{x(1 - x)},$$

so $x = 0$ and $x = 1$ are the only singular points on the x-axis. Also,

$$xP(x) = \frac{c - (a + b + 1)x}{1 - x} = [c - (a + b + 1)x](1 + x + x^2 + \cdots)$$
$$= c + [c - (a + b + 1)]x + \cdots$$

and

$$x^2 Q(x) = \frac{-abx}{1 - x} = -abx(1 + x + x^2 + \cdots)$$
$$= -abx - abx^2 - \cdots,$$

so $x = 0$ (and similarly $x = 1$) is a regular singular point. These expansions show that $p_0 = c$ and $q_0 = 0$, so the indicial equation is

$$m(m - 1) + mc = 0 \quad \text{or} \quad m[m - (1 - c)] = 0$$

and the exponents are $m_1 = 0$ and $m_2 = 1 - c$. If $1 - c$ is not a positive integer, that is, if c is not zero or a negative integer, then Theorem 30-A guarantees that (1) has a solution of the form

$$y = x^0 \sum_{n=0}^{\infty} a_n x^n = a_0 + a_1 x + a_2 x^2 + \cdots, \tag{2}$$

where a_0 is a nonzero constant. On the substituting this into (1) and equating to zero the coefficient of x^n, we obtain the following recursion formula for the a_n:

$$a_{n+1} = \frac{(a + n)(b + n)}{(n + 1)(c + n)} a_n. \tag{3}$$

We now set $a_0 = 1$ and calculate the other a_n in succession:

$$a_1 = \frac{ab}{1 \cdot c}, \qquad a_2 = \frac{a(a + 1)b(b + 1)}{1 \cdot 2c(c + 1)},$$

$$a_3 = \frac{a(a + 1)(a + 2)b(b + 1)(b + 2)}{1 \cdot 2 \cdot 3c(c + 1)(c + 2)}, \ldots$$

With these coefficients, (2) becomes

$$y = 1 + \frac{ab}{1 \cdot c}x + \frac{a(a + 1)b(b + 1)}{1 \cdot 2c(c + 1)}x^2$$

$$+ \frac{a(a + 1)(a + 2)b(b + 1)(b + 2)}{1 \cdot 2 \cdot 3c(c + 1)(c + 2)}x^3 + \cdots$$

$$= 1 + \sum_{n=1}^{\infty} \frac{a(a + 1)\cdots(a + n - 1)b(b + 1)\cdots(b + n - 1)}{n!c(c + 1)\cdots(c + n - 1)}x^n. \quad (4)$$

This is known as the *hypergeometric series,* and is denoted by the symbol $F(a,b,c,x)$. It is called this because it generalizes the familiar geometric series as follows: when $a = 1$ and $c = b$, we obtain

$$F(1,b,b,x) = 1 + x + x^2 + \cdots = \frac{1}{1 - x}.$$

If a or b is zero or a negative integer, the series (4) breaks off and is a polynomial; otherwise the ratio test shows that it converges for $|x| < 1$, since (3) gives

$$\left| \frac{a_{n+1}x^{n+1}}{a_n x^n} \right| = \left| \frac{(a + n)(b + n)}{(n + 1)(c + n)} \right| |x| \to |x| \quad \text{as } n \to \infty.$$

This convergence behavior could also have been predicted from the fact that the singular point closest to the origin is $x = 1$. Accordingly, when c is not zero or a negative integer, $F(a,b,c,x)$ is an analytic function—called the *hypergeometric function*—on the interval $|x| < 1$. It is the simplest particular solution of the hypergeometric equation. The hypergeometric function has a great many properties, of which the most obvious is that it is unaltered when a and b are interchanged: $F(a,b,c,x) = F(b,a,c,x)$.[8]

　　If $1 - c$ is not zero or a negative integer—which means that c is not a positive integer—then Theorem 30-A also tells us that there is a second independent solution of (1) near $x = 0$ with exponent $m_2 = 1 - c$. This

---

[8] A summary of some of its other properties can be found in A. Erdélyi (ed.), *Higher Transcendental Functions*, Vol. I, pp. 56–119, McGraw-Hill, New York, 1953.

solution can be found directly, by substituting

$$y = x^{1-c}(a_0 + a_1 x + a_2 x^2 + \cdots)$$

into (1) and calculating the coefficients. It is more instructive, however, to change the dependent variable in (1) from y to z by writing

$$y = x^{1-c}z.$$

When the necessary computations are performed—students should do this work themselves—equation (1) becomes

$$x(1 - x)z'' + [(2 - c) - ([a - c + 1] + [b - c + 1] + 1)x]z'$$
$$- (a - c + 1)(b - c + 1)z = 0, \quad (5)$$

which is the hypergeometric equation with the constants a, b, and c replaced by $a - c + 1$, $b - c + 1$, and $2 - c$. We already know that (5) has the power series solution

$$z = F(a - c + 1, b - c + 1, 2 - c, x)$$

near the origin, so our desired second solution is

$$y = x^{1-c}F(a - c + 1, b - c + 1, 2 - c, x).$$

Accordingly, when c is not an integer, we have

$$y = c_1 F(a,b,c,x) + c_2 x^{1-c}F(a - c + 1, b - c + 1, 2 - c, x) \quad (6)$$

as the general solution of the hypergeometric equation near the singular point $x = 0$.

In general, the above solution is only valid near the origin. We now solve (1) near the singular point $x = 1$. The simplest procedure is to obtain this solution from the one already found, by introducing a new independent variable $t = 1 - x$. This makes $x = 1$ correspond to $t = 0$ and transforms (1) into

$$t(1 - t)y'' + [(a + b - c + 1) - (a + b + 1)t]y' - aby = 0,$$

where the primes signify derivatives with respect to t. Since this is a hypergeometric equation, its general solution near $t = 0$ can be written down at once from (6), by replacing x by t and c by $a + b - c + 1$; and when t is replaced by $1 - x$, we see that the general solution of (1) near $x = 1$ is

$$y = c_1 F(a,b, a + b - c + 1, 1 - x)$$
$$+ c_2(1 - x)^{c-a-b}F(c - b, c - a, c - a - b + 1, 1 - x). \quad (7)$$

In this case it is necessary to assume that $c - a - b$ is not an integer.

Formulas (6) and (7) show that the adaptability of the constants in equation (1) makes it possible to express the general solution of this equation near each of its singular points in terms of the single function F.

Much more than this is true, for these ideas are applicable to a wide class of differential equations. The key is to notice the following general features of the hypergeometric equation: that the coefficients of y'', y', and y are polynomials of degrees 2, 1, and 0, and also that the first of these polynomials has distinct real zeros. Any differential equation with these characteristics can be brought into the hypergeometric form by a linear change of the independent variable, and hence can be solved near its singular points in terms of the hypergeometric function.

To make these remarks somewhat more concrete, we briefly consider the general equation of this type,

$$(x - A)(x - B)y'' + (C + Dx)y' + Ey = 0, \tag{8}$$

where $A \neq B$. If we change the independent variable from x to t by means of

$$t = \frac{x - A}{B - A},$$

then $x = A$ corresponds to $t = 0$, and $x = B$ to $t = 1$. With a little calculation, equation (8) assumes the form

$$t(1 - t)y'' + (F + Gt)y' + Hy = 0,$$

where F, G, and H are certain combinations of the constants in (8) and the primes indicate derivatives with respect to t. This is a hypergeometric equation with a, b, and c defined by

$$F = c, \qquad G = -(a + b + 1), \qquad H = -ab,$$

and can therefore be solved near $t = 0$ and $t = 1$ in terms of the hypergeometric function. But this means that (8) can be solved in terms of the same function near $x = A$ and $x = B$.

The above ideas suggest the protean versatility of the hypergeometric function $F(a,b,c,x)$ in the field of differential equations. We will also see (in Problem 1) that the flexibility afforded by the three constants a, b, and c allows the hypergeometric function to include as special cases most of the familiar functions of elementary analysis. This function was known to Euler, who discovered a number of its properties; but it was first studied systematically in the context of the hypergeometric equation by Gauss, who in this connection gave the earliest satisfactory treatment of the convergence of an infinite series. Gauss's work was of great historical importance because it initiated far-reaching developments in many branches of analysis—not only in infinite series, but also in the general theories of linear differential equations and functions of a complex variable. The hypergeometric function has retained its significance in modern mathematics because of its powerful unifying influence, since many of the principal special functions of higher analysis are also related to it.[9]

PROBLEMS

1. Verify each of the following by examining the series expansions of the functions on the left sides:
   (a) $(1 + x)^p = F(-p,b,b,-x)$;
   (b) $\log(1 + x) = xF(1,1,2,-x)$;

   (c) $\sin^{-1} x = xF\left(\frac{1}{2}, \frac{1}{2}, \frac{3}{2}, x^2\right)$;

   (d) $\tan^{-1} x = xF\left(\frac{1}{2}, 1, \frac{3}{2}, -x^2\right)$.

   It is also true that

   (e) $e^x = \lim_{b \to \infty} F\left(a,b,a,\frac{x}{b}\right)$;

   (f) $\sin x = x\left[\lim_{a \to \infty} F\left(a,a,\frac{3}{2},\frac{-x^2}{4a^2}\right)\right]$;

   (g) $\cos x = \lim_{a \to \infty} F\left(a,a,\frac{1}{2},\frac{-x^2}{4a^2}\right)$.

   Satisfy yourself of the validity of these statements without attempting to justify the limit processes involved.

2. Find the general solution of each of the following differential equations near the indicated singular point:

   (a) $x(1 - x)y'' + \left(\frac{3}{2} - 2x\right)y' + 2y = 0$, $x = 0$;
   (b) $(2x^2 + 2x)y'' + (1 + 5x)y' + y = 0$, $x = 0$;
   (c) $(x^2 - 1)y'' + (5x + 4)y' + 4y = 0$, $x = -1$;
   (d) $(x^2 - x - 6)y'' + (5 + 3x)y' + y = 0$, $x = 3$.

3. In Problem 28-6 we discussed Chebyshev's equation

$$(1 - x^2)y'' - xy' + p^2 y = 0,$$

   where p is a nonnegative constant. Transform it into a hypergeometric equation by replacing x by $t = \frac{1}{2}(1 - x)$, and show that its general solution near $x = 1$ is

$$y = c_1 F\left(p, -p, \frac{1}{2}, \frac{1-x}{2}\right) + c_2\left(\frac{1-x}{2}\right)^{1/2} F\left(p + \frac{1}{2}, -p + \frac{1}{2}, \frac{3}{2}, \frac{1-x}{2}\right).$$

---

[9] A brief account of Gauss and his scientific work is given in Appendix C.

4. Consider the differential equation

$$x(1 - x)y'' + [p - (p + 2)x]y' - py = 0,$$

where p is a constant.
   (a) If p is not an integer, find the general solution near $x = 0$ in terms of hypergeometric functions.
   (b) Write the general solution found in (a) in terms of elementary functions.
   (c) When $p = 1$, the differential equation becomes

$$x(1 - x)y'' + (1 - 3x)y' - y = 0,$$

   and the solution in (b) is no longer the general solution. Find the general solution in this case by the method of Section 16.

5. Some differential equations are of the hypergeometric type even though they may not appear to be so. Find the general solution of

$$(1 - e^x)y'' + \frac{1}{2}y' + e^x y = 0$$

near the singular point $x = 0$ by changing the independent variable to $t = e^x$.

6. (a) Show that $F'(a,b,c,x) = \dfrac{ab}{c} F(a + 1, b + 1, c + 1, x)$.

   (b) By applying the differentiation formula in (a) to the result of Problem 3, show that the only solutions of Chebyshev's equation whose derivatives are bounded near $x = 1$ are $y = c_1 F\left(p, -p, \dfrac{1}{2}, \dfrac{1 - x}{2}\right)$. Conclude that the only polynomial solutions of Chebyshev's equation are constant multiples of $F\left(n, -n, \dfrac{1}{2}, \dfrac{1 - x}{2}\right)$ where n is a non-negative integer.

The *Chebyshev polynomial* of degree n is denoted by $T_n(x)$ and defined by $T_n(x) = F\left(n, -n, \dfrac{1}{2}, \dfrac{1 - x}{2}\right)$.[10] An interesting application of these polynomials to the theory of approximation is discussed in Appendix D,

32   THE POINT AT INFINITY

It is often desirable, in both physics and pure mathematics, to study the solutions of

$$y'' + P(x)y' + Q(x)y = 0 \tag{1}$$

for large values of the independent variable. For instance, if the variable is time, we may want to know how the physical system described by (1) behaves in the distant future, when transient disturbances have faded away.

---

[10] The notation $T_n(x)$ is used because Chebyshev's name was formerly transliterated as Tchebychev, Tchebycheff, or Tschebycheff.

We can adapt our previous ideas to this broader purpose by studying solutions near the *point at infinity*. The procedure is quite simple, for if we change the independent variable from x to

$$t = \frac{1}{x}, \tag{2}$$

then large x's correspond to small t's. Consequently, if we apply (2) to (1), solve the transformed equation near $t = 0$, and then replace t by $1/x$ in these solutions, we have solutions of (1) that are valid for large values of x. To carry out this program, we need the formulas

$$y' = \frac{dy}{dx} = \frac{dy}{dt}\frac{dt}{dx} = \frac{dy}{dt}\left(-\frac{1}{x^2}\right) = -t^2\frac{dy}{dt} \tag{3}$$

and

$$y'' = \frac{d}{dx}\left(\frac{dy}{dx}\right) = \frac{d}{dt}\left(\frac{dy}{dx}\right)\frac{dt}{dx} = \left(-t^2\frac{d^2y}{dt^2} - 2t\frac{dy}{dt}\right)(-t^2). \tag{4}$$

When these expressions are inserted in (1), and primes are used to denote derivatives with respect to t, then (1) becomes

$$y'' + \left[\frac{2}{t} - \frac{P(1/t)}{t^2}\right]y' + \frac{Q(1/t)}{t^4}y = 0. \tag{5}$$

We say that equation (1) has $x = \infty$ as an ordinary point, a regular singular point with exponents m_1 and m_2, or an irregular singular point, if the point $t = 0$ has the corresponding character for the transformed equation (5).

As a simple illustration, consider the Euler equation

$$y'' + \frac{4}{x}y' + \frac{2}{x^2}y = 0. \tag{6}$$

A comparison of (6) with (5) shows that the transformed equation is

$$y'' - \frac{2}{t}y' + \frac{2}{t^2}y = 0. \tag{7}$$

It is clear that $t = 0$ is a regular singular point for (7), with indicial equation

$$m(m - 1) - 2m + 2 = 0$$

and exponents $m_1 = 2$ and $m_2 = 1$. This means that (6) has $x = \infty$ as a regular singular point with exponents 2 and 1.

Our main example is the hypergeometric equation

$$x(1 - x)y'' + [c - (a + b + 1)x]y' - aby = 0. \tag{8}$$

We already know that (8) has two finite regular singular points: $x = 0$ with exponents 0 and $1 - c$; and $x = 1$ with exponents 0 and $c - a - b$. To determine the nature of the point $x = \infty$, we substitute (3) and (4) directly into (8). After a little rearrangement, we find that the transformed equation is

$$y'' + \left[\frac{(1 - a - b) - (2 - c)t}{t(1 - t)} \right] y' + \frac{ab}{t^2(1 - t)} y = 0. \tag{9}$$

This equation has $t = 0$ as a regular singular point with indicial equation

$$m(m - 1) + (1 - a - b)m + ab = 0$$

or

$$(m - a)(m - b) = 0.$$

This shows that the exponents of equation (9) at $t = 0$ are a and b, so equation (8) has $x = \infty$ as a regular singular point with exponents a and b. We conclude that the hypergeometric equation (8) has precisely three regular singular points: 0, 1, and ∞ with corresponding exponents 0 and $1 - c$, 0 and $c - a - b$, and a and b. In Appendix E we demonstrate that the form of the hypergeometric equation is completely determined by the specification of these three regular singular points together with the added requirement that at least one exponent must be zero at each of the points $x = 0$ and $x = 1$.

Another classical differential equation of considerable importance is the *confluent hypergeometric equation*

$$xy'' + (c - x)y' - ay = 0. \tag{10}$$

To understand where this equation comes from and why it bears this name, we consider the ordinary hypergeometric equation (8) in the form

$$s(1 - s) \frac{d^2y}{ds^2} + [c - (a + b + 1)s] \frac{dy}{ds} - aby = 0. \tag{11}$$

If the independent variable is changed from s to $x = bs$, then we have

$$\frac{dy}{ds} = \frac{dy}{dx} \frac{dx}{ds} = b \frac{dy}{dx} \quad \text{and} \quad \frac{d^2y}{ds^2} = b^2 \frac{d^2y}{dx^2},$$

and (11) becomes

$$x\left(1 - \frac{x}{b}\right)y'' + \left[(c - x) - \frac{(a + 1)x}{b}\right]y' - ay = 0, \tag{12}$$

where the primes denote derivatives with respect to x. Equation (12) has regular singular points at $x = 0$, $x = b$, and $x = \infty$; it differs from (11) in that the singular point $x = b$ is now mobile. If we let $b \to \infty$, then (12) becomes (10). The singular point at b has evidently coalesced with the one at ∞, and this confluence of two regular singular points at ∞ is easily seen to produce an irregular singular point there (Problem 3).

PROBLEMS

1. Use (3) and (4) to determine the nature of the point $x = \infty$ for
   (a) Legendre's equation $(1 - x^2)y'' - 2xy' + p(p + 1)y = 0$;
   (b) Bessel's equation $x^2y'' + xy' + (x^2 - p^2)y = 0$.

2. Show that the change of dependent variable defined by $y = t^a w$ transforms equation (9) into the hypergeometric equation

$$t(1 - t)w'' + \{(1 + a - b) - [a + (1 + a - c) + 1]t\}w'$$
$$- a(1 + a - c)w = 0.$$

If a and b are not equal and do not differ by an integer, conclude that the hypergeometric equation (8) has the following independent solutions for large values of x:

$$y_1 = \frac{1}{x^a} F\left(a, 1 + a - c, 1 + a - b, \frac{1}{x}\right)$$

and

$$y_2 = \frac{1}{x^b} F\left(b, 1 + b - c, 1 + b - a, \frac{1}{x}\right).$$

3. Verify that the confluent hypergeometric equation (10) has $x = \infty$ as an irregular singular point.

4. Verify that the confluent hypergeometric equation (10) has $x = 0$ as a regular singular point with exponents 0 and $1 - c$. If c is not zero or a negative integer, show that the Frobenius series solution corresponding to the exponent 0 is

$$1 + \sum_{n=1}^{\infty} \frac{a(a + 1)\cdots(a + n - 1)}{n!c(c + 1)\cdots(c + n - 1)} x^n.$$

The function defined by this series is known as the *confluent hypergeometric function*, and is often denoted by the symbol $F(a,c,x)$.

5. *Laguerre's equation* is

$$xy'' + (1 - x)y' + py = 0,$$

where p is a constant.[11] Use Problem 4 to show that the only solutions bounded near the origin are constant multiples of $F(-p,1,x)$, and also that these solutions are polynomials if p is a nonnegative integer. The functions

---

[11] Edmond Laguerre (1834–1886) was a professor at the Collège de France in Paris, and worked primarily in geometry and the theory of equations. He was one of the first to point out that a "reasonable" distance function (metric) can be imposed on the coordinate plane of analytic geometry in more than one way.

$L_n(x) = F(-n,1,x)$, where $n = 0, 1, 2, \ldots$, are called *Laguerre polynomials*; they have important applications in the quantum mechanics of the hydrogen atom.

APPENDIX A. TWO CONVERGENCE PROOFS

Proof of Theorem 28-A (conclusion). Our assumption is that the series

$$P(x) = \sum_{n=0}^{\infty} p_n x^n \quad \text{and} \quad Q(x) = \sum_{n=0}^{\infty} q_n x^n \tag{1}$$

converge for $|x| < R$, $R > 0$. We must prove that the series

$$y = \sum_{n=0}^{\infty} a_n x^n \tag{2}$$

converges at least on the same interval if a_0 and a_1 are arbitrary and if a_{n+2} is defined recursively for $n \geq 0$ by

$$(n + 1)(n + 2)a_{n+2} = -\sum_{k=0}^{n} [(k + 1)p_{n-k}a_{k+1} + q_{n-k}a_k]. \tag{3}$$

Let r be a positive number such that $r < R$. Since the series (1) converge for $x = r$, and the terms of a convergent series approach zero and are therefore bounded, there exists a constant $M > 0$ such that

$$|p_n|r^n \leq M \quad \text{and} \quad |q_n| r^n \leq M$$

for all n. Using these inequalities in (3), we find that

$$(n + 1)(n + 2) |a_{n+2}| \leq \frac{M}{r^n} \sum_{k=0}^{n} [(k + 1) |a_{k+1}| + |a_k|]r^k$$

$$\leq \frac{M}{r^n} \sum_{k=0}^{n} [(k + 1) |a_{k+1}| + |a_k|]r^k + M |a_{n+1}| r,$$

where the term $M |a_{n+1}| r$ is inserted because it will be needed below. We now define $b_0 = |a_0|$, $b_1 = |a_1|$, and b_{n+2} (for $n \geq 0$) by

$$(n + 1)(n + 2)b_{n+2} = \frac{M}{r^n} \sum_{k=0}^{n} [(k + 1)b_{k+1} + b_k]r^k + Mb_{n+1}r. \tag{4}$$

It is clear that $0 \leq |a_n| \leq b_n$ for every n. We now try to learn something about the values of x for which the series

$$\sum_{n=0}^{\infty} b_n x^n \tag{5}$$

converges, and for this we need information about the behavior of the ratio b_{n+1}/b_n as $n \to \infty$. We acquire this information at follows. Replacing n in (4) first by $n - 1$ and then by $n - 2$ yields

$$n(n + 1)b_{n+1} = \frac{M}{r^{n-1}} \sum_{k=0}^{n-1} [(k + 1)b_{k+1} + b_k]r^k + Mb_n r$$

and

$$(n - 1)nb_n = \frac{M}{r^{n-2}} \sum_{k=0}^{n-2} [(k + 1)b_{k+1} + b_k]r^k + Mb_{n-1}r.$$

By multiplying the first of these equations by r and using the second, we obtain

$$
\begin{aligned}
rn(n + 1)b_{n+1} &= \frac{M}{r^{n-2}} \sum_{k=0}^{n-2} [(k + 1)b_{k+1} + b_k]r^k \\
&\quad + rM(nb_n + b_{n-1}) + Mb_n r^2 \\
&= (n - 1)nb_n - Mb_{n-1}r + rM(nb_n + b_{n-1}) + Mb_n r^2 \\
&= [(n - 1)n + rMn + Mr^2]b_n,
\end{aligned}
$$

so

$$\frac{b_{n+1}}{b_n} = \frac{(n - 1)n + rMn + Mr^2}{rn(n + 1)}.$$

This tells us that

$$\left| \frac{b_{n+1}x^{n+1}}{b_n x^n} \right| = \frac{b_{n+1}}{b_n} |x| \to \frac{|x|}{r}.$$

The series (5) therefore converges for $|x| < r$, so by the inequality $|a_n| \le b_n$ and the comparison test, the series (2) also converges for $|x| < r$. Since r was an arbitrary positive number smaller than R, we conclude that (2) converges for $|x| < R$, and the proof is complete.

Proof of Theorem 30-A (conclusion). The argument is similar to that just given for Theorem 28-A, but is sufficiently different in its details to merit separate consideration. We assume that the series

$$xP(x) = \sum_{n=0}^{\infty} p_n x^n \qquad \text{and} \qquad x^2Q(x) = \sum_{n=0}^{\infty} q_n x^n \tag{6}$$

converge for $|x| < R$, $R > 0$. The indicial equation is

$$f(m) = m(m - 1) + mp_0 + q_0 = 0, \tag{7}$$

and we consider only the case in which (7) has two real roots m_1 and m_2 with $m_2 \le m_1$. The series whose convergence behavior we must examine is

$$\sum_{n=0}^{\infty} a_n x^n, \tag{8}$$

where a_0 is an arbitrary nonzero constant and the other a_n are defined recursively in terms of a_0 by

$$f(m + n)a_n = -\sum_{k=0}^{n-1} a_k[(m + k)p_{n-k} + q_{n-k}]. \tag{9}$$

Our task is to prove that the series (8) converges for $|x| < R$ if $m = m_1$, and also if $m = m_2$ and $m_1 - m_2$ is not a positive integer.

We begin by observing that $f(m)$ can be written in the form

$$f(m) = (m - m_1)(m - m_2) = m^2 - (m_1 + m_2)m + m_1m_2.$$

With a little calculation, this enables us to write

$$f(m_1 + n) = n(n + m_1 - m_2)$$

and

$$f(m_2 + n) = n(n + m_2 - m_1);$$

and consequently

$$|f(m_1 + n)| \geq n(n - |m_1 - m_2|) \tag{10}$$

and

$$|f(m_2 + n)| \geq n(n - |m_2 - m_1|). \tag{11}$$

Let r be a positive number such that $r < R$. Since the series (6) converge for $x = r$, there exists a constant $M > 0$ with the property that

$$|p_n| r^n \leq M \quad \text{and} \quad |q_n| r^n \leq M \tag{12}$$

for all n. If we put $m = m_1$ in (9) and use (10) and (12), we obtain

$$n(n - |m_1 - m_2|) |a_n| \leq M \sum_{k=0}^{n-1} \frac{|a_k|}{r^{n-k}} (|m_1| + k + 1).$$

We now define a sequence $\{b_n\}$ by writing

$$b_n = |a_n| \quad \text{for} \quad 0 \leq n \leq |m_1 - m_2|$$

and

$$n(n - |m_1 - m_2|)b_n = M \sum_{k=0}^{n-1} \frac{b_k}{r^{n-k}} (|m_1| + k + 1) \tag{13}$$

for $n > |m_1 - m_2|$. It is clear that $0 \leq |a_n| \leq b_n$ for every n. We shall prove that the series

$$\sum_{n=0}^{\infty} b_n x^n \tag{14}$$

converges for $|x| < r$, and to achieve this we seek a convenient expression for the ratio b_{n+1}/b_n. By replacing n by $n + 1$ in (13), multiplying by r, and using (13) to simplify the result, we obtain

$$r(n + 1)(n + 1 - |m_1 - m_2|)b_{n+1}$$
$$= n(n - |m_1 - m_2|)b_n + Mb_n(|m_1| + n + 1),$$

so

$$\frac{b_{n+1}}{b_n} = \frac{n(n - |m_1 - m_2|) + M(|m_1| + n + 1)}{r(n + 1)(n + 1 - |m_1 - m_2|)}.$$

This tells us that

$$\left| \frac{b_{n+1}x^{n+1}}{b_n x^n} \right| = \frac{b_{n+1}}{b_n} |x| \to \frac{|x|}{r},$$

so (14) converges for $|x| < r$. It now follows from $0 \le |a_n| \le b_n$ that (8) also converges for $|x| < r$; and since r was taken to be an arbitrary positive number smaller than R, we conclude that (8) converges for $|x| < R$. If m_1 is everywhere replaced by m_2 and (11) is used instead of (10), then the same calculations prove that in this case the series (8) also converges for $|x| < R$—assuming, of course, that $m_1 - m_2$ is not a positive integer so that the series (8) is well defined.

APPENDIX B. HERMITE POLYNOMIALS AND QUANTUM MECHANICS

The most important single application of the Hermite polynomials is to the theory of the linear harmonic oscillator in quantum mechanics. A differential equation that arises in this theory and is closely related to Hermite's equation (Problem 28-7) is

$$\frac{d^2w}{dx^2} + (2p + 1 - x^2)w = 0, \tag{1}$$

where p is a constant. For reasons discussed at the end of this appendix, physicists are interested only in solutions of (1) that approach zero as $|x| \to \infty$. If we try to solve (1) directly by power series, we get a three-term recursion formula for the coefficients, and this is too inconvenient to merit further consideration. To simplify the problem, we introduce a new dependent varible y by means of

$$w = ye^{-x^2/2}. \tag{2}$$

This transforms (1) into

$$\frac{d^2y}{dx^2} - 2x\frac{dy}{dx} + 2py = 0, \tag{3}$$

which is Hermite's equation. The desired solutions of (1) therefore correspond to the solutions of (3) that grow in magnitude (as $|x| \to \infty$) less rapidly than $e^{x^2/2}$, and we shall see that these are essentially the Hermite polynomials.

Physicists motivate the transformation (2) by the following ingenious argument. When x is large, the constant $2p + 1$ in equation (1) is negligible compared with x^2, so (1) is approximately

$$\frac{d^2w}{dx^2} = x^2 w.$$

It is not too outrageous to guess that the functions $w = e^{\pm x^2/2}$ might be solutions of this equation. We now observe that

$$w' = \pm x e^{\pm x^2/2} \quad \text{and} \quad w'' = x^2 e^{\pm x^2/2} \pm e^{\pm x^2/2};$$

and since for large x the second term of w'' can be neglected compared with the first, it appears that $w = e^{x^2/2}$ and $w = e^{-x^2/2}$ are indeed "approximate solutions" of (1). The first of these is now discarded because it does not approach zero as $|x| \to \infty$. It is therefore reasonable to suppose that the exact solution of (1) has the form (2), where we hope that the function $y(x)$ has a simpler structure than $w(x)$.

Whatever one thinks of this reasoning, it works. For we have seen in Problem 28-7 that Hermite's equation (3) has a two-term recursion formula

$$a_{n+2} = -\frac{2(p-n)}{(n+1)(n+2)} a_n, \tag{4}$$

and also that this formula generates two independent series solutions

$$y_1(x) = 1 - \frac{2p}{2!}x^2 + \frac{2^2 p(p-2)}{4!}x^4 - \frac{2^3 p(p-2)(p-4)}{6!}x^6 + \cdots \tag{5}$$

and

$$y_2(x) = x - \frac{2(p-1)}{3!}x^3 + \frac{2^2(p-1)(p-3)}{5!}x^5$$
$$- \frac{2^3(p-1)(p-3)(p-5)}{7!}x^7 + \cdots \tag{6}$$

that converge for all x.

We now compare the rates of growth of the functions $y_1(x)$ and $e^{x^2/2}$. Our purpose is to prove that

$$\frac{y_1(x)}{e^{x^2/2}} \to 0 \qquad \text{as } |x| \to \infty$$

if and only if the series for $y_1(x)$ breaks off and is a polynomial, that is, if and only if the parameter p has one of the values $0, 2, 4, \ldots$. The "if" part is clear by l'Hospital's rule. To prove the "only if" part, we assume that $p \neq 0, 2, 4, \ldots$, and show that in this case the above quotient does not approach zero. To do this, we use the fact that $y_1(x)$ has the form $y_1(x) = \sum a_{2n} x^{2n}$ with its coefficients determined by (4) and the condition $a_0 = 1$, and also that $e^{x^2/2}$ has the series expansion $e^{x^2/2} = \sum b_{2n} x^{2n}$ where

$b_{2n} = 1/(2^n n!)$, so

$$\frac{y_1(x)}{e^{x^2/2}} = \frac{a_0 + a_2 x^2 + a_4 x^4 + \cdots + a_{2n} x^{2n} + \cdots}{b_0 + b_2 x^2 + b_4 x^4 + \cdots + b_{2n} x^{2n} + \cdots}.$$

Formula (4) tells us that all coefficients in the numerator with sufficiently large subscripts have the same sign, so without loss of generality these coefficients may be assumed to be positive. To prove that our quotient does not approach zero as $|x| \to \infty$, it therefore suffices to show that $a_{2n} > b_{2n}$ if n is large enough. To establish this, we begin by observing that

$$\frac{a_{2n+2}}{a_{2n}} = -\frac{2(p - 2n)}{(2n + 1)(2n + 2)} \quad \text{and} \quad \frac{b_{2n+2}}{b_{2n}} = \frac{1}{2(n + 1)},$$

so

$$\frac{a_{2n+2}/a_{2n}}{b_{2n+2}/b_{2n}} = -\frac{2(p - 2n)2(n + 1)}{(2n + 1)(2n + 2)} \to 2.$$

This implies that

$$\frac{a_{2n+2}}{b_{2n+2}} > \frac{3}{2} \cdot \frac{a_{2n}}{b_{2n}}$$

for all sufficiently large n's. If N is any one of these n's, then repeated application of this inequality shows that

$$\frac{a_{2N+2k}}{b_{2N+2k}} > \left(\frac{3}{2}\right)^k \frac{a_{2N}}{b_{2N}} > 1$$

for all sufficiently large k's, so $a_{2n}/b_{2n} > 1$ or $a_{2n} > b_{2n}$ if n is large enough. The above argument proves that $y_1(x)e^{-x^2/2} \to 0$ as $|x| \to \infty$ if and only if the parameter p has one of the values $0, 2, 4, \ldots$. Similar reasoning yields the same conclusion for $y_2(x)e^{-x^2/2}$ (with $p = 1, 3, 5, \ldots$), so the desired solutions of Hermite's equation are constant multiples of the Hermite polynomials $H_0(x)$, $H_1(x)$, $H_2(x)$, $\ldots$ defined in Problem 28-7.

The generating function and Rodrigues' formula. We have seen how the Hermite polynomials arise, and we now turn to a consideration of their most useful properties. The significance of these properties will become clear at the end of this appendix.

These polynomials are often defined by means of the following power series expansion:

$$e^{2xt-t^2} = \sum_{n=0}^{\infty} \frac{H_n(x)}{n!} t^n = H_0(x) + H_1(x)t + \frac{H_2(x)}{2!} t^2 + \cdots. \quad (7)$$

The function e^{2xt-t^2} is called the *generating function* of the Hermite polynomials. This definition has the advantage of efficiency for deducing properties of the $H_n(x)$, and the obvious weakness of being totally unmotivated. We shall therefore derive (7) from the series solutions (5) and (6).

All polynomial solutions of (3) are obtained from these series by replacing p by an integer $n \geq 0$ and multiplying by an arbitrary constant. They all have the form

$$h_n(x) = \cdots + a_{n-6}x^{n-6} + a_{n-4}x^{n-4} + a_{n-2}x^{n-2} + a_n x^n$$

$$= a_n x^n + a_{n-2}x^{n-2} + a_{n-4}x^{n-4} + a_{n-6}x^{n-6} + \cdots,$$

where the sum last written ends with a_0 or $a_1 x$ according as n is even or odd and its coefficients are related by

$$a_{k+2} = -\frac{2(n-k)}{(k+1)(k+2)} a_k. \tag{8}$$

We shall find $a_{n-2}, a_{n-4}, \ldots$ in terms of a_n, and to this end we replace k in (8) by $k-2$ and get

$$a_k = -\frac{2(n-k+2)}{(k-1)k} a_{k-2}$$

or

$$a_{k-2} = -\frac{k(k-1)}{2(n-k+2)} a_k.$$

Letting k be n, $n-2$, $n-4$, etc., yields

$$a_{n-2} = -\frac{n(n-1)}{2 \cdot 2} a_n,$$

$$a_{n-4} = -\frac{(n-2)(n-3)}{2 \cdot 4} a_{n-2}$$

$$= \frac{n(n-1)(n-2)(n-3)}{2^2 \cdot 2 \cdot 4} a_n,$$

$$a_{n-6} = -\frac{(n-4)(n-5)}{2 \cdot 6} a_{n-4}$$

$$= -\frac{n(n-1)(n-2)(n-3)(n-4)(n-5)}{2^3 \cdot 2 \cdot 4 \cdot 6} a_n,$$

and so on, so

$$h_n(x) = a_n\left[x^n - \frac{n(n-1)}{2\cdot 2}x^{n-2} + \frac{n(n-1)(n-2)(n-3)}{2^2\cdot 2\cdot 4}x^{n-4}\right.$$

$$- \frac{n(n-1)(n-2)(n-3)(n-4)(n-5)}{2^3\cdot 2\cdot 4\cdot 6}x^{n-6} + \cdots$$

$$\left. + (-1)^k\frac{n(n-1)\cdots(n-2k+1)}{2^k\cdot 2\cdot 4\cdots(2k)}x^{n-2k} + \cdots\right].$$

This expression can be written in the form

$$h_n(x) = a_n\sum_{k=0}^{[n/2]}(-1)^k\frac{n!}{2^{2k}k!(n-2k)!}x^{n-2k},$$

where $[n/2]$ is the standard notation fo the greatest integer $\leq n/2$. To get the nth Hermite polynomial $H_n(x)$, we put $a_n = 2^n$ and obtain

$$H_n(x) = \sum_{k=0}^{[n/2]}(-1)^k\frac{n!}{k!(n-2k)!}(2x)^{n-2k}. \tag{9}$$

This choice for the value of a_n is purely a matter of convenience; it has the effect of simplifying the formulas expressing the various properties of the Hermite polynomials.

In order to make the transition from (9) to (7), we digress briefly. The defining formula for the product of two power series,

$$\left(\sum_{n=0}^{\infty}a_nt^n\right)\left(\sum_{n=0}^{\infty}b_nt^n\right) = \sum_{n=0}^{\infty}\left(\sum_{k=0}^{n}a_kb_{n-k}\right)t^n,$$

is awkward to use when the first series contains only even powers of t:

$$\left(\sum_{n=0}^{\infty}a_nt^{2n}\right)\left(\sum_{n=0}^{\infty}b_nt^n\right) = ?.$$

What we want to do here is gather together the nth powers of t from all possible products $a_kt^{2k}b_jt^j$, so $2k + j = n$ and the terms we consider are $a_kt^{2k}b_{n-2k}t^{n-2k}$. The restrictions are $k \geq 0$ and $n - 2k \geq 0$, so $0 \leq k \leq n/2$; and for each $n \geq 0$ we see that k varies from 0 to the greatest integer $\leq n/2$. This yields the product formula

$$\left(\sum_{n=0}^{\infty}a_nt^{2n}\right)\left(\sum_{n=0}^{\infty}b_nt^n\right) = \sum_{n=0}^{\infty}\left(\sum_{k=0}^{[n/2]}a_kb_{n-2k}\right)t^n. \tag{10}$$

If we now insert (9) into the right side of (7) and use (10), we obtain

$$\sum_{n=0}^{\infty} \frac{H_n(x)}{n!} t^n = \sum_{n=0}^{\infty} \left[\sum_{k=0}^{[n/2]} \frac{(-1)^k (2x)^{n-2k}}{k!(n-2k)!} \right] t^n$$

$$= \left[\sum_{n=0}^{\infty} \frac{(-1)^n}{n!} t^{2n} \right] \left[\sum_{n=0}^{\infty} \frac{(2x)^n}{n!} t^n \right]$$

$$= \left[\sum_{n=0}^{\infty} \frac{(-t^2)^n}{n!} \right] \left[\sum_{n=0}^{\infty} \frac{(2xt)^n}{n!} \right]$$

$$= e^{-t^2} e^{2xt} = e^{2xt-t^2},$$

which establishes (7).

As an application of (7) we prove *Rodrigues' formula* for the Hermite polynomials:

$$H_n(x) = (-1)^n e^{x^2} \frac{d^n}{dx^n} e^{-x^2}. \tag{11}$$

In view of formula 26-(9) for the coefficients of a power series, (7) yields

$$H_n(x) = \left(\frac{\partial^n}{\partial t^n} e^{2xt-t^2} \right)_{t=0} = e^{x^2} \left(\frac{\partial^n}{\partial t^n} e^{-(x-t)^2} \right)_{t=0}.$$

If we introduce a new variable $z = x - t$ and use the fact that $\partial/\partial t = -(\partial/\partial z)$, then since $t = 0$ corresponds to $z = x$, the expression last written becomes

$$(-1)^n e^{x^2} \left(\frac{d^n}{dz^n} e^{-z^2} \right)_{z=x} = (-1)^n e^{x^2} \frac{d^n}{dx^n} e^{-x^2},$$

and the proof is complete.

Orthogonality. We know that for each nonnegative integer n the function

$$w_n(x) = e^{-x^2/2} H_n(x), \tag{12}$$

called the *Hermite function of order n*, approaches zero as $|x| \to \infty$ and is a solution of the differential equation

$$w_n'' + (2n + 1 - x^2) w_n = 0. \tag{13}$$

An important property of these functions is the fact that

$$\int_{-\infty}^{\infty} w_m w_n \, dx = \int_{-\infty}^{\infty} e^{-x^2} H_m(x) H_n(x) \, dx = 0 \qquad \text{if } m \neq n. \tag{14}$$

This relation is often expressed by saying that the Hermite functions are *orthogonal* on the interval $(-\infty, \infty)$.

To prove (14) we begin by writing down the equation satisfied by $w_m(x)$,

$$w_m'' + (2m + 1 - x^2)w_m = 0. \tag{15}$$

Now, multiplying (13) by w_m and (15) by w_n and subtracting, we obtain

$$\frac{d}{dx}(w_n'w_m - w_m'w_n) + 2(n - m)w_mw_n = 0.$$

If we integrate this equation from $-\infty$ to ∞ and use the fact that $w_n'w_m - w_m'w_n$ vanishes at both limits, we see that

$$2(n - m)\int_{-\infty}^{\infty} w_mw_n \, dx = 0,$$

which implies (14)

We will also need to know that the value of the integral in (14) when $m = n$ is

$$\int_{-\infty}^{\infty} e^{-x^2}[H_n(x)]^2 \, dx = 2^n n! \sqrt{\pi}. \tag{16}$$

To establish this, we use Rodrigues' formula (11) and integrate

$$\int_{-\infty}^{\infty} e^{-x^2}H_n(x)H_n(x) \, dx = (-1)^n \int_{-\infty}^{\infty} H_n(x)\frac{d^n}{dx^n}e^{-x^2} \, dx$$

by parts, with

$$u = H_n(x), \qquad du = H_n'(x) \, dx,$$

$$dv = \frac{d^n}{dx^n}e^{-x^2} \, dx, \qquad v = \frac{d^{n-1}}{dx^{n-1}}e^{-x^2}.$$

Since uv is the product of e^{-x^2} and a polynomial, it vanishes at both limits and

$$\int_{-\infty}^{\infty} e^{-x^2}[H_n(x)]^2 \, dx = (-1)^{n+1}\int_{-\infty}^{\infty} H_n'(x)\frac{d^{n-1}}{dx^{n-1}}e^{-x^2} \, dx$$

$$= (-1)^{n+2}\int_{-\infty}^{\infty} H_n''(x)\frac{d^{n-2}}{dx^{n-2}}e^{-x^2} \, dx$$

$$= \cdots = (-1)^{2n}\int_{-\infty}^{\infty} H_n^{(n)}(x)e^{-x^2} \, dx.$$

Now the term containing the highest power of x in $H_n(x)$ is $2^n x^n$, so

$H_n^{(n)}(x) = 2^n n!$ and the last integral is

$$2^n n! \int_{-\infty}^{\infty} e^{-x^2} dx = (2^n n!)2 \int_0^{\infty} e^{-x^2} dx = 2^n n! \sqrt{\pi},$$

which is the desired result.[12]

These orthogonality properties can be used to expand an "arbitrary" function $f(x)$ in a *Hermite series*:

$$f(x) = \sum_{n=0}^{\infty} a_n H_n(x). \tag{17}$$

If we proceed formally, the coefficients a_n can be found by multiplying (17) by $e^{-x^2} H_m(x)$ and integrating term by term from $-\infty$ to ∞. By (14) and (16) this gives

$$\int_{-\infty}^{\infty} e^{-x^2} H_m(x)f(x)\, dx = \sum_{n=0}^{\infty} a_n \int_{-\infty}^{\infty} e^{-x^2} H_m(x)H_n(x)\, dx = a_m 2^m m! \sqrt{\pi},$$

so (replacing m by n)

$$a_n = \frac{1}{2^n n! \sqrt{\pi}} \int_{-\infty}^{\infty} e^{-x^2} H_n(x)f(x)\, dx. \tag{18}$$

This formal procedure suggests the mathematical problem of determining conditions on the function $f(x)$ that guarantee that (17) is valid when the a_n's are defined by (18). Problems of this kind are part of the general theory of orthogonal functions. Some direct physical applications of orthogonal expansions like (17) are discussed in Appendices A and B of Chapter 8.

The harmonic oscillator. As we stated at the beginning, the mathematical ideas developed above have their main application in quantum mechanics. An adequate discussion of the underlying physical concepts is clearly beyond the scope of this appendix. Nevertheless, it is quite easy to understand the role played by the Hermite polynomials $H_n(x)$ and the corresponding Hermite functions $e^{-x^2/2} H_n(x)$.

In Section 20 we analyzed the classical harmonic oscillator, which can be thought of as a particle of mass m constrained to move along the x-axis and bound to the equilibrium position $x = 0$ by a restoring force $-kx$. The equation of motion is

$$m \frac{d^2 x}{dt^2} = -kx;$$

---

[12] The fact that the integral of e^{-x^2} from 0 to ∞ is $\sqrt{\pi}/2$ is often proved in elementary calculus. See Problem 46-3.

and with suitable initial conditions, we found that its solution is the harmonic oscillation

$$x = x_0 \cos \sqrt{\frac{k}{m}}\, t,$$

where x_0 is the amplitude. We also recall that the period T is given by $T = 2\pi\sqrt{m/k}$; and since the vibrational frequency v is the reciprocal of the period, we have $k = 4\pi^2 m v^2$. Furthermore, since the kinetic energy is $\frac{1}{2}m(dx/dt)^2$ and the potential energy is $\frac{1}{2}kx^2$, an easy calculation shows that the total energy of the system is $E = \frac{1}{2}kx_0^2$, a constant. This total energy may clearly take any positive value whatever.

In quantum mechanics, the *Schrödinger wave equation* for the harmonic oscillator described above is

$$\frac{d^2\psi}{dx^2} + \frac{8\pi^2 m}{h^2}\left(E - \frac{1}{2}kx^2\right)\psi = 0, \tag{19}$$

where E is again the total energy, h is Planck's constant, and satisfactory solutions $\psi(x)$ are known as *Schrödinger wave functions*.[13] If we use the equation $k = 4\pi^2 m v^2$ to eliminate the force constant k, then (19) can be written in the form

$$\frac{d^2\psi}{dx^2} + \frac{8\pi^2 m}{h^2}(E - 2\pi^2 m v^2 x^2)\psi = 0. \tag{20}$$

The physically admissible (or "civilized") solutions of this equation are those satisfying the conditions

$$\psi \to 0 \text{ as } |x| \to \infty \quad \text{and} \quad \int_{-\infty}^{\infty} |\psi|^2\, dx = 1. \tag{2}$$

These solutions—the Schrödinger wave functions—are also called the *eigenfunctions* of the problem, and we shall see that they exist only when E has certain special values called *eigenvalues*.

If we change the independent variable to

$$u = 2\pi\sqrt{\frac{vm}{h}}\, x, \tag{22}$$

---

[13] Erwin Schrödinger (1887–1961) was an Austrian theoretical physicist who shared the 1933 Nobel Prize with Dirac. His scientific work can be appreciated only by experts, but he was a man of broad cultural interests and was a brilliant and lucid writer in the tradition of Poincaré. He liked to write pregnant little books on big themes: *What Is Life?*, *Science and Humanism*, *Nature and the Greeks*, Cambridge University Press, New York, 1944, 1952, 1954, respectively.

then (20) becomes

$$\frac{d^2\psi}{du^2} + \left(\frac{2E}{hv} - u^2\right)\psi = 0 \tag{23}$$

and conditions (21) become

$$\psi \to 0 \text{ as } |u| \to \infty \quad \text{and} \quad \int_{-\infty}^{\infty} |\psi|^2 \, du = 2\pi\sqrt{\frac{vm}{h}}. \tag{24}$$

Except for notation, equation (23) has exactly the form of equation (1), so we know that it has solutions satisfying the first condition of (24) if and only if $2E/hv = 2n + 1$ or

$$E = hv\left(n + \frac{1}{2}\right) \tag{25}$$

for some non-negative integer n. We also know that in this case these solutions of (23) have the form

$$\psi = ce^{-u^2/2}H_n(u)$$

where c is a constant. If we now impose the second condition of (24) and use (16), then it follows that

$$c = \left[\frac{4\pi vm}{2^{2n}(n!)^2h}\right]^{1/4}.$$

The eigenfunction corresponding to the eigenvalue (25) is therefore

$$\psi = \left[\frac{4\pi vm}{2^{2n}(n!)^2h}\right]^{1/4} e^{-u^2/2}H_n(u), \tag{26}$$

where (22) gives u in terms of x.

Physicists have a deep professional interest in the detailed properties of these eigenfunctions. For us, however, the problem is only an illustration of the occurrence of the Hermite polynomials, so we will not pursue the matter any further—beyond pointing out that formula (25) yields the so-called *quantized energy levels* of the harmonic oscillator. This means that the energy E may assume only these discrete values, which of course is very different from the corresponding classical situation described above. The simplest concrete application of these ideas is to the vibrational motion of the atoms in a diatomic molecule. When this phenomenon is studied experimentally, the observed energies are found to be precisely in accord with (25).

NOTE ON HERMITE. Charles Hermite (1822–1901), one of the most eminent French mathematicians of the nineteenth century, was particularly distinguished for the elegance and high artistic quality of his work. As a student, he courted disaster by neglecting his routine assigned work to study the classic masters of mathematics; and though he nearly failed his examinations, he became a first-rate creative mathematician himself while still in his early twenties. In 1870 he was appointed to a professorship at the Sorbonne, where he trained a whole generation of well known French mathematicians, including Picard, Borel, and Poincaré.

The unusual character of his mind is suggested by the following remark of Poincaré: "Talk with M. Hermite. He never evokes a concrete image, yet you soon perceive that the most abstract entities are to him like living creatures." He disliked geometry, but was strongly attracted to number theory and analysis, and his favorite subject was elliptic functions, where these two fields touch in many remarkable ways. The reader may be aware that Abel had proved many years before that the general polynomial equation of the fifth degree cannot be solved by functions involving only rational operations and root extractions. One of Hermite's most surprising achievements (in 1858) was to show that this equation can be solved by elliptic functions. His 1873 proof of the transcendence of e was another high point of his career.

Several of his purely mathematical discoveries had unexpected applications many years later to mathematical physics. For example, the Hermitian forms and matrices he invented in connection with certain problems of number theory turned out to be crucial for Heisenberg's 1925 formulation of quantum mechanics, and we have seen that Hermite polynomials and Hermite functions are useful in solving Schrödinger's wave equation. The reason is not clear, but it seems to be true that mathematicians do some of their most valuable practical work when thinking about problems that appear to have nothing whatever to do with physical reality.

APPENDIX C. GAUSS

Carl Friedrich Gauss (1777–1855) was the greatest of all mathematicians and perhaps the most richly gifted genius of whom there is any record. This gigantic figure, towering at the beginning of the nineteenth century, separates the modern era in mathematics from all that went before. His visionary insight and originality, the extraordinary range and depth of his achievements, his repeated demonstrations of almost superhuman power and tenacity—all these qualities combined in a single individual present an enigma as baffling to us as it was to his contemporaries.

Gauss was born in the city of Brunswick in northern Germany. His exceptional skill with numbers was clear at a very early age, and in later life he joked that he knew how to count before he could talk. It is said that Goethe wrote and directed little plays for a puppet theater when he was six, and that Mozart composed his first childish minuets when he was five, but Gauss corrected an error in his father's payroll accounts at the age of three.[14] His father was a gardener and bricklayer without either

---

[14] See W. Sartorius von Waltershausen, "Gauss zum Gedächtniss." These personal recollections appeared in 1856, and a translation by Helen W. Gauss (the mathematician's great-granddaughter) was privately printed in Colorado Springs in 1966.

the means or the inclination to help develop the talents of his son. Fortunately, however, Gauss's remarkable abilities in mental computation attracted the interest of several influential men in the community, and eventually brought him to the attention of the Duke of Brunswick. The Duke was impressed with the boy and undertook to support his further education, first at the Caroline College in Brunswick (1792–1795) and later at the University of Göttingen (1795–1798).

At the Caroline College, Gauss completed his mastery of the classical languages and explored the works of Newton, Euler, and Lagrange. Early in this period—perhaps at the age of fourteen or fifteen—he discovered the prime number theorem, which was finally proved in 1896 after great efforts by many mathematicians (see our notes on Chebyshev and Riemann). He also invented the method of least squares for minimizing the errors inherent in observational data, and conceived the Gaussian (or normal) law of distribution in the theory of probability.

At the university, Gauss was attracted by philology but repelled by the mathematics courses, and for a time the direction of his future was uncertain. However, at the age of eighteen he made a wonderful geometric discovery that caused him to decide in favor of mathematics and gave him great pleasure to the end of his life. The ancient Greeks had known ruler-and-compass constructions for regular polygons of 3, 4, 5, and 15 sides, and for all others obtainable from these by bisecting angles. But this was all, and there the matter rested for 2000 years, until Gauss solved the problem completely. He proved that a regular polygon with n sides is constructible if and only if n is the product of a power of 2 and distinct prime numbers of the form $p_k = 2^{2^k} + 1$. In particular, when $k = 0,1,2,3$, we see that each of the corresponding numbers $p_k = 3,5,17,257$ is prime, so regular polygons with these numbers of sides are constructible.[15]

During these years Gauss was almost overwhelmed by the torrent of ideas which flooded his mind. He began the brief notes of his scientific diary in an effort to record his discoveries, since there were far too many to work out in detail at that time The first entry, dated March 30, 1796, states the constructibility of the regular polygon with 17 sides, but even earlier than this he was penetrating deeply into several unexplored continents in the theory of numbers. In 1795 he discovered the law of quadratic reciprocity, and as he later wrote, "For a whole year this theorem tormented me and absorbed my greatest efforts, until at last I

---

[15] Details of some of these constructions are given in H. Tietze, *Famous Problems of Mathematics*, chap. IX, Graylock Press, New York, 1965.

found a proof,"[16] At that time Gauss was unaware that the theorem had already been imperfectly stated without proof by Euler, and correctly stated with an incorrect proof by Legendre. It is the core of the central part of his famous treatise *Disquisitiones Arithmeticae,* which was published in 1801 although completed in 1798.[17] Apart from a few fragmentary results of earlier mathematicians, this great work was wholly original. It is usually considered to mark the true beginning of modern number theory, to which it is related in much the same way as Newton's *Principia* is to physics and astronomy. In the introductory pages Gauss develops his method of congruences for the study of divisibility problems and gives the first proof of the fundamental theorem of arithmetic (also called the unique factorization theorem), which asserts that every integer $n > 1$ can be expressed uniquely as a product of primes. The central part is devoted mainly to quadratic congruences, forms, and residues. The last section presents his complete theory of the cyclotomic (circle-dividing) equation, with its applications to the constructibility of regular polygons. The entire work was a gargantuan feast of pure mathematics, which his successors were able to digest only slowly and with difficulty.

In his *Disquisitiones* Gauss also created the modern rigorous approach to mathematics. He had become thoroughly impatient with the loose writing and sloppy proofs of his predecessors, and resolved that his own works would be beyond criticism in this respect. As he wrote to a friend, "I mean the word proof not in the sense of the lawyers, who set two half proofs equal to a whole one, but in the sense of the mathematician, where 1/2 proof = 0 and it is demanded for proof that every doubt becomes impossible." The *Disquisitiones* was composed in this spirit and in Gauss's mature style, which is terse, rigorous, devoid of motivation, and in many places so carefully polished that it is almost unintelligible. In another letter he said, "You know that I write slowly. This is chiefly because I am never satisfied until I have said as much as possible in a few words, and writing briefly takes far more time than writing at length." One of the effects of this habit is that his publications concealed almost as much as they revealed, for he worked very hard at removing every trace of the train of thought that led him to his discoveries. Abel remarked, "He is like the fox, who effaces his tracks in the sand with his tail." Gauss replied to such criticisms by saying that no

---

[16] See D. W. Smith, *A Source Book in Mathematics,* pp. 112–118, McGraw-Hill, New York, 1929. This selection includes a statement of the theorem and the fifth of eight proofs that Gauss found over a period of many years. There are probably more than 50 known today.

[17] There is a translation by Arthur A. Clarke (Yale University Press, New Haven, Conn., 1966).

self-respecting architect leaves the scaffolding in place after completing his building. Nevertheless, the difficulty of reading his works greatly hindered the diffusion of his ideas.

Gauss's doctoral dissertation (1799) was another milestone in the history of mathematics. After several abortive attempts by earlier mathematicians—d'Alembert, Euler, Lagrange, Laplace—the fundamental theorem of algebra was here given its first satisfactory proof. This theorem asserts the existence of a real or complex root for any polynomial equation with real or complex coefficients. Gauss's success inaugurated the age of existence proofs, which ever since have played an important part in pure mathematics. Furthermore, in this first proof (he gave four altogether) Gauss appears as the earliest mathematician to use complex numbers and the geometry of the complex plane with complete confidence.[18]

The next period of Gauss's life was heavily weighted toward applied mathematics, and with a few exceptions the great wealth of ideas in his diary and notebooks lay in suspended animation.

In the last decades of the eighteenth century, many astronomers were searching for a new planet between the orbits of Mars and Jupiter, where Bode's law (1772) suggested that there ought to be one. The first and largest of the numerous minor planets known as asteroids was discovered in that region in 1801, and was named Ceres. This discovery ironically coincided with an astonishing publication by the philosopher Hegel, who jeered at astronomers for ignoring philosophy: this science (he said) could have saved them from wasting their efforts by demonstrating that no new planet could possibly exist.[19] Hegel continued his career in a similar vein, and later rose to even greater heights of clumsy obfuscation. Unfortunately the tiny new planet was difficult to see under the best of circumstances, and it was soon lost in the light of the sky near the sun. The sparse observational data posed the problem of calculating the orbit with sufficient accuracy to locate Ceres again after it had moved away from the sun. The astronomers of Europe attempted this task without success for many months. Finally, Gauss was attracted by the challenge; and with the aid of his method of least squares and his unparalleled skill at numerical computation he determined the orbit, told the astronomers where to look with their telescopes, and there it was. He had succeeded in rediscovering Ceres after all the experts had failed.

---

[18] The idea of this proof is very clearly explained by F. Klein, *Elementary Mathematics from an Advanced Standpoint*, pp. 101–104, Dover, New York, 1945.

[19] See the last few pages of "De Orbitis Planetarum," vol. I of Georg Wilhelm Hegel's *Sämtliche Werke*, Frommann Verlag, Stuttgart, 1965.

This achievement brought him fame, an increase in his pension from the Duke, and in 1807 an appointment as professor of astronomy and first director of the new observatory at Göttingen. He carried out his duties with his customary thoroughness, but as it turned out, he disliked administrative chores, committee meetings, and all the tedious red tape involved in the business of being a professor. He also had little enthusiasm for teaching, which he regarded as a waste of his time and as essentially useless (for different reasons) for both talented and untalented students. However, when teaching was unavoidable he apparently did it superbly. One of his students was the eminent algebraist Richard Dedekind, for whom Gauss's lectures after the passage of 50 years remained "unforgettable in memory as among the finest which I have ever heard."[20] Gauss had many opportunities to leave Göttingen, but he refused all offers and remained there for the rest of his life, living quietly and simply, traveling rarely, and working with immense energy on a wide variety of problems in mathematics and its applications. Apart from science and his family—he had two wives and six children, two of whom emigrated to America—his main interests were history and world literature, international politics, and public finance. He owned a large library of about 6000 volumes in many languages, including Greek, Latin, English, French, Russian, Danish, and of course German. His acuteness in handling his own financial affairs is shown by the fact that although he started with virtually nothing, he left an estate over a hundred times as great as his average annual income during the last half of his life.

In the first two decades of the nineteenth century Gauss produced a steady stream of works on astronomical subjects, of which the most important was the treatise *Theoria Motus Corporum Coelestium* (1809). This remained the bible of planetary astronomers for over a century. Its methods for dealing with perturbations later led to the discovery of Neptune. Gauss thought of astronomy as his profession and pure mathematics as his recreation, and from time to time he published a few of the fruits of his private research. His great work on the hypergeometric series (1812) belongs to this period. This was a typical Gaussian effort, packed with new ideas in analysis that have kept mathematicians busy ever since.

Around 1820 he was asked by the government of Hanover to

---

[20] Dedekind's detailed recollections of this course are given in G. Waldo Dunnington, *Carl Friedrich Gauss: Titan of Science*, pp. 259–261, Hafner, New York, 1955. This book is useful mainly for its many quotations, its bibliography of Gauss's publications, and its list of the courses he offered (but often did not teach) from 1808 to 1854.

supervise a geodetic survey of the kingdom, and various aspects of this task—including extensive field work and many tedious triangulations—occupied him for a number of years. It is natural to suppose that a mind like his would have been wasted on such an assignment, but the great ideas of science are born in many strange ways. These apparently unrewarding labors resulted in one of his deepest and most far-reaching contributions to pure mathematics, without which Einstein's general theory of relativity would have been quite impossible.

Gauss's geodetic work was concerned with the precise measurement of large triangles on the earth's surface. This provided the stimulus that led him to the ideas of his paper *Disquisitiones generales circa superficies curvas* (1827), in which he founded the intrinsic differential geometry of general curved surfaces.[21] In this work he introduced curvilinear coordinates u and v on a surface; he obtained the fundamental quadratic differential form $ds^2 = E\,du^2 + 2F\,du\,dv + G\,dv^2$ for the element of arc length ds, which makes it possible to determine geodesic curves; and he formulated the concepts of Gaussian curvature and integral curvature.[22] His main specific results were the famous *theorema egregium*, which states that the Gaussian curvature depends only on E, F, and G, and is therefore invariant under bending; and the Gauss–Bonnet theorem on integral curvature for the case of a geodesic triangle, which in its general form is the central fact of modern differential geometry in the large. Apart from his detailed discoveries, the crux of Gauss's insight lies in the word *intrinsic*, for he showed how to study the geometry of a surface by operating only on the surface itself and paying no attention to the surrounding space in which it lies. To make this more concrete, let us imagine an intelligent two-dimensional creature who inhabits a surface but has no awareness of a third dimension or of anything not on the surface. If this creature is capable of moving about, measuring distances along the surface, and determining the shortest path (geodesic) from one point to another, then he is also capable of measuring the Gaussian curvature at any point and of creating a rich geometry on the surface—and this geometry will be Euclidean (flat) if and only if the Gaussian curvature is everywhere zero. When these conceptions are generalized to more than two dimensions, then they open the door to Riemannian geometry, tensor analysis, and the ideas of Einstein.

Another great work of this period was his 1831 paper on biquadratic

---

[21] A translation by A. Hiltebeitel and J. Morehead was published under the title *General Investigations of Curved Surfaces* by the Raven Press, Hewlett, New York, in 1965.

[22] These ideas are explained in nontechnical language in C. Lanczos, *Albert Einstein and the Cosmic World Order*, chap. 4, Interscience-Wiley, New York, 1965.

residues. Here he extended some of his early discoveries in number theory with the aid of a new method, his purely algebraic approach to complex numbers. He defined these numbers as ordered pairs of real numbers with suitable definitions for the algebraic operations, and in so doing laid to rest the confusion that still surrounded the subject and prepared the way for the later algebra and geometry of n-dimensional spaces. But this was only incidental to his main purpose, which was to broaden the ideas of number theory into the complex domain. He defined complex integers (now called Gaussian integers) as complex numbers $a + ib$ with a and b ordinary integers; he introduced a new concept of prime numbers, in which 3 remains prime but $5 = (1 + 2i)(1 - 2i)$ does not; and he proved the unique factorization theorem for these integers and primes. The ideas of this paper inaugurated algebraic number theory, which has grown steadily from that day to this.[23]

From the 1830s on, Gauss was increasingly occupied with physics, and he enriched every branch of the subject he touched. In the theory of surface tension, he developed the fundamental idea of conservation of energy and solved the earliest problem in the calculus of variations involving a double integral with variable limits. In optics, he introduced the concept of the focal length of a system of lenses and invented the Gauss wide-angle lens (which is relatively free of chromatic aberration) for telescope and camera objectives. He virtually created the science of geomagnetism, and in collaboration with his friend and colleague Wilhelm Weber he built and operated an iron-free magnetic observatory, founded the Magnetic Union for collecting and publishing observations from many places in the world, and invented the electromagnetic telegraph and the bifilar magnetometer. There are many references to his work in James Clerk Maxwell's famous *Treatise on Electricity and Magnetism* (1873). In his preface, Maxwell says that Gauss "brought his powerful intellect to bear on the theory of magnetism and on the methods of observing it, and he not only added greatly to our knowledge of the theory of attractions, but reconstructed the whole of magnetic science as regards the instruments used, the methods of observation, and the calculation of results, so that his memoirs on Terrestrial Magnetism may be taken as models of physical research by all those who are engaged in the measurement of any of the forces in nature." In 1839 Gauss published his fundamental paper on the general theory of inverse square forces, which established potential theory as a coherent branch of

---

[23] See E. T. Bell, "Gauss and the Early Development of Algebraic Numbers," *National Math. Mag.*, vol. 18, pp. 188–204, 219–233 (1944).

mathematics.[24] As usual, he had been thinking about these matters for many years; and among his discoveries were the divergence theorem (also called Gauss's theorem) of modern vector analysis, the basic mean value theorem for harmonic functions, and the very powerful statement which later became known as "Dirichlet's principle" and was finally proved by Hilbert in 1899.

We have discussed the published portion of Gauss's total achievement, but the unpublished and private part was almost equally impressive. Much of this came to light only after his death, when a great quantity of material from his notebooks and scientific correspondence was carefully analyzed and included in his collected works. His scientific diary has already been mentioned. This little booklet of 19 pages, one of the most precious documents in the history of mathematics, was unknown until 1898, when it was found among family papers in the possession of one of Gauss's grandsons. It extends from 1796 to 1814 and consists of 146 very concise statements of the results of his investigations, which often occupied him for weeks or months.[25] All of this material makes it abundantly clear that the ideas Gauss conceived and worked out in considerable detail, but kept to himself, would have made him the greatest mathematician of his time if he had published them and done nothing else.

For example, the theory of functions of a complex variable was one of the major accomplishments of nineteenth century mathematics, and the central facts of this discipline are Cauchy's integral theorem (1827) and the Taylor and Laurent expansions of an analytic function (1831, 1843). In a letter written to his friend Bessel in 1811, Gauss explicitly states Cauchy's theorem and then remarks, "This is a very beautiful theorem whose fairly simple proof I will give on a suitable occasion. It is connected with other beautiful truths which are concerned with series expansions."[26] Thus, many years in advance of those officially credited with these important discoveries, he knew Cauchy's theorem and probably knew both series expansions. However, for some reason the "suitable occasion" for publication did not arise. A possible explanation for this is suggested by his comments in a letter to Wolfgang Bolyai, a close friend from his university years with whom he maintained a lifelong correspondence: "It is not knowledge but the act of learning, not possession but the act of getting there, which grants the greatest

---

[24] George Green's "Essay on the Application of Mathematical Analysis to the Theories of Electricity and Magnetism" (1828) was neglected and almost completely unknown until it was reprinted in 1846.

[25] See Gauss's *Werke*, vol. X, pp. 483–574, 1917.

[26] *Werke*, vol. VIII, p. 91, 1900.

enjoyment. When I have clarified and exhausted a subject, then I turn away from it in order to go into darkness again." His was the temperament of an explorer, who is reluctant to take the time to write an account of his last expedition when he could be starting another. As it was, Gauss wrote a great deal; but to publish every fundamental discovery he made in a form satisfactory to himself would have required several long lifetimes.

Another prime example is non-Euclidean geometry, which has been compared with the Copernican revolution in astronomy for its impact on the minds of civilized men. From the time of Euclid to the boyhood of Gauss, the postulates of Euclidean geometry were universally regarded as necessities of thought. Yet there was a flaw in the Euclidean structure that had long been a focus of attention: the so-called parallel postulate, stating that through a point not on a line there exists a single line parallel to the given line. This postulate was thought not to be independent of the others, and many had tried without success to prove it as a theorem. We now know that Gauss joined in these efforts at the age of fifteen, and he also failed. But he failed with a difference, for he soon came to the shattering conclusion—which had escaped all his predecessors—that the Euclidean form of geometry is not the only one possible. He worked intermittently on these ideas for many years, and by 1820 he was in full possession of the main theorems of non-Euclidean geometry (the name is due to him).[27] But he did not reveal his conclusions, and in 1829 and 1832 Lobachevsky and Johann Bolyai (son of Wolfgang) published their own independent work on the subject. One reason for Gauss's silence in this case is quite simple. The intellectual climate of the time in Germany was totally dominated by the philosophy of Kant, and one of the basic tenets of his system was the idea that Euclidean geometry is the only possible way of thinking about space. Gauss knew that this idea was totally false and that the Kantian system was a structure built on sand. However, he valued his privacy and quiet life, and held his peace in order to avoid wasting his time on disputes with the philosophers. In 1829 he wrote as follows to Bessel: "I shall probably not put my very extensive investigations on this subject [the foundations of geometry] into publishable form for a long time, perhaps not in my lifetime, for I dread the shrieks we would hear from the Boeotians if I were to express myself fully on this matter."[28]

The same thing happened again in the theory of elliptic functions, a

---

[27] Everything he is known to have written about the foundations of geometry was published in his *Werke*, vol. VIII, pp. 159–268, 1900.

[28] *Werke*, vol. VIII, p. 200. The Boeotians were a dull-witted tribe of the ancient Greeks.

very rich field of analysis that was launched primarily by Abel in 1827 and also by Jacobi in 1828–1829. Gauss had published nothing on this subject, and claimed nothing, so the mathematical world was filled with astonishment when it gradually became known that he had found many of the results of Abel and Jacobi before these men were born. Abel was spared this devastating knowledge by his early death in 1829, at the age of twenty-six, but Jacobi was compelled to swallow his disappointment and go on with his work. The facts became known partly through Jacobi himself. His attention was caught by a cryptic passage in the *Disquisitiones* (Article 335), whose meaning can only be understood if one knows something about elliptic functions. He visited Gauss on several occasions to verify his suspicions and tell him about his own most recent discoveries, and each time Gauss pulled 30-year-old manuscripts out of his desk and showed Jacobi what Jacobi had just shown him. The depth of Jacobi's chagrin can readily be imagined. At this point in his life Gauss was indifferent to fame and was actually pleased to be relieved of the burden of preparing the treatise on the subject which he had long planned. After a week's visit with Gauss in 1840, Jacobi wrote to his brother, "Mathematics would be in a very different position if practical astronomy had not diverted this colossal genius from his glorious career."

Such was Gauss, the supreme mathematician. He surpassed the levels of achievement possible for ordinary men of genius in so many ways that one sometimes has the eerie feeling that he belonged to a higher species.

APPENDIX D. CHEBYSHEV POLYNOMIALS AND THE MINIMAX PROPERTY

In Problem 31-6 we defined the Chebyshev polynomials $T_n(x)$ in terms of the hypergeometric function by $T_n(x) = F\left(n, -n, \frac{1}{2}, \frac{1-x}{2}\right)$, where $n = 0, 1, 2, \ldots$. Needless to say, this definition by itself tells us practically nothing, for the question that matters is: what purpose do these polynomials serve? We will now try to answer this question.

It is convenient to begin by adopting a different definition for the polynomials $T_n(x)$. We will see later that the two definitions agree. Our starting point is the fact that if n is a nonnegative integer, then de Moivre's formula from the theory of complex numbers gives

$$\cos n\theta + i \sin n\theta = (\cos \theta + i \sin \theta)^n$$
$$= \cos^n \theta + n \cos^{n-1} \theta(i \sin \theta)$$
$$+ \frac{n(n-1)}{2} \cos^{n-2} \theta(i \sin \theta)^2 + \cdots + (i \sin \theta)^n, \quad (1)$$

so $\cos n\theta$ is the real part of the sum on the right. Now the real terms in this sum are precisely those that contain even powers of $i \sin \theta$; and since $\sin^2 \theta = 1 - \cos^2 \theta$, it is apparent that $\cos n\theta$ is a polynomial function of $\cos \theta$. We use this as the definition of the nth Chebyshev polynomial: $T_n(x)$ is that polynomial for which

$$\cos n\theta = T_n(\cos \theta). \tag{2}$$

Since $T_n(x)$ is a polynomial, it is defined for all values of x. However, if x is restricted to lie in the interval $-1 \le x \le 1$ and we write $x = \cos \theta$ where $0 \le \theta \le \pi$, then (2) yields

$$T_n(x) = \cos (n \cos^{-1} x). \tag{3}$$

With the same restrictions, we can obtain another curious expression for $T_n(x)$. For on adding the two formulas

$$\cos n\theta \pm i \sin n\theta = (\cos \theta \pm i \sin \theta)^n,$$

we get

$$\cos n\theta = \frac{1}{2}[(\cos \theta + i \sin \theta)^n + (\cos \theta - i \sin \theta)^n]$$

$$= \frac{1}{2}[(\cos \theta + i\sqrt{1 - \cos^2 \theta})^n + (\cos \theta - i\sqrt{1 - \cos^2 \theta})^n]$$

$$= \frac{1}{2}[(\cos \theta + \sqrt{\cos^2 \theta - 1})^n + (\cos \theta - \sqrt{\cos^2 \theta - 1})^n],$$

so

$$T_n(x) = \frac{1}{2}[(x + \sqrt{x^2 - 1})^n + (x - \sqrt{x^2 - 1})^n]. \tag{4}$$

Another explicit expression for $T_n(x)$ can be found by using the binomial formula to write (1) as

$$\cos n\theta + i \sin n\theta = \sum_{m=0}^{n} \binom{n}{m} \cos^{n-m}\theta(i \sin \theta)^m.$$

We have remarked that the real terms in this sum correspond to the even values of m, that is, to $m = 2k$ where $k = 0,1,2, \ldots , [n/2]$.[29] Since

$$(i \sin \theta)^m = (i \sin \theta)^{2k} = (-1)^k(1 - \cos^2 \theta)^k = (\cos^2 \theta - 1)^k,$$

we have

$$\cos n\theta = \sum_{k=0}^{[n/2]} \binom{n}{2k} \cos^{n-2k} \theta(\cos^2 \theta - 1)^k,$$

---

[29] The symbol $[n/2]$ is the standard notation for the greatest integer $\le n/2$.

and therefore

$$T_n(x) = \sum_{k=0}^{[n/2]} \frac{n!}{(2k)!(n-2k)!} x^{n-2k}(x^2 - 1)^k. \tag{5}$$

It is clear from (4) that $T_0(x) = 1$ and $T_1(x) = x$; but for higher values of n, $T_n(x)$ is most easily computed from a recursion formula. If we write

$$\cos n\theta = \cos[\theta + (n-1)\theta] = \cos\theta\cos(n-1)\theta - \sin\theta\sin(n-1)\theta$$

and

$$\cos(n-2)\theta = \cos[-\theta + (n-1)\theta]$$
$$= \cos\theta\cos(n-1)\theta + \sin\theta\sin(n-1)\theta,$$

then it follows that

$$\cos n\theta + \cos(n-2)\theta = 2\cos\theta\cos(n-1)\theta.$$

If we use (2) and replace $\cos\theta$ by x, then this trigonometric identity gives the desired recursion formula:

$$T_n(x) + T_{n-2}(x) = 2xT_{n-1}(x). \tag{6}$$

By starting with $T_0(x) = 1$ and $T_1(x) = x$, we find from (6) that $T_2(x) = 2x^2 - 1$, $T_3(x) = 4x^3 - 3x$, $T_4(x) = 8x^4 - 8x^2 + 1$, and so on.

The hypergeometric form. To establish a connection between Chebyshev's differential equation and the Chebyshev polynomials as we have just defined them, we use the fact that the polynomial $y = T_n(x)$ becomes the function $y = \cos n\theta$ when the variable is changed from x to θ by means of $x = \cos\theta$. Now the function $y = \cos n\theta$ is clearly a solution of the differential equation

$$\frac{d^2y}{d\theta^2} + n^2y = 0, \tag{7}$$

and an easy calculation shows that changing the variable from θ back to x transforms (7) into Chebyshev's equation

$$(1 - x^2)\frac{d^2y}{dx^2} - x\frac{dy}{dx} + n^2y = 0. \tag{8}$$

We therefore know that $y = T_n(x)$ is a polynomial solution of (8). But Problem 31-6 tells us that the only polynomial solutions of (8) have the form $cF\left(n, -n, \frac{1}{2}, \frac{1-x}{2}\right)$; and since (4) implies that $T_n(1) = 1$ for every n, and $cF\left(n, -n, \frac{1}{2}, \frac{1-1}{2}\right) = c$, we conclude that

$$T_n(x) = F\left(n, -n, \frac{1}{2}, \frac{1-x}{2}\right). \tag{9}$$

Orthogonality. One of the most important properties of the functions $y_n(\theta) = \cos n\theta$ for different values of n is their orthogonality on the interval $0 \leq \theta \leq \pi$, that is, the fact that

$$\int_0^\pi y_m y_n \, d\theta = \int_0^\pi \cos m\theta \cos n\theta \, d\theta = 0 \qquad \text{if } m \neq n. \tag{10}$$

To prove this, we write down the differential equations satisfied by $y_m = \cos m\theta$ and $y_n = \cos n\theta$:

$$y_m'' + m^2 y_m = 0 \qquad \text{and} \qquad y_n'' + n^2 y_n = 0.$$

On multiplying the first of these equations by y_n and the second by y_m, and subtracting, we obtain

$$\frac{d}{d\theta}(y_m' y_n - y_n' y_m) + (m^2 - n^2) y_m y_n = 0;$$

and (10) follows at once by integrating each term of this equation from 0 to π, since y_m' and y_n' both vanish at the endpoints and $m^2 - n^2 \neq 0$.

When the variable in (10) is changed from θ to $x = \cos\theta$, (10) becomes

$$\int_{-1}^1 \frac{T_m(x) T_n(x)}{\sqrt{1 - x^2}} \, dx = 0 \qquad \text{if } m \neq n. \tag{11}$$

This fact is usually expressed by saying that the Chebyshev polynomials are *orthogonal* on the interval $-1 \leq x \leq 1$ with respect to the *weight function* $(1 - x^2)^{-1/2}$. When $m = n$ in (11), we have

$$\int_{-1}^1 \frac{[T_n(x)]^2}{\sqrt{1 - x^2}} \, dx = \begin{cases} \dfrac{\pi}{2} & \text{for } n \neq 0, \\[2mm] \pi & \text{for } n = 0. \end{cases} \tag{12}$$

These additional statements follow from

$$\int_0^\pi \cos^2 n\theta \, d\theta = \begin{cases} \dfrac{\pi}{2} & \text{for } n \neq 0, \\[2mm] \pi & \text{for } n = 0, \end{cases}$$

which are easy to establish by direct integration.

Just as in the case of the Hermite polynomials discussed in Appendix B, the orthogonality properties (1) and (12) can be used to expand an "arbitrary" function $f(x)$ in a *Chebyshev series*:

$$f(x) = \sum_{n=0}^\infty a_n T_n(x). \tag{13}$$

The same formal procedure as before yields the coefficients

$$a_0 = \frac{1}{\pi} \int_{-1}^{1} \frac{f(x)}{\sqrt{1 - x^2}} dx \tag{14}$$

and

$$a_n = \frac{2}{\pi} \int_{-1}^{1} \frac{T_n(x)f(x)}{\sqrt{1 - x^2}} dx \tag{15}$$

for $n > 0$. And again the true mathematical issue is the problem of finding conditions under which the series (13)—with the a_n defined by (14) and (15)—actually converges to $f(x)$.

The minimax property. The Chebyshev problem we now consider is to see how closely the function x^n can be approximated on the interval $-1 \le x \le 1$ by polynomials $a_{n-1}x^{n-1} + \cdots + a_1x + a_0$ of degree $n - 1$; that is, to see how small the number

$$\max_{-1 \le x \le 1} |x^n - a_{n-1}x^{n-1} - \cdots - a_1x - a_0|$$

can be made by an appropriate choice of the coefficients. This in turn is equivalent to the following problem: among all polynomials $P(x) = x^n + a_{n-1}x^{n-1} + \cdots + a_1x + a_0$ of degree n with leading coefficient 1, to minimize the number

$$\max_{-1 \le x \le 1} |P(x)|,$$

and if possible to find a polynomial that attains this minimum value.

It is clear from $T_1(x) = x$ and the recursion formula (6) that when $n > 0$ the coefficient of x^n in $T_n(x)$ is 2^{n-1}, so $2^{1-n}T_n(x)$ has leading coefficient 1. These polynomials completely solve Chebyshev's problem, in the sense that they have the following remarkable property.

Minimax property. *Among all polynomials $P(x)$ of degree $n > 0$ with leading coefficient 1, $2^{1-n}T_n(x)$ deviates least from zero in the interval $-1 \le x \le 1$:*

$$\max_{-1 \le x \le 1} |P(x)| \ge \max_{-1 \le x \le 1} |2^{1-n}T_n(x)| = 2^{1-n}. \tag{16}$$

Proof. First, the equality in (16) follows at once from

$$\max_{-1 \le x \le 1} |T_n(x)| = \max_{0 \le \theta \le \pi} |\cos n\theta| = 1.$$

To complete the argument, we assume that $P(x)$ is a polynomial of the stated type for which

$$\max_{-1 \le x \le 1} |P(x)| < 2^{1-n}, \tag{17}$$

and we deduce a contradiction from this hypothesis. We begin by noticing that the polynomial $2^{1-n}T_n(x) = 2^{1-n}\cos n\theta$ has the alternately positive and negative values 2^{1-n}, -2^{1-n}, 2^{1-n}, ..., $\pm 2^{1-n}$ at the $n + 1$ points x that correspond to $\theta = 0$, π/n, $2\pi/n$, ..., $n\pi/n = \pi$. By assumption (17), $Q(x) = 2^{1-n}T_n(x) - P(x)$ has the same sign as $2^{1-n}T_n(x)$ at these points, and must therefore have at least n zeros in the interval $-1 \le x \le 1$. But this is impossible since $Q(x)$ is a polynomial of degree at most $n - 1$ which is not identically zero.

In this very brief treatment the minimax property unfortunately seems to appear out of nowhere, with no motivation and no hint as to why the Chebyshev polynomials behave in this extraordinary way. We hope the reader will accept our assurance that in the broader context of Chebyshev's original ideas this surprising property is really quite natural.[30] For those who like their mathematics to have concrete applications, it should be added that the minimax property is closely related to the important place Chebyshev polynomials occupy in contemporary numerical analysis.

NOTE ON CHEBYSHEV. Pafnuty Lvovich Chebyshev (1821–1894) was the most eminent Russian mathematician of the nineteenth century. He was a contemporary of the famous geometer Lobachevsky (1793–1856), but his work had a much deeper influence throughout Western Europe and he is considered the founder of the great school of mathematics that has been flourishing in Russia for the past century.

As a boy he was fascinated by mechanical toys, and apparently was first attracted to mathematics when he saw the importance of geometry for understanding machines. After his student years in Moscow, he became professor of mathematics at the University of St. Petersburg, a position he held until his retirement. His father was a member of the Russian nobility, but after the famine of 1840 the family estates were so diminished that for the rest of his life Chebyshev was forced to live very frugally and he never married. He spent much of his small income on mechanical models and occasional journeys to Western Europe, where he particularly enjoyed seeing windmills, steam engines, and the like.

Chebyshev was a remarkably versatile mathematician with a rare talent for solving difficult problems by using elementary methods. Most of his effort went into pure mathematics, but he also valued practical applications of his subject, as the following remark suggests: "To isolate mathematics from the practical

---

[30] Those readers who are blessed with indomitable skepticism, and rightly refuse to accept assurances of this kind without personal investigation, are invited to consult N. I. Achieser, *Theory of Approximation,* Ungar, New York, 1956; E. W. Cheney, *Introduction to Approximation Theory,* McGraw-Hill, New York, 1966; or G. G. Lorentz, *Approximation of Functions,* Holt, New York, 1966.

demands of the sciences is to invite the sterility of a cow shut away from the bulls." He worked in many fields, but his most important achievements were in probability, the theory of numbers, and the approximation of functions (to which he was led by his interest in mechanisms).

In probability, he introduced the concepts of mathematical expectation and variance for sums and arithmetic means of random variables, gave a beautifully simple proof of the law of large numbers based on what is now known as Chebyshev's inequality, and worked extensively on the central limit theorem. He is regarded as the intellectual father of a long series of well-known Russian scientists who contributed to the mathematical theory of probability, including A. A. Markov, S. N. Bernstein, A. N. Kolmogorov, A. Y. Khinchin, and others.

In the late 1840s Chebyshev helped to prepare an edition of some of the works of Euler. It appears that this task caused him to turn his attention to the theory of numbers, particularly to the very difficult problem of the distribution of primes. As the reader probably knows, a *prime number* is an integer $p > 1$ that has no positive divisors except 1 and p. The first few are easily seen to be 2, 3, 5, 7, 11, 13, 17, 19, 23, 29, 31, 37, 41, 43, ... It is clear that the primes are distributed among all the positive integers in a rather irregular way; for as we move out, they seem to occur less and less frequently, and yet there are many adjoining pairs separated by a single even number. The problem of discovering the law governing their occurrence—and of understanding the reasons for it—is one that has challenged the curiosity of men for hundreds of years. In 1751 Euler expressed his own bafflement in these words: "Mathematicians have tried in vain to this day to discover some order in the sequence of prime numbers, and we have reason to believe that it is a mystery into which the human mind will never penetrate."

Many attempts have been made to find simple formulas for the nth prime and for the exact number of primes among the first n positive integers. All such efforts have failed, and real progress was achieved only when mathematicians started instead to look for information about the average distribution of the primes among the positive integers. It is customary to denote by $\pi(x)$ the number of primes less than or equal to a positive number x. Thus $\pi(1) = 0$, $\pi(2) = 1$, $\pi(3) = 2$, $\pi(\pi) = 2$, $\pi(4) = 2$, and so on. In his early youth Gauss studied $\pi(x)$ empirically, with the aim of finding a simple function that seems to approximate it with a small relative error for large x. On the basis of his observations he conjectured (perhaps at the age of fourteen or fifteen) that $x/\log x$ is a good approximating function, in the sense that

$$\lim_{x \to \infty} \frac{\pi(x)}{x/\log x} = 1. \tag{18}$$

This statement is the famous *prime number theorem*; and as far as anyone knows, Gauss was never able to support his guess with even a fragment of proof.

Chebyshev, unaware of Gauss's conjecture, was the first mathematician to establish any firm conclusions about this question. In 1848 and 1850 he proved that

$$0.9213 \ldots < \frac{\pi(x)}{x/\log x} < 1.1055 \ldots \tag{19}$$

for all sufficiently large x, and also that if the limit in (18) exists, then its value must be 1.[31] As a by-product of this work, he also proved Bertrand's postulate: for every integer $n \geq 1$ there is a prime p such that $n < p \leq 2n$. Chebyshev's efforts did not bring him to a final proof of the prime number theorem (this came in 1896), but they did stimulate many other mathematicians to continue working on the problem. We shall return to this subject in Appendix E, in our note on Riemann.

APPENDIX E. RIEMANN'S EQUATION

Our purpose in this appendix is to understand the structure of Gauss's hypergeometric equation

$$x(1 - x)y'' + [c - (a + b + 1)x]y' - aby = 0. \tag{1}$$

In Sections 31 and 32 we saw that this equation has exactly three regular singular points $x = 0$, $x = 1$, and $x = \infty$, and also that at least one exponent has the value 0 at each of the points $x = 0$ and $x = 1$. We shall prove that (1) is fully determined by these properties, in the sense that if we make these assumptions about the general equation

$$y'' + P(x)y' + Q(x)y = 0, \tag{2}$$

then (2) necessarily has the form (1).

We begin by recalling from Section 32 that if the independent variable in (2) is changed from x to $t = 1/x$, then (2) becomes

$$y'' + \left[\frac{2}{t} - \frac{P(1/t)}{t^2} \right]y' + \frac{Q(1/t)}{t^4} y = 0, \tag{3}$$

where the primes denote derivatives with respect to t. It is clear from (3) that the point $x = \infty$ is a regular singular point of (2) if it is not an ordinary point and the functions

$$\frac{1}{t} P\left(\frac{1}{t}\right) \quad \text{and} \quad \frac{1}{t^2} Q\left(\frac{1}{t}\right)$$

are both analytic at $t = 0$.

We now explicitly assume that (2) has $x = 0$, $x = 1$, and $x = \infty$ as regular singular points and that all other points are ordinary. It follows that $xP(x)$ is analytic at $x = 0$, that $(x - 1)P(x)$ is analytic at $x = 1$, and that $x(x - 1)P(x)$ is analytic for all finite values of x:

$$x(x - 1)P(x) = \sum_{n=0}^{\infty} a_n x^n. \tag{4}$$

---

[31] The number on the left side of (19) is $A = \log 2^{\frac{1}{2}} 3^{\frac{1}{3}} 5^{\frac{1}{5}} 30^{-\frac{1}{30}}$, and that on the right is $\frac{6}{5}A$.

If we substitute $x = 1/t$, then (4) becomes

$$\frac{1}{t}\left(\frac{1}{t} - 1\right)P\left(\frac{1}{t}\right) = \sum_{n=0}^{\infty} a_n\left(\frac{1}{t}\right)^n,$$

so

$$\frac{1}{t}P\left(\frac{1}{t}\right) = \frac{t}{1-t}\sum_{n=0}^{\infty} a_n\left(\frac{1}{t}\right)^n = \frac{1}{1-t}\left(a_0 t + a_1 + \frac{a_2}{t} + \cdots\right).$$

Since $x = \infty$ is a regular singular point of (2), this function must be analytic at $t = 0$. We conclude that $a_2 = a_3 = \cdots = 0$, so (4) yields

$$P(x) = \frac{a_0 + a_1 x}{x(x-1)} = \frac{A}{x} + \frac{B}{x-1} \tag{5}$$

for certain constants A and B. Similarly, $x^2(x-1)^2 Q(x)$ is analytic for all finite values of x, so

$$x^2(x-1)^2 Q(x) = \sum_{n=0}^{\infty} b_n x^n,$$

$$\frac{1}{t^2}\left(\frac{1}{t} - 1\right)^2 Q\left(\frac{1}{t}\right) = \sum_{n=0}^{\infty} b_n\left(\frac{1}{t}\right)^n,$$

and

$$\frac{1}{t^2}Q\left(\frac{1}{t}\right) = \frac{t^2}{(1-t)^2}\sum_{n=0}^{\infty} b_n\left(\frac{1}{t}\right)^n$$

$$= \frac{1}{(1-t)^2}\left(b_0 t^2 + b_1 t + b_2 + \frac{b_3}{t} + \cdots\right). \tag{6}$$

As before, the assumption that $x = \infty$ is a regular singular point of (2) implies that (6) must be analytic at $t = 0$, so $b_3 = b_4 = \cdots = 0$ and

$$Q(x) = \frac{b_0 + b_1 x + b_2 x^2}{x^2(x-1)^2} = \frac{C}{x} + \frac{D}{x^2} + \frac{E}{x-1} + \frac{F}{(x-1)^2}. \tag{7}$$

Now the fact that (6) is bounded near $t = 0$ means that $x^2 Q(x)$ is bounded for large x, so

$$x^2\left(\frac{C}{x} + \frac{E}{x-1}\right) = x^2\left[\frac{(C+E)x - C}{x(x-1)}\right]$$

is also bounded and $C + E = 0$. This enables us to write (7) as

$$Q(x) = \frac{D}{x^2} + \frac{F}{(x-1)^2} - \frac{C}{x(x-1)}; \tag{8}$$

and in view of (5) and (8), equation (2) takes the form

$$y'' + \left(\frac{A}{x} + \frac{B}{x-1}\right)y' + \left[\frac{D}{x^2} + \frac{F}{(x-1)^2} - \frac{C}{x(x-1)}\right]y = 0. \tag{9}$$

Let the exponents belonging to the regular singular points 0, 1, and ∞ be denoted by α_1 and α_2, β_1 and β_2, γ_1 and γ_2, respectively. These numbers are the roots of the indicial equations at these three points:

$$m(m - 1) + Am + D = 0,$$
$$m(m - 1) + Bm + F = 0,$$
$$m(m - 1) + (2 - A - B)m + (D + F - C) = 0.$$

The first two of these equations can be written down directly by inspecting (9), but the third requires a little calculation based on (3). If we write these equtions as

$$m^2 + (A - 1)m + D = 0,$$
$$m^2 + (B - 1)m + F = 0,$$
$$m^2 + (1 - A - B)m + (D + F - C) = 0,$$

then by the well-known relations connecting the roots of a quadratic equation with its coefficients, we obtain

$$
\begin{array}{ll}
\alpha_1 + \alpha_2 = 1 - A, & \alpha_1\alpha_2 = D, \\
\beta_1 + \beta_2 = 1 - B, & \beta_1\beta_2 = F, \\
\gamma_1 + \gamma_2 = A + B - 1, & \gamma_1\gamma_2 = D + F - C.
\end{array}
\qquad (10)
$$

It is clear from the first column that

$$\alpha_1 + \alpha_2 + \beta_1 + \beta_2 + \gamma_1 + \gamma_2 = 1; \qquad (11)$$

and by using (10), we can write (9) in the form

$$y'' + \left(\frac{1 - \alpha_1 - \alpha_2}{x} + \frac{1 - \beta_1 - \beta_2}{x - 1}\right)y'$$
$$+ \left[\frac{\alpha_1\alpha_2}{x^2} + \frac{\beta_1\beta_2}{(x - 1)^2} + \frac{\gamma_1\gamma_2 - \alpha_1\alpha_2 - \beta_1\beta_2}{x(x - 1)}\right]y = 0. \quad (12)$$

This is called *Riemann's equation*, and (11) is known as *Riemann's identity*.

The qualitative content of this remarkable conclusion can be expressed as follows: the precise form of (2) is completely determined by requiring that it have only three regular singular points $x = 0$, $x = 1$, and $x = \infty$ and by specifying the values of its exponents at each of these points.

Let us now impose the additional condition that at least one exponent must have the value 0 at each of the points $x = 0$ and $x = 1$, say $\alpha_1 = \beta_1 = 0$. Then with a little simplification and the aid of (11), Riemann's equation reduces to

$$x(1 - x)y'' + [(1 - \alpha_2) - (\gamma_1 + \gamma_2 + 1)x]y' - \gamma_1\gamma_2 y = 0,$$

which clearly becomes Gauss's equation (1) if we introduce the customary notation $a = \gamma_1$, $b = \gamma_2$, $c = 1 - \alpha_2$. For this reason, equation (12) is sometimes called the *generalized hypergeometric equation*.

These results are merely the first few steps in a far-reaching theory of differential equations initiated by Riemann. One of the aims of this theory is to characterize in as simple a manner as possible all differential equations whose solutions are expressible in terms of Gauss's hypergeometric function. Another is to achieve a systematic classification of all differential equations with rational coefficients according to the number and nature of their singular points. One surprising fact that emerges from this classification is that virtually all such equations arising in mathematical physics can be generated by confluence from a single equation with five regular singular points in which the difference between the exponents at each point is $1/2$.[32]

NOTE ON RIEMANN. No great mind of the past has exerted a deeper influence on the mathematics of the twentieth century than Bernhard Riemann (1826–1866), the son of a poor country minister in northern Germany. He studied the works of Euler and Legendre while he was still in secondary school, and it is said that he mastered Legendre's treatise on the theory of numbers is less than a week. But he was shy and modest, with little awareness of his own extraordinary abilities, so at the age of nineteen he went to the University of Göttingen with the aim of pleasing his father by studying theology and becoming a minister himself. Fortunately this worthy purpose soon stuck in his throat, and with his father's willing permission he switched to mathematics.

The presence of the legendary Gauss automatically made Göttingen the center of the mathematical world. But Gauss was remote and unapproachable—particularly to beginning students—and after only a year Riemann left this unsatisfying environment and went to the University of Berlin. There he attracted the friendly interest of Dirichlet and Jacobi, and learned a great deal from both men. Two years later he returned to Göttingen, where he obtained his doctor's degree in 1851. During the next eight years he endured debilitating poverty and created his greatest works. In 1854 he was appointed Privatdozent (unpaid lecturer), which at that time was the necessary first step on the academic ladder. Gauss died in 1855, and Dirichlet was called to Göttingen at his successor. Dirichlet helped Riemann in every way he could, first with a small salary (about one-tenth of that paid to a full professor) and then with a promotion to an assistant professorship. In 1859 he also died, and Riemann was appointed as a full professor to replace him. Riemann's years of poverty were over, but his health

---

[32] A full understanding of these further developments requires a grasp of the main principles of complex analysis. Nevertheless, a reader without this equipment can glean a few useful impressions from E. T. Whittaker and G. N. Watson, *Modern Analysis*, pp. 203–208, Cambridge University Press, London, 1935; or E. D. Rainville, *Intermediate Differential Equations*, chap. 6, Macmillan, New York, 1964.

was broken. At the age of thirty-nine he died of tuberculosis in Italy, on the last of several trips he undertook in order to escape the cold, wet climate of northern Germany. Riemann had a short life and published comparatively little, but his works permanently altered the course of mathematics in analysis, geometry, and number theory.[33]

His first published paper was his celebrated dissertation of 1851 on the general theory of functions of a complex variable.[34] Riemann's fundamental aim here was to free the concept of an analytic function from any dependence on explicit expressions such as power series, and to concentrate instead on general principles and geometric ideas. He founded his theory on what are now called the Cauchy–Riemann equations, created the ingenious device of Riemann surfaces for clarifying the nature of multiple-valued functions, and was led to the Riemann mapping theorem. Gauss was rarely enthusiastic about the mathematical achievements of his contemporaries, but in his official report to the faculty he warmly praised Riemann's work: "The dissertation submitted by Herr Riemann offers convincing evidence of the author's thorough and penetrating investigations in those parts of the subject treated in the dissertation, of a creative, active, truly mathematical mind, and of a gloriously fertile originality."

Riemann later applied these ideas to the study of hypergeometric and Abelian functions. In his work on Abelian functions he relied on a remarkable combination of geometric reasoning and physical insight, the latter in the form of Dirichlet's principle from potential theory. He used Riemann surfaces to build a bridge between analysis and geometry which made it possible to give geometric expression to the deepest analytic properties of functions. His powerful intuition often enabled him to discover such properties—for instance, his version of the Riemann–Roch theorem—by simply thinking about possible configurations of closed surfaces and performing imaginary physical experiments on these surfaces. Riemann's geometric methods in complex analysis constituted the true beginning of topology, a rich field of geometry concerned with those properties of figures that are unchanged by continuous deformations.

In 1854 he was required to submit a probationary essay in order to be admitted to the position of Privatdozent, and his response was another pregnant work whose influence is indelibly stamped on the mathematics of our own time.[35] The problem he set himself was to analyze Dirichlet's conditions (1829) for the representability of a function by its Fourier series. One of these conditions was that the function must be integrable. But what does this mean? Dirichlet had used Cauchy's definition of integrability, which applies only to functions that are continuous or have at most a finite number of points of discontinuity. Certain

---

[33] His *Gesammelte Mathematische Werke* (reprinted by Dover in 1953) occupy only a single volume, of which two-thirds consists of posthumously published material. Of the nine papers Riemann published himself, only five deal with pure mathematics.

[34] Grundlagen für eine allgemeine Theorie der Functionen einer veränderlichen complexen Grösse, in *Werke*, pp. 3–43.

[35] Ueber die Darstellbarkeit einer Function durch eine trigonometrische Reihe, in *Werke*, pp. 227–264.

functions that arise in number theory suggested to Riemann that this definition should be broadened. He developed the concept of the Riemann integral as it now appears in most textbooks on calculus, established necessary and sufficient conditions for the existence of such an integral, and generalized Dirichlet's criteria for the validity of Fourier expansions. Cantor's famous theory of sets was directly inspired by a problem raised in this paper, and these ideas led in turn to the concept of the Lebesgue integral and even more general types of integration. Riemann's pioneering investigations were therefore the first steps in another new branch of mathematics, the theory of functions of a real variable.

The Riemann rearrangement theorem in the theory of infinite series was an incidental result in the paper just described. He was familiar with Dirichlet's example showing that the sum of a conditionally convergent series can be changed by altering the order of its terms:

$$1 - \frac{1}{2} + \frac{1}{3} - \frac{1}{4} + \frac{1}{5} - \frac{1}{6} + \frac{1}{7} - \frac{1}{8} + \cdots = \log 2, \tag{13}$$

$$1 + \frac{1}{3} - \frac{1}{2} + \frac{1}{5} + \frac{1}{7} - \frac{1}{4} + \cdots = \frac{3}{2}\log 2. \tag{14}$$

It is apparent that these two series have different sums but the same terms; for in (14) the first two positive terms in (13) are followed by the first negative term, then the next two positive terms are followed by the second negative term, and so on. Riemann proved that it is possible to rearrange the terms of any conditionally convergent series in such a manner that the new series will converge to an arbitrary preassigned sum, or diverge to ∞ or $-\infty$.

In addition to his probationary essay, Riemann was also required to present a trial lecture to the faculty before he could be appointed to his unpaid lectureship. It was the custom for the candidate to offer three titles, and the head of his department usually accepted the first. However, Riemann rashly listed as his third topic the foundations of geometry, a profound subject on which he was unprepared but which Gauss had been turning over in his mind for 60 years. Naturally, Gauss was curious to see how this particular candidate's "gloriously fertile originality" would cope with such a challenge, and to Riemann's dismay he designated this as the subject of the lecture. Riemann quickly tore himself away from his other interests at the time—"my investigtions of the connection between electricity, magnetism, light, and gravitation"—and wrote his lecture in the next two months. The result was one of the great classical masterpieces of mathematics, and probably the most important scientific lecture ever given.[36] It is recorded that even Gauss was surprised and enthusiastic.

Riemann's lecture presented in nontechnical language a vast generalization of all known geometries, both Euclidean and non-Euclidean. This field is now called Riemannian geometry; and apart from its great importance in pure

---

[36] Ueber die Hypothesen, Welche der Geometrie zu Grunde liegen, in *Werke*, pp. 272–286. There is a translation in D. E. Smith, *A Source Book in Mathematics*, McGraw-Hill, New York, 1929.

mathematics, it turned out 60 years later to be exactly the right framework for Einstein's general theory of relativity. Like most of the great ideas of science, Riemannian geometry is quite easy to understand if we set aside the technical details and concentrate on its essential features. Let us recall the intrinsic differential geometry of curved surfaces which Gauss had discovered 25 years earlier. If a surface imbedded in three dimensional space is defined parametrically by three functions $x = x(u,v)$, $y = y(u,v)$, and $z = z(u,v)$, then u and v can be interpreted as the coordinates of points on the surface. The distance ds along the surface between two nearby points (u,v) and $(u + du, v + dv)$ is given by Gauss's quadratic differential form

$$ds^2 = E\,du^2 + 2F\,du\,dv + G\,dv^2,$$

where E, F, and G are certain functions of u and v. This differential form makes it possible to calculate the lengths of curves on the surface, to find the geodesic (or shortest) curves, and to compute the Gaussian curvature of the surface at any point—all in total disregard of the surrounding space. Riemann generalized this by discarding the idea of a surrounding Euclidean space and introducing the concept of a continuous n-dimensional manifold of points $(x_1, x_2, \ldots, x_n)$. He then imposed an arbitrarily given distance (or metric) ds between nearby points

$$(x_1, x_2, \ldots, x_n) \quad \text{and} \quad (x_1 + dx_1, x_2 + dx_2, \ldots, x_n + dx_n)$$

by means of a quadratic differential form

$$ds^2 = \sum_{i,j=1}^{n} g_{ij}\,dx_i\,dx_j, \tag{15}$$

where the g_{ij} are suitable functions of $x_1, x_2, \ldots, x_n$ and different systems of g_{ij} define different Riemannian geometries on the manifold under discussion. His next steps were to examine the idea of curvature for these Riemannian manifolds and to investigate the special case of constant curvature. All of this depends on massive computational machinery, which Riemann mercifully omitted from his lecture but included in a posthumous paper on heat conduction. In that paper he explicitly introduced the Riemann curvature tensor, which reduces to the Gaussian curvature when $n = 2$ and whose vanishing he showed to be necessary and sufficient for the given quadratic metric to be equivalent to a Euclidean metric. From this point of view, the curvature tensor measures the deviation of the Riemannian geometry defined by formula (15) from Euclidean geometry. Einstein has summarized these ideas in a single statement: "Riemann's geometry of an n-dimensional space bears the same relation to Euclidean geometry of an n-dimensional space as the general geometry of curved surfaces bears to the geometry of the plane."

The physical significance of geodesics appears in its simplest form as the following consequence of Hamilton's principle in the calculus of variations: if a particle is constrained to move on a curved surface, and if no force acts on it, then it glides along a geodesic.[37] A direct extension of this idea is the heart of the

---

[37] This is proved in Appendix B of Chapter 12.

general theory of relativity, which is essentially a theory of gravitation. Einstein conceived the geometry of space as a Riemannian geometry in which the curvature and geodesics are determined by the distribution of matter; in this curved space, planets move in their orbits around the sun by simply coasting along geodesics instead of being pulled into curved paths by a mysterious force of gravity whose nature no one has ever really understood.

In 1859 Riemann published his only work on the theory of numbers, a brief but exceedingly profound paper of less than 10 pages devoted to the prime number theorem.[38] This mighty effort started tidal waves in several branches of pure mathematics, and its influence will probably still be felt a thousand years from now. His starting point was a remarkable identity discovered by Euler over a century earlier: if s is a real number greater than 1, then

$$\sum_{n=1}^{\infty} \frac{1}{n^s} = \prod_{p} \frac{1}{1 - (1/p^s)}, \tag{16}$$

where the expression on the right denotes the product of the numbers $(1 - p^{-s})^{-1}$ for all primes p. To understand how this identity arises, we note that $1/(1 - x) = 1 + x + x^2 + \cdots$ for $|x| < 1$, so for each p we have

$$\frac{1}{1 - (1/p^s)} = 1 + \frac{1}{p^s} + \frac{1}{p^{2s}} + \cdots.$$

On multiplying these series for all primes p and recalling that each integer $n > 1$ is uniquely expressible as a product of powers of different primes, we see that

$$\prod_{p} \frac{1}{1 - (1/p^s)} = \prod_{p} \left(1 + \frac{1}{p^s} + \frac{1}{p^{2s}} + \cdots \right)$$

$$= 1 + \frac{1}{2^s} + \frac{1}{3^s} + \cdots + \frac{1}{n^s} + \cdots$$

$$= \sum_{n=1}^{\infty} \frac{1}{n^s},$$

which is the identity (16). The sum of the series on the left of (16) is evidently a function of the real variable $s > 1$, and the identity establishes a connection between the behavior of this function and properties of the primes. Euler himself exploited this connection in several ways, but Riemann perceived that access to the deeper features of the distribution of primes can only be gained by allowing s to be a complex variable. He denoted the resulting function by $\zeta(s)$, and it has since been known as the *Riemann zeta function*:

$$\zeta(s) = 1 + \frac{1}{2^s} + \frac{1}{3^s} + \cdots, \qquad s = \sigma + it.$$

---

[38] Ueber die Anzahl der Primzahlen unter einer gegebenen Grösse, in *Werke*, pp. 145–153. See the statement of the prime number theorem in our note on Chebyshev in Appendix D.

In his paper he proved several important properties of this function, and in a sovereign way simply stated a number of others without proof. During the century since his death, many of the finest mathematicians in the world have exerted their strongest efforts and created rich new branches of analysis in attempts to prove these statements. The first success was achieved in 1893 by J. Hadamard, and with one exception every statement has since been settled in the sense Riemann expected.[39] This exception is the famous Riemann hypothesis: that all the zeros of $\zeta(s)$ in the strip $0 \le \sigma \le 1$ lie on the central line $\sigma = \frac{1}{2}$. It stands today as the most important unsolved problem of mathematics, and is probably the most difficult problem that the mind of man has ever conceived. In a fragmentary note found among his posthumous papers, Riemann wrote that these theorems "follow from an expression for the function $\zeta(s)$ which I have not yet simplified enough to publish."[40] Writing about this fragment in 1944, Hadamard remarked with justified exasperation, "We still have not the slightest idea of what the expression could be."[41] He adds the further comment: "In general, Riemann's intuition is highly geometrical; but this is not the case for his memoir on prime numbers, the one in which that intuition is the most powerful and mysterious."

---

[39] Hadamard's work led him to his 1896 proof of the prime number theorem. See E. C. Titchmarsh, *The Theory of the Riemann Zeta Function*, chap. 3, Oxford University Press, London, 1951. This treatise has a bibliography of 326 items.

[40] *Werke*, p. 154.

[41] *The Psychology of Invention in the Mathematical Field*, p. 118, Dover, New York, 1954.

CHAPTER
6

FOURIER SERIES AND ORTHOGONAL FUNCTIONS

33 THE FOURIER COEFFICIENTS

Trigonometric series of the form

$$f(x) = \frac{1}{2}a_0 + \sum_{n=1}^{\infty} (a_n \cos nx + b_n \sin nx) \qquad (1)$$

are needed in the treatment of many physical problems that lead to partial differential equations, for instance, in the theory of sound, heat conduction, electromagnetic waves, and mechanical vibrations.[1] We shall examine some of these applications in the next chapter. The representation of functions by power series is familiar to us from calculus and also from our work in the preceding chapter. An important advantage of the series (1) is that it can represent very general functions with many discontinuities—like the discontinuous "impulse" functions of electrical engineering—whereas power series can represent only continuous functions that have derivatives of all orders.

---

[1] It is only for reasons of convenience that the constant term in (1) is written $\frac{1}{2}a_0$ instead of a_0. This will become clear below.

Aside from the great practical value of trigonometric series for solving problems in physics and engineering, the purely theoretical part of this subject has had a profound influence on the general development of mathematical analysis over the past 250 years. Specifically, it provided the main driving force behind the evolution of the modern notion of function, which in all its ramifications is certainly the central concept of mathematics; it led Riemann and Lebesgue to create their successively more powerful theories of integration, and Cantor his theory of sets; it led Weierstrass to his critical study of the real number system and the properties of continuity and differentiability for functions; and it provided the context within which the geometric idea of orthogonality (perpendicularity) was able to develop into one of the major unifying concepts of modern analysis. We shall comment further on all of these matters throughout this chapter.

We begin our treatment with some classical calculations that were first performed by Euler. Our point of view is that the function $f(x)$ in (1) is defined on the closed interval $-\pi \leq x \leq \pi$, and we must find the coefficients a_n and b_n in the series expansion. It is convenient to assume, temporarily, that the series is uniformly convergent, because this implies that the series can be integrated term by term from $-\pi$ to π.[2]

Since

$$\int_{-\pi}^{\pi} \cos nx \, dx = 0 \quad \text{and} \quad \int_{-\pi}^{\pi} \sin nx \, dx = 0 \tag{2}$$

for $n = 1, 2, \ldots$, the term-by-term integration yields

$$\int_{-\pi}^{\pi} f(x) \, dx = a_0 \pi,$$

so

$$a_0 = \frac{1}{\pi} \int_{-\pi}^{\pi} f(x) \, dx. \tag{3}$$

It is worth noticing here that formula (3) shows that the constant term $\frac{1}{2} a_0$ in (1) is simply the average value of $f(x)$ over the interval. The coefficient a_n is found in a similar way. Thus, if we multiply (1) by $\cos nx$ the result is

$$f(x) \cos nx = \tfrac{1}{2} a_0 \cos nx + \cdots + a_n \cos^2 nx + \cdots, \tag{4}$$

---

[2] Readers who are not acquainted with the concept of uniform convergence can freely integrate the series term by term anyway—as Euler and his contemporaries did without a qualm—as long as they realize that this operation is not always legitimate and ultimately needs theoretical justification.

where the terms not written contain products of the form $\sin mx \cos nx$ or of the form $\cos mx \cos nx$ with $m \neq n$. At this point it is necessary to recall the trigonometric identities

$$\sin mx \cos nx = \tfrac{1}{2}[\sin (m + n)x + \sin (m - n)x],$$
$$\cos mx \cos nx = \tfrac{1}{2}[\cos (m + n)x + \cos (m - n)x],$$
$$\sin mx \sin nx = \tfrac{1}{2}[\cos (m - n)x - \cos (m + n)x],$$

which follow directly from the addition and subtraction formulas for the sine and cosine. It is now easy to verify that for integral values of m and $n \geq 1$ we have

$$\int_{-\pi}^{\pi} \sin mx \cos nx \, dx = 0 \tag{5}$$

and

$$\int_{-\pi}^{\pi} \cos mx \cos nx \, dx = 0, \qquad m \neq n. \tag{6}$$

These facts enable us to integrate (4) term by term and obtain

$$\int_{-\pi}^{\pi} f(x) \cos nx \, dx = a_n \int_{-\pi}^{\pi} \cos^2 nx \, dx = a_n\pi,$$

so

$$a_n = \frac{1}{\pi} \int_{-\pi}^{\pi} f(x) \cos nx \, dx. \tag{7}$$

By (3), formula (7) is also valid for $n = 0$; this is the reason for writing the constant term in (1) as $\tfrac{1}{2}a_0$ rather than a_0. We get the corresponding formula for b_n by essentially the same procedure—we multiply (1) through by $\sin nx$, integrate term by term, and use the additional fact that

$$\int_{-\pi}^{\pi} \sin mx \sin nx \, dx = 0, \qquad m \neq n. \tag{8}$$

This yields

$$\int_{-\pi}^{\pi} f(x) \sin nx \, dx = b_n \int_{-\pi}^{\pi} \sin^2 nx \, dx = b_n\pi,$$

so

$$b_n = \frac{1}{\pi} \int_{-\pi}^{\pi} f(x) \sin nx \, dx. \tag{9}$$

These calculations show that if the series (1) is uniformly convergent, then the coefficients a_n and b_n can be obtained from the sum $f(x)$ by means of the above formulas. However, this situation is too restricted to be of much practical value, because how do we know whether a given $f(x)$ admits an expansion as a uniformly convergent trigonometric series?

We don't—and for this reason it is better to set aside the idea of *finding* the coefficients a_n and b_n in an expansion (1) that may or may not exist, and instead use formulas (7) and (9) to *define* certain numbers a_n and b_n that are then used to construct the trigonometric series (1). When this is done, these a_n and b_n are called the *Fourier coefficients* of the function $f(x)$, and the series (1) is called the *Fourier series* of $f(x)$. A Fourier series is thus a special kind of trigonometric series—one whose coefficients are obtained by applying formulas (7) and (9) to some given function $f(x)$. In order to form this series, it is not necessary to assume that $f(x)$ is continuous, but only that the integrals (7) and (9) exist; and for this it suffices to assume that $f(x)$ is integrable on the interval $-\pi \le x \le \pi$.[3]

Of course, we hope that the Fourier series of $f(x)$ will converge and have $f(x)$ for its sum, and that therefore (1) will constitute a valid representation or expansion of this function. Unfortunately, however, this is not always true, for there exist many integrable—even continuous—functions whose Fourier series diverge at one or more points. Advanced treatises on Fourier series usually replace the equals sign in (1) by the symbol $\sim$, in order to emphasize that the series on the right is the Fourier series of the function on the left but that the series is not necessarily convergent. We shall continue to use the equals sign because the series obtained in this book actually do converge for every value of x.

Just as being a Fourier series does not imply convergence, convergence for a trigonometric series does not imply that it is a Fourier series. For example, it is known that

$$\sum_{n=1}^{\infty} \frac{\sin nx}{\log (1 + n)} \tag{10}$$

converges for every value of x, and yet this series is known not to be a Fourier series.[4] This means that the coefficients in (10) cannot be obtained by applying formulas (7) and (9) to *any* integrable function $f(x)$, not even if we make the obvious choice and take $f(x)$ to be the function that is the sum of the series.

These surprising phenomena prevent the theory of Fourier series

---

[3] In this context "integrable" means "Riemann integrable," which is defined in terms of upper sums and lower sums and is the standard concept used in most calculus courses.

[4] For convergence, see Problem 2(a) in Appendix C.12 of George F. Simmons, *Calculus With Analytic Geometry*, McGraw-Hill, New York, 1985. The fact that (10) is not a Fourier series is a consequence of the remarkable theorem that the term-by-term integral of any Fourier series (whether convergent or not) must converge for all x—and this is not true for (10).

from being at all simple or straightforward, but they also render it extraordinarily fascinating to mathematicians. The fundamental problem of the subject is clearly to discover properties of an integrable function that guarantee that its Fourier series not only converges, but also converges to the function. We shall state such properties in the next section, but first it is desirable to gain some direct, hands-on experience with the calculation of Fourier series for particular functions.

Example 1. Find the Fourier series of the function $f(x) = x$, $-\pi \le x \le \pi$. First, by (3) we have

$$a_0 = \frac{1}{\pi} \int_{-\pi}^{\pi} x \, dx = \frac{1}{\pi} \cdot \frac{x^2}{2} \Big]_{-\pi}^{\pi} = 0.$$

If $n \ge 1$, then we find a_n by using (7) and integrating by parts with $u = x$, $dv = \cos nx \, dx$,

$$a_n = \frac{1}{\pi} \int_{-\pi}^{\pi} x \cos nx \, dx = \frac{1}{\pi} \left[\frac{x \sin nx}{n} + \frac{\cos nx}{n^2} \right]_{-\pi}^{\pi}$$
$$= 0;$$

and using (9) with $u = x$, $dv = \sin nx \, dx$ gives

$$b_n = \frac{1}{\pi} \int_{-\pi}^{\pi} x \sin nx \, dx = \frac{1}{\pi} \left[-\frac{x \cos nx}{n} + \frac{\sin nx}{n^2} \right]_{-\pi}^{\pi}$$
$$= \frac{1}{\pi} \left[-\frac{\pi \cos n\pi}{n} - \frac{\pi \cos(-n\pi)}{n} \right]$$
$$= -\frac{2}{n} \cos n\pi = \frac{2}{n} (-1)^{n+1},$$

since $\cos n\pi = (-1)^n$. Now, substituting these results in (1) suggests that

$$x = 2 \left(\sin x - \frac{\sin 2x}{2} + \frac{\sin 3x}{3} - \cdots \right). \tag{11}$$

It should be clearly understood that the use of the equals sign here is an expression of hope rather than definite knowledge.

In Appendix A we prove that the series (11) converges to x for $-\pi < x < \pi$. To discuss the convergence behavior of the series outside this interval, we introduce the concept of periodicity. A function $f(x)$ is said to be *periodic* if $f(x + p) = f(x)$ for all values of x, where p is a positive constant.[5] Any positive number p with this property is called a *period* of $f(x)$; for instance, $\sin x$ in (11) has periods $2\pi, 4\pi, \ldots$, and $\sin 2x$ has periods $\pi, 2\pi, \ldots$.

---

[5] It follows that we also have $f(x - p) = f(x)$, as can be seen by replacing x by $x - p$ in the above equation.

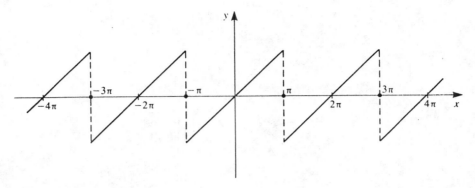

FIGURE 33

It is easy to see that each term of the series (11) has period 2π—in fact, 2π is the smallest period common to all the terms—so the sum also has period 2π. This means that the known graph of the sum between $-\pi$ and π is simply repeated on each successive interval of length 2π to the right and left. The graph of the sum therefore has the sawtooth appearance shown in Fig. 33. It is clear from this that the sum of the series is equal to x only on the interval $-\pi < x < \pi$, and not on the entire real line $-\infty < x < \infty$.

It remains to describe what happens at the points $x = \pm\pi$, $\pm 3\pi, \ldots$, where the sum of the series as shown in the figure has a sudden jump from $-\pi$ to $+\pi$. By putting $x = \pm\pi, \pm 3\pi, \ldots$ in (11), we see that every term of the series is zero. Therefore the sum is also zero, and we show this fact in the figure by putting a dot at these points.

The first four terms of the series (11) are

$$2 \sin x, \qquad -\sin 2x, \qquad \tfrac{2}{3} \sin 3x, \qquad -\tfrac{1}{2} \sin 4x.$$

These and the next two terms are sketched as the numbered curves in Fig. 34. The sum of the four terms listed above is

$$y = 2 \sin x - \sin 2x + \tfrac{2}{3} \sin 3x - \tfrac{1}{2} \sin 4x. \tag{12}$$

Since this is a partial sum of the Fourier series, and the series converges to x for $-\pi < x < \pi$, we expect the partial sum (12) to approximate the function $y = x$ on this interval. The accuracy of the approximation is indicated by the upper curves in Fig. 34, which show this partial sum of four terms and also the sums of six and ten terms. As the number of terms increases, the approximating curves approach $y = x$ for each fixed x on the interval $-\pi < x < \pi$, but not for $x = \pm\pi$.

Example 2. Find the Fourier series of the function defined by

$$f(x) = 0, \qquad -\pi \le x < 0;$$
$$f(x) = \pi, \qquad 0 \le x \le \pi.$$

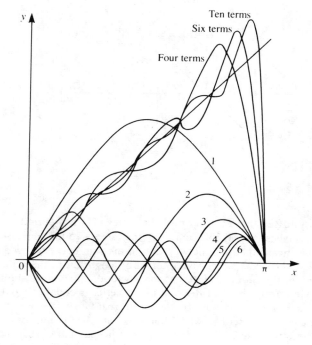

FIGURE 34

By (3), (7) and (9) we have

$$a_0 = \frac{1}{\pi}\left[\int_{-\pi}^{0} 0\, dx + \int_{0}^{\pi} \pi\, dx\right] = \pi;$$

$$a_n = \frac{1}{\pi}\int_{0}^{\pi} \pi \cos nx\, dx = 0, \qquad n \geq 1;$$

$$b_n = \frac{1}{\pi}\int_{0}^{\pi} \pi \sin nx\, dx = \frac{1}{n}(1 - \cos n\pi)$$

$$= \frac{1}{n}[1 - (-1)^n].$$

Since the nth even number is $2n$ and the nth odd number is $2n - 1$, the last of these formulas tells us that

$$b_{2n} = 0, \qquad b_{2n-1} = \frac{2}{2n - 1}.$$

By substituting in (1) we obtain the required Fourier series,

$$f(x) = \frac{\pi}{2} + 2\left(\sin x + \frac{\sin 3x}{3} + \frac{\sin 5x}{5} + \cdots\right). \tag{13}$$

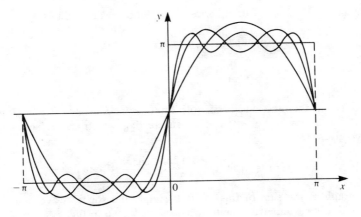

FIGURE 35

The successive partial sums are

$$y = \frac{\pi}{2}, \qquad y = \frac{\pi}{2} + 2 \sin x, \qquad y = \frac{\pi}{2} + 2 \sin x + \frac{2}{3} \sin 3x, \ldots$$

The first four of these are sketched in Fig. 35, together with the graph of $y = f(x)$.

    We will see in the next section that the series (13) converges to the function $f(x)$ on the subintervals $-\pi < x < 0$ and $0 < x < \pi$, but not at the points 0, π, $-\pi$. The sum of the series (13) is clearly periodic with period 2π, and therefore the graph of this sum has the square wave appearance shown in Fig. 36, with a jump from 0 to π at each point $x = 0, \pm\pi, \pm 2\pi, \ldots$. Further, this sum evidently has the value $\pi/2$ at each of these points of discontinuity, and we indicate this fact in the figure as we did before, by placing a dot at each of the points in question. And just as before, each dot is halfway between the limit of the function as we approach the point of discontinuity from the left and the limit from the right.

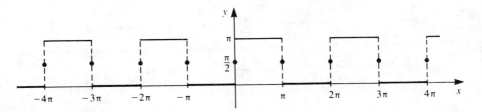

FIGURE 36

Example 3. Find the Fourier series of the function defined by

$$f(x) = -\frac{\pi}{2}, \qquad -\pi \le x < 0;$$

$$f(x) = \frac{\pi}{2}, \qquad 0 \le x \le \pi.$$

This is the function in Example 2 minus the constant $\pi/2$. Its Fourier series can therefore be obtained by subtracting $\pi/2$ from the series (13), which gives

$$f(x) = 2\left(\sin x + \frac{\sin 3x}{3} + \frac{\sin 5x}{5} + \cdots\right). \tag{14}$$

The graph of the sum of this series is simply the square wave in Fig. 36 lowered to be symmetric about the x-axis, as shown in Fig. 37.

Example 4. Find the Fourier series of the function defined by

$$f(x) = -\frac{\pi}{2} - \frac{1}{2}x, \qquad -\pi \le x < 0;$$

$$f(x) = \frac{\pi}{2} - \frac{1}{2}x, \qquad 0 \le x \le \pi.$$

This is the function defined in Example 3 minus one-half the function in Example 1. The Fourier series can therefore be obtained by subtracting one-half the series (11) term by term from the series (14):

$$\begin{aligned}
f(x) &= 2\left(\sin x + \frac{\sin 3x}{3} + \frac{\sin 5x}{5} + \cdots\right) \\
&\quad - \left(\sin x - \frac{\sin 2x}{2} + \frac{\sin 3x}{3} - \cdots\right) \\
&= \sin x + \frac{\sin 2x}{2} + \frac{\sin 3x}{3} + \cdots = \sum_{n=1}^{\infty} \frac{\sin nx}{n}.
\end{aligned} \tag{15}$$

The graph of the sum of this series is the sawtooth wave shown in Fig. 38.

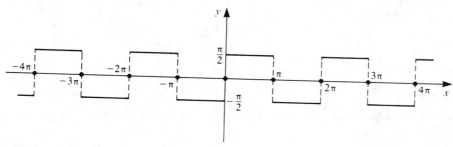

FIGURE 37

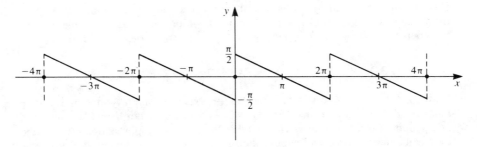

FIGURE 38

The validity of the procedures used in Examples 3 and 4 depends on the easily verified fact that the operation of forming the Fourier coefficients is linear; that is, the coefficients for the sum $f(x) + g(x)$ are the sums of the respective coefficients for $f(x)$ and for $g(x)$, and if c is any constant, then the coefficients for $cf(x)$ are c times the coefficients for $f(x)$. Also, the Fourier series of a constant function is simply the constant itself.

Remark 1. In Section 36 we show how the interval $-\pi \leq x \leq \pi$ of length 2π can be replaced by an interval of arbitrary length, with no difficulty except for a slight loss of simplicity in the formulas. This extension of the ideas is necessary for many of the applications to science.

Remark 2. Our work in this section—and throughout this chapter—rests on the property of *orthogonality* for the system of functions

$$1, \qquad \cos nx, \qquad \sin nx \qquad (n = 1, 2, \dots)$$

over the interval $-\pi \leq x \leq \pi$. This means that the integral of the product of any two of these functions over the interval is zero—which is precisely the substance of equations (2), (5), (6) and (8). We shall return to this concept in Sections 37 and 38 and use it to give a simple and satisfying geometric structure to the theory of Fourier series.

NOTE ON FOURIER. Jean Baptiste Joseph Fourier (1768–1830), an excellent mathematical physicist, was a friend of Napoleon (so far as such people have friends) and accompanied his master to Egypt in 1798. On his return he became prefect of the district of Isère in southeastern France, and in this capacity built the first real road from Grenoble to Turin. He also befriended the boy Champollion, who later deciphered the Rosetta Stone as the first long step toward understanding the hieroglyphic writing of the ancient Egyptians.

During these years he worked on the theory of the conduction of heat, and in 1822 published his famous *Théorie Analytique de la Chaleur,* in which he made extensive use of the series that now bear his name. These series were of profound significance in connection with the evolution of the concept of a function. The general attitude at that time was co call $f(x)$ a function if it could be represented by a single expression like a polynomial, a finite combination of elementary functions, a power series $\sum_{n=0}^{\infty} a_n x^n$, or a trigonometric series of the form

$$\tfrac{1}{2}a_0 + \sum_{n=1}^{\infty} (a_n \cos nx + b_n \sin nx).$$

If the graph of $f(x)$ were "arbitrary"—for example, a polygonal line with a number of corners and even a few gaps—then $f(x)$ would not have been accepted as a genuine function. Fourier claimed that "arbitrary" graphs can be represented by trigonometric series and should therefore be treated as legitimate functions, and it came as a shock to many that he turned out to be right. It was a long time before these issues were completely clarified, and it was no accident that the definition of a function that is now almost universally used was first formulated by Dirichlet in 1837 in a research paper on the theory of Fourier series. Also, the classical definition of the definite integral due to Riemann was first given in his fundamental paper of 1854 on the subject of Fourier series. Indeed, many of the most important mathematical discoveries of the nineteenth century are directly linked to the theory of Fourier series, and the applications of this subject to mathematical physics have been scarcely less profound.

Fourier himself is one of the fortunate few: his name has become rooted in all civilized languages as an adjective that is well known to physical scientists and mathematicians in every part of the world.

PROBLEMS

1. Find the Fourier series for the function defined by

$$f(x) = \pi, \qquad -\pi \le x \le \frac{\pi}{2};$$

$$f(x) = 0, \qquad \frac{\pi}{2} < x \le \pi.$$

2. Find the Fourier series for the function defined by

$$f(x) = \begin{cases} 0, & -\pi \le x < 0; \\ 1, & 0 \le x \le \frac{\pi}{2}; \\ 0, & \frac{\pi}{2} < x \le \pi. \end{cases}$$

3. Find the Fourier series for the function defined by

$$f(x) = 0, \qquad -\pi \le x < 0;$$
$$f(x) = \sin x, \qquad 0 \le x \le \pi.$$

4. Solve Problem 3 with $\sin x$ replaced by $\cos x$.

5. Find the Fourier series for the function defined by
  (a) $f(x) = \pi$, $-\pi \le x \le \pi$;
  (b) $f(x) = \sin x$, $-\pi \le x \le \pi$;
  (c) $f(x) = \cos x$, $-\pi \le x \le \pi$;
  (d) $f(x) = \pi + \sin x + \cos x$, $-\pi \le x \le \pi$.
  Pay special attention to the reasoning used to establish your conclusions, including the possibility of alternate lines of thought.

Solve Problems 6 and 7 by using the methods of Examples 3 and 4, without actually calculating the Fourier coefficients.

6. Find the Fourier series for the function defined by
  (a) $f(x) = -a$, $-\pi \le x < 0$ and $f(x) = a$, $0 \le x \le \pi$ (*a* is a positive number);
  (b) $f(x) = -1$, $-\pi \le x < 0$ and $f(x) = 1$, $0 \le x \le \pi$;
  (c) $f(x) = -\dfrac{\pi}{4}$, $-\pi \le x < 0$ and $f(x) = \dfrac{\pi}{4}$, $0 \le x \le \pi$;
  (d) $f(x) = -1$, $-\pi \le x < 0$ and $f(x) = 2$, $0 \le x \le \pi$;
  (e) $f(x) = 1$, $-\pi \le x < 0$ and $f(x) = 2$, $0 \le x \le \pi$.
7. Obtain the Fourier series for the function in Problem 2 from the result of Problem 1. *Hint*: Begin by forming $\pi - $ (the function in Example 2).
8. Without using Fourier series at all, show graphically that the sawtooth wave of Fig. 33 can be represented as the sum of a sawtooth wave of period π and a square wave of period 2π.

34 THE PROBLEM OF CONVERGENCE

The examples and problems in Section 33 illustrate several features that are characteristic of Fourier series in general and which we now discuss from a general point of view. Our purpose is to attain a good understanding of a useful set of conditions that will guarantee that the Fourier series of a function not only converges, but also converges to the function.

We begin by pointing out that each term of the series

$$f(x) = \frac{1}{2}a_0 + \sum_1^\infty (a_n \cos nx + b_n \sin nx) \tag{1}$$

has period 2π, and therefore, if the function $f(x)$ is to be represented by the sum, $f(x)$ must also have period 2π. Whenever we consider a series like (1), we shall assume that $f(x)$ is initially given on the basic interval $-\pi \le x < \pi$ or $-\pi < x \le \pi$, and that for other values of x, $f(x)$ is defined by the periodicity condition

$$f(x + 2\pi) = f(x). \tag{2}$$

In particular, (2) requires that we must always have $f(\pi) = f(-\pi)$.

Accordingly, the complete function we consider is the so-called "periodic extension" of the originally given part to the successive intervals of length 2π that lie to the right and left of the basic interval.

The phrase *simple discontinuity* (or often *jump discontinuity*) is used to describe the situation where a function has a finite jump at a point $x = x_0$. This means that $f(x)$ approaches finite but different limits from the left side of x_0 and from the right side, as shown in Fig. 39. We can express this behavior by writing

$$\lim_{\epsilon \to 0} f(x_0 - \epsilon) \neq \lim_{\epsilon \to 0} f(x_0 + \epsilon), \qquad \epsilon > 0,$$

where it is understood that both limits exist and are finite. It will be convenient to denote these limits by the simpler symbols $f(x_0-)$ and $f(x_0+)$, so that the above inequality can be written as

$$f(x_0-) \neq f(x_0+).$$

A function $f(x)$ is said to be *bounded* if an inequality of the form

$$|f(x)| \leq M$$

holds for some constant M and all x under consideration. For example, the functions x^2, e^x and $\sin x$ are bounded on $-\pi \leq x < \pi$, but $f(x) = 1/(\pi - x)$ is not. It can be proved (see Problem 7 below) that if a bounded function $f(x)$ has only a finite number of discontinuities and only a finite number of maxima and minima, then all its discontinuities

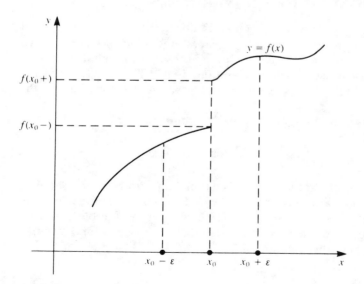

FIGURE 39

are simple. This means that $f(x-)$ and $f(x+)$ exist at every point x, and points of continuity are those for which $f(x-) = f(x+)$.

Each of the functions shown in Figs. 33, 36, 37 and 38 satisfies these conditions on every finite interval. However, the function defined by

$$f(x) = \sin\frac{1}{x} \qquad (x \neq 0), \qquad f(0) = 0$$

has infinitely many maxima near $x = 0$, and the discontinuity at $x = 0$ is not simple [Fig. 40(a)]. The functions defined by

$$g(x) = x\sin\frac{1}{x} \qquad (x \neq 0), \qquad g(0) = 0$$

and

$$h(x) = x^2\sin\frac{1}{x} \qquad (x \neq 0), \qquad h(0) = 0$$

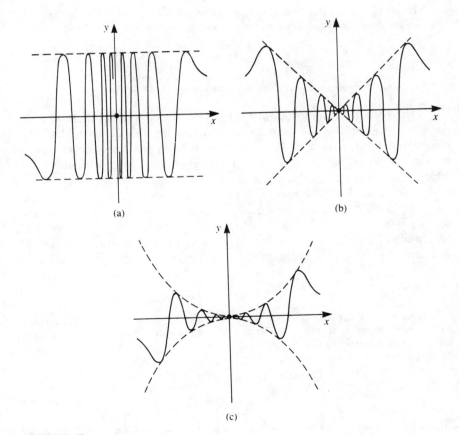

(a)

(b)

(c)

FIGURE 40

also have infinitely many maxima near $x = 0$ [Figs. 40(b) and 40(c)], but both are continuous at $x = 0$ whereas only $h(x)$ is differentiable at this point.

We are now in a position to state the following theorem, which establishes the desired convergence behavior for a very large class of functions.

Dirichlet's Theorem. *Assume that $f(x)$ is defined and bounded for $-\pi \le x < \pi$, and also that it has only a finite number of discontinuities and only a finite number of maxima and minima on this interval. Let $f(x)$ be defined for other values of x by the periodicity condition $f(x + 2\pi) = f(x)$. Then the Fourier series of $f(x)$ converges to*

$$\frac{1}{2}[f(x-) + f(x+)]$$

at every point x, and therefore it converges to $f(x)$ at every point of continuity of the function. Thus, if at every point of discontinuity the value of the function is redefined as the average of its two one-sided limits there,

$$f(x) = \frac{1}{2}[f(x-) + f(x+)],$$

then the Fourier series represents the function everywhere.[6]

The conditions imposed on $f(x)$ in this theorem are called *Dirichlet conditions*, after the German mathematician P. G. L. Dirichlet who discovered the theorem in 1829. In Appendix A we establish the same conclusion under slightly different hypotheses—piecewise smoothness— which are still sufficiently weak to cover almost all applications.[7]

The general situation is as follows: The continuity of a function is not *sufficient* for the convergence of its Fourier series to the function, and neither is it necessary.[8] That is, it is quite possible for a discontinuous function to be represented everywhere by its Fourier series, provided its discontinuities are relatively mild, and provided it is relatively well-

---

[6] We remind the reader that the value of an integrable function can be redefined at any finite number of points without changing the value of its integral, and therefore without changing the Fourier series of the function.

[7] Proofs of Dirichlet's theorem in a slightly more general form can be found in E. C. Titchmarsh, *The Theory of Functions*, 2d ed., Oxford University Press, 1950, pp. 406–407; in W. Rogosinski, *Fourier Series*, Chelsea, New York, 1950, pp. 72–74; and in Béla Sz.-Nagy, *Introduction to Real Functions and Orthogonal Expansions*, Oxford University Press, 1965, pp. 399–402.

[8] It is a major unsolved problem of mathematics to find conditions that are both necessary and sufficient.

behaved between the points of discontinuity. In Dirichlet's theorem above, the discontinuities are simple and the graph consists of a finite number of increasing or decreasing continuous pieces; and in the theorem we prove in Appendix A, the discontinuities are again simple and the graph consists of a finite number of continuous pieces with continuously turning tangents.

Example. Find the Fourier series of the periodic function defined by

$$f(x) = 0, \qquad -\pi \le x < 0;$$
$$f(x) = x, \qquad 0 \le x < \pi.$$

First, we have

$$a_0 = \frac{1}{\pi} \int_0^\pi x \, dx = \frac{1}{\pi} \cdot \frac{x^2}{2}\bigg]_0^\pi = \frac{\pi}{2}.$$

For $n \ge 1$, we integrate by parts to obtain

$$a_n = \frac{1}{\pi} \int_0^\pi x \cos nx \, dx = \frac{1}{\pi}\left[\frac{x \sin nx}{n} + \frac{\cos nx}{n^2} \right]_0^\pi$$
$$= \frac{1}{\pi n^2}(\cos n\pi - 1) = \frac{1}{\pi n^2}[(-1)^n - 1],$$

so

$$a_{2n} = 0 \qquad \text{and} \qquad a_{2n-1} = -\frac{2}{\pi(2n-1)}.$$

Similarly,

$$b_n = \frac{1}{\pi} \int_0^\pi x \sin nx \, dx = \frac{1}{\pi}\left[-\frac{x \cos nx}{n} + \frac{\sin nx}{n^2} \right]_0^\pi$$
$$= \frac{1}{\pi}\left[-\frac{\pi \cos n\pi}{n} \right] = \frac{(-1)^{n+1}}{n}.$$

The Fourier series is therefore

$$f(x) = \frac{\pi}{4} - \frac{2}{\pi} \sum_1^\infty \frac{\cos(2n-1)x}{(2n-1)^2} + \sum_1^\infty (-1)^{n+1} \frac{\sin nx}{n}. \tag{3}$$

By Dirichlet's theorem this equation is valid at all points of continuity, since $f(x)$ is understood to be the periodic extension of the initially given part (see Fig. 41). At the point of discontinuity $x = \pi$, the series converges to

$$\frac{1}{2}[f(\pi-) + f(\pi+)] = \frac{\pi}{2}.$$

When $x = \pi$ is substituted in (3), this yields the following interesting sum of the reciprocals of the squares of the odd numbers,

$$\sum_1^\infty \frac{1}{(2n-1)^2} = 1 + \frac{1}{3^2} + \frac{1}{5^2} + \frac{1}{7^2} + \cdots = \frac{\pi^2}{8}. \tag{4}$$

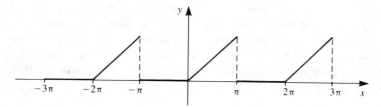

FIGURE 41

The same sum is obtained by substituting the point of continuity $x = 0$ into (3). Further, we can use (4) to find the sum of the reciprocals of the squares of *all* the positive integers,

$$\sum_{1}^{\infty} \frac{1}{n^2} = 1 + \frac{1}{2^2} + \frac{1}{3^2} + \frac{1}{4^2} + \cdots = \frac{\pi^2}{6}. \tag{5}$$

All that is needed to establish this is to write

$$\sum \frac{1}{n^2} = \sum \frac{1}{(2n)^2} + \sum \frac{1}{(2n-1)^2} = \frac{1}{4} \sum \frac{1}{n^2} + \frac{\pi^2}{8},$$

$$\frac{3}{4} \sum \frac{1}{n^2} = \frac{\pi^2}{8}, \quad \text{and} \quad \sum \frac{1}{n^2} = \frac{4}{3} \cdot \frac{\pi^2}{8} = \frac{\pi^2}{6}.$$

The sum (5) was found by Euler in 1736, and is one of the most memorable discoveries in the early history of infinite series.[9]

NOTE ON DIRICHLET. Peter Gustav Lejeune Dirichlet (1805–1859) was a German mathematician who made many contributions of lasting value to analysis and number theory. As a young man he was drawn to Paris by the reputations of Cauchy, Fourier, and Legendre, but he was most deeply influenced by his encounter and lifelong contact with Gauss's *Disquisitiones Arithmeticae* (1801). This prodigious but cryptic work contained many of the great master's far-reaching discoveries in number theory, but it was understood by very few mathematicians at that time. As Kummer later said, "Dirichlet was not satisfied to study Gauss's *Disquisitiones* once or several times, but continued throughout his life to keep in close touch with the wealth of deep mathematical thoughts which it contains by perusing it again and again. For this reason the book was never put on the shelf but had an abiding place on the table at which he worked. Dirichlet was the first one who not only fully understood this work, but also made it accessible to others." In later life Dirichlet became a friend and disciple of Gauss, and also a friend and advisor of Riemann, whom he helped in a small way with his doctoral dissertation. In 1855, after lecturing at Berlin for many years, he succeeded Gauss in the professorship at Göttingen.

---

[9] For Euler's own wonderfully ingenious way of discovering (5), see Appendix A.12 in the Simmons book cited in footnote 4.

One of the Dirichlet's earliest achievements was a milestone in analysis: In 1829 he gave the first satisfactory proof that certain specific types of functions are actually the sums of their Fourier series. Previous work in this field had consisted wholly of the uncritical manipulation of formulas; Dirichlet transformed the subject into genuine mathematics in the modern sense. As a byproduct of this research, he also contributed greatly to the correct understanding of the nature of a function, and gave the definition which is now most often used, namely, that y is a function of x when to each value of x in a given interval there corresponds a unique value of y. He added that it does not matter whether y depends on x according to some "formula" or "law" or "mathematical operation," and he emphasized this by giving the example of the function of x which has the value 1 for all rational x's and the value 0 for all irrational x's.

Perhaps his greatest works were two long memoirs of 1837 and 1839 in which he made very remarkable applications of analysis to the theory of numbers. It was in the first of these that he proved his wonderful theorem that there are an infinite number of primes in any arithmetic progression of the form $a + nb$, where a and b are positive integers with no common factor. His discoveries about absolutely convergent series also appeared in 1837. His convergence test, referred to in footnote 4 in Section 33, was published posthumously in his *Vorlesungen über Zahlentheorie* (1863). These lectures went through many editions and had a very wide influence.

He was also interested in mathematical physics, and formulated the so-called Dirichlet principle of potential theory, which asserts the existence of harmonic functions (functions that satisfy Laplace's equation) with prescribed boundary values. Riemann—who gave the principle its name—used it with great effect in some of his profoundest researches. Hilbert gave a rigorous proof of Dirichlet's principle in the early twentieth century.

PROBLEMS

1. In Problems 1, 2, 3, 4, 6 of Section 33, sketch the graph of the sum of each Fourier series on the interval $-5\pi \le x \le 5\pi$.
2. Use the example in the text to write down without calculation the Fourier series for the function defined by

$$f(x) = -x, \qquad -\pi < x \le 0;$$

$$f(x) = 0, \qquad 0 < x \le \pi.$$

Sketch the graph of the sum of this series on the interval $-5\pi \le x \le 5\pi$.
3. Find the Fourier series for the periodic function defined by

$$f(x) = -\pi, \qquad -\pi \le x < 0;$$

$$f(x) = x, \qquad 0 \le x < \pi.$$

Sketch the graph of the sum of this series on the interval $-5\pi \le x \le 5\pi$ and find what numerical sums are implied by the convergence behavior at the points of discontinuity $x = 0$ and $x = \pi$.

4. (a) Show that the Fourier series for the periodic function defined by $f(x) = 0$, $-\pi \le x < 0$ and $f(x) = x^2$, $0 \le x < \pi$ is

$$f(x) = \frac{\pi^2}{6} + 2 \sum_{1}^{\infty} (-1)^n \frac{\cos nx}{n^2}$$

$$+ \pi \sum_{1}^{\infty} (-1)^{n+1} \frac{\sin nx}{n} - \frac{4}{\pi} \sum_{1}^{\infty} \frac{\sin (2n - 1)x}{(2n - 1)^3}.$$

(b) Sketch the graph of the sum of this series on the interval $-5\pi \le x \le 5\pi$.

(c) Use the series in (a) with $x = 0$ and π to obtain the sums

$$1 - \frac{1}{2^2} + \frac{1}{3^2} - \frac{1}{4^2} + \cdots = \frac{\pi^2}{12}$$

and

$$1 + \frac{1}{2^2} + \frac{1}{3^2} + \frac{1}{4^2} + \cdots = \frac{\pi^2}{6}.$$

(d) Derive the second sum in (c) from the first. *Hint*: Add $2 \sum \left(\frac{1}{2n}\right)^2$ to both sides.

5. (a) Find the Fourier series for the periodic function defined by $f(x) = e^x$, $-\pi \le x < \pi$. *Hint*: Recall that $\sinh x = (e^x - e^{-x})/2$.

(b) Sketch the graph of the sum of this series on the interval $-5\pi \le x \le 5\pi$.

(c) Use the series in (a) to establish the sums

$$\sum_{1}^{\infty} \frac{1}{n^2 + 1} = \frac{1}{2}\left[\frac{\pi}{\tanh \pi} - 1\right]$$

and

$$\sum_{1}^{\infty} \frac{(-1)^n}{n^2 + 1} = \frac{1}{2}\left[\frac{\pi}{\sinh \pi} - 1\right].$$

6. Mathematicians prefer the classes of functions they study to be linear spaces, that is, to be closed under the operations of addition and multiplication by scalars. Unfortunately this is not true for the class of functions defined on the interval $-\pi \le x < \pi$ that satisfy the Dirichlet conditions. Verify this statement by examining the functions

$$f(x) = x^2 \sin \frac{1}{x} + 2x \qquad (x \ne 0), \qquad f(0) = 0$$

and

$$g(x) = -2x.$$

7. If $f(x)$ is defined on the interval $-\pi \le x < \pi$ and satisfies the Dirichlet conditions there, prove that $f(x-)$ and $f(x+)$ exist at every interior point, and also that $f(x+)$ exists at the left endpoint and $f(x-)$ exists at the right endpoint. *Hint*: Each interior point of discontinuity is isolated from other such points, in the sense that the function is continuous at all nearby points; also, on each side of such a point and near enough to it, the function does not oscillate, and is therefore increasing or decreasing.

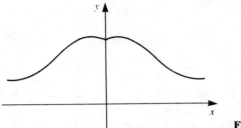

FIGURE 42

35   EVEN AND ODD FUNCTIONS.
COSINE AND SINE SERIES

In principle, our work in the preceding sections could have been based on any interval of length 2π, for instance, on the interval $0 \le x \le 2\pi$. However, the symmetrically placed interval $-\pi \le x \le \pi$ has substantial advantages for the exploitation of symmetry properties of functions, as we now show.

A function $f(x)$ defined on this interval (or on any symmetrically placed interval) is said to be *even* if

$$f(-x) = f(x), \tag{1}$$

and $f(x)$ is said to be *odd* if

$$f(-x) = -f(x). \tag{2}$$

For example, x^2 and $\cos x$ are even, and x^3 and $\sin x$ are odd. The graph of an even function is symmetric about the y-axis, as shown in Fig. 42, and the graph of an odd function is skew-symmetric (Fig. 43). By putting $x = 0$ in (2), we see that an odd function always has the property that

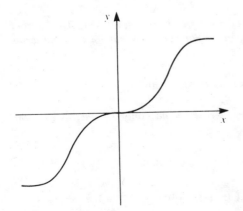

FIGURE 43

$f(0) = 0$. It is clear from the figures that

$$\int_{-a}^{a} f(x)\, dx = 2 \int_{0}^{a} f(x)\, dx \qquad \text{if } f(x) \text{ is even,} \tag{3}$$

and

$$\int_{-a}^{a} f(x)\, dx = 0 \qquad \text{if } f(x) \text{ is odd,} \tag{4}$$

because the integrals represent the algebraic (signed) areas under the curves. These facts can also be established by analytic reasoning based on the definitions (1) and (2) [see Problem 3 below].

Products of even and odd functions have the simple properties

$$(\text{even})(\text{even}) = \text{even}, \qquad (\text{even})(\text{odd}) = \text{odd}, \qquad (\text{odd})(\text{odd}) = \text{even},$$

which correspond to the familiar rules

$$(+1)(+1) = +1, \qquad (+1)(-1) = -1, \qquad (-1)(-1) = +1.$$

For instance, to prove the second property we consider the function $F(x) = f(x)g(x)$, where $f(x)$ is even and $g(x)$ is odd. Then

$$F(-x) = f(-x)g(-x) = f(x)[-g(x)] = -f(x)g(x) = -F(x),$$

which shows that the product $f(x)g(x)$ is odd. The other two properties can be proved similarly. As an example, we know that $x^3 \cos nx$ is odd because x^3 is odd and $\cos nx$ is even, so (4) tells us at once that

$$\int_{-\pi}^{\pi} x^3 \cos nx \, dx = 0,$$

without the need for detailed integrations by parts.

The following simple theorem clarifies the significance of these ideas for the study of Fourier series.

Theorem. Let $f(x)$ be an integrable function defined on the interval $-\pi \le x \le \pi$. If $f(x)$ is even, then its Fourier series has only cosine terms and the coefficients are given by

$$a_n = \frac{2}{\pi} \int_{0}^{\pi} f(x) \cos nx \, dx, \qquad b_n = 0. \tag{5}$$

And if $f(x)$ is odd, then its Fourier series has only sine terms and the coefficients are given by

$$a_n = 0, \qquad b_n = \frac{2}{\pi} \int_{0}^{\pi} f(x) \sin nx \, dx. \tag{6}$$

To prove this, we assume first that $f(x)$ is even. Then $f(x) \cos nx$ is even (even times even) and by (3) we have

$$a_n = \frac{1}{\pi} \int_{-\pi}^{\pi} f(x) \cos nx \, dx = \frac{2}{\pi} \int_0^{\pi} f(x) \cos nx \, dx.$$

On the other hand, $f(x) \sin nx$ is odd (even times odd), so (4) tells us that

$$b_n = \frac{1}{\pi} \int_{-\pi}^{\pi} f(x) \sin nx \, dx = 0,$$

which completes the argument for (5). It is easy to establish (6) by similar reasoning.

Example 1. (a) First, we briefly consider the function $f(x) = x$ on the interval $-\pi \le x \le \pi$. Since this is an odd function, its Fourier series is automatically a sine series, and therefore it is not necessary to bother calculating the cosine coefficients. We found in Section 33 that the Fourier series is

$$x = 2\left(\sin x - \frac{\sin 2x}{2} + \frac{\sin 3x}{3} - \cdots\right), \tag{7}$$

and we know that this expansion is valid only on the open interval $-\pi < x < \pi$ and not at the endpoints $x = \pm\pi$, because any series of sines converges to zero at these points.

(b) Next, we consider the function $f(x) = |x|$ on the interval $-\pi \le x \le \pi$ (Fig. 44). Since this is an even function, its Fourier series reduces to a cosine series, and by (5) we have

$$a_n = \frac{1}{\pi} \int_{-\pi}^{\pi} |x| \cos nx \, dx$$

$$= \frac{2}{\pi} \int_0^{\pi} x \cos nx \, dx.$$

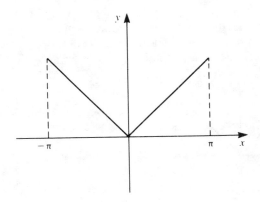

FIGURE 44

It is easy to see that $a_0 = \pi$, and for $n \geq 1$ an integration by parts gives

$$a_n = \frac{2}{\pi n^2}(\cos n\pi - 1) = \frac{2}{\pi n^2}[(-1)^n - 1].$$

This tells us that

$$a_{2n} = 0 \quad \text{and} \quad a_{2n-1} = -\frac{4}{\pi(2n-1)^2},$$

so we have the expansion

$$|x| = \frac{\pi}{2} - \frac{4}{\pi}\left(\cos x + \frac{\cos 3x}{3^2} + \frac{\cos 5x}{5^2} + \cdots\right). \tag{8}$$

The periodic extension of the initially given function is shown in Fig. 45. We see at once from the ideas of Section 34 that the series in (8) converges to this extension for all x, and therefore the expansion (8) is valid on the closed interval $-\pi \leq x \leq \pi$.

Since $|x| = x$ for $x \geq 0$, the two series (7) and (8) are both expansions of the same function $f(x) = x$ on the interval $0 \leq x \leq \pi$. The first series (7) is called the *Fourier sine series* for x, and (8) is called the *Fourier cosine series* for x. Similarly, any function $f(x)$ defined on the interval $0 \leq x \leq \pi$ that satisfies the Dirichlet conditions there can be expanded in both a sine series and a cosine series on this interval—with the proviso that the sine series cannot converge to $f(x)$ at the endpoints $x = 0$ and $x = \pi$ unless $f(x)$ has the value 0 at these points.

To obtain the sine series for $f(x)$, we redefine the function (if necessary) to have the value 0 at $x = 0$, and then we extend it over the interval $-\pi \leq x < 0$ in such a way that the extended function is odd. That is, we define $f(x)$ for $-\pi \leq x < 0$ by putting $f(x) = -f(-x)$. The extended function is clearly odd, so its Fourier series contains sine terms only, and its coefficients are given by (6). Similarly, we obtain the cosine series for $f(x)$ by extending $f(x)$ to be an even function on the interval $-\pi \leq x \leq \pi$ and using (5) to calculate the coefficients. With respect to the sine and cosine series described here, we emphasize particularly that the original function $f(x)$ is not assumed in advance to be odd, or even, or periodic, or defined elsewhere at all; it is intended to be an essentially

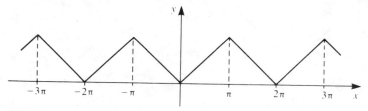

FIGURE 45

arbitrary function on the interval $0 \le x \le \pi$—within the very weak restrictions imposed by the Dirichlet conditions.

Example 2. Find the sine series, and also the cosine series, for the function $f(x) = \cos x$, $0 \le x \le \pi$.

For the sine series, (6) gives

$$a_n = 0 \quad \text{and} \quad b_n = \frac{2}{\pi} \int_0^\pi \cos x \sin nx \, dx.$$

For $n = 1$ we have $b_1 = 0$, and for $n > 1$ a short calculation yields

$$b_n = \frac{2n}{\pi} \left[\frac{1 + (-1)^n}{n^2 - 1} \right].$$

We therefore have

$$b_{2n-1} = 0 \quad \text{and} \quad b_{2n} = \frac{8n}{\pi(4n^2 - 1)},$$

so the sine series is

$$\cos x = \frac{8}{\pi} \sum_1^\infty \frac{n \sin 2nx}{4n^2 - 1}, \qquad 0 < x < \pi.$$

To obtain the cosine series, we observe that (5) gives $b_n = 0$ and

$$a_n = \frac{2}{\pi} \int_0^\pi \cos x \sin nx \, dx = \begin{cases} 1 & \text{for } n = 1 \\ 0 & \text{for } n \ne 1. \end{cases}$$

Therefore the cosine series for $\cos x$ is simply $\cos x$, just as we would have expected. This conclusion also follows directly from the equation $\cos x = \cos x$, because our work in Section 33 shows that any finite trigonometric series (the right side) is automatically the Fourier series of its sum (the left side).

PROBLEMS

1. Determine whether each of the following functions is even, odd, or neither:
$x^5 \sin x$, $x^2 \sin 2x$, e^x, $(\sin x)^3$, $\sin x^2$, $\cos (x + x^3)$, $x + x^2 + x^3$, $\log \dfrac{1 + x}{1 - x}$.

2. Show that any function $f(x)$ defined on a symmetrically placed interval can be written as the sum of an even function and an odd function. *Hint*: $f(x) = \frac{1}{2}[f(x) + f(-x)] + \frac{1}{2}[f(x) - f(-x)]$.

3. Prove properties (3) and (4) analytically, by making the substitution $x = -t$ in the part of the integral from $-a$ to 0 and using the definitions (1) and (2).

4. Show that the sine series of the constant function $f(x) = \pi/4$ is

$$\frac{\pi}{4} = \sin x + \frac{\sin 3x}{3} + \frac{\sin 5x}{5} + \cdots, \qquad 0 < x < \pi.$$

What sum is obtained by putting $x = \pi/2$? What is the cosine series of this function?

5. Find the Fourier series for the function of period 2π defined by $f(x) = \cos\frac{1}{2}x$, $-\pi \le x \le \pi$. Sketch the graph of the sum of this series on the interval $-5\pi \le x \le 5\pi$.

6. Find the sine and cosine series for $\sin x$.

7. Find the Fourier series for the function of period 2π defined by

$$f(x) = x + \frac{\pi}{2}, \qquad -\pi \le x < 0;$$

$$f(x) = -x + \frac{\pi}{2}, \qquad 0 \le x \le \pi$$

(a) by computing the Fourier coefficients;
(b) directly from the expansion (8).
Sketch the graph of the sum of this series (a triangular wave) on the interval $-5\pi \le x \le 5\pi$.

8. For the function $f(x) = \pi - x$, find
(a) its Fourier series on the interval $-\pi < x < \pi$;
(b) its cosine series on the interval $0 \le x \le \pi$;
(c) its sine series on the interval $0 < x \le \pi$.
Sketch the graph of the sum of each of these series on the interval $-5\pi \le x \le 5\pi$.

9. If $f(x) = x$ for $0 \le x \le \pi/2$ and $f(x) = \pi - x$ for $\pi/2 < x \le \pi$, show that the cosine series for this function is

$$f(x) = \frac{\pi}{4} - \frac{2}{\pi}\sum_{1}^{\infty}\frac{\cos 2(2n-1)x}{(2n-1)^2}.$$

Sketch the graph of the sum of this series on the interval $-5\pi \le x \le 5\pi$.

10. (a) Show that the cosine series for x^2 is

$$x^2 = \frac{\pi^2}{3} + 4\sum_{1}^{\infty}(-1)^n\frac{\cos nx}{n^2}, \qquad -\pi \le x \le \pi.$$

(b) Find the sine series for x^2, and use this expansion together with formula (7) to obtain the sum

$$1 - \frac{1}{3^3} + \frac{1}{5^3} - \frac{1}{7^3} + \cdots = \frac{\pi^3}{32}.$$

(c) Denote by s the sum of the reciprocals of the cubes of the odd numbers,

$$1 + \frac{1}{3^3} + \frac{1}{5^3} + \frac{1}{7^3} + \cdots = s,$$

and show that then

$$\sum_{1}^{\infty}\frac{1}{n^3} = 1 + \frac{1}{2^3} + \frac{1}{3^3} + \frac{1}{4^3} + \cdots = \frac{8}{7}s.$$

The exact numerical value of the latter sum has been one of unsolved mysteries of mathematics since Euler first raised the question in 1736.

11. (a) Show that the cosine series for x^3 is

$$x^3 = \frac{\pi^3}{4} + 6\pi \sum_1^\infty (-1)^n \frac{\cos nx}{n^2} + \frac{24}{\pi} \sum_1^\infty \frac{\cos (2n-1)x}{(2n-1)^4},$$

$0 \le x \le \pi$.

(b) Use the series in (a) to obtain, in this order, the sums

$$\sum_1^\infty \frac{1}{(2n-1)^4} = \frac{\pi^4}{96} \quad \text{and} \quad \sum_1^\infty \frac{1}{n^4} = \frac{\pi^4}{90}.$$

12. (a) Show that the cosine series for x^4 is

$$x^4 = \frac{\pi^4}{5} + 8 \sum_1^\infty (-1)^n \frac{\pi^2 n^2 - 6}{n^4} \cos nx,$$

$-\pi \le x \le \pi$.

(b) Use the series in (a) to obtain again the second sum in Problem 11(b).

13. (a) If α is not an integer, show that

$$\cos \alpha x = \frac{\sin \alpha\pi}{\alpha\pi} + \frac{2\alpha \sin \alpha\pi}{\pi} \sum_1^\infty (-1)^n \frac{\cos nx}{\alpha^2 - n^2}$$

for $-\pi \le x \le \pi$.

(b) Use the series in (a) to obtain the formula

$$\pi \cot \alpha\pi = \frac{1}{\alpha} + 2\alpha \sum_1^\infty \frac{1}{\alpha^2 - n^2}.$$

This is called *Euler's partial fractions expansion of the cotangent*.

(c) Rewrite the expansion in (b) in the form

$$\pi \cot \pi t - \frac{\pi}{\pi t} = \sum_1^\infty \frac{-2t}{n^2 - t^2},$$

and by integrating term by term from $t = 0$ to $t = x$ $(0 < x < 1)$ obtain

$$\log \left(\frac{\sin \pi x}{\pi x} \right) = \sum_1^\infty \log \left(1 - \frac{x^2}{n^2} \right)$$

or

$$\frac{\sin \pi x}{\pi x} = \left(1 - \frac{x^2}{1^2} \right)\left(1 - \frac{x^2}{2^2} \right)\left(1 - \frac{x^2}{3^2} \right) \cdots.$$

If x is replaced by x/π, this infinite product takes the equivalent form

$$\frac{\sin x}{x} = \left(1 - \frac{x^2}{\pi^2} \right)\left(1 - \frac{x^2}{4\pi^2} \right)\left(1 - \frac{x^2}{9\pi^2} \right) \cdots,$$

which is called *Euler's infinite product for the sine*. Observe that this formula displays the nonzero roots $x = \pm\pi, \pm2\pi, \pm3\pi, \ldots$ of the transcendental equation $\sin x = 0$.

14. The functions $\sin^2 x$ and $\cos^2 x$ are both even. Show briefly, without calculation, that the identities

$$\sin^2 x = \frac{1}{2}(1 - \cos 2x) = \frac{1}{2} - \frac{1}{2}\cos 2x$$

and

$$\cos^2 x = \frac{1}{2}(1 + \cos 2x) = \frac{1}{2} + \frac{1}{2}\cos 2x$$

are the Fourier series expansions of these functions.

15. Find the sine series of the functions in Problem 14, and verify that these expansions satisfy the identity $\sin^2 x + \cos^2 x = 1$.

16. Prove the trigonometric identities

$$\sin^3 x = \frac{3}{4}\sin x - \frac{1}{4}\sin 3x \quad \text{and} \quad \cos^3 x = \frac{3}{4}x + \frac{1}{4}\cos 3x,$$

and show briefly, without calculation that these are the Fourier series expansions of the functions on the left.

36 EXTENSION TO ARBITRARY INTERVALS

The standard form of a Fourier series is the one we have worked with in the preceding sections, where the function under consideration is defined on the interval $-\pi \le x < \pi$. In many applications it is desirable to adapt the form of a Fourier series to a function $f(x)$ defined on an interval $-L \le x < L$, where L is a positive number different from π. This is done by a change of variable that amounts to a change of scale on the horizontal axis.

We introduce a new variable t that runs from $-\pi$ to π as x runs from $-L$ to L. This is easy to remember as a statement about proportions:

$$\frac{t}{\pi} = \frac{x}{L}, \quad \text{so} \quad t = \frac{\pi x}{L} \quad \text{and} \quad x = \frac{Lt}{\pi}. \tag{1}$$

The function $f(x)$ is thereby transformed into a function of t,

$$f(x) = f\left(\frac{Lt}{\pi}\right) = g(t), \quad -\pi \le t < \pi,$$

and if we assume that $f(x)$ satisfies the Dirichlet conditions, then so does $g(t)$. We can therefore expand $g(t)$ in a Fourier series of the usual form,

$$g(t) = \frac{1}{2}a_0 + \sum_{1}^{\infty}(a_n \cos nt + b_n \sin nt), \tag{2}$$

where we use the familiar formulas for the coefficients,

$$a_n = \frac{1}{\pi} \int_{-\pi}^{\pi} g(t) \cos nt \, dt \quad \text{and} \quad b_n = \frac{1}{\pi} \int_{-\pi}^{\pi} g(t) \sin nt \, dt. \quad (3)$$

Having found the expansion (2), we now use (1) to transform this back into a solution of our original problem, namely, to find an expansion of $f(x)$ on the interval $-L \leq x < L$:

$$f(x) = \frac{1}{2} a_0 + \sum_{1}^{\infty} \left(a_n \cos \frac{n\pi x}{L} + b_n \sin \frac{n\pi x}{L} \right). \quad (4)$$

Of course, we can also transform formulas (3) into integrals with respect to x,

$$a_n = \frac{1}{L} \int_{-L}^{L} f(x) \cos \frac{n\pi x}{L} \, dx \quad \text{and} \quad b_n = \frac{1}{L} \int_{-L}^{L} f(x) \sin \frac{n\pi x}{L} \, dx.$$

$$(5)$$

We can use formulas (5) directly if we wish to do so, but changing the variable to t usually makes the work easier because it simplifies the calculations.

Example. Expand $f(x)$ in a Fourier series on the interval $-2 \leq x < 2$ if $f(x) = 0$ for $-2 \leq x < 0$ and $f(x) = 1$ for $0 \leq x < 2$.

Here we introduce t by writing

$$\frac{t}{\pi} = \frac{x}{2}, \quad \text{so} \quad t = \frac{\pi x}{2} \quad \text{and} \quad x = \frac{2t}{\pi}.$$

Then $g(t) = 0$ for $-\pi \leq t < 0$ and $g(t) = 1$ for $0 \leq t < \pi$, and we have

$$a_0 = \frac{1}{\pi} \left[\int_{-\pi}^{0} 0 \, dt + \int_{0}^{\pi} 1 \, dt \right] = 1;$$

$$a_n = \frac{1}{\pi} \int_{0}^{\pi} \cos nt \, dt = 0, \quad n \geq 1;$$

$$b_n = \frac{1}{\pi} \int_{0}^{\pi} \sin nt \, dt = \frac{1}{n\pi} [1 - (-1)^n].$$

The last of these formulas tells us that

$$b_{2n} = 0 \quad \text{and} \quad b_{2n-1} = \frac{2}{(2n-1)\pi}.$$

We therefore have

$$g(t) = \frac{1}{2} + \frac{2}{\pi} \sum_{1}^{\infty} \frac{\sin (2n-1)t}{2n-1},$$

so the desired expansion is

$$f(x) = \frac{1}{2} + \frac{2}{\pi} \sum_{1}^{\infty} \frac{1}{2n-1} \sin(2n-1)\frac{\pi x}{2}.$$

Further, we know that this series converges to the periodic extension of $f(x)$ [with period 4] at all points x except the points of discontinuity $x = 0, \pm 2, \pm 4, \ldots$, and at these points it converges to the sum $1/2$, which is the average of the two one-sided limits.

PROBLEMS

1. For the function defined by

$$f(x) = -3, \qquad -2 \leq x < 0 \qquad \text{and} \qquad f(x) = 3, \qquad 0 \leq x < 2,$$

   write down its Fourier expansion directly from the example in the text, without calculation.
2. Find the Fourier series for the functions defined by
   (a) $f(x) = 1 + x$, $-1 \leq x < 0$ and $f(x) = 1 - x$, $0 \leq x \leq 1$;
   (b) $f(x) = |x|$, $-2 \leq x \leq 2$.
3. Show that

$$\frac{1}{2}L - x = \frac{L}{\pi} \sum_{1}^{\infty} \frac{1}{n} \sin\frac{2n\pi x}{L}, \qquad 0 < x < L.$$

4. Find the cosine series for the function defined on the interval $0 \leq x \leq 1$ by $f(x) = x^2 - x + \frac{1}{6}$. (In the context of Problem 9 below, this function is the Bernoulli polynomial $B_2(x)$, and the series found here is the simplest special case of the expansion in Problem 10.)
5. Find the cosine series for the function defined by

$$f(x) = 2, \qquad 0 \leq x \leq 1 \qquad \text{and} \qquad f(x) = 0, \qquad 1 < x \leq 2.$$

6. Expand $f(x) = \cos \pi x$ in a Fourier series on the interval $-1 \leq x \leq 1$.
7. Find the cosine series for the function defined by

$$f(x) = \frac{1}{4} - x, \qquad 0 \leq x < \frac{1}{2} \qquad \text{and} \qquad f(x) = x - \frac{3}{4}, \qquad \frac{1}{2} \leq x \leq 1.$$

8. (This problem and the next are necessary preliminaries for the Fourier series problem that follows them, and this in turn is aimed at obtaining the remarkable formulas in Problem 11.) Since

$$\frac{e^x - 1}{x} = 1 + \frac{x}{2!} + \frac{x^2}{3!} + \cdots$$

   for $x \neq 0$, and this power series has the value 1 at $x = 0$, the reciprocal function $x/(e^x - 1)$ has a power series expansion valid in some neighborhood of the origin if the value of this function is defined to be 1 at $x = 0$:

$$\frac{x}{e^x - 1} = \sum_0^\infty \frac{B_n}{n!} x^n = B_0 + B_1 x + \frac{B_2}{2!} x^2 + \cdots . \qquad (*)$$

The numbers B_n defined in this way are called *Bernoulli numbers,* and play an important role in the theory of infinite series.[10] Evidently $B_0 = 1$.

(a) By writing

$$\frac{x}{e^x - 1} = \frac{x}{2}\left(\frac{e^x + 1}{e^x - 1} - 1\right) = -\frac{x}{2} + \frac{x}{2} \cdot \frac{e^x + 1}{e^x - 1}$$

and noticing that the second term on the right is an even function, conclude that $B_1 = -\frac{1}{2}$ and $B_n = 0$ if n is odd and > 1.

(b) By writing $(*)$ in the form

$$\left(\frac{B_0}{0!} + \frac{B_1}{1!} x + \frac{B_2}{2!} x^2 + \cdots\right)\left(\frac{1}{1!} + \frac{x}{2!} + \frac{x^2}{3!} + \cdots\right) = 1$$

and multiplying the two power series on the left, conclude by examining the coefficient of x^{n-1} that

$$\binom{n}{0} B_0 + \binom{n}{1} B_1 + \binom{n}{2} B_2 + \cdots + \binom{n}{n-1} B_{n-1} = 0 \quad (**)$$

for $n \geq 2$, where $\binom{n}{k}$ is the binomial coefficient $n!/[k! \, (n-k)!]$.

(c) By taking $n = 3, 5, 7, 9, 11$ in $(**)$, show that

$$B_2 = \frac{1}{6}, \qquad B_4 = -\frac{1}{30}, \qquad B_6 = \frac{1}{42}, \qquad B_8 = -\frac{1}{30}, \qquad B_{10} = \frac{5}{66}.$$

From the recursive mode of calculation, all the Bernoulli numbers can be considered as known (even though considerable labor may be required to make any particular one of them visibly present) and all of them are rational.

9. The *Bernoulli polynomials* $B_0(x), B_1(x), B_2(x), \ldots$ are defined by the resulting coefficients in the following product of two power series (see the preceding problem):

$$e^{xt} \cdot \frac{t}{e^t - 1} = \left(\sum_0^\infty \frac{(xt)^n}{n!}\right)\left(\sum_0^\infty \frac{B_n}{n!} t^n\right) = \sum_0^\infty \frac{B_n(x)}{n!} t^n.$$

---

[10] For instance, it can be proved that the power series expansion of $\tan x$ is

$$\tan x = \sum_1^\infty (-1)^{n+1} \frac{2^{2n}(2^{2n} - 1)B_{2n}}{(2n)!} x^{2n-1}.$$

See Appendix A.18 in the Simmons book mentioned in footnote 4.

(a) Show that $B_n(x)$ is a polynomial of degree n that is given by the formula

$$B_n(x) = \binom{n}{0} B_0 x^n + \binom{n}{1} B_1 x^{n-1} + \cdots + \binom{n}{n-1} B_{n-1} x + \binom{n}{n} B_n.$$

(b) Show that $B_n(0) = B_n$ for $n \geq 0$, and by using (**) in the preceding problem, show that also $B_n(1) = B_n$ for $n \geq 2$.

(c) Show that

$$B'_{n+1}(x) = (n + 1)B_n(x),$$

and deduce from this that

$$B_{n+1}(x) = B_{n+1} + (n + 1) \int_0^x B_n(t)\, dt$$

and (if $n \geq 1$)

$$\int_0^1 B_n(x)\, dx = 0.$$

(d) Show that

$$B_0(x) = 1, \qquad B_1(x) = x - \frac{1}{2}, \qquad B_2(x) = x^2 - x + \frac{1}{6},$$

$$B_3(x) = x^3 - \frac{3}{2}x^2 + \frac{1}{2}x, \qquad B_4(x) = x^4 - 2x^3 + x^2 - \frac{1}{30}.$$

10. Show that the cosine series for the Bernoulli polynomial $B_{2n}(x)$ on the interval $0 \leq x \leq 1$ is

$$B_{2n}(x) = (-1)^{n+1} \frac{2(2n)!}{(2\pi)^{2n}} \sum_{k=1}^{\infty} \frac{\cos 2k\pi x}{k^{2n}}, \qquad n \geq 1.$$

11. Use the expansion in Problem 10 to show that

$$\sum_{n=1}^{\infty} \frac{1}{n^{2p}} = (-1)^{p+1} \frac{2^{2p} B_{2p}}{2(2p)!} \pi^{2p},$$

where p is a positive integer. Use the results of Problem 8 to obtain the special sums corresponding to $p = 1, 2, 3, 4, 5$:

$$\sum_1^{\infty} \frac{1}{n^2} = \frac{\pi^2}{6}, \qquad \sum_1^{\infty} \frac{1}{n^4} = \frac{\pi^4}{90}, \qquad \sum_1^{\infty} \frac{1}{n^6} = \frac{\pi^6}{945},$$

$$\sum_1^{\infty} \frac{1}{n^8} = \frac{\pi^8}{9450}, \qquad \sum_1^{\infty} \frac{1}{n^{10}} = \frac{\pi^{10}}{93555}.$$

These discoveries are all due to Euler.[11]

---

[11] For more information on the background of these formulas, see the article by Raymond Ayoub, "Euler and the Zeta Function," *American Mathematical Monthly*, vol. 81 (1974), pp. 1067–1086.

37 ORTHOGONAL FUNCTIONS

A sequence of functions $\theta_n(x)$, $n = 1, 2, 3, \ldots$, is said to be *orthogonal* on the interval $[a,b]$[12] if

$$\int_a^b \theta_m(x)\theta_n(x)\,dx \begin{cases} = 0 & \text{for } m \neq n, \\ \neq 0 & \text{for } m = n. \end{cases} \tag{1}$$

For example, the sequence

$$\theta_1(x) = \sin x, \qquad \theta_2(x) = \sin 2x, \qquad \ldots, \qquad \theta_n(x) = \sin nx, \qquad \ldots$$

is orthogonal on $[0,\pi]$ because

$$\int_0^\pi \theta_m(x)\theta_n(x)\,dx = \int_0^\pi \sin mx \sin nx\,dx$$

$$= \int_0^\pi \tfrac{1}{2}[\cos(m-n)x - \cos(m+n)x]\,dx \begin{cases} = 0 & \text{for } m \neq n, \\ = \dfrac{\pi}{2} & \text{for } m = n. \end{cases}$$

We pointed out in Section 33 that the sequence

$$1, \qquad \cos x, \qquad \sin x, \qquad \cos 2x, \qquad \sin 2x, \qquad \ldots \tag{2}$$

is orthogonal on $[-\pi,\pi]$, but it is not orthogonal on $[0,\pi]$ because

$$\int_0^\pi 1 \cdot \sin x\,dx = 2 \neq 0.$$

In the preceding sections of this chapter the trigonometric sequence (2) was used for the formation of Fourier series. During the nineteenth and early twentieth centuries many mathematicians and physicists became aware that one can form series similar to Fourier series by using any orthogonal sequence of functions. These generalized Fourier series turned out to be indispensable tools in many branches of mathematical physics, especially in quantum mechanics. They are also of central importance in several major areas of twentieth century mathematics, in connection with such topics as function spaces and theories of integration.[13]

The formula for the generalized Fourier coefficients is particularly simple if the integral (1) has the value 1 for $m = n$. In this case the functions $\theta_n(x)$ are said to be *normalized*, and $\{\theta_n(x)\}$ is called an

---

[12] As usual, this notation designates the *closed* interval $a \leq x \leq b$.

[13] See, for example, the excellent book by Béla Sz.-Nagy, *Introduction to Real Functions and Orthogonal Expansions*, Oxford University Press, 1965.

orthonormal sequence. On the other hand, if

$$\int_a^b [\theta_n(x)]^2 \, dx = \alpha_n \neq 1$$

in (1), then it is easy to see that the functions

$$\phi_n(x) = \frac{\theta_n(x)}{\sqrt{\alpha_n}}$$

are orthonormal, that is,

$$\int_a^b \phi_m(x)\phi_n(x) \, dx \begin{cases} = 0 & \text{for } m \neq n, \\ = 1 & \text{for } m = n. \end{cases} \tag{3}$$

For example, since

$$\int_{-\pi}^{\pi} 1 \, dx = 2\pi, \qquad \int_{-\pi}^{\pi} \cos^2 nx \, dx = \pi, \qquad \int_{-\pi}^{\pi} \sin^2 nx \, dx = \pi \tag{4}$$

for $n \geq 1$, the orthonormal sequence corresponding to the orthogonal sequence (2) is

$$\frac{1}{\sqrt{2\pi}}, \quad \frac{\cos x}{\sqrt{\pi}}, \quad \frac{\sin x}{\sqrt{\pi}}, \quad \frac{\cos 2x}{\sqrt{\pi}}, \quad \frac{\sin 2x}{\sqrt{\pi}}, \dots. \tag{5}$$

Now let $\{\phi_n(x)\}$ be an orthonormal sequence of functions on $[a,b]$, and suppose that we are trying to expand another function $f(x)$ in a series of the form

$$f(x) = a_1\phi_1(x) + a_2\phi_2(x) + \cdots + a_n\phi_n(x) + \cdots. \tag{6}$$

To determine the coefficients a_n we multiply both sides of (6) by $\phi_n(x)$. This gives

$$f(x)\phi_n(x) = a_1\phi_1(x)\phi_n(x) + \cdots + a_n[\phi_n(x)]^2 + \cdots, \tag{7}$$

where the terms not written contain products $\phi_m(x)\phi_n(x)$ with $m \neq n$. If we assume that term-by-term integration of (7) is valid, then by carrying out this integration and using (3) we find that most of the terms disappear and all that remains is

$$\int_a^b f(x)\phi_n(x) \, dx = \int_a^b a_n[\phi_n(x)]^2 \, dx = a_n,$$

so

$$a_n = \int_a^b f(x)\phi_n(x) \, dx. \tag{8}$$

In deriving formula (8) for the coefficients in the expansion (6), we made two very large assumptions. First, we assumed that the function $f(x)$ can be represented by a series of the form (6). Second, we assumed that the term-by-term integration of the series (7) is permissible. Unfortunately, we have no reason whatever—apart from wishful thinking—for believing that either assumption is legitimate. To express this somewhat differently, we have no guarantee at all that the series (6) with coefficients defined by (8) will even converge, let alone converge to the function $f(x)$. Nevertheless, the numbers (8) are called the *Fourier coefficients* of $f(x)$ with respect to the orthonormal sequence $\{\phi_n(x)\}$, and the resulting series (6) is called the *Fourier series* of $f(x)$ with respect to $\{\phi_n(x)\}$.[14] When these ideas are applied to the orthonormal sequence (5), they yield the ordinary Fourier series as described in the preceding sections (see Problem 2 below).

We also point out, as we did in Section 33, that the term-by-term integration of (7) that leads to (8) is legal if the functions are continuous and the series is uniformly convergent. However, in the next section formula (8) will be obtained in an entirely different manner, having nothing to do with uniform convergence. It will then be clear that there is no need to feel uneasy because formula (8) seems to have been derived by faulty reasoning. The truth is, that we can use whatever reasoning we please as *motivation* for the definitions of the Fourier coefficients and Fourier series, and we then turn to the problem of discovering conditions under which the Fourier series (6) is a valid expansion of the function $f(x)$.

Most orthogonal sequences of functions are obtained by solving differential equations, as suggested in the following example. A broader discussion of this topic is given in Section 43.

> **Example.** Use the differential equation $y'' + \lambda y = 0$, or equivalently $y'' = -\lambda y$, to show that the trigonometric sequence (2) is orthogonal on $[-\pi, \pi]$.
>
> Let m and n be positive integers. If $y_m = \sin mx$ or $\cos mx$ and $y_n = \sin nx$ or $\cos nx$, then
>
> $$y''_m = -m^2 y_m \quad \text{and} \quad y''_n = -n^2 y_n.$$
>
> If the first equation is multiplied by y_n, the second by y_m, and the resulting equations are subtracted, the result is
>
> $$y_n y''_m - y_m y''_n = (n^2 - m^2) y_m y_n.$$
>
> We now notice that the left side of this is the derivative of $y_n y'_m - y_m y'_n$, so

---

[14] Some writers make consistent use of the terms *generalized Fourier coefficients* and *generalized Fourier series*. We prefer to simplify the terminology by omitting the adjective "generalized," and to rely on the context to tell us whether we are dealing with generalized or ordinary Fourier series.

integrating from $-\pi$ to π gives

$$(y_n y'_m - y_m y'_n)]^{\pi}_{-\pi} = (n^2 - m^2) \int_{-\pi}^{\pi} y_m y_n \, dx. \tag{9}$$

The function $y_n y'_m - y_m y'_n$ is periodic with period 2π and therefore has the same values at $-\pi$ and π, so the left side of (9) is zero. This yields the orthogonality property

$$\int_{-\pi}^{\pi} y_m y_n \, dx = 0,$$

except in the case $m = n$. In this case, however, the relevant integral is easy to evaluate:

$$\int_{-\pi}^{\pi} \sin nx \cos nx \, dx = \frac{1}{2n} \sin^2 nx \Big]_{-\pi}^{\pi} = 0.$$

All that remains is to notice that the function 1 in the sequence (2) is orthogonal to all the others, that is,

$$\int_{-\pi}^{\pi} 1 \cdot y_n \, dx = 0$$

for every n, and this completes the argument.

There is a very suggestive analogy between Fourier series and vectors that should be mentioned here. Let us briefly consider ordinary three-dimensional Euclidean space. In this space $\mathbf{i}$, $\mathbf{j}$, $\mathbf{k}$ are familiar mutually perpendicular unit vectors in the coordinate directions, and other vectors can be written in the form

$$\mathbf{A} = a_1 \mathbf{i} + a_2 \mathbf{j} + a_3 \mathbf{k}$$

and

$$\mathbf{B} = b_1 \mathbf{i} + b_2 \mathbf{j} + b_3 \mathbf{k}.$$

Let us denote the "dot product" $\mathbf{A} \cdot \mathbf{B}$ of $\mathbf{A}$ and $\mathbf{B}$ by the symbol $(\mathbf{A},\mathbf{B})$, so that

$$(\mathbf{A},\mathbf{B}) = a_1 b_1 + a_2 b_2 + a_3 b_3. \tag{10}$$

In the present context we prefer to call this quantity the *inner product* of $\mathbf{A}$ and $\mathbf{B}$, and our purpose is to point out that this inner product is closely connected with the most important geometric features of the space.

First, two vectors $\mathbf{A}$ and $\mathbf{B}$ are *orthogonal* (or perpendicular) if their inner product is zero, that is, if

$$(\mathbf{A},\mathbf{B}) = a_1 b_1 + a_2 b_2 + a_3 b_3 = 0. \tag{11}$$

Next, the inner product underlies the concept of the *norm*, or length, of a vector $\mathbf{A}$: if we denote the norm by $\|\mathbf{A}\|$—a symbol that resembles, but differs from, the absolute value sign—then

$$\|\mathbf{A}\| = \sqrt{a_1^2 + a_2^2 + a_3^2} = \sqrt{(\mathbf{A},\mathbf{A})}. \tag{12}$$

This norm in turn gives rise to the concept of the *distance* between any two points in the space, or equivalently, the distance between the tips of any two vectors,

$$d(\mathbf{A},\mathbf{B}) = \|\mathbf{A} - \mathbf{B}\|. \tag{13}$$

As our final bit of review, we recall that if $\mathbf{u}_1$, $\mathbf{u}_2$, $\mathbf{u}_3$ are any three mutually orthogonal unit vectors, then every vector $\mathbf{V}$ can be expressed in the form

$$\mathbf{V} = \alpha_1\mathbf{u}_1 + \alpha_2\mathbf{u}_2 + \alpha_3\mathbf{u}_3, \tag{14}$$

where α_1, α_2, α_3 are constants. In order to determine these constant coefficients for a given vector $\mathbf{V}$, we form the inner product of both sides of (14) with $\mathbf{u}_k$, where $k = 1$, 2, or 3. This yields

$$(\mathbf{V},\mathbf{u}_k) = \alpha_1(\mathbf{u}_1,\mathbf{u}_k) + \alpha_2(\mathbf{u}_2,\mathbf{u}_k) + \alpha_3(\mathbf{u}_3,\mathbf{u}_k);$$

and since the vectors $\mathbf{u}_1$, $\mathbf{u}_2$, $\mathbf{u}_3$ are mutually orthogonal and have length 1, the sum on the right collapses to a single term,

$$(\mathbf{V},\mathbf{u}_k) = \alpha_k.$$

The formula for the coefficients is therefore

$$\alpha_k = (\mathbf{V}, \mathbf{u}_k). \tag{15}$$

Equations (14) and (15) should be compared with (6) and (8), because their meanings are very similar. In essence, the α_k are the "Fourier coefficients" of the vector $\mathbf{V}$, and (14) is its expansion in a "Fourier series."

In the case of genuine Fourier series, we work with functions defined on an interval $[a,b]$ instead of with vectors. We speak of a "function space" instead of a three-dimensional "vector space." This function space is infinite-dimensional, in the sense that we need an infinite orthonormal sequence to represent an arbitrary function. Life is somewhat more complicated in this infinite-dimensional space than it is in the three-dimensional space described above. First, it turns out that only special kinds of orthonormal sequences are capable of representing "arbitrary" functions. And second, it is necessary to introduce restrictions that remove the vagueness from the expression "arbitrary function" and precisely define the class of functions that are to be represented by their Fourier series. We begin this precise discussion in the next few paragraphs, and continue it in the next section.

The function space we consider is denoted by R and consists of all functions $f(x)$ that are defined and Riemann integrable on the interval $[a,b]$. Since the inner product (10) is the sum of products of components, and since the values of a function can be thought of as its components, it is natural to define the *inner product* (f,g) of two functions in R by

$$(f,g) = \int_a^b f(x)g(x)\,dx. \tag{16}$$

Clearly,

$$(f_1 + f_2,g) = (f_1,g) + (f_2,g),$$
$$(cf,g) = c(f,g) \quad \text{and} \quad (f,g) = (g,f).$$

With (11) as our guide, we say that f and g are *orthogonal* if their inner product is zero, that is, if

$$(f,g) = 0.$$

This is precisely the meaning of orthogonality as given in Section 33,

$$\int_a^b f(x)g(x)\,dx = 0.$$

By the definition at the beginning of this section, an *orthogonal sequence* in R is a sequence with the property that each function is orthogonal to every other and no function is orthogonal to itself. Continuing the analogy, the *norm* of a function f is defined by

$$\|f\| = \sqrt{(f,f)} = \left[\int_a^b [f(x)]^2\,dx\right]^{1/2}, \tag{17}$$

so that

$$\|f\|^2 = (f,f).$$

A function f is called a *null function* if

$$\|f\| = 0 \quad \text{or, equivalently, if} \quad \int_a^b [f(x)]^2\,dx = 0.$$

A null function need not be identically zero. For example, if $f(x) = 0$ on $[-\pi,\pi]$ except at the points $x = 1, \frac{1}{2}, \frac{1}{3}, \ldots$, but $f(x) = 1$ at these points, then f is a null function. In the present context it is convenient to consider a null function as being essentially equal to zero, so that two functions are considered to be equal if their difference is a null function. With this understanding, the norm has the simple properties

$$\|cf\| = |c|\,\|f\|, \qquad \|f\| \geq 0,$$
$$\|f\| = 0 \quad \text{if and only if} \quad f = 0. \tag{18}$$

Two properties that are not so simple are

$$|(f,g)| \leq \|f\|\,\|g\| \tag{19}$$

and

$$\|f + g\| \leq \|f\| + \|g\|. \tag{20}$$

The inequality (19) is called the *Schwarz inequality*. By using (16) and (17), it can be written out as follows [in the form $(f,g)^2 \leq \|f\|^2 \|g\|^2$]:

$$\left[\int_a^b f(x)g(x)\, dx \right]^2 \leq \int_a^b [f(x)]^2\, dx \cdot \int_a^b [g(x)]^2\, dx.$$

The inequality (20) is called the *Minkowski inequality*; its written-out form is

$$\left[\int_a^b [f(x) + g(x)]^2\, dx \right]^{1/2} \leq \left[\int_a^b [f(x)]^2\, dx \right]^{1/2} + \left[\int_a^b [g(x)]^2\, dx \right]^{1/2}.$$

The integral versions of these inequalities have a formidable appearance, and one might think that probably they cannot be established except by the use of complicated reasoning. In fact, however, there exists a simple but ingenious proof of (19) which we ask readers to think through for themselves (Problem 3 below); and (20) follows quite easily from (19) by an argument that we give here. Thus, by Schwarz's inequality we have

$$\begin{aligned}
\|f + g\|^2 = (f + g, f + g) &= (f,f) + 2(f,g) + (g,g) \\
&= \|f\|^2 + 2(f,g) + \|g\|^2 \\
&\leq \|f\|^2 + 2\,|(f,g)| + \|g\|^2 \\
&\leq \|f\|^2 + 2\,\|f\|\,\|g\| + \|g\|^2 \\
&= (\|f\| + \|g\|)^2,
\end{aligned}$$

and we now obtain (20) by taking square roots.

By using the concept of the norm of a function, we are now able to define the *distance* $d(f,g)$ between two functions f and g in R:

$$d(f,g) = \|f - g\| = \left[\int_a^b [f(x) - g(x)]^2\, dx \right]^{1/2}. \tag{21}$$

We also speak of $d(f,g)$ as the distance from f to g, or the distance of g from f. It is easy to see from (18) and (20) that distance has the following properties:

$d(f,g) \geq 0$, and $d(f,g) = 0$ if and only if $f = g$;

$d(f,g) = d(g,f)$ [*symmetry*];

$d(f,g) \leq d(f,h) + d(h,g)$ [*triangle inequality*].

A space (of vectors, functions, or any objects whatever) with a distance function possessing these properties is called a *metric space*. With the understanding that functions in R are considered to be equal if they differ by a null function, R is a metric space whose structure we continue to investigate in the next section.

NOTE ON MINKOWSKI. At the age of 18 the Russian–German mathematician Hermann Minkowski (1864–1909) won the Grand Prize of the Academy of Sciences in Paris for his brilliant research on quadratic forms, starting from a problem about the representation of an integer as the sum of five squares. This work later led to the creation of a whole new branch of number theory now called the Geometry of Numbers, which in turn is based on his highly original ideas about the properties of convex bodies in n-dimensional space. In this connection he introduced the abstract concept of distance, analyzed the notions of volume and surface, and established the important inequality that bears his name. In the years 1907–1908 Minkowski became the mathematician of relativity by geometrizing the new subject. He created the concept of four-dimensional space–time as the proper mathematical setting for Einstein's essentially physical (and nonmathematical) way of thinking about special relativity. In a now-famous lecture of 1908 he began with a sentence that is not easily forgotten: "From now on space by itself, and time by itself, are doomed to fade away into mere shadows, and only a kind of union of the two will retain an independent existence."

NOTE ON SCHWARZ. Hermann Amadeus Schwarz (1843–1921), a pupil of Weierstrass whom he succeeded in Berlin, made substantial contributions to the theory of minimal surfaces in geometry and to conformal mapping, potential theory, hypergeometric functions, and other topics in analysis. In conformal mapping, he rescued and rigorously nailed down some of Riemann's very important but rather intuitive discoveries, especially the basic Riemann mapping theorem. In minimal surfaces, he gave the first rigorous proof that a sphere has a smaller surface area than any other body of the same volume. He also discovered and proved the "pedal triangle" theorem of elementary geometry: In any acute-angled triangle, the inscribed triangle with smallest perimeter is the one whose vertices are the three feet of the altitudes of the given triangle.[15]

PROBLEMS

1. One of the important consquences of the orthogonality properties of the trigonometric sequence (2) [namely, equations (4) in this section and (2), (5), (6), (8) in Section 33] is *Bessel's inequality*: If $f(x)$ is any function integrable on $[-\pi, \pi]$, its ordinary Fourier coefficients satisfy the inequality

$$\frac{1}{2} a_0^2 + \sum_{k=1}^{\infty} (a_k^2 + b_k^2) \le \frac{1}{\pi} \int_{-\pi}^{\pi} [f(x)]^2 \, dx. \qquad (*)$$

Prove this by the following steps:
(a) For any $n \ge 1$, define

$$s_n(x) = \frac{1}{2} a_0 + \sum_{k=1}^{n} (a_k \cos kx + b_k \sin kx)$$

---

[15] For details, see Chapter 5 of H. Rademacher and O. Toeplitz, *The Enjoyment of Mathematics*, Princeton University Press, 1957; or R. Courant and H. Robbins, *What Is Mathematics?*, Oxford University Press, 1941, pp. 346–51.

and show that

$$\frac{1}{\pi} \int_{-\pi}^{\pi} f(x)s_n(x)\, dx = \frac{1}{2}a_0^2 + \sum_{k=1}^{n} (a_k^2 + b_k^2).$$

(b) By considering all possible products in the multiplication of $s_n(x)$ by itself, show that

$$\frac{1}{\pi} \int_{-\pi}^{\pi} [s_n(x)]^2\, dx = \frac{1}{2}a_0^2 + \sum_{k=1}^{n} (a_k^2 + b_k^2).$$

(c) By writing

$$\frac{1}{\pi} \int_{-\pi}^{\pi} [f(x) - s_n(x)]^2\, dx$$

$$= \frac{1}{\pi} \int_{-\pi}^{\pi} [f(x)]^2\, dx - \frac{2}{\pi} \int_{-\pi}^{\pi} f(x)s_n(x)\, dx + \frac{1}{\pi} \int_{-\pi}^{\pi} [s_n(x)]^2\, dx$$

$$= \frac{1}{\pi} \int_{-\pi}^{\pi} [f(x)]^2\, dx - \frac{1}{2}a_0^2 - \sum_{k=1}^{n} (a_k^2 + b_k^2),$$

conclude that

$$\frac{1}{2}a_0^2 + \sum_{k=1}^{n} (a_k^2 + b_k^2) \le \frac{1}{\pi} \int_{-\pi}^{\pi} [f(x)]^2\, dx,$$

and from this complete the proof.

Observe that the convergence of the series on the left side of (*) implies the following corollary of Bessel's inequality: If a_n and b_n are the ordinary Fourier coefficients of $f(x)$, then $a_n \to 0$ and $b_n \to 0$ as $n \to \infty$.

2. In the case of the orthonormal sequence (5), verify in detail that the Fourier coefficients (8) are slightly different from the ordinary Fourier coefficients, but that the Fourier series (6) is exactly the same as the ordinary Fourier series.

3. Prove the Schwarz inequality (19). *Hint:* If $\|g\| \ne 0$, then the function $F(\alpha) = \|f + \alpha g\|^2$ is a second degree polynomial in α that has no negative values; examine the discriminant.

4. A well-known theorem of elementary geometry states that the sum of the squares of the sides of a parallelogram equals the sum of the squares of its diagonals. Prove that this so-called *parallelogram law* is true for the norm in R:

$$2\|f\|^2 + 2\|g\|^2 = \|f + g\|^2 + \|f - g\|^2.$$

5. Prove the *Pythagorean theorem* and its converse in R: f is orthogonal to g if and only if $\|f - g\|^2 = \|f\|^2 + \|g\|^2$.

6. Show that a null function is zero at each point of continuity, so that a continuous null function is identically zero.

38 THE MEAN CONVERGENCE OF FOURIER SERIES

Consider a function $f(x)$ and a sequence of functions $p_n(x)$, all defined and integrable on the interval $[a,b]$. There are different ways in which

$p_n(x)$ can converge to $f(x)$, and these are best understood in terms of the problem of approximating $f(x)$ by $p_n(x)$.

If we try to approximate $f(x)$ by $p_n(x)$, then each of the numbers

$$|f(x) - p_n(x)| \quad \text{and} \quad [f(x) - p_n(x)]^2 \tag{1}$$

gives a measure of the error in the approximation at the point x. It is clear that if one of these numbers is small, then so is the other. The usual definition of convergence amounts to the statement that the sequence of functions $p_n(x)$ converges to the function $f(x)$ if for each point x either of the expressions (1) approaches zero as $n \to \infty$. This is the familiar concept used in Sections 33 to 36, and for obvious reasons it is called *pointwise convergence.*

On the other hand, we might prefer to use a measure of error that refers to the whole interval $[a,b]$ simultaneously, instead of point by point. We can obtain such a measure by integrating the expressions (1) from a to b,

$$\int_a^b |f(x) - p_n(x)| \, dx \quad \text{and} \quad \int_a^b [f(x) - p_n(x)]^2 \, dx.$$

The second integral here is a better choice than the first, for two reasons: it avoids the awkward absolute value sign in the first integral; and the exponent 2 makes many of the necessary calculations very convenient to carry out, as we will see below. The measure of error we adopt is therefore

$$E_n = \int_a^b [f(x) - p_n(x)]^2 \, dx. \tag{2}$$

This quantity is called the *mean square error.* The terminology is appropriate because if the integral (2) is divided by $b - a$, the result is exactly the mean value of the square error $[f(x) - p_n(x)]^2$. If (2) approaches zero as $n \to \infty$, the sequence $\{p_n(x)\}$ is said to *converge in the mean* to $f(x)$, and this concept is called *mean convergence.* We sometimes symbolize this mode of convergence by writing

$$f(x) = \underset{n\to\infty}{\text{l.i.m.}} \, p_n(x),$$

where "l.i.m." stands for "limit in the mean." Our discussion in the rest of this section will show that in the case of Fourier series mean convergence is much easier to work with than ordinary pointwise convergence.

We assumed at the beginning that the functions $f(x)$ and $p_n(x)$ belong to the function space R described in the preceding section. We now point out that the mean square error (2) is precisely the square of the norm of $f - p_n$ in R,

$$E_n = \int_a^b [f(x) - p_n(x)]^2 \, dx = \|f - p_n\|^2. \tag{3}$$

The mean convergence of $p_n(x)$ to $f(x)$ is therefore completely equivalent to the convergence of the sequence $\{p_n\}$ to the limit f in the metric space R, namely,

$$d(f,p_n) = \|f - p_n\| \to 0 \qquad \text{as } n \to \infty.$$

As indicated here, we will often use f and p_n as abbreviations for $f(x)$ and $p_n(x)$, in order to simplify the notation.

We now come to the main business of this section. Let $\{\phi_n(x)\}$ be an orthonormal sequence of integrable functions on $[a,b]$, so that

$$\int_a^b \phi_m(x)\phi_n(x) \, dx = \begin{cases} 0 & \text{for } m \neq n, \\ 1 & \text{for } m = n. \end{cases} \tag{4}$$

We consider the first n of these functions,

$$\phi_1(x), \qquad \phi_2(x), \qquad \ldots, \qquad \phi_n(x), \tag{5}$$

and we seek to approximate a given integrable function $f(x)$ by a linear combination of the functions (5),

$$p_n(x) = b_1\phi_1(x) + b_2\phi_2(x) + \cdots + b_n\phi_n(x).$$

Our purpose is to minimize the mean square error (2),

$$E_n = \int_a^b [f - p_n]^2 \, dx = \int_a^b [f - (b_1\phi_1 + \cdots + b_n\phi_n)]^2 \, dx, \tag{6}$$

by making a suitable choice of the coefficients $b_1, \ldots, b_n$.

Our first step is to expand the term in brackets in (6), which yields

$$E_n = \int_a^b f^2 \, dx - 2\int_a^b (b_1\phi_1 + \cdots + b_n\phi_n)f \, dx$$

$$+ \int_a^b (b_1\phi_1 + \cdots + b_n\phi_n)^2 \, dx. \tag{7}$$

If the Fourier coefficients of f with respect to the orthonormal sequence $\{\phi_k\}$ are denoted by

$$a_k = \int_a^b f\phi_k \, dx,$$

as in Section 37, then the second integral in (7) is

$$\int_a^b (b_1\phi_1 + \cdots + b_n\phi_n)f \, dx = a_1b_1 + \cdots + a_nb_n.$$

The third integral in (7) can be written

$$\int_a^b (b_1\phi_1 + \cdots + b_n\phi_n)(b_1\phi_1 + \cdots + b_n\phi_n)\, dx$$

$$= \int_a^b (b_1^2\phi_1^2 + \cdots + b_n^2\phi_n^2 + \cdots)\, dx$$

$$= b_1^2 + \cdots + b_n^2,$$

where the second group of terms " $+ \cdots$" contains products $\phi_i\phi_j$ with $i \neq j$ and the final value results from using (4). These considerations enable us to write the mean square error (7) as

$$E_n = \int_a^b f^2\, dx - 2\sum_{k=1}^n a_k b_k + \sum_{k=1}^n b_k^2. \tag{8}$$

If we now notice that

$$-2a_k b_k + b_k^2 = -a_k^2 + (b_k - a_k)^2,$$

then the formula for E_n takes its final form,

$$E_n = \int_a^b f^2\, dx - \sum_{k=1}^n a_k^2 + \sum_{k=1}^n (b_k - a_k)^2. \tag{9}$$

Formula (9) for the mean square error E_n has a number of important consequences that follow by very simple reasoning. First, the terms $(b_k - a_k)^2$ in (9) are positive unless $b_k = a_k$, in which case they are zero. Therefore the choice of the b_k that minimizes E_n is obviously $b_k = a_k$, and we have

Theorem 1. *For each positive integer n, the nth partial sum of the Fourier series of f, namely,*

$$\sum_{k=1}^n a_k\phi_k = a_1\phi_1 + \cdots + a_n\phi_n,$$

gives a smaller mean square error $E_n = \int_a^b (f - p_n)^2\, dx$ than is given by any other linear combination $p_n = b_1\phi_1 + \cdots + b_n\phi_n$. Further, this minimum value of the error is

$$\min E_n = \int_a^b f^2\, dx - \sum_{k=1}^n a_k^2. \tag{10}$$

Formula (6) tells us that we always have $E_n \geq 0$, because the integrand in (6), being a square, is nonnegative. Since $E_n \geq 0$ for all choices of the b_k, it is clear that the minimum value of E_n (which arises when $b_k = a_k$) is also ≥ 0. Therefore (10) implies that

$$\int_a^b f^2\, dx - \sum_{k=1}^n a_k^2 \geq 0 \qquad \text{or} \qquad \sum_{k=1}^n a_k^2 \leq \int_a^b f^2\, dx.$$

By letting $n \to \infty$ we at once obtain

Theorem 2. *If the numbers* $a_n = \int_a^b f\phi_n \, dx$ *are the Fourier coefficients of* f *with respect to the orthonormal sequence* $\{\phi_n\}$, *then the series* $\sum a_n^2$ *converges and satisfies Bessel's inequality,*

$$\sum_{n=1}^{\infty} a_n^2 \le \int_a^b [f(x)]^2 \, dx. \tag{11}$$

Since the nth term of a convergent series must approach zero, Theorem 2 implies

Theorem 3. *If the numbers* $a_n = \int_a^b f\phi_n \, dx$ *are the Fourier coefficients of* f *with respect to the orthonormal sequence* $\{\phi_n\}$, *then* $a_n \to 0$ *as* $n \to \infty$.

Theorems 2 and 3 are obtained for ordinary Fourier series in Problem 37-1. Here they are seen to be true for generalized Fourier series with respect to arbitrary orthonormal sequences.

For applications it is important to know whether or not the Fourier series of f is a valid expansion of f in the sense of mean convergence. This is equivalent to asking whether or not the partial sums of the Fourier series of f converge in the mean to f, that is, whether or not

$$f = \underset{n \to \infty}{\text{l.i.m.}} \sum_{k=1}^{n} a_k \phi_k. \tag{12}$$

In view of Theorem 1 it is evident that we do have a valid expansion of f if and only if

$$\min E_n \to 0 \qquad \text{as } n \to \infty,$$

and by formula (10) we see that this happens if and only if *Parseval's equation* holds:

$$\int_a^b f^2 \, dx - \sum_{k=1}^{\infty} a_k^2 = 0.$$

We summarize these observations in the following theorem.

Theorem 4. *The representation of* f *by its Fourier series, namely,*

$$f = a_1\phi_1 + a_2\phi_2 + \cdots + a_n\phi_n + \cdots, \tag{13}$$

is valid in the sense of mean convergence if and only if Bessel's inequality (11) *becomes Parseval's equation,*

$$\sum_{n=1}^{\infty} a_n^2 = \int_a^b [f(x)]^2 \, dx. \tag{14}$$

If a Fourier expansion of the form (13) is valid (in the sense of mean convergence) for every function $f(x)$ in R, then the orthonormal sequence $\{\phi_n(x)\}$ is said to be *complete*. A complete sequence, then, is a sequence $\{\phi_n\}$ that can be used for mean square approximations of the form (12) for arbitrary functions f in R. It can be proved that the trigonometric sequence

$$\frac{1}{\sqrt{2\pi}}, \quad \frac{\cos x}{\sqrt{\pi}}, \quad \frac{\sin x}{\sqrt{\pi}}, \quad \frac{\cos 2x}{\sqrt{\pi}}, \quad \frac{\sin 2x}{\sqrt{\pi}}, \cdots \quad (15)$$

is complete on $[-\pi, \pi]$.

Remark 1. The proof of the theorem just stated about the trigonometric sequence (15) is long and would take us much too far afield.[16] However, if we recall Problem 2 in Section 37, then we see that this theorem immediately yields the following major conclusion, which can be interpreted as sweeping away all the difficulties that arise in the theory of pointwise convergence for Fourier series.

Theorem 5. *If $f(x)$ is any function defined and integrable on $[-\pi, \pi]$, then $f(x)$ is represented by its ordinary Fourier series in the sense of mean convergence,*

$$f(x) = \frac{1}{2} a_0 + \sum_{n=1}^{\infty} (a_n \cos nx + b_n \sin nx), \quad (16)$$

where the a_n and b_n are the ordinary Fourier coefficients of $f(x)$.

To appreciate the clean simplicity of this statement, it helps to recall from our previous work that this representation theorem is false if (16) is interpreted in the sense of pointwise convergence; further, the representation even fails for some continuous functions.

Remark 2. In Problem 6 below we ask the student to show that if we specialize to the interval $[-\pi, \pi]$ and use the ordinary Fourier coefficients, then Parseval's equation (14) takes the form

$$\frac{1}{\pi} \int_{-\pi}^{\pi} [f(x)]^2 \, dx = \frac{1}{2} a_0^2 + \sum_{1}^{\infty} (a_n^2 + b_n^2). \quad (17)$$

The function $f(x)$ in this equation is assumed to belong to R, that is, to be Riemann integrable on $[-\pi, \pi]$, and for any such function its square

---

[16]The basic tools for the proof we have in mind are two major theorems of classical analysis, *Fejer's summability theorem* and the *Weierstrass approximation theorem*.

$[f(x)]^2$ is also automatically integrable. It therefore follows from (17) that for this function the Fourier coefficients $a_0, a_1, b_1, a_2, b_2, \ldots$ have the property that the series $\sum (a_n^2 + b_n^2)$ converges. Of course, we already knew this from Problem 1 in Section 37.

However, if the Riemann integral is replaced by its more powerful cousin the Lebesgue integral, then this statement has a converse that was proved by F. Riesz and E. Fischer in 1907. The famous *Riesz–Fischer theorem*, one of the great achievements of the Lebesgue theory of integration, states that given any sequence of numbers $a_0, a_1, b_1, a_2, b_2, \ldots$ such that the series $\sum (a_n^2 + b_n^2)$ converges, there exists a unique square-integrable function $f(x)$ with these numbers as its Fourier coefficients.

It is customary to use the symbol L_2 to denote the space of functions $f(x)$ that are square-integrable on $[-\pi, \pi]$ in the sense of Lebesgue, where as usual two functions are considered to be equal if they differ by a null function.[17] When Parseval's equation (17) and the Riesz–Fischer theorem are taken together, we see from this discussion that they give a very simple characterization of the functions in L_2 in terms of their Fourier coefficients. It is remarkable that no other important class of functions has a characterization of comparable simplicity and completeness—a fact that delights the souls of mathematicians.

NOTE ON PARSEVAL. Marc-Antoine Parseval des Chênes (1755–1836), member of an aristocratic French family and ardent royalist, poet, and amateur mathematician, managed to survive the French Revolution with his head still on his shoulders, but was imprisoned briefly in 1792 and luckily fled the country when Napoleon ordered his arrest for publishing poetry attacking the regime. He published very little mathematics—and none of any distinction—but this little included (in 1799) a rough statement that only slightly resembles Parseval's equation as it is known to mathematicians today throughout the world . . . and for this his name is immortal.

PROBLEMS

1. Consider the sequence of functions $f_n(x)$, $n = 1, 2, 3, \ldots$, defined on the interval $[0,1]$ by

$$f_n(x) = \begin{cases} 0, & 0 \le x \le 1/n, \\ \sqrt{n}, & 1/n < x < 2/n, \\ 0, & 2/n \le x \le 1. \end{cases}$$

---

[17] It should be pointed out that L_2 contains R and many other functions as well, and that whenever the Lebesgue integral is applied to a function in R, it yields the same numerical result as the Riemann integral.

(a) Show that the sequence $\{f_n(x)\}$ converges pointwise to the zero function on the interval $[0,1]$.

(b) Show that the sequence $\{f_n(x)\}$ does *not* converge in the mean to the zero function on the interval $[0,1]$.

2. Consider the following sequence of closed subintervals of $[0,1]$: $[0,\frac{1}{2}]$, $[\frac{1}{2},1]$, $[0,\frac{1}{4}]$, $[\frac{1}{4},\frac{1}{2}]$, $[\frac{1}{2},\frac{3}{4}]$, $[\frac{3}{4},1]$, $[0,\frac{1}{8}]$, $[\frac{1}{8},\frac{1}{4}]$, ..., and denote the nth subinterval by I_n. Now define a sequence of functions $f_n(x)$ on $[0,1]$ by

$$f_n(x) = \begin{cases} 1 & \text{for } x \text{ in } I_n, \\ 0 & \text{for } x \text{ not in } I_n. \end{cases}$$

(a) Show that the sequence $\{f_n(x)\}$ converges in the mean to the zero function on the interval $[0,1]$.

(b) Show that the sequence $\{f_n(x)\}$ does *not* converge pointwise at any point of the interval $[0,1]$.

3. Obtain the formula $b_k = a_k$ from both (8) and (9), by using the fact that $\partial E_n/\partial b_k = 0$ when E_n has a minimum value.

4. The function $f(x) = 1$ is to be approximated on $[0,\pi]$ by $p(x) = b_1 \sin x + b_2 \sin 2x + b_3 \sin 3x + b_4 \sin 4x + b_5 \sin 5x$ in such a way that $\int_0^\pi [1 - p(x)]^2 \, dx$ is minimized. What values should the coefficients b_k have?

5. The function $f(x) = x$ to be approximated on $[0,\pi]$ by

$$p(x) = b_1 \sin x + b_2 \sin 2x + b_3 \sin 3x$$

in such a way that $\int_0^\pi [x - p(x)]^2 \, dx$ is minimized. What values should the coefficients b_k have?

6. Show that Parseval's equation (14) has the form (17) when the orthonormal sequence $\{\phi_n(x)\}$ is the trigonometric sequence (15).

7. Obtain the sums

$$\sum_1^\infty \frac{1}{n^2} = \frac{\pi^2}{6} \quad \text{and} \quad \sum_1^\infty \frac{1}{n^4} = \frac{\pi^4}{90}$$

by applying Parseval's equation in the preceding problem to the two Fourier series

$$x = 2\left(\sin x - \frac{\sin 2x}{2} + \frac{\sin 3x}{3} - \cdots\right)$$

and

$$x^2 = \frac{\pi^2}{3} + 4\sum_1^\infty (-1)^n \frac{\cos nx}{n^2}.$$

[These series are found in Example 33-1 and Problem 35-10(a).]

8. Use the method and results of Problem 7 to obtain the sum

$$\sum_1^\infty \frac{1}{n^6} = \frac{\pi^6}{945}$$

from the sine series for x^2 [Problem 35-10(b)].

9. Use the method and results of Problems 7 and 8 to obtain the sum

$$\sum_{1}^{\infty} \frac{1}{n^8} = \frac{\pi^8}{9450}$$

from the cosine series for x^4 [Problem 35-12(a)].

APPENDIX A. A POINTWISE CONVERGENCE THEOREM

We divide the work of stating and proving the theorem into stages, for easier comprehension.

1. Our first purpose is to obtain a convenient explicit formula for the difference between a function and the nth partial sum of its Fourier series. This formula will enable us to prove pointwise convergence for a large class of functions that includes all the examples given in this chapter.

To develop this formula, we begin by assuming only that $f(x)$ is an integrable function of period 2π. The nth partial sum of its Fourier series is then

$$s_n(x) = \frac{1}{2} a_0 + \sum_{k=1}^{n} (a_k \cos kx + b_k \sin kx), \tag{1}$$

where

$$a_k = \frac{1}{\pi} \int_{-\pi}^{\pi} f(t) \cos kt\, dt \quad \text{and} \quad b_k = \frac{1}{\pi} \int_{-\pi}^{\pi} f(t) \sin kt\, dt. \tag{2}$$

By substituting (2) into (1) we obtain

$$s_n(x) = \frac{1}{\pi} \int_{-\pi}^{\pi} f(t) \left[\frac{1}{2} + \sum_{k=1}^{n} (\cos kt \cos kx + \sin kt \sin kx) \right] dt$$

$$= \frac{1}{\pi} \int_{-\pi}^{\pi} f(t) \left[\frac{1}{2} + \sum_{k=1}^{n} \cos k(t - x) \right] dt. \tag{3}$$

If we define the *Dirichlet kernel* by

$$D_n(u) = \frac{1}{2} + \sum_{k=1}^{n} \cos ku, \tag{4}$$

then (3) can be put in the more compact form

$$s_n(x) = \frac{1}{\pi} \int_{-\pi}^{\pi} f(t) D_n(t - x)\, dt. \tag{5}$$

Putting $u = t - x$ in (5) yields

$$s_n(x) = \frac{1}{\pi} \int_{-\pi-x}^{\pi-x} f(x + u)D_n(u)\, du. \tag{6}$$

By the definition (4), $D_n(u)$ has period 2π; and as a function of u, $f(x + u)$ also has period 2π. Therefore the integral of $f(x + u)D_n(u)$ over any interval of length 2π equals the integral over any other interval of length 2π, and (6) can be written

$$s_n(x) = \frac{1}{\pi} \int_{-\pi}^{\pi} f(x + u)D_n(u)\, du. \tag{7}$$

Since $D_n(-u) = D_n(u)$, we can replace u by $-u$ in (7) to obtain

$$s_n(x) = -\frac{1}{\pi} \int_{\pi}^{-\pi} f(x - u)D_n(u)\, du$$

$$= \frac{1}{\pi} \int_{-\pi}^{\pi} f(x - u)D_n(u)\, du, \tag{8}$$

and adding (7) and (8) yields

$$2s_n(x) = \frac{1}{\pi} \int_{-\pi}^{\pi} [f(x + u) + f(x - u)]D_n(u)\, du.$$

The integrand here is an even function of u, so the integral from $-\pi$ to π is twice the integral from 0 to π, and we have

$$s_n(x) = \frac{1}{\pi} \int_{0}^{\pi} [f(x + u) + f(x - u)]D_n(u)\, du. \tag{9}$$

To bring $f(x)$ into our discussion and put the difference $s_n(x) - f(x)$ into a convenient form, we notice that

$$\frac{1}{\pi} \int_{0}^{\pi} D_n(u)\, du = \frac{1}{2},$$

since the terms $\cos ku$ in (4) integrate to zero. If we now multiply this by $2f(x)$ we obtain

$$f(x) = \frac{1}{\pi} \int_{0}^{\pi} 2f(x)D_n(u)\, du, \tag{10}$$

and subtracting (10) from (9) yields

$$s_n(x) - f(x) = \frac{1}{\pi} \int_{0}^{\pi} [f(x + u) + f(x - u) - 2f(x)]D_n(u)\, du. \tag{11}$$

This formula is our fundamental tool for studying the convergence of $s_n(x)$ to $f(x)$.

2. At this point we need the following closed formula for the Dirichlet kernel (4),

$$D_n(u) = \frac{1}{2} + \sum_{k=1}^{n} \cos ku = \frac{\sin(n + \frac{1}{2})u}{2 \sin \frac{1}{2}u}, \tag{12}$$

if $\sin \frac{1}{2}u \neq 0.$[18] This enables us to write (11) in the form

$$s_n(x) - f(x) = \frac{1}{\pi} \int_0^\pi g(u) \sin(n + \frac{1}{2})u \, du, \tag{13}$$

where

$$g(u) = \frac{f(x + u) + f(x - u) - 2f(x)}{2 \sin \frac{1}{2}u}. \tag{14}$$

Of course, $g(u)$ is really a function of both u and x. However, we are going to be examining $g(u)$ with x fixed and u variable, and this notation helps to avoid confusion. In view of (13), to prove that $s_n(x) \rightarrow f(x)$ as $n \rightarrow \infty$, we must prove that

$$\lim_{n \to \infty} \int_0^\pi g(u) \sin(n + \frac{1}{2})u \, du = 0. \tag{15}$$

Our task is to give a rigorous proof of (15) with appropriate, understandable, and clearly stated assumptions about the behavior of the function $f(x)$.

3. As a preliminary to the proof of the main convergence theorem stated below, we need the following lemma.

Lemma. *If $\phi(u)$ is integrable on the interval $[0,\pi]$, then*

$$\lim_{n \to \infty} \int_0^\pi \phi(u) \sin(n + \frac{1}{2})u \, du = 0. \tag{16}$$

Proof. By the addition formula for the sine, this integral can be broken up into

$$\int_0^\pi \phi(u) \cos \frac{1}{2}u \cdot \sin nu \, du + \int_0^\pi \phi(u) \sin \frac{1}{2}u \cdot \cos nu \, du.$$

---

[18] This formula can easily be proved by writing down the identity $2 \cos A \sin B = \sin(A + B) - \sin(A - B)$ n times, with $A = u, 2u, 3u, \ldots, nu$ and $B = u/2$, and adding the results to obtain

$$2 \sin \frac{1}{2}u(\cos u + \cos 2u + \cdots + \cos nu) = \sin(n + \frac{1}{2}u) - \sin \frac{1}{2}u.$$

If we write

$$A_n = \frac{2}{\pi} \int_0^\pi \phi(u) \sin \tfrac{1}{2}u \cdot \cos nu \, du$$

and

$$B_n = \frac{2}{\pi} \int_0^\pi \phi(u) \cos \tfrac{1}{2}u \cdot \sin nu \, du,$$

then the integral (16) is

$$\frac{\pi}{2}(A_n + B_n).$$

It is easy to see that A_n is the nth coefficient in the cosine series for $\phi(u) \sin \tfrac{1}{2}u$, and B_n is the nth coefficient in the sine series for $\phi(u) \cos \tfrac{1}{2}u$. Since $\phi(u)$ is integrable, each of these functions is also integrable. It now follows from the corollary to Bessel's inequality stated at the end of Problem 37-1 that $A_n \to 0$ and $B_n \to 0$ as $n \to \infty$, and the proof of (16) is complete.

4. In view of condition (15) and the lemma, all that remains is to formulate assumptions sufficient to guarantee that the function $g(u)$ defined by (14) is integrable on $[0,\pi]$.

So far, we have only the general requirements that $f(x)$ is integrable on $[-\pi,\pi]$ and periodic with period 2π. We now make the further assumption that $f(x)$ is *piecewise smooth* on $[-\pi,\pi]$. This means that the graph on $[-\pi,\pi]$ consists of a finite number of continuous curves on each of which $f'(x)$ exists and is continuous. It also means that the derivative exists at the endpoints of these curves, in the sense of

$$\lim_{u \to 0+} \frac{f(x + u) - f(x+)}{u} \quad \text{and} \quad \lim_{u \to 0+} \frac{f(x - u) - f(x-)}{-u}. \quad (17)$$

In this way, the function $f(x)$ is guaranteed to have a right derivative and a left derivative at every point x—including points of discontinuity—which we denote by $f'_+(x)$ and $f'_-(x)$.

Of course, the function $f(x)$ is allowed to have a finite number of jump discontinuities on $[-\pi,\pi]$. However, since the Fourier coefficients are not changed if $f(x)$ is redefined at a finite number of points, we may assume without loss of generality that

$$f(x) = \frac{f(x-) + f(x+)}{2} \quad (18)$$

at every point x, whether $f(x)$ is continuous at x or not.

Our pointwise convergence theorem can now be stated as follows.

Theorem. *If $f(x)$ is piecewise smooth on $[-\pi,\pi]$, is periodic with period 2π, and is defined at points of discontinuity by (18), then the Fourier series of $f(x)$ converges to $f(x)$ at every point x.*

5. To prove this theorem, let x be any fixed point. We wish to establish the correctness of (15), and in view of the lemma, it suffices to show that the function

$$g(u) = \frac{f(x + u) + f(x - u) - 2f(x)}{2 \sin \frac{1}{2}u} \qquad (19)$$

is integrable on $[0,\pi]$. It is clear that the only doubt about integrability arises from the fact that $\sin \frac{1}{2}u = 0$ when $u = 0$—for elsewhere in the interval, $\sin \frac{1}{2}u$ is continuous and positive, and the numerator of (19) is certainly an integrable function of u on $[0,\pi]$. We see from these remarks that $g(u)$ will be integrable on $[0,\pi]$ if we can show that $g(u)$ approaches a finite limit as $u \to 0+$.

By using (18) we can write

$$g(u) = \frac{f(x + u) + f(x - u) - f(x-) - f(x+)}{2 \sin \frac{1}{2}u}$$

$$= \left[\frac{f(x + u) - f(x+)}{u} + \frac{f(x - u) - f(x-)}{u} \right] \cdot \frac{\frac{1}{2}u}{\sin \frac{1}{2}u}.$$

But as $u \to 0+$, (17) tells us that

$$\frac{f(x + u) - f(x+)}{u} \to f'_+(x) \qquad \text{and} \qquad \frac{f(x - u) - f(x-)}{u} \to -f'_-(x),$$

and we know that

$$\frac{\frac{1}{2}u}{\sin \frac{1}{2}u} \to 1.$$

It therefore follows that

$$g(u) \to f'_+(x) - f'_-(x),$$

so $g(u)$ is integrable on $[0,\pi]$ and the proof is complete.

CHAPTER
7

PARTIAL DIFFERENTIAL EQUATIONS AND BOUNDARY VALUE PROBLEMS

39 INTRODUCTION. HISTORICAL REMARKS

The theory of Fourier series discussed in the preceding chapter had its historical origin in the middle of the eighteenth century, when several mathematicians were studying the vibrations of stretched strings. The mathematical theory of these vibrations amounts to the problem of solving the partial differential equation

$$a^2 \frac{\partial^2 y}{\partial x^2} = \frac{\partial^2 y}{\partial t^2}, \tag{1}$$

where a is a positive constant. This *one-dimensional wave equation* has many solutions, and the problem, for a particular vibrating string, is to find the solution that satisfies certain preliminary conditions associated with this string, such as its initial shape, its initial velocity, etc. The solution then describes the subsequent motion of the string as it vibrates under tension. The equilibrium position of the string is assumed to be along the x-axis, and if $y = y(x,t)$ is the desired solution of (1), then for a fixed value of $t \geq 0$ the curve $y = y(x,t)$ gives the shape of the displaced string at that moment (see the dashed curve in Fig. 46), and this shape changes from moment to moment.

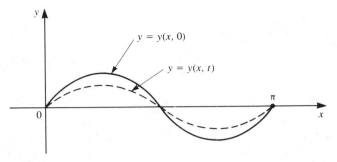

FIGURE 46

For the case of a string stretched between the points $x = 0$ and $x = \pi$, and then deformed into an arbitrary shape and released at the moment $t = 0$, Daniel Bernoulli (in 1753) gave the solution of (1) as a series of the form

$$y = b_1 \sin x \cos at + b_2 \sin 2x \cos 2at + \cdots. \tag{2}$$

It is easy to verify by inspection that a typical term of this series, $b_n \sin nx \cos nat$, is a solution of equation (1). Further, every finite sum of such terms is a solution, and the series (2) will also be a solution if term-by-term differentiation of the series is justified.[1] When $t = 0$, the series (2) reduces to

$$y = b_1 \sin x + b_2 \sin 2x + \cdots.$$

This should give the initial shape of the string, that is, the curve $y = y(x,0)$ into which the string is deformed at the moment $t = 0$ when the string is released and the vibrations begin (see the solid curve in Fig. 46).

However, d'Alembert (in 1747) and Euler (in 1748) had already published solutions of the problem which, for the case stated above, have the form

$$y = \frac{1}{2}[f(x + at) + f(x - at)]. \tag{3}$$

Here the curve $y = f(x)$ is assumed to be the shape of the string at time $t = 0$; also, the function $f(x)$ is assumed to be defined outside the interval $[0,\pi]$ by the requirement that it is an odd function of period 2π,

---

[1] In Bernoulli's time no mathematicians had any doubt that infinite series of functions can be differentiated freely term-by-term. Such doubt was the product of a later, more skeptical, and more sophisticated age.

that is,

$$f(-x) = -f(x) \qquad \text{and} \qquad f(x + 2\pi) = f(x).$$

If we compare the solution of Bernoulli with that of d'Alembert and Euler, then we see at once that we ought to have

$$f(x) = b_1 \sin x + b_2 \sin 2x + \cdots, \qquad (4)$$

because this is what we get if the solutions (2) and (3) agree at time $t = 0$. Therefore, as a result of mathematically analyzing this physical problem, Bernoulli arrived at an idea that has had very far-reaching influence on the history of mathematics and physical science, namely, the possibility that a function as general as the shape of an arbitrarily deformed taut string can be expanded in a trigonometric series of the form (4).

Both d'Alembert and Euler rejected Bernoulli's idea, and for essentially the same reason. It is clear on physical grounds that there is a great amount of freedom in the way the string can be constrained in its initial position. For example, if the string is plucked aside at a single point, then the shape will be a broken line (Fig. 47(a)); and if it is pushed aside by using a circular object of some kind, then the shape will be partly a straight line, partly an arc of a circle, and partly another straight line, as in Fig. 47(b). It is reasonable to expect that the single "formula" or "analytic expression" (4) could represent a straight line on part of the interval $[0,\pi]$, a circle on another part, and a second straight line on still another part? To the mathematicians of that time (except Bernoulli) this seemed absurd. To d'Alembert the curve in Fig. 47(b) would have represented three separate graphs of three distinct functions, merely pieced together. To Euler it would have been a single graph, but of three functions rather than a single function. Both dismissed the possibility that such a graph could be represented by a single "reasonable" function like the series (4). The controversy bubbled on for many years, and in the absence of mathematical proofs, no one converted anyone else to his way of thinking.

The more general form of a trigonometric series containing both

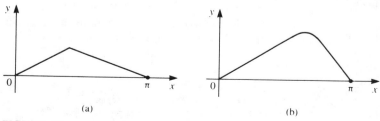

(a)   (b)

FIGURE 47

sines and cosines, namely,

$$f(x) = \frac{1}{2} a_0 + \sum_{n=1}^{\infty} (a_n \cos nx + b_n \sin nx), \tag{5}$$

arises naturally in another physical problem, that of the conduction of heat. In 1807 the French physicist–mathematician Fourier announced in this connection that an "arbitrary function" $f(x)$ can be represented in the form (5), with coefficients given by the formulas

$$a_n = \frac{1}{\pi} \int_{-\pi}^{\pi} f(x) \cos nx \, dx \quad \text{and} \quad b_n = \frac{1}{\pi} \int_{-\pi}^{\pi} f(x) \sin nx \, dx. \tag{6}$$

No one believed him, and for the next 15 years he labored at the task of accumulating empirical evidence to support his assertion. The results were presented in his classic treatise, *Théorie Analytique de la Chaleur* (1822). He supplied no proofs, but instead heaped up the evidence of many solved problems and many convincing specific expansions—so many, indeed, that the mathematicians of the time began to spend more effort on proving, rather than disproving, his conjecture. The first major result of this shift in the winds of opinion was the classical paper of Dirichlet in 1829, in which he proved with full mathematical rigor that the series (5) actually does converge to the function $f(x)$ for all continuous functions whose graphs consist of a finite number of increasing or decreasing pieces—in particular, for the functions illustrated in Fig. 47. Thus were Bernoulli and Fourier vindicated. We must add, however, that Euler found formulas (6) in 1777, but believed them to be valid only in the case of functions $f(x)$ already known to be represented in the form (5).

As we know from Chapter 6, in recognition of Fourier's pioneering tenacity a trigonometric series of the form (5) is called a *Fourier series* if its coefficients are calculated by formulas (6) from some given integrable function $f(x)$.

Those readers who would like a more detailed description of these memorable events in our intellectual history are urged to consult any (or all) of the following masterly accounts: Philip J. Davis and Reuben Hersh, *The Mathematical Experience*, Houghton Mifflin Co., Boston, 1982, pp. 255–270; Béla Sz.-Nagy, *Introduction to Real Functions and Orthogonal Expansions*, Oxford University Press, 1965, pp. 375–380; and particularly Bernhard Riemann, in *A Source Book In Classical Analysis*, ed. Garrett Birkhoff, Harvard University Press, 1973, pp. 16–21.

In the next section and its problems we present an organized exposition of the theory of the vibrating string sketched above; and in the sections after that we turn to other applications of Fourier series in physics and mathematics.

NOTE ON d'ALEMBERT. Jean le Rond d'Alembert (1717–1783) was a French physicist, mathematician, and man of letters. In science he is remembered for *d'Alembert's principle* in mechanics and his solution of the wave equation. The main work of his life was his collaboration with Diderot in preparing the latter's famous *Encyclopédie,* which played a major role in the French Enlightenment by emphasizing science and literature and attacking the forces of reaction in church and state. D'Alembert was a valued friend of Euler, Lagrange, and Laplace.

40 EIGENVALUES, EIGENFUNCTIONS, AND THE VIBRATING STRING

We begin by seeking a nontrivial solution $y(x)$ of the equation

$$y'' + \lambda y = 0 \tag{1}$$

that satisfies the *boundary conditions*

$$y(0) = 0 \quad \text{and} \quad y(\pi) = 0. \tag{2}$$

The parameter λ in (1) is free to assume any real value whatever, and part of our task is to discover the λ's for which the problem can be solved. In our previous work we have considered only *initial value problems,* in which the solution of a second order equation is sought that satisfies two conditions at a single value of the independent variable. Here we have an entirely different situation, for we wish to satisfy one condition at each of two distinct values of x. Problems of this kind are called *boundary value problems,* and in general they are more difficult and far-reaching—in both theory and practice—than initial value problems.

In the problem posed by (1) and (2), however, there are no difficulties. If λ is negative, then Theorem 24-B tells us that only the trivial solution of (1) can satisfy (2); and if $\lambda = 0$, then the general solution of (1) is $y(x) = c_1 x + c_2$, and we have the same conclusion. We are thus restricted to the case in which λ is positive, where the general solution of (1) is

$$y(x) = c_1 \sin \sqrt{\lambda}\, x + c_2 \cos \sqrt{\lambda}\, x;$$

and since $y(0)$ must be 0, this reduces to

$$y(x) = c_1 \sin \sqrt{\lambda}\, x. \tag{3}$$

Thus, if our problem has a solution, it must be of the form (3). For the second boundary condition $y(\pi) = 0$ to be satisfied, it is clear that $\sqrt{\lambda}\, \pi$ must equal $n\pi$ for some positive integer n, so $\lambda = n^2$. In other words, λ must equal one of the numbers $1, 4, 9, \ldots$. These values of λ are called the *eigenvalues* of the problem, and corresponding solutions

$$\sin x, \quad \sin 2x, \quad \sin 3x, \quad \ldots \tag{4}$$

are called *eigenfunctions*. It is clear that the eigenvalues are uniquely determined by the problem, but that the eigenfunctions are not; for any nonzero constant multiples of (4), say $a_1 \sin x$, $a_2 \sin 2x$, $a_3 \sin 3x$, . . . , will serve just as well and are also eigenfunctions. For future reference we notice two facts: the eigenvalues form an increasing sequence of positive numbers that approaches ∞; and the nth eigenfunction, $\sin nx$, vanishes at the endpoints of the interval $[0,\pi]$ and has exactly $n - 1$ zeros inside this interval.

We now examine the classical problem of mathematical physics described in the preceding section—that of the vibrating string. Our purpose is to understand how eigenvalues and eigenfunctions arise. Suppose that a flexible string is pulled taut on the x-axis and fastened at two points that for convenience we take to be $x = 0$ and $x = \pi$. The string is then drawn aside into a certain curve $y = f(x)$ in the xy-plane (Fig. 48) and released. In order to obtain the equation of motion, we make several simplifying assumptions, the first of which is that the subsequent vibration is entirely transverse. This means that each point of the string has constant x-coordinate, so that its y-coordinate depends only on x and the time t. Accordingly, the displacement of the string from its equilibrium position is given by some function $y = y(x,t)$, and the time derivatives $\partial y/\partial t$ and $\partial^2 y/\partial t^2$ represent the string's velocity and acceleration. We consider the motion of a small piece which in its equilibrium position has length Δx. If the linear mass density of the string is $m = m(x)$, so that the mass of the piece is $m\,\Delta x$, then by Newton's second law of motion the transverse force F acting on it is given by

$$F = m\,\Delta x\,\frac{\partial^2 y}{\partial t^2}. \tag{5}$$

Since the string is flexible, the tension $T = T(x)$ at any point is directed along the tangent (see Fig. 48) and has $T \sin \theta$ as its y-component. We next assume that the motion of the string is due solely to the tension in it.

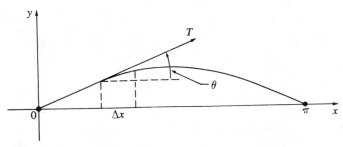

FIGURE 48

As a consequence, F is the difference between the values of $T \sin \theta$ at the ends of our piece, namely $\Delta(T \sin \theta)$, so (5) becomes

$$\Delta(T \sin \theta) = m \, \Delta x \frac{\partial^2 y}{\partial t^2}. \tag{6}$$

If the vibrations are relatively small, so that θ is small and $\sin \theta$ is approximately equal to $\tan \theta = \partial y / \partial x$, then (6) yields

$$\frac{\Delta(T \, \partial y / \partial x)}{\Delta x} = m \frac{\partial^2 y}{\partial t^2};$$

and when Δx is allowed to approach 0, we obtain

$$\frac{\partial}{\partial x}\left(T \frac{\partial y}{\partial x}\right) = m \frac{\partial^2 y}{\partial t^2}. \tag{7}$$

Our present interest in this equation is confined to the case in which both m and T are constant, so that the equation can be written

$$a^2 \frac{\partial^2 y}{\partial x^2} = \frac{\partial^2 y}{\partial t^2} \tag{8}$$

with $a = \sqrt{T/m}$. For reasons that will emerge in the Problems, equation (8) is called the *one-dimensional wave equation*. We seek a solution $y(x,t)$ that satisfies the boundary conditions

$$y(0,t) = 0 \tag{9}$$

and

$$y(\pi,t) = 0, \tag{10}$$

and the initial conditions

$$\left.\frac{\partial y}{\partial t}\right]_{t=0} = 0 \tag{11}$$

and

$$y(x,0) = f(x). \tag{12}$$

Conditions (9) and (10) express the assumption that the ends of the string are permanently fixed at the points $x = 0$ and $x = \pi$; and (11) and (12) assert that the string is motionless when it is released and that $y = f(x)$ is its shape at that moment. We note explicitly, however, that none of these conditions are in any way connected with the derivation of (7) and (8).

We shall give a formal solution of (8) by the method of *separation of variables*. This amounts to looking for solutions of the form

$$y(x,t) = u(x)v(t), \tag{13}$$

which are factorable into a product of functions each of which depends on only one of the independent variables. When (13) is substituted into

(8), we get

$$a^2 u''(x)v(t) = u(x)v''(t)$$

or

$$\frac{u''(x)}{u(x)} = \frac{1}{a^2}\frac{v''(t)}{v(t)}. \tag{14}$$

Since the left side is a function only of x and the right side is a function only of t, equation (14) can hold only if both sides are constant. If we denote this constant by $-\lambda$, then (14) splits into two ordinary differential equations for $u(x)$ and $v(t)$:

$$u'' + \lambda u = 0 \tag{15}$$

and

$$v'' + \lambda a^2 v = 0. \tag{16}$$

It is possible to satisfy (9) and (10) by solving (15) with the boundary conditions $u(0) = u(\pi) = 0$. We have already seen that this problem has a nontrivial solution if and only if $\lambda = n^2$ for some positive integer n, and that corresponding solutions (the eigenfunctions) are

$$u_n(x) = \sin nx.$$

Similarly, for these λ's (the eigenvalues) the general solution of (16) is

$$v(t) = c_1 \sin nat + c_2 \cos nat;$$

and if we impose the requirement that $v'(0) = 0$, so that (11) is satisfied, then $c_1 = 0$ and we have solutions

$$v_n(t) = \cos nat.$$

The corresponding products of the form (13) are therefore

$$y_n(x,\,t) = \sin nx \cos nat.$$

Each of these functions, for $n = 1, 2, \ldots$, satisfies equation (8) and conditions (9), (10), and (11); and it is easily verified that the same is true for any finite sum of constant multiples of the y_n:

$$b_1 \sin x \cos at + b_2 \sin 2x \cos 2at + \cdots + b_n \sin nx \cos nat.$$

If we proceed formally—that is, ignoring all questions of convergence, term-by-term differentiability, and the like—then any infinite series of the form

$$y(x,t) = \sum_{n=1}^{\infty} b_n \sin nx \cos nat = b_1 \sin x \cos at$$
$$+ b_2 \sin 2x \cos 2at + \cdots + b_n \sin nx \cos nat + \cdots \tag{17}$$

is also a solution that satisfies (9), (10), and (11). This brings us to the final condition (12), namely, that for $t = 0$ our solution (17) should yield the initial shape of the string:

$$f(x) = b_1 \sin x + b_2 \sin 2x + \cdots + b_n \sin nx + \cdots. \qquad (18)$$

As we said in the preceding section, when these formulas were developed by Daniel Bernoulli in 1753, it seemed to many mathematicians that (18) ought to be impossible unless $f(x)$ were a function of some very special type. During the next century it became clear that this opinion was mistaken, and that in reality expressions of the form (18) are valid for very wide classes of functions $f(x)$ that vanish at 0 and π. Assuming that this is true, the problem remained for Bernoulli and his contemporaries of finding the coefficients b_n when the function $f(x)$ is given. This problem was solved by Euler in 1777, and his solution launched the vast subject of Fourier series. We know how to find these coefficients from our work in Section 35, but we shall find them again by methods that fit into a broader pattern of ideas.

The eigenfunctions $u_m(x)$ and $u_n(x)$, that is, $\sin mx$ and $\sin nx$, satisfy the equations

$$u_m'' = -m^2 u_m \qquad \text{and} \qquad u_n'' = -n^2 u_n.$$

If the first equation is multiplied by u_n and the second by u_m, then the difference of the resulting equations is

$$u_n u_m'' - u_m u_n'' = (n^2 - m^2) u_m u_n$$

or

$$(u_n u_m' - u_m u_n')' = (n^2 - m^2) u_m u_n. \qquad (19)$$

On integrating both sides of (19) from 0 to π and using the fact that $u_m(x) = \sin mx$ and $u_n(x) = \sin nx$ both vanish at 0 and π, we obtain

$$(n^2 - m^2) \int_0^\pi u_m(x) u_n(x)\, dx = [u_n(x) u_m'(x) - u_m(x) u_n'(x)]_0^\pi = 0,$$

so

$$\int_0^\pi \sin mx \sin nx\, dx = 0 \qquad \text{when } m \ne n. \qquad (20)$$

This result suggests multiplying (18) through by $\sin nx$ and integrating the result term by term from 0 to π. When these operations are carried out, (20) produces a wholesale disappearance of terms, leaving only

$$\int_0^\pi f(x) \sin nx\, dx = b_n \int_0^\pi \sin^2 nx\, dx;$$

and since

$$\int_0^\pi \sin^2 nx\, dx = \frac{1}{2} \int_0^\pi (1 - \cos 2nx)\, dx = \frac{\pi}{2},$$

we have

$$b_n = \frac{2}{\pi} \int_0^\pi f(x) \sin nx\, dx. \qquad (21)$$

These b_n are very familiar to us and are called the *Fourier coefficients of* $f(x)$. With these coefficients, (18) is the *Fourier sine series* of $f(x)$ or the *eigenfunction expansion* of $f(x)$ in terms of the eigenfunctions sin nx, and (17) is called *Bernoulli's solution* of the wave equation.

The above "solution" of the wave equation is clearly riddled with doubtful procedures and unanswered questions, so much so, indeed, that from a strictly rigorous point of view it cannot be regarded as having more than a suggestive value. But even this much is well worth the effort, for some of the questions that arise—especially those about the meaning and validity of (18)—are exceedingly fruitful. For instance, if the b_n are computed by means of (21) and used to form the series on the right of (18), under what circumstances will this series converge? And if it converges at a point x, does it necessarily converge to $f(x)$? We give the following brief statement of one answer to these questions that is fully covered by the theorem proved in Appendix A at the end of the preceding chapter.

The function $f(x)$ under consideration is defined on the interval $[0,\pi]$ and vanishes at the endpoints. Suppose that $f(x)$ is continuous on the entire interval, and also that its derivative is continuous with the possible exception of a finite number of *jump discontinuities*, where the derivative approaches finite but different limits from the left and from the right. In geometric language, the graph of such a function is a continuous curve with the property that the direction of the tangent changes continuously as it moves along the curve, except possibly at a finite number of "corners" where its direction changes abruptly. Under these hypotheses the expansion (18) is valid; that is, if the b_n are defined by (21), then the series on the right converges at every point to the value of the function at that point. The need for a carefully constructed theory can be seen from the fact that if $f(x)$ is merely assumed to be continuous, and nothing is said about its derivative, then it is known to be possible for the series on the right of (18) to diverge at some points.[2]

Another line of investigation considers the possiblity of eigenfunction expansions like (18) for other boundary value problems. If we put aside the issue of the validity of such expansions, then the main problem becomes that of showing in other cases that we have an adequate supply of suitable building materials, i.e., a sequence of eigenvalues with corresponding eigenfunctions that satisfy some condition similar to (20).

Suppose, for instance, we consider the vibrating string studied above with one significant difference: the string is *nonhomogeneous*, in

---

[2] It has been known since 1966 that there even exists a continuous function whose Fourier series diverges at *every* rational point in $[0,\pi]$.

the sense that its density $m = m(x)$ may vary from point to point. In this situation, (8) is replaced by

$$\frac{\partial^2 y}{\partial x^2} = \frac{m(x)}{T} \frac{\partial^2 y}{\partial t^2}.$$
(22)

If we again seek a solution of the form (13), then (22) becomes

$$\frac{u''(x)}{m(x)u(x)} = \frac{1}{T} \frac{v''(t)}{v(t)};$$

and as before, we are led to the following boundary value problem:

$$u'' + \lambda m(x)u = 0, \qquad u(0) = u(\pi) = 0.$$
(23)

What are the eigenvalues and eigenfunctions in this case? Needless to say, we cannot give precise answers without knowing something definite about the density function $m(x)$. But at least we can prove that these eigenvalues and eigenfunctions exist. The details of this argument are given in Appendix A at the end of this chapter.

PROBLEMS

1. Find the eigenvalues λ_n and eigenfunctions $y_n(x)$ for the equation $y'' + \lambda y = 0$ in each of the following cases:
   (a) $y(0) = 0$, $y(\pi/2) = 0$;
   (b) $y(0) = 0$, $y(2\pi) = 0$;
   (c) $y(0) = 0$, $y(1) = 0$;
   (d) $y(0) = 0$, $y(L) = 0$ when $L > 0$;
   (e) $y(-L) = 0$, $y(L) = 0$ when $L > 0$;
   (f) $y(a) = 0$, $y(b) = 0$ when $a < b$.

Solve the following two problems formally, i.e., without considering such purely mathematical issues as the differentiability of functions and the convergence of series.

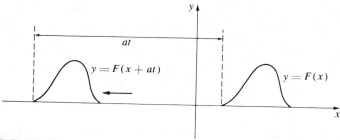

FIGURE 49

2. If $y = F(x)$ is an arbitrary function, then $y = F(x + at)$ represents a wave of fixed shape that moves to the left along the x-axis with velocity a (Fig. 49). Similarly, if $y = G(x)$ is another arbitrary function, then $y = G(x - at)$ is a wave moving to the right, and the most general one-dimensional wave with velocity a is

$$y(x,t) = F(x + at) + G(x - at). \tag{*}$$

(a) Show that (*) satisfies the wave equation (8).
(b) It is easy to see that the constant a in equation (8) has the dimensions of velocity. Also, it is intuitively clear that if a stretched string is disturbed, then waves will move in both directions away from the source of the disturbance. These considerations suggest introducing the new variables $\alpha = x + at$ and $\beta = x - at$. Show that with these independent variables, equation (8) becomes

$$\frac{\partial^2 y}{\partial \alpha\, \partial \beta} = 0,$$

and from this derive (*) by integration. Formula (*) is called *d'Alembert's solution* of the wave equation. It was also obtained by Euler, independently of d'Alembert but slightly later.

3. Consider an infinite string stretched taut on the x-axis from $-\infty$ to ∞. Let the string be drawn aside into a curve $y = f(x)$ and released, and assume that its subsequent motion is described by the wave equation (8).
(a) Use (*) to show that the string's displacement is given by *d'Alembert's formula*,

$$y(x,t) = \frac{1}{2}[f(x + at) + f(x - at)]. \tag{**}$$

Hint: Remember the initial conditions (11) and (12).
(b) Assume further that the string remains motionless at the points $x = 0$ and $x = \pi$ (such points are called *nodes*), so that $y(0,t) = y(\pi,t) = 0$, and use (**) to show that $f(x)$ is an odd function that is periodic with period 2π [that is, $f(-x) = -f(x)$ and $f(x + 2\pi) = f(x)$].
(c) Show that since $f(x)$ is odd and periodic with period 2π, it necessarily vanishes at 0 and π.
(d) Show that Bernoulli's solution (17) can be written in the form of (**). *Hint*: $2 \sin nx \cos nat = \sin[n(x + at)] + \sin[n(x - at)]$.

4. Consider a uniform flexible chain of constant mass density m_0 hanging freely from one end. If a coordinate system is established as in Fig. 50, then the lateral vibrations of the chain, when it is disturbed, are governed by equation (7). In this case, the tension T at any point is the weight of the chain below that point, and is therefore given by $T = m_0 x g$, where g is the acceleration due to gravity. When m_0 is canceled, (7) becomes

$$\frac{\partial}{\partial x}\left(gx\, \frac{\partial y}{\partial x}\right) = \frac{\partial^2 y}{\partial t^2}.$$

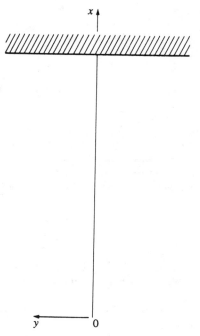

FIGURE 50

(a) Assume that this partial differential equation has a solution of the form $y(x,t) = u(x)v(t)$, and show as a consequence that $u(x)$ satisfies the following ordinary differential equation:

$$\frac{d}{dx}\left(gx\frac{du}{dx}\right) + \lambda u = 0. \qquad (***)$$

(b) If the independent variable is changed from x to $z = 2\sqrt{\lambda x/g}$, show that equation (***) becomes

$$z\frac{d^2u}{dz^2} + \frac{du}{dz} + zu = 0,$$

which (apart from notation) is Bessel's equation 1-(9) for the special case in which $p = 0$.

5. Solve the vibrating string problem in the text if the initial shape (12) is given by the function

(a) $f(x) = \begin{cases} 2cx/\pi, & 0 \le x \le \pi/2, \\ 2c(\pi - x)/\pi, & \pi/2 \le x \le \pi; \end{cases}$

(b) $f(x) = \dfrac{1}{\pi}x(\pi - x);$

(c) $f(x) = \begin{cases} x, & 0 \le x \le \pi/4, \\ \pi/4, & \pi/4 \le x \le 3\pi/4, \\ \pi - x, & 3\pi/4 \le x \le \pi. \end{cases}$

In each case, sketch the initial shape of the string.

6. Solve the vibrating string problem in the text if the initial shape (12) is that of a single arch of a sine curve, $f(x) = c \sin x$. Show that the moving string always has the same general shape. Do the same for functions of the form $f(x) = c \sin nx$. Show, in particular, that there are $n - 1$ points between $x = 0$ and $x = \pi$ at which the string remains motionless; these points are called *nodes*, and these solutions are called *standing waves*. Draw sketches to illustrate the movement of the standing waves.

7. The problem of the *struck string* is that of solving equation (8) with the boundary conditions (9) and (10) and the initial conditions

$$\frac{\partial y}{\partial t}\bigg]_{t=0} = g(x) \quad \text{and} \quad y(x,0) = 0.$$

(These initial conditions mean that the string is initially in the equilibrium position, and has an initial velocity $g(x)$ at the point x as a result of being struck.) By separating variables and proceeding formally, obtain the solution

$$y(x,t) = \sum_{1}^{\infty} c_n \sin nx \sin nat$$

where

$$c_n = \frac{2}{\pi na} \int_0^\pi g(x) \sin nx \, dx.$$

41 THE HEAT EQUATION

When we study the flow of heat in thermally conducting bodies, we encounter an entirely different type of problem leading to a partial differential equation.

In the interior of a body where heat is flowing from one region to another, the temperature generally varies from point to point at any one time, and from time to time at any one point. Thus, the temperature w is a function of the space coordinates x, y, z and the time t, say $w = w(x,y,z,t)$. The precise form of this function naturally depends on the shape of the body, the thermal characteristics of its material, the initial distribution of temperature, and the conditions maintained on the surface of the body. The French physicist–mathematician Fourier studied this problem in his classic treatise of 1822, *Théorie Analytique de la Chaleur*. He used physical principles to show that the temperature function w must satisfy the *heat equation*

$$a^2\left(\frac{\partial^2 w}{\partial x^2} + \frac{\partial^2 w}{\partial y^2} + \frac{\partial^2 w}{\partial z^2}\right) = \frac{\partial w}{\partial t}. \tag{1}$$

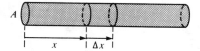

FIGURE 51

We shall retrace his reasoning in a simple one-dimensional situation, and thereby derive the one-dimensional heat equation.

The following are the physical principles that will be needed:

(a) Heat flows in the direction of decreasing temperature, that is, from hot regions to cold regions.

(b) The rate at which heat flows across an area is proportional to the area and to the rate of change of temperature with respect to distance in a direction perpendicular to the area. (This proportionality factor is denoted by k and called the *thermal conductivity* of the substance.)

(c) The quantity of heat gained or lost by a body when its temperature changes, that is, the change in its thermal energy, is proportional to the mass of the body and to the change of temperature. (This proportionality factor is denoted by c and called the *specific heat* of the substance.)

We now consider the flow of heat in a thin cylindrical rod of cross-sectional area A (Fig. 51) whose lateral surface is perfectly insulated so that no heat flows through it. This use of the word "thin" means that the temperature is assumed to be uniform on any cross section, and is therefore a function only of the time and the position of the cross section, say $w = w(x,t)$. We examine the rate of change of the heat contained in a thin slice of the rod between the positions x and $x + \Delta x$.

If ρ is the density of the rod, that is, its mass per unit volume, then the mass of the slice is

$$\Delta m = \rho A \, \Delta x.$$

Furthermore, if Δw is the temperature change at the point x in a small time interval Δt, then (c) tells us that the quantity of heat stored in the slice in this time interval is

$$\Delta H = c \, \Delta m \, \Delta w = c\rho A \, \Delta x \, \Delta w,$$

so the rate at which heat is being stored is approximately

$$\frac{\Delta H}{\Delta t} = c\rho A \, \Delta x \frac{\Delta w}{\Delta t}. \tag{2}$$

We assume that no heat is generated inside the slice—for instance, by chemical or electrical processes—so that the slice gains heat only by means of the flow of heat through its faces. By (b) the rate at which heat flows into the slice through the left face is

$$-kA \left.\frac{\partial w}{\partial x}\right|_x.$$

The negative sign here is chosen in accordance with (a), so that this quantity will be positive if $\partial w/\partial x$ is negative. Similarly, the rate at which heat flows into the slice through the right face is

$$kA \left.\frac{\partial w}{\partial x}\right|_{x+\Delta x},$$

so the total rate at which heat flows into the slice is

$$kA \left.\frac{\partial w}{\partial x}\right|_{x+\Delta x} - kA \left.\frac{\partial w}{\partial x}\right|_x. \tag{3}$$

If we equate the expressions (2) and (3), the result is

$$kA \left.\frac{\partial w}{\partial x}\right|_{x+\Delta x} - kA \left.\frac{\partial w}{\partial x}\right|_x = c\rho A \, \Delta x \frac{\Delta w}{\Delta t},$$

or

$$\frac{k}{c\rho} \left[\frac{\partial w/\partial x|_{x+\Delta x} - \partial w/\partial x|_x}{\Delta x} \right] = \frac{\Delta w}{\Delta t}.$$

Finally, by letting Δx and $\Delta t \to 0$ we obtain the desired equation,

$$a^2 \frac{\partial^2 w}{\partial x^2} = \frac{\partial w}{\partial t}, \tag{4}$$

where $a^2 = k/c\rho$. This is the physical reasoning that leads to the one-dimensional heat equation. The three-dimensional equation (1) can be derived in essentially the same way.

We now solve the one-dimensional heat equation (4), subject to the following set of conditions: the rod is π units long and lies along the x-axis between $x = 0$ and $x = \pi$; the initial temperature is a prescribed function $f(x)$, so that

$$w(x,0) = f(x); \tag{5}$$

and the ends of the rod have the constant temperature zero for all values of $t \geq 0$,

$$w(0,t) = 0 \quad \text{and} \quad w(\pi,t) = 0. \tag{6}$$

We try for a solution of this boundary value problem by the method of separation of variables that worked so well in the case of the wave equation; that is, we seek a solution of (4) having the form

$$w(x,t) = u(x)v(t). \tag{7}$$

When this expression is substituted in (4), the result can be written

$$\frac{u''(x)}{u(x)} = \frac{1}{a^2}\frac{v'(t)}{v(t)}. \tag{8}$$

Since each side of this equation depends on only one of the variables, both sides must be constant, and if we denote this common constant value by $-\lambda$, then (8) splits into the two ordinary differential equations

$$u'' + \lambda u = 0 \tag{9}$$

and

$$v' + \lambda a^2 v = 0. \tag{10}$$

Just as in Section 40, we solve (9) and satisfy the boundary conditions (6) by setting $\lambda = n^2$ for any positive integer n, and the corresponding eigenfunction is

$$u_n(x) = \sin nx.$$

With this value of λ, equation (10) becomes

$$v' + n^2 a^2 v = 0,$$

which has the easy solution

$$v_n(t) = e^{-n^2 a^2 t}.$$

The resulting products of the form (7) are therefore

$$w_n(x,t) = e^{-n^2 a^2 t}\sin nx, \qquad n = 1, 2, 3, \ldots. \tag{11}$$

This brings us to the point where we know that each of the functions (11) satisfies equation (4) and the boundary conditions (6), and it is clear that the same is true for any finite linear combination of the w_n:

$$b_1 e^{-a^2 t}\sin x + b_2 e^{-4a^2 t}\sin 2x + \cdots + b_n e^{-n^2 a^2 t}\sin nx. \tag{12}$$

Without dwelling on the important mathematical issues of convergence and term-by-term differentiability, we now pass from (12) to the corresponding infinite series,

$$w(x,t) = \sum_{n=1}^{\infty} b_n e^{-n^2 a^2 t}\sin nx. \tag{13}$$

This will be a solution of our original boundary value problem if it allows us to satisfy the initial condition (5), that is, if (13) reduces to the initial temperature distribution $f(x)$ when $t = 0$:

$$f(x) = \sum_{n=1}^{\infty} b_n \sin nx. \tag{14}$$

To finish this part of our work and make the solution (13) completely explicit, all that remains is to determine the b_n as the Fourier coefficients in the expansion (14) of $f(x)$ in a Fourier sine series,

$$b_n = \frac{2}{\pi} \int_0^{\pi} f(x) \sin nx \, dx. \tag{15}$$

Example 1. Suppose that the thin rod discussed above is first immersed in boiling water so that its temperature is 100°C throughout, and then removed from the water at time $t = 0$ with its ends immediately put in ice so that these ends are kept at temperature 0°C. Find the temperature $w = w(x,t)$ under these circumstances.

Solution. This is the special case of the above discussion in which the initial temperature distribution is given by the constant function

$$f(x) = 100, \qquad 0 < x < \pi.$$

We must therefore find the sine series of this function, which we can either calculate from scratch by using (15) or obtain in some other way (see Problem 35-4),

$$f(x) = \frac{400}{\pi} \left(\sin x + \frac{\sin 3x}{3} + \frac{\sin 5x}{5} + \cdots \right).$$

By referring to formula (13), we now see that the desired temperature function is

$$w(x,t) = \frac{400}{\pi} \left[e^{-a^2 t} \sin x + \frac{1}{3} e^{-9a^2 t} \sin 3x + \frac{1}{5} e^{-25a^2 t} \sin 5x + \cdots \right].$$

Example 2. Find the steady-state temperature of the thin rod discussed above if the fixed temperatures at the ends $x = 0$ and $x = \pi$ are w_1 and w_2, respectively.

Solution. "Steady-state" means that $\partial w / \partial t = 0$, so the heat equation (4) reduces to $\partial^2 w / \partial x^2 = 0$ or $d^2 w / dx^2 = 0$. The general solution is therefore $w = c_1 x + c_2$, and by using the boundary conditions we easily determine these constants of integration and obtain the desired solution,

$$w = w_1 + \frac{1}{\pi} (w_2 - w_1) x.$$

The steady-state version of the three-dimensional heat equation (1) is

$$\frac{\partial^2 w}{\partial x^2} + \frac{\partial^2 w}{\partial y^2} + \frac{\partial^2 w}{\partial z^2} = 0; \tag{16}$$

it is called *Laplace's equation.* The study of this equation and its solutions and uses—there are many applications in the theory of gravitation—is a rich branch of mathematics called *potential theory.* This topic is continued in Appendix A at the end of the next chapter. The corresponding equation in two dimensions is

$$\frac{\partial^2 w}{\partial x^2} + \frac{\partial^2 w}{\partial y^2} = 0; \tag{17}$$

this is a valuable tool if plane problems are under consideration. Equation (17) also has a special significance of its own in complex analysis.

PROBLEMS

1. Derive the three-dimensional heat equation (1) by adapting the reasoning in the text to the case of a small box with edges Δx, Δy, Δz contained in a region R in xyz-space where the temperature function $w(x,y,z,t)$ is sought. *Hint:* Consider the flow of heat through two opposite faces of the box, first perpendicular to the x-axis, then the y-axis, and finally the z-axis.
2. Solve the boundary value problem in the text if the conditions are altered from (5) and (6) to

$$w(x,0) = f(x) \qquad \text{and} \qquad w(0,t) = w_1, \qquad w(\pi,t) = w_2.$$

   Hint: Write $w(x,t) = W(x,t) + g(x)$ and remember Example 2.
3. Suppose that the lateral surface of the thin rod in the text is not insulated, but instead radiates heat into the surroundings. If Newton's law of cooling applies, show that the one-dimensional heat equation becomes

$$a^2 \frac{\partial^2 w}{\partial x^2} = \frac{\partial w}{\partial t} + c(w - w_0),$$

   where c is a positive constant and w_0 is the temperature of the surroundings.
4. In the preceding problem, find $w(x,t)$ if the ends of the rod are kept at $0°C$, $w_0 = 0°C$, and the initial temperature distribution is $f(x)$.
5. In Example 1, suppose the ends of the rod are insulated instead of being kept at $0°C$. What are the new boundary conditions? Find the temperature $w(x,t)$ in this case by using only common sense.
6. Solve the problem of finding $w(x,t)$ for the rod with insulated ends at $x = 0$ and $x = \pi$ (see the preceding problem) if the initial temperature distribution is given by $w(x,0) = f(x)$.

7. The two-dimensional heat equation is

$$a^2\left(\frac{\partial^2 w}{\partial x^2} + \frac{\partial^2 w}{\partial y^2}\right) = \frac{\partial w}{\partial t}.$$

Use the method of separation of variables to find a steady-state solution of this equation in the infinite strip of the xy-plane bounded by the lines $x = 0$, $x = \pi$, and $y = 0$ if the following conditions are satisfied:

$$w(0,y) = 0, \qquad\qquad w(\pi,y) = 0,$$

$$w(x,0) = f(x), \qquad \lim_{y\to\infty} w(x,y) = 0.$$

42 THE DIRICHLET PROBLEM FOR A CIRCLE. POISSON'S INTEGRAL

We continue our overall program in this chapter of acquainting the student with important mathematical problems related to both partial differential equations and Fourier series. Even though we cannot treat these problems in the depth they deserve within the limitations of the present book, at least it is possible to convey an impression of what these problems are and briefly describe some of the standard methods for dealing with them.

We begin with the two-dimensional Laplace equation mentioned at the end of Section 41. In rectangular coordinates (x,y) it is

$$\frac{\partial^2 w}{\partial x^2} + \frac{\partial^2 w}{\partial y^2} = 0; \tag{1}$$

and in polar coordinates (r,θ) it is

$$\frac{\partial^2 w}{\partial r^2} + \frac{1}{r}\frac{\partial w}{\partial r} + \frac{1}{r^2}\frac{\partial^2 w}{\partial \theta^2} = 0. \tag{2}$$

It is an exercise in the use of the chain rule for partial derivatives to transform these equations into one another (see Problem 1 below). Many types of physical problems require solutions of Laplace's equation, and there exists a wide variety of solutions containing many different kinds of functions. However, just as in the preceding sections, a specific physical problem usually asks for a solution that is defined in a certain region and satisfies a given condition on the boundary of that region.

There is a famous problem in analysis called the *Dirichlet problem*, one version of which can be stated as follows: Given a region R in the plane bounded by a simple closed curve C, and given a function $f(P)$ defined and continuous for points P on C, it is required to find a function $w(P)$ continuous in R and on C, such that $w(P)$ satisfies Laplace's equation in R and equals $f(P)$ on the boundary C.

We shall consider the special case in which R is the interior of the unit circle $x^2 + y^2 = 1$, and we use polar coordinates as the geometry suggests. Let $w = w(r, \theta)$ be a function continuous inside and on this circle. The values of this function when $r = 1$ are called its *boundary values* for the circular region. The function $w(1, \theta)$ is evidently a continuous function of θ with period 2π. The Dirichlet problem for this circular region is then the following: Let $f(\theta)$ be any given continuous function of θ with period 2π. It is required to find a function $w = w(r, \theta)$ that satisfies Laplace's equation (2) for $0 \le r < 1$, and has the further property that $w(1, \theta) = f(\theta)$ for each value of θ. In some versions of the Dirichlet problem the condition that $f(\theta)$ must be continuous is relaxed and the condition $w(1, \theta) = f(\theta)$ is expressed in a different form; we shall comment further on these matters below.

If w is understood to be temperature, then we know from our work in Section 41 that the Dirichlet problem for a circle is the problem of finding the steady-state temperature throughout a thin circular plate when the temperature along the edge is prescribed in advance. Solutions of Laplace's equation are often called *harmonic functions*. Using this language, the Dirichlet problem is the problem of finding a function that is harmonic in the circular region and assumes preassigned continuous values on the boundary.

Now for the details of solving this problem. We begin by ignoring the boundary function $f(\theta)$ and seeking solutions of Laplace's equation (2) that have the form $w = w(r, \theta) = u(r)v(\theta)$, that is, that can be written as the product of a function of r alone and a function of θ alone. Thus, we make yet another application of the method of separation of variables. When this function is substituted in equation (2) we obtain

$$u''(r)v(\theta) + \frac{1}{r}u'(r)v(\theta) + \frac{1}{r^2}u(r)v''(\theta) = 0$$

or

$$\frac{r^2 u''(r) + r u'(r)}{u(r)} = -\frac{v''(\theta)}{v(\theta)}. \tag{3}$$

The left side of (3) is independent of θ, and the right side is independent of r, so both sides must be constant; and if we denote this common constant value by λ, then (3) splits into the two equations

$$v'' + \lambda v = 0 \tag{4}$$

and

$$r^2 u'' + r u' - \lambda u = 0. \tag{5}$$

We want $v(\theta)$ to be continuous and periodic with period 2π—and, of course, not identically zero. This requires us to conclude that the constant λ in (4) must be of the form $\lambda = n^2$ with $n = 0, 1, 2, 3, \ldots$. For

$n = 0$ the only suitable solution is $v =$ a constant, and for $n = 1, 2, 3, \ldots$ the solutions of (4) are linear combinations of $\cos n\theta$ and $\sin n\theta$,

$$v_n(\theta) = a_n \cos n\theta + b_n \sin n\theta.$$

We next set $\lambda = n^2$ in equation (5), which then becomes

$$r^2 \frac{d^2u}{dr^2} + r \frac{du}{dr} - n^2 u = 0.$$

This is Euler's equidimensional equation (Problem 17-5), with solutions

$$u(r) = A + B \log r \qquad \text{if } n = 0,$$
$$u(r) = Ar^n + Br^{-n} \qquad \text{if } n = 1, 2, 3, \ldots,$$

where A and B are constants. We want $u(r)$ to be continuous at $r = 0$, so we take $B = 0$ in all cases, and we therefore have

$$u_n(r) = r^n.$$

If we now write down all the solutions $w = u_n(r)v_n(\theta)$ in sequential order, the result is as follows:

$$n = 0, \qquad w = \text{a constant } \tfrac{1}{2}a_0;$$
$$n = 1, \qquad w = r(a_1 \cos \theta + b_1 \sin \theta);$$
$$n = 2, \qquad w = r^2(a_2 \cos 2\theta + b_2 \sin 2\theta);$$
$$n = 3, \qquad w = r^3(a_3 \cos 3\theta + b_3 \sin 3\theta);$$
$$\cdots.$$

It is easy to see that any finite sum of solutions of Laplace's equation is also a solution, and the same is true for an infinite series of solutions if the series has suitable convergence properties. This leads us to the solution

$$w = w(r, \theta) = \frac{1}{2}a_0 + \sum_{n=1}^{\infty} r^n(a_n \cos n\theta + b_n \sin n\theta). \tag{6}$$

If we put $r = 1$ in (6) and remember that we want to satisfy the boundary condition $w(1, \theta) = f(\theta)$, then we obtain

$$f(\theta) = \frac{1}{2}a_0 + \sum_{n=1}^{\infty} (a_n \cos n\theta + b_n \sin n\theta). \tag{7}$$

It is now clear what must be done to solve the Dirichlet problem for the unit circle: start with the given boundary function $f(\theta)$ and find its Fourier series (7); then form the solution (6) by merely inserting the factor r^n in front of the expression in parentheses in (7). Of course, the constant term in (6) is written as $\tfrac{1}{2}a_0$ for the sake of agreement with the standard notation for Fourier series.

Example. Solve the Dirichlet problem for the unit circle if $f(\theta) = 1$ on the top half of the circle $(0 < \theta < \pi)$ and $f(\theta) = -1$ on the bottom half of the circle $(-\pi < \theta < 0)$, with $f(0) = f(\pm \pi) = 0$.

Solution. We know from Problem 35-4 that the Fourier series for $f(\theta)$ is

$$f(\theta) = \frac{4}{\pi}\left(\sin\theta + \frac{\sin 3\theta}{3} + \frac{\sin 5\theta}{5} + \cdots\right).$$

The solution of the Dirichlet problem is therefore

$$w(r,\theta) = \frac{4}{\pi}\left(r\sin\theta + \frac{1}{3}r^3\sin 3\theta + \frac{1}{5}r^5\sin 5\theta + \cdots\right).$$

The discussion given above is concerned mostly with formal procedures and not with delicate questions of convergence. However, we state without proof that if the a_n and b_n are the Fourier coefficients of $f(\theta)$, then the series (6) converges for $0 \le r < 1$ and its sum $w(r,\theta)$ is a solution of Laplace's equation in this region. For this to be true it is not necessary to assume that $f(\theta)$ is continuous, or even that its Fourier series converges. It is enough to assume that $f(\theta)$ is integrable. Furthermore, even with this weak hypothesis it turns out that $f(\theta)$ is the boundary value of $w(r,\theta)$, in the sense that

$$\lim_{r \to 1} w(r,\theta) = f(\theta)$$

at every point of continuity of the function $f(\theta)$. These remarkable facts have emerged from careful theoretical studies of the Poisson integral, which we now briefly describe.[3]

The Poisson integral. The Dirichlet problem for the unit circle is now solved, at least formally. However, a simpler expression for this solution can be found as follows, if we don't mind a bit of calculating with complex numbers. As we know, the coefficients in (6) are given by the formulas

$$a_n = \frac{1}{\pi}\int_{-\pi}^{\pi} f(\phi)\cos n\phi \, d\phi, \qquad b_n = \frac{1}{\pi}\int_{-\pi}^{\pi} f(\phi)\sin n\phi \, d\phi.$$

When these are substituted in (6), then by using the identity

$$\cos(\theta - \phi) = \cos\theta\cos\phi + \sin\theta\sin\phi$$

---

[3] More details on these interesting matters of theory can be found in H. S. Carslaw, *Introduction to the Theory of Fourier's Series and Integrals*, 3d ed., Macmillan, London, 1930, pp. 250–254; R. T. Seeley, *An Introduction to Fourier Series and Integrals*, W. A. Benjamin, New York, 1966, pp. 16–19; or pp. 436–442 of the book of Sz.-Nagy mentioned in Section 39.

and interchanging the order of integration and summation, we obtain

$$w(r, \theta) = \frac{1}{\pi} \int_{-\pi}^{\pi} f(\phi) \left[\frac{1}{2} + \sum_{1}^{\infty} r^n \cos n(\theta - \phi) \right] d\phi. \tag{8}$$

To sum the series in brackets, we put $\alpha = \theta - \phi$ and let $z = re^{i\alpha} = r(\cos \alpha + i \sin \alpha)$. Then $z^n = r^n e^{in\alpha} = r^n (\cos n\alpha + i \sin n\alpha)$ and

$$\frac{1}{2} + \sum_{1}^{\infty} r^n \cos n\alpha = \text{real part} \left[\frac{1}{2} + \sum_{1}^{\infty} z^n \right]$$

$$= \text{real part} \left[-\frac{1}{2} + \frac{1}{1 - z} \right]$$

$$= \text{real part} \left[\frac{1 + z}{2(1 - z)} \right]$$

$$= \text{real part} \left[\frac{(1 + z)(1 - \bar{z})}{2 |1 - z|^2} \right]$$

$$= \frac{1 - |z|^2}{2 |1 - z|^2} = \frac{1 - r^2}{2(1 - 2r \cos \alpha + r^2)}.$$

By substituting this in (8) we obtain

$$w(r, \theta) = \frac{1}{2\pi} \int_{-\pi}^{\pi} \frac{1 - r^2}{1 - 2r \cos(\theta - \phi) + r^2} f(\phi) \, d\phi. \tag{9}$$

This remarkable formula for the solution of the Dirichlet problem is called the *Poisson integral*; it expresses the value of the harmonic function $w(r, \theta)$ at all points *inside* the circle in terms of its values on the *circumference* of the circle. It should also be observed that for $r = 0$ formula (9) yields

$$w(0, \theta) = \frac{1}{2\pi} \int_{-\pi}^{\pi} f(\phi) \, d\phi.$$

This shows that the value of the harmonic function w at the center of the circle is the average of its values on the circumference.

NOTE ON POISSON. Siméon Denis Poisson (1781–1840), a very eminent French mathematician and physicist, succeeded Fourier in 1806 as full professor at the École Polytechnique. In physics, Poisson's equation describes the variation of potential inside continuous distributions of mass or electric charge, just as Laplace's equation does in empty space. He also made important theoretical contributions to the study of elasticity, magnetism, heat, and capillary action. In pure mathematics, the Poisson summation formula is a major tool in analytic number theory, and the Poisson integral pointed the way to many important developments in Fourier analysis. In addition, he worked extensively in probability. It was he who named the law of large numbers; and the Poisson distribution—or law of small numbers—has many applications to such phenomena as the distribution of blood cells on a microscope slide, of automobiles on a highway, of customers at a theater ticket office, etc. According to Abel, Poisson was a short, plump man. His family tried to encourage him in many directions, from being a doctor to being a lawyer, this last on the theory that perhaps he was fit for nothing better, but at last he found his niche as a scientist and produced over 300 works in a relatively short lifetime. "La vie, c'est le travail (Life is work)," he said—and he had good reason to know.

PROBLEMS

1. If $w = F(x,y) = G(r,\theta)$ with $x = r \cos \theta$ and $y = r \sin \theta$, show that

$$\frac{\partial^2 w}{\partial x^2} + \frac{\partial^2 w}{\partial y^2} = \frac{1}{r}\left[\frac{\partial}{\partial r}\left(r\frac{\partial w}{\partial r}\right) + \frac{1}{r}\frac{\partial^2 w}{\partial \theta^2}\right]$$

$$= \frac{\partial^2 w}{\partial r^2} + \frac{1}{r}\frac{\partial w}{\partial r} + \frac{1}{r^2}\frac{\partial^2 w}{\partial \theta^2}.$$

Hint:

$$\frac{\partial w}{\partial r} = \frac{\partial w}{\partial x}\cos\theta + \frac{\partial w}{\partial y}\sin\theta \quad \text{and} \quad \frac{\partial w}{\partial \theta} = \frac{\partial w}{\partial x}(-r\sin\theta) + \frac{\partial w}{\partial y}(r\cos\theta).$$

Similarly, compute $\dfrac{\partial}{\partial r}\left(r\dfrac{\partial w}{\partial r}\right)$ and $\dfrac{\partial^2 w}{\partial \theta^2}$.

2. Solve the Dirichlet problem for the unit circle if the boundary function $f(\theta)$ is defined by
(a) $f(\theta) = \cos \frac{1}{2}\theta$, $-\pi \le \theta \le \pi$;
(b) $f(\theta) = \theta$, $-\pi < \theta < \pi$;
(c) $f(\theta) = 0$ for $-\pi \le \theta < 0$, $f(\theta) = \sin\theta$ for $0 \le \theta \le \pi$;
(d) $f(\theta) = 0$ for $-\pi < \theta < 0$, $f(\theta) = 1$ for $0 \le \theta \le \pi$;
(e) $f(\theta) = \frac{1}{4}\theta^2$, $-\pi \le \theta \le \pi$.

3. Show that the Dirichlet problem for the circle $x^2 + y^2 = R^2$, where $f(\theta)$ is the boundary function, has the solution

$$w(r,\theta) = \frac{1}{2}a_0 + \sum_{1}^{\infty}\left(\frac{r}{R}\right)^n(a_n \cos n\theta + b_n \sin n\theta),$$

where a_n and b_n are the Fourier coefficients of $f(\theta)$. Show also that the Poisson integral for this more general case is

$$w(r,\theta) = \frac{1}{2\pi}\int_{-\pi}^{\pi}\frac{R^2 - r^2}{R^2 - 2Rr\cos(\theta - \phi) + r^2}f(\phi)\,d\phi.$$

4. Let $w(P)$ be harmonic in a plane region, and let C be any circle entirely contained in this region. Prove that the value of w at the center of C is the average of its values on the circumference. (This is a major theorem of potential theory due to Gauss.)

43 STURM–LIOUVILLE PROBLEMS

We return briefly to the discussion of eigenvalues and eigenfunctions at the beginning of Section 40. Our purpose here is to place these ideas in a broader context that will help make an easier transition to the topics of the next chapter.

As we know, a sequence of functions $y_n(x)$ with the property that

$$\int_a^b y_m(x)y_n(x)\, dx = \begin{cases} 0 & \text{if } m \neq n, \\ \alpha_n \neq 0 & \text{if } m = n, \end{cases} \tag{1}$$

is said to be *orthogonal* on the interval $[a,b]$. If $\alpha_n = 1$ for all n, the functions are said to be *normalized*, and we speak of an *orthonormal sequence*. A more general type of orthogonality is defined by the property

$$\int_a^b q(x)y_m(x)y_n(x)\, dx = \begin{cases} 0 & \text{if } m \neq n, \\ \alpha_n \neq 0 & \text{if } m = n. \end{cases} \tag{2}$$

In this case the sequence is said to be *orthogonal with respect to the weight function* $q(x)$. Orthogonality properties of this kind are possessed by the eigenfunctions associated with a wide variety of boundary value problems.

Consider a differential equation of the form

$$\frac{d}{dx}\left[p(x)\frac{dy}{dx}\right] + [\lambda q(x) + r(x)]y = 0, \tag{3}$$

for which we are interested in solutions valid on the interval $[a,b]$. We know from Theorem A in Section 14 that if $p(x)$, $p'(x)$, $q(x)$, and $r(x)$ are continuous on this interval, and if $p(x)$ does not vanish there, then there is one and only one solution $y(x)$ for the initial value problem in which we arbitrarily assign prescribed values to both $y(a)$ and $y'(a)$. Suppose, however, that we wish to assign prescribed values to both $y(a)$ and $y(b)$, that is, to $y(x)$ at two different points, rather than to $y(x)$ and $y'(x)$ at the same point. We examine the circumstances under which this *boundary value problem* has a nontrivial solution.

Example 1. At the beginning of Section 40 we considered the special case of (3) in which $p(x) = q(x) = 1$ and $r(x) = 0$, so that the equation is

$$y'' + \lambda y = 0.$$

The interval was taken to be $[0,\pi]$ and the boundary conditions were

$$y(0) = 0 \quad \text{and} \quad y(\pi) = 0.$$

We found that for this problem to be solvable λ must have one of the values

$$\lambda_n = n^2, \quad n = 1, 2, 3, \ldots,$$

and that corresponding solutions are

$$y_n(x) = \sin nx.$$

We called the λ_n the *eigenvalues* of the problem, and the $y_n(x)$ are corresponding *eigenfunctions.*

In the case of the more general equation (3), it turns out that if the functions $p(x)$ and $q(x)$ are restricted in a reasonable way—specifically, if $p(x) > 0$ and $q(x) > 0$ on $[a,b]$—then we will also be able to obtain nontrivial solutions satisfying suitable boundary conditions at the two distinct points a and b if and only if the parameter λ takes on certain specific values. These are the *eigenvalues* of the boundary value problem; they are real numbers that can be arranged in an increasing sequence

$$\lambda_1 < \lambda_2 < \lambda_3 < \cdots < \lambda_n < \lambda_{n+1} < \cdots, \tag{4}$$

and furthermore,

$$\lambda_n \to \infty \quad \text{as} \quad n \to \infty.$$

This ordering is desirable because it enables us to arrange the corresponding *eigenfunctions*

$$y_1(x), \qquad y_2(x), \qquad \ldots, \qquad y_n(x), \qquad \ldots \tag{5}$$

in their own natural order. As in the case of Example 1, the eigenfunctions are not unique, but with the boundary conditions we will be interested in, they are determined up to a nonzero constant factor.

We now look for possible orthogonality properties of the sequence of eigenfunctions (5), and in the process of doing this, we will discover what types of boundary conditions are "suitable." Consider the differential equation (3) written down for two different eigenvalues λ_m and λ_n, with y_m and y_n the corresponding eigenfunctions:

$$\frac{d}{dx}\left[p\frac{dy_m}{dx}\right] + [\lambda_m q + r]y_m = 0$$

and

$$\frac{d}{dx}\left[p\frac{dy_n}{dx}\right] + [\lambda_n q + r]y_n = 0.$$

If we shift to the more compact prime notation for derivatives, then on multiplying the first equation by y_n and the second by y_m, and subtracting, we find that

$$y_n(py_m')' - y_m(py_n')' + (\lambda_m - \lambda_n)qy_m y_n = 0.$$

We now move the first two terms to the right and integrate from a to b, using integration by parts, to obtain

$$(\lambda_m - \lambda_n) \int_a^b q y_m y_n \, dx$$

$$= \int_a^b y_m (p y_n')' \, dx - \int_a^b y_n (p y_m')' \, dx$$

$$= [y_m (p y_n')]_a^b - \int_a^b y_m' (p y_n') \, dx - [y_n (p y_m')]_a^b + \int_a^b y_n' (p y_m') \, dx$$

$$= p(b)[y_m(b) y_n'(b) - y_n(b) y_m'(b)] - p(a)[y_m(a) y_n'(a) - y_n(a) y_m'(a)].$$

$$(6)$$

If we denote by $W(x)$ the Wronskian determinant of the solutions $y_m(x)$ and $y_n(x)$, which is defined by

$$W(x) = \begin{vmatrix} y_m(x) & y_m'(x) \\ y_n(x) & y_n'(x) \end{vmatrix} = y_m(x) y_n'(x) - y_n(x) y_m'(x),$$

then (6) can be written in the convenient form

$$(\lambda_m - \lambda_n) \int_a^b q y_m y_n \, dx = p(b) W(b) - p(a) W(a). \qquad (7)$$

We point out particularly that the integrations by parts in the calculation (6), and the consequent cancellations, are possible only because of the special form of the first term in the differential equation (3).[4]

We want the right side of (6) or (7) to vanish, so that we can obtain the orthogonality property

$$\int_a^b q y_m y_n \, dx = 0 \qquad \text{for } m \neq n. \qquad (8)$$

By looking at the right side of (6), we see that this will certainly happen if the boundary conditions required of a nontrivial solution of (3) are

$$y(a) = 0 \qquad \text{and} \qquad y(b) = 0$$

or

$$y'(a) = 0 \qquad \text{and} \qquad y'(b) = 0.$$

Each of these is a special case of the more general boundary conditions

$$c_1 y(a) + c_2 y'(a) = 0 \qquad \text{and} \qquad d_1 y(b) + d_2 y'(b) = 0, \qquad (9)$$

where c_1 or $c_2 \neq 0$ and d_1 or $d_2 \neq 0$. To see that these boundary

---

[4] Differential equations having this special form are called *self-adjoint*. See the problems below for an explanation of this terminology.

conditions really do make the right side of (7) vanish, suppose that the solutions $y_m(x)$ and $y_n(x)$ both satisfy the first condition (9), so that

$$c_1 y_m(a) + c_2 y_m'(a) = 0,$$
$$c_1 y_n(a) + c_2 y_n'(a) = 0.$$

Since this system has a nontrivial solution c_1, c_2, the coefficient determinant must vanish:

$$\begin{vmatrix} y_m(a) & y_m'(a) \\ y_n(a) & y_n'(a) \end{vmatrix} = W(a) = 0.$$

Similarly $W(b) = 0$, and it follows from this that the right side of (7) vanishes.

Boundary conditions of the form (9) are called *homogeneous boundary conditions*. Their special feature is the fact that any sum of solutions of equation (3) that individually satisfy such boundary conditions will also satisfy the same boundary conditions. Any differential equation of the form (3) with homogeneous boundary conditions is called a *Sturm–Liouville problem*.

The significance of these ideas is that the orthogonality property (8) gives us a formal method for finding series expansions of functions $f(x)$ in terms of the eigenfunctions of such a Sturm–Liouville problem. Formally, we are led to the following procedure. We assume that $f(x)$ can be written in the form

$$f(x) = a_1 y_1(x) + a_2 y_2(x) + \cdots + a_n y_n(x) + \cdots. \tag{10}$$

Multiplying both sides of this by $q(x)y_n(x)$ and integrating term by term from a to b yields

$$\int_a^b f(x)q(x)y_n(x)\, dx$$

$$= a_1 \int_a^b q(x)y_1(x)y_n(x)\, dx + \cdots + a_n \int_a^b q(x)[y_n(x)]^2\, dx + \cdots$$

$$= a_n \int_a^b q(x)[y_n(x)]^2\, dx, \tag{11}$$

because of (8). With the coefficients a_n determined by (11), formula (10) is called an *eigenfunction expansion* of $f(x)$.

A very important mathematical question now arises that is familiar to us from Chapter 6 and the earlier sections of this chapter—how do we know that the series (10) with coefficients determined by (11) really represents $f(x)$? And what does "represents" mean? Does it mean in the sense of pointwise convergence? Or mean convergence? Or perhaps some other concept altogether? We have seen in Chapter 6 how difficult some of these theoretical problems are for ordinary Fourier series, which are the simplest of all eigenfunction expansions. Two further special cases that are particularly important for applications to physics are concerned with the orthogonal sequences of the Legendre polynomials and the Bessel functions. These two sequences of functions, and their properties, and the associated eigenfunction expansions, are the subject of the next chapter.

Self-adjoint boundary value problems of the kind described above are called *regular*, because the interval $[a,b]$ is finite and the functions $p(x)$ and $q(x)$ are positive and continuous on the entire interval. *Singular* problems are those in which one of these functions vanishes or becomes infinite at an endpoint, or the interval itself is infinite. Unfortunately, many of the more important problems are singular, and the theory must be correspondingly more complicated to cope with them.[5]

Example 2. Consider the important Legendre equation in its self-adjoint form,

$$\frac{d}{dx}\left[(1 - x^2)\frac{dy}{dx}\right] + \lambda y = 0, \qquad -1 \le x \le 1.$$

Here the function $p(x) = 1 - x^2$ vanishes at both endpoints. No boundary conditions of the usual kind are imposed at the endpoints $x = \pm 1$, but it is required that the solutions remain bounded near these points. It turns out that this happens only when $\lambda = n(n + 1)$ for $n = 0, 1, 2, \ldots$, and the corresponding solutions are the Legendre polynomials $P_n(x)$. The details of this singular self-adjoint boundary value problem are found in Chapter 8.

Remark. We have done little more in this section than acquaint the student with some of the issues in this subject, and we have certainly not provided any substantive proofs. One of the first questions about any self-adjoint boundary value problem—Sturm–Liouville or otherwise—is this: Does there exists an adequate supply of eigenvalues and corresponding eigenfunctions? For the reader who is interested in these theoretical matters, a full and rigorous proof of this existence theorem is given in Appendix A, but only for a somewhat special case of the regular Sturm–Liouville problem described above.

NOTE ON LIOUVILLE. Joseph Liouville (1809–1882) was a highly respected professor at the Collège de France in Paris and the founder and editor of the *Journal des Mathématiques Pures et Appliquées*, a famous periodical that played

---

[5] Full treatments can be found in E. C. Titchmarsh, *Eigenfunction Expansions*, 2 vols, Oxford University Press, 1946 and 1958; and in E. A. Coddington and N. Levinson, *Theory of Ordinary Differential Equations*, McGraw-Hill, New York, 1955.

an important role in French mathematical life throughout the nineteenth century. For some reason, however, his own remarkable achievements as a creative mathematician have not received the appreciation they deserve. The fact that his collected works have never been published is an unfortunate and rather surprising oversight on the part of his countrymen.

He was the first to solve a boundary value problem by solving an equivalent integral equation, a method developed by Fredholm and Hilbert in the early 1900s into one of the major fields of modern analysis. His ingenious theory of fractional differentiation answered the long-standing question of what reasonable meaning can be assigned to the symbol $d^n y/dx^n$ when n is not a positive integer. He discovered the fundamental result in complex analysis now known as *Liouville's theorem*—that a bounded entire function is necessarily a constant—and used it as the basis for his own theory of elliptic functions. There is also a well-known Liouville theorem in Hamiltonian mechanics, which states that volume integrals are time-invariant in phase space. His theory of the integrals of elementary functions was perhaps the most original of all his achievements, for in it he proved that such integrals as

$$\int e^{-x^2}\,dx, \qquad \int \frac{e^x}{x}\,dx, \qquad \int \frac{\sin x}{x}\,dx, \qquad \int \frac{dx}{\log x},$$

as well as the elliptic integrals of the first and second kinds, cannot be expresed in terms of a finite number of elementary functions.[6]

The fascinating and difficult theory of transcendental numbers is another important branch of mathematics that originated in Liouville's work. The irrationality of π and e—that is, the fact that these numbers are not roots of any linear equation $ax + b = 0$ whose coefficients are integers—had been proved in the eighteenth century by Lambert and Euler. In 1844 Liouville showed that e is also not a root of any quadratic equation with integral coefficients. This led him to conjecture that e is *transcendental*, which means that it does not satisfy any polynomial equation

$$a_n x^n + a_{n-1} x^{n-1} + \cdots + a_1 x + a_0 = 0$$

with integral coefficients. His efforts to prove this failed, but his ideas contributed to Hermite's success in 1873 and then to Lindemann's 1882 proof that π is also transcendental. Lindemann's result showed at last that the age-old problem of squaring the circle by a ruler-and-compass construction is impossible. One of the great mathematical achievements of modern times was Gelfond's 1929 proof that e^π is transcendental, but nothing is yet known about the nature of any of the numbers $\pi + e$, πe or π^e. Liouville also discovered a sufficient condition for transcendence and used it in 1844 to produce the first examples of real numbers

---

[6] See D. G. Mead, "Integration," *Am. Math. Monthly*, vol. 68, pp. 152–156 (1961). For additional details, see G. H. Hardy, *The Integration of Functions of a Single Variable*, Cambridge University Press, London, 1916; or J. F. Ritt, *Integration in Finite Terms*, Columbia University Press, New York, 1948.

that are provably transcendental. One of these is

$$\sum_{n=1}^{\infty} \frac{1}{10^{n!}} = \frac{1}{10^1} + \frac{1}{10^2} + \frac{1}{10^6} + \cdots = 0.11000100\cdots.$$

His methods here have also led to extensive further research in the twentieth century.[7]

PROBLEMS

1. The differential equation $P(x)y'' + Q(x)y' + R(x)y = 0$ is called *exact* if it can be written in the form $[P(x)y']' + [S(x)y]' = 0$ for some function $S(x)$. In this case the second equation can be integrated at once to give the first order linear equation $P(x)y' + S(x)y = c_1$, which can then be solved by the method of Section 10. By equating coefficients and eliminating $S(x)$, show that a necessary and sufficient condition for exactness is $P''(x) - Q'(x) + R(x) = 0$.

2. Consider the Euler equidimensional equation that arose in Section 42,

$$x^2 y'' + xy' - n^2 y = 0,$$

where n is a positive integer. Find the values of n for which this equation is exact, and for these values find the general solution by the method suggested in Problem 1.

3. If the equation in Problem 1 is not exact, it can be made exact by multiplying by a suitable integrating factor $\mu(x)$. Thus, $\mu(x)$ must satisfy the condition that the equation $\mu(x)P(x)y'' + \mu(x)Q(x)y' + \mu(x)R(x)y = 0$ is expressible in the form $[\mu(x)P(x)y']' + [S(x)y]' = 0$ for some function $S(x)$. Show that $\mu(x)$ must be a solution of the *adjoint equation*

$$P(x)\mu'' + [2P'(x) - Q(x)]\mu' + [P''(x) - Q'(x) + R(x)]\mu = 0.$$

In general (but not always) the adjoint equation is just as difficult to solve as the original equation. Find the adjoint equation in each of the following cases:
   (a) Legendre's equation: $(1 - x^2)y'' - 2xy' + p(p + 1)y = 0$;
   (b) Bessel's equation: $x^2 y'' + xy' + (x^2 - p^2)y = 0$;
   (c) Chebyshev's equation: $(1 - x^2)y'' - xy' + p^2 y = 0$;
   (d) Hermite's equation: $y'' - 2xy' + 2py = 0$;
   (e) Airy's equation: $y'' + xy = 0$;
   (f) Laguerre's equation: $xy'' + (1 - x)y' + py = 0$.

4. Solve the equation

$$y'' - \left(2x + \frac{3}{x}\right)y' - 4y = 0$$

by finding a simple solution of the adjoint equation by inspection.

---

[7] An impression of the depth and complexity of this subject can be gained by looking into A. O. Gelfond, *Transcendental and Algebraic Numbers*, Dover, New York, 1960.

5. Show that the adjoint of the adjoint of the equation $P(x)y'' + Q(x)y' + R(x)y = 0$ is the original equation.

6. The equation $P(x)y'' + Q(x)y' + R(x)y = 0$ is called *self-adjoint* if its adjoint is the same equation (except for notation).

   (a) Show that this equation is self-adjoint if and only if $P'(x) = Q(x)$. In this case the equation becomes

   $$P(x)y'' + P'(x)y' + R(x) = 0$$

   or

   $$[P(x)y']' + R(x)y = 0,$$

   which is the standard form of a self-adjoint equation.

   (b) Which of the equations in Problem 3 are self-adjoint?

7. Show that any equation $P(x)y'' + Q(x)y' + R(x)y = 0$ can be made self-adjoint by multiplying through by

   $$\frac{1}{P}e^{\int (Q/P)\,dx}.$$

8. Using Problem 7 when necessary, put each equation in Problem 3 into the standard self-adjoint form described in Problem 6.

9. Consider the regular Sturm–Liouville problem consisting of equation (3) with the boundary conditions (9). Prove that every eigenfunction is unique except for a constant factor. *Hint*: Let $y = u(x)$ and $y = v(x)$ be eigenfunctions corresponding to a single eigenvalue λ, and use their Wronskian to show that they are linearly dependent on $[a,b]$.

10. Consider the following self-adjoint boundary value problem on $[a,b]$:

    $$\frac{d}{dx}\left[p(x)\frac{dy}{dx}\right] + [\lambda q(x) + r(x)]y = 0,$$

    $$y(a) = y(b) \qquad \text{and} \qquad y'(a) = y'(b),$$

    where $p(a) = p(b)$. It is assumed that $p(x)$, $p'(x)$, $q(x)$, and $r(x)$ are continuous and that $p(x) > 0$ and $q(x) > 0$ for $a \le x \le b$. This problem is then said to have *periodic boundary conditions*. It can be proved that there exists a sequence of eigenvalues

    $$\lambda_0 < \lambda_1 < \lambda_2 < \cdots < \lambda_n < \lambda_{n+1} < \cdots$$

    such that

    $$\lim_{n\to\infty} \lambda_n = \infty.$$

    (a) By examining the calculation (6), show that eigenfunctions corresponding to distinct eigenvalues are orthogonal with respect to the weight function $q(x)$.

    (b) In this case, however, to each eigenvalue there may correspond either one or two linearly independent eigenfunctions. Verify this by finding the eigenvalues and corresponding eigenfunctions for the problem $y'' + \lambda y = 0$, where $y(-\pi) = y(\pi)$ and $y'(-\pi) = y'(\pi)$.

    (c) Why can this problem not have more than two independent eigenfunctions associated with a particular eigenvalue?

APPENDIX A. THE EXISTENCE OF EIGENVALUES AND EIGENFUNCTIONS

The general theory of eigenvalues, eigenfunctions, and eigenfunction expansions is one of the deepest and richest parts of modern mathematics. In this appendix we confine our attention to a small but significant fragment of this broad subject. Our primary purpose is to prove that any boundary value problem of the form 40-(23)—which arose in connection with the nonhomogeneous vibrating string—has eigenvalues and eigenfunctions with properties similar to those encountered in Section 40. Once this is accomplished, we will find that a simple change of variable allows us to extend this result to a considerably more general class of problems.

We begin with several easy consequences of the Sturm comparison theorem.

Lemma 1. *Let $y(x)$ and $z(x)$ be nontrivial solutions of*

$$y'' + q(x)y = 0$$

and

$$z'' + r(x)z = 0,$$

where $q(x)$ and $r(x)$ are positive continuous functions such that $q(x) > r(x)$. Suppose that $y(x)$ and $z(x)$ both vanish at a point b_0, and that $z(x)$ has a finite or infinite number of successive zeros $b_1, b_2, \ldots, b_n, \ldots$ to the right of b_0. Then $y(x)$ has at least as many zeros as $z(x)$ on every closed interval $[b_0, b_n]$; and if the successive zeros of $y(x)$ to the right of b_0 are $a_1, a_2, \ldots, a_n, \ldots$, then $a_n < b_n$ for every n.

Proof. By the Sturm comparison theorem (Theorem 25-B), $y(x)$ has at least one zero in each of the open intervals $(b_0, b_1), (b_1, b_2), \ldots, (b_{n-1}, b_n)$, and both statements follow at once from this.

Lemma 2. *Let $q(x)$ be a positive continuous function that satisfies the inequalities*

$$0 < m^2 < q(x) < M^2$$

on a closed interval $[a, b]$. If $y(x)$ is a nontrivial solution of $y'' + q(x)y = 0$ on this interval, and if x_1 and x_2 are successive zeros of $y(x)$, then

$$\frac{\pi}{M} < x_2 - x_1 < \frac{\pi}{m}. \tag{1}$$

Furthermore, if $y(x)$ vanishes at a and b, and at $n - 1$ points in the open interval (a, b), then

$$\frac{m(b - a)}{\pi} < n < \frac{M(b - a)}{\pi}. \tag{2}$$

Proof. To prove (1), we begin by comparing the given equation with $z'' + m^2 z = 0$. A nontrivial solution of this that vanishes at x_1 is $z(x) = \sin m(x - x_1)$. Since the next zero of $z(x)$ is $x_1 + \pi/m$, and Theorem 25-B tells us that x_2 must occur before this, we have $x_2 < x_1 + \pi/m$ or $x_2 - x_1 < \pi/m$. A similar argument gives the other inequality in (1).

To prove (2), we first observe that there are n subintervals between the $n + 1$ zeros, so by (1) we have $b - a =$ the sum of the lengths of the n subintervals $< n(\pi/m)$, and therefore $m(b - a)/\pi < n$. In the same way we see that $b - a > n(\pi/M)$, so $n < M(b - a)/\pi$.

Our main preliminary result is the next lemma.

Lemma 3. *Let $q(x)$ be a positive continuous function and consider the differential equation*

$$y'' + \lambda q(x)y = 0 \tag{3}$$

on a closed interval $[a,b]$. For each λ, let $y_\lambda(x)$ be the unique solution of equation (3) which satisfies the initial conditions $y_\lambda(a) = 0$ and $y'_\lambda(a) = 1$. Then there exists an increasing sequence of positive numbers

$$\lambda_1 < \lambda_2 < \cdots < \lambda_n < \cdots$$

that approaches ∞ and has the property that $y_\lambda(b) = 0$ if and only if λ equals one of the λ_n. Furthermore, the function $y_{\lambda_n}(x)$ has exactly $n - 1$ zeros in the open interval (a,b).

Proof. It is clear by Theorem 24-B that $y_\lambda(x)$ has no zeros to the right of a when $\lambda \le 0$. Our plan is to watch the oscillation behavior of $y_\lambda(x)$ as λ increases from 0. We begin with the observation that by the continuity of $q(x)$ there exist positive numbers m and M such that on $[a,b]$ we have $0 < m^2 < q(x) < M^2$. Thus, in the sense made precise in Section 25, $y_\lambda(x)$ oscillates more rapidly on $[a,b]$ than solutions of

$$y'' + \lambda m^2 y = 0,$$

and less rapidly than solutions of

$$y'' + \lambda M^2 y = 0.$$

By Lemma 2, when λ is positive and small (so small that $\pi/\sqrt{\lambda}\, M \ge b - a$) the function $y_\lambda(x)$ has no zeros in $[a,b]$ to the right of a; and when λ increases to the point where $\pi/\sqrt{\lambda}\, m \le b - a$, then $y_\lambda(x)$ has at least one such zero. Similarly, as λ increases to ∞, the number of zeros of $y_\lambda(x)$ in $[a,b]$ tends toward ∞. It follows from Lemma 1 that the nth zero of $y_\lambda(x)$ to the right of a moves to the left as λ increases, and we shall take it for granted (it can be proved) that this zero moves continuously. Consequently, as λ starts at 0 and increases to ∞, there are infinitely many values $\lambda_1, \lambda_2, \ldots, \lambda_n, \ldots$ for which a zero of $y_\lambda(x)$ reaches b and subsequently enters the interval, so that $y_{\lambda_n}(x)$ vanishes at a and b and has $n - 1$ zeros in (a,b). To show that the sequence $\lambda_1, \lambda_2, \ldots, \lambda_n, \ldots$ approaches ∞, we appeal to the inequalities (2), which in this case become

$$\frac{\sqrt{\lambda_n}\, m(b - a)}{\pi} < n < \frac{\sqrt{\lambda_n}\, M(b - a)}{\pi}$$

or

$$\frac{n^2\pi^2}{M^2(b - a)^2} < \lambda_n < \frac{n^2\pi^2}{m^2(b - a)^2}.$$

Equation (3) is the special case of the *Sturm–Liouville equation*

$$\frac{d}{dx}\left[p(x)\frac{dy}{dx}\right] + \lambda q(x)y = 0 \tag{4}$$

in which $p(x) = 1$. We assume here that $p(x)$ and $q(x)$ are positive continuous functions on $[a,b]$, and also that $p(x)$ has a continuous derivative on this interval. If we change the independent variable in (4) from x to a new variable w defined by

$$w(x) = \int_a^x \frac{dt}{p(t)},$$

so that

$$\frac{dw}{dx} = \frac{1}{p(x)} \quad \text{and} \quad \frac{dy}{dx} = \frac{dy}{dw}\frac{dw}{dx} = \frac{1}{p(x)}\frac{dy}{dw},$$

then (4) takes the form

$$\frac{d^2y}{dw^2} + \lambda q_1(w)y = 0, \tag{5}$$

where $q_1(w)$ is positive and continuous on the transformed interval $0 \le w \le c = w(b)$. On applying Lemma 3 to equation (5), we immediately obtain the following statement about (4).

Theorem A. *Consider the boundary value problem*

$$\frac{d}{dx}\left[p(x)\frac{dy}{dx}\right] + \lambda q(x)y = 0, \qquad y(a) = y(b) = 0, \tag{6}$$

where $p(x)$ and $q(x)$ satisfy the conditions stated above. Then there exists an increasing sequence of positive numbers

$$\lambda_1 < \lambda_2 < \cdots < \lambda_n < \cdots$$

that approaches ∞ and has the property that (6) has a nontrivial solution if and only if λ equals one of the λ_n. The solution corresponding to $\lambda = \lambda_n$ is unique except for an arbitrary constant factor, and has exactly $n - 1$ zeros in the open interval (a,b).

One final remark is in order. As pointed out in Section 43, we usually refer to (6) as a *regular* Sturm–Liouville problem because the interval is finite and the functions $p(x)$ and $q(x)$ are positive and continuous on the entire interval. *Singular* problems arise when the interval is infinite, or when it is finite and $p(x)$ or $q(x)$ vanishes or is discontinuous at one or both endpoints. These problems are considerably more difficult, and of course are not covered by our discussion in this appendix. Unfortunately, many of the most interesting differential equations are singular in this sense. We mention *Legendre's equation*

$$\frac{d}{dx}\left[(1 - x^2)\frac{dy}{dx}\right] + \lambda y = 0, \qquad -1 \le x \le 1;$$

Chebyshev's equation

$$\frac{d}{dx}\left[(1 - x^2)^{1/2}\frac{dy}{dx}\right] + \lambda(1 - x^2)^{-1/2}y = 0, \qquad -1 < x < 1;$$

Hermite's equation

$$\frac{d}{dx}\left[e^{-x^2}\frac{dy}{dx}\right] + \lambda e^{-x^2}y = 0, \qquad -\infty < x < \infty;$$

and *Laguerre's equation*

$$\frac{d}{dx}\left[xe^{-x}\frac{dy}{dx}\right] + \lambda e^{-x}y = 0, \qquad 0 \le x < \infty.$$

These equations appeared in Chapter 5, where they were studied from an entirely different point of view.

CHAPTER
8

SOME SPECIAL FUNCTIONS OF MATHEMATICAL PHYSICS

44 LEGENDRE POLYNOMIALS

This section and the next are entirely devoted to the technical task of defining the Legendre polynomials and establishing a number of their special properties. It is natural to wonder about the purpose of this elaborate machinery, and more generally, why we care about Legendre polynomials at all. The simplest answer is that the Legendre polynomials have many important applications to mathematical physics, and these applications depend on this machinery. For the benefit of readers who wish to see for themselves, the physical background and several typical applications are discussed in Appendix A. There is another answer, however, which is less utilitarian and applies equally to our subsequent treatment of Bessel functions. It is that the study of specific classical functions and their individual properties provides a healthy counterpoise to the abstract ideas that sometimes seem to dominate contemporary mathematics. In addition, we mention several items that arise naturally in the context of this chapter which we hope will be of interest to all students of mathematics: the gamma function and the formula $(-\frac{1}{2})! =$

$\sqrt{\pi}$; Lambert's continued fraction for the tangent,

$$\tan x = \cfrac{1}{\cfrac{1}{x} - \cfrac{1}{\cfrac{3}{x} - \cfrac{1}{\cfrac{5}{x} - \cdots}}};$$

and the famous series

$$\sum_{n=1}^{\infty} \frac{1}{n^2} = \frac{\pi^2}{6} \quad \text{and} \quad \sum_{n=1}^{\infty} \frac{1}{n^4} = \frac{\pi^4}{90},$$

whose sums were discovered by Euler in the early eighteenth century and which appear again in a surprising way in connection with the zeros of Bessel functions.

Now for the Legendre polynomials themselves, which we approach by way of the hypergeometric equation.[1]

In Section 28 we used Legendre's equation to illustrate the technique of finding power series solutions at ordinary points. For reasons explained in Appendix A, we now write this equation in the form

$$(1 - x^2)y'' - 2xy' + n(n + 1)y = 0, \tag{1}$$

where n is understood to be a non-negative integer. The reader will recall that all the solutions of (1) found in Section 28 are analytic on the interval $-1 < x < 1$. However, the solutions most useful in the applications are those bounded near $x = 1$, and for convenience in singling these out we change the independent variable from x to $t = \frac{1}{2}(1 - x)$. This makes $x = 1$ correspond to $t = 0$ and transforms (1) into

$$t(1 - t)y'' + (1 - 2t)y' + n(n + 1)y = 0, \tag{2}$$

where the primes signify derivatives with respect to t. This is a hypergeometric equation with $a = -n$, $b = n + 1$, and $c = 1$, so it has the following polynomial solution near $t = 0$:

$$y_1 = F(-n, n + 1, 1, t). \tag{3}$$

---

[1] Adrien Marie Legendre (1752–1833) encountered his polynomials in his research on the gravitational attraction of ellipsoids. He was a very good French mathematician who had the misfortune of seeing most of his best work—in elliptic integrals, number theory, and the method of least squares—superseded by the achievements of younger and abler men. For instance, he devoted 40 years to his research on elliptic integrals, and his two-volume treatise on the subject had scarcely appeared in print when the discoveries of Abel and Jacobi revolutionized the field completely. He was very remarkable for the generous spirit with which he repeatedly welcomed newer and better work that made his own obsolete.

Since the exponents of (2) at the origin are both zero ($m_1 = 0$ and $m_2 = 1 - c = 0$), we seek a second solution by the method of Section 16. This second solution is $y_2 = v y_1$, where

$$v' = \frac{1}{y_1^2} e^{-\int P \, dt} = \frac{1}{y_1^2} e^{\int (2t-1)/t(1-t) \, dt}$$

$$= \frac{1}{y_1^2 t(1 - t)} = \frac{1}{t} \left[\frac{1}{y_1^2(1 - t)} \right]$$

by an elementary integration. Since y_1^2 is a polynomial with constant term 1, the bracketed expression on the right is an analytic function of the form $1 + a_1 t + a_2 t^2 + \cdots$, and we have

$$v' = \frac{1}{t} + a_1 + a_2 t + \cdots.$$

This yields $v = \log t + a_1 t + \cdots$, so

$$y_2 = y_1(\log t + a_1 t + \cdots)$$

and the general solution of (2) near the origin is

$$y = c_1 y_1 + c_2 y_2. \tag{4}$$

Because of the presence of the term $\log t$ in y_2, it is clear that (4) is bounded near $t = 0$ if and only if $c_2 = 0$. If we replace t in (3) by $\frac{1}{2}(1 - x)$, it follows that the solutions of (1) bounded near $x = 1$ are precisely constant multiples of the polynomial $F[-n, n + 1, 1, \frac{1}{2}(1 - x)]$.

   This brings us to the fundamental definition. The nth *Legendre polynomial* is denoted by $P_n(x)$ and defined by

$$P_n(x) = F\left[-n, n + 1, 1, \frac{1}{2}(1 - x)\right] = 1 + \frac{(-n)(n + 1)}{(1!)^2}\left(\frac{1 - x}{2}\right)$$

$$+ \frac{(-n)(-n + 1)(n + 1)(n + 2)}{(2!)^2}\left(\frac{1 - x}{2}\right)^2 + \cdots$$

$$+ \frac{(-n)(-n + 1)\cdots[-n + (n - 1)](n + 1)(n + 2)\cdots(2n)}{(n!)^2}$$

$$\times \left(\frac{1 - x}{2}\right)^n$$

$$= 1 + \frac{n(n + 1)}{(1!)^2 2}(x - 1) + \frac{n(n - 1)(n + 1)(n + 2)}{(2!)^2 2^2}(x - 1)^2$$

$$+ \cdots + \frac{(2n)!}{(n!)^2 2^n}(x - 1)^n. \tag{5}$$

We know from our work in Section 28 that $P_n(x)$ is a polynomial of degree n that contains only even or only odd powers of x according as n is even or odd. It can therefore be written in the form

$$P_n(x) = a_n x^n + a_{n-2} x^{n-2} + a_{n-4} x^{n-4} + \cdots, \tag{6}$$

where this sum ends with a_0 if n is even and $a_1 x$ if n is odd. It is clear from (5) that $P_n(1) = 1$ for every n, and in view of (6) we also have $P_n(-1) = (-1)^n$.

As it stands, formula (5) is a very inconvenient tool to use in studying $P_n(x)$, so we look for something simpler. We could expand each term in (5), collect like powers of x, and arrange the result in the form (6), but this would be unnecessarily laborious. What we shall do is notice from (5) that $a_n = (2n)!/(n!)^2 2^n$ and calculate $a_{n-2}, a_{n-4}, \ldots$ recursively in terms of a_n. What is needed here is formula 28-(9) with p replaced by n and n by $k - 2$:

$$a_k = -\frac{(n - k + 2)(n + k - 1)}{(k - 1)k} a_{k-2}$$

or

$$a_{k-2} = -\frac{k(k - 1)}{(n - k + 2)(n + k - 1)} a_k.$$

When $k = n, n - 2, \ldots$, this yields

$$a_{n-2} = -\frac{n(n - 1)}{2(2n - 1)} a_n,$$

$$a_{n-4} = -\frac{(n - 2)(n - 3)}{4(2n - 3)} a_{n-2}$$

$$= \frac{n(n - 1)(n - 2)(n - 3)}{2 \cdot 4(2n - 1)(2n - 3)} a_n,$$

and so on, so (6) becomes

$$P_n(x) = \frac{(2n)!}{(n!)^2 2^n} \left[x_n - \frac{n(n - 1)}{2(2n - 1)} x^{n-2} \right.$$

$$+ \frac{n(n - 1)(n - 2)(n - 3)}{2 \cdot 4(2n - 1)(2n - 3)} x^{n-4} + \cdots$$

$$\left. + (-1)^k \frac{n(n - 1) \cdots (n - 2k + 1)}{2^k k! \, (2n - 1)(2n - 3) \cdots (2n - 2k + 1)} x^{n-2k} + \cdots \right].$$

$$\tag{7}$$

Since

$$n(n - 1) \cdots (n - 2k + 1) = \frac{n!}{(n - 2k)!}$$

and

$(2n - 2k + 1)(2n - 2k + 3) \cdots (2n - 3)(2n - 1)$

$$= \frac{(2n - 2k + 1)(2n - 2k + 2)(2n - 2k + 3) \cdots (2n - 3)(2n - 2)(2n - 1)2n}{(2n - 2k + 2) \cdots (2n - 2)2n}$$

$$= \frac{(2n)!}{(2n - 2k)!\, 2^k (n - k + 1) \cdots (n - 1)n} \frac{1}{} = \frac{(2n)!\,(n - k)!}{(2n - 2k)!\, 2^k n!},$$

the coefficient of x^{n-2k} in (7) is

$$(-1)^k \frac{n!}{2^k k!\,(n - 2k)!} \frac{(2n - 2k)!\, 2^k n!}{(2n)!\,(n - k)!} = (-1)^k \frac{(n!)^2 (2n - 2k)!}{k!\,(2n)!\,(n - k)!\,(n - 2k)!}.$$

This enables us to write (7) as

$$P_n(x) = \sum_{k=0}^{[n/2]} (-1)^k \frac{(2n - 2k)!}{2^n k!\,(n - k)!\,(n - 2k)!}\, x^{n-2k}, \tag{8}$$

where $[n/2]$ is the usual symbol for the greatest integer $\leq n/2$. We continue toward an even more concise form by observing that

$$P_n(x) = \sum_{k=0}^{[n/2]} \frac{(-1)^k}{2^n k!\,(n - k)!} \frac{(2n - 2k)!}{(n - 2k)!}\, x^{n-2k}$$

$$= \sum_{k=0}^{[n/2]} \frac{(-1)^k}{2^n k!\,(n - k)!} \frac{d^n}{dx^n} x^{2n-2k}$$

$$= \frac{1}{2^n n!} \frac{d^n}{dx^n} \sum_{k=0}^{[n/2]} \frac{n!}{k!\,(n - k)!} (x^2)^{n-k} (-1)^k.$$

If we extend the range of this sum by letting k vary from 0 to n—which changes nothing since the new terms are of degree $< n$ and their nth derivatives are zero—then we get

$$P_n(x) = \frac{1}{2^n n!} \frac{d^n}{dx^n} \sum_{k=0}^{n} \binom{n}{k} (x^2)^{n-k} (-1)^k;$$

and the binomial formula yields

$$P_n(x) = \frac{1}{2^n n!} \frac{d^n}{dx^n} (x^2 - 1)^n. \tag{9}$$

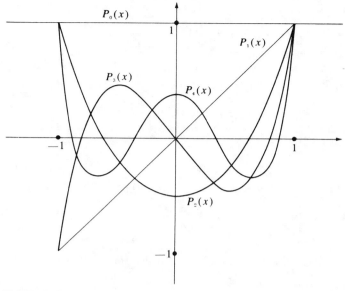

FIGURE 52

This expression for $P_n(x)$ is called *Rodrigues' formula.*[2] It provides a relatively easy method for computing the successive Legendre polynomials, of which the first few (Fig. 52) are

$$P_0(x) = 1, \qquad\qquad P_1(x) = x,$$

$$P_2(x) = \frac{1}{2}(3x^2 - 1), \qquad P_3(x) = \frac{1}{2}(5x^3 - 3x). \qquad (10)$$

An even easier procedure is suggested in Problem 2, and a more significant application of (9) will appear in the next section.

PROBLEMS

1. The function on the left side of

$$\frac{1}{\sqrt{1 - 2xt + t^2}} = P_0(x) + P_1(x)t + P_2(x)t^2 + \cdots + P_n(x)t^n + \cdots$$

---

[2] Olinde Rodrigues (1794–1851) was a French banker who came to the aid of Claude Henri Saint-Simon (the founder of socialism) in his destitute old age, supported him during the last years of his life, and became one of his earliest disciples. He discovered the above formula in 1816, but soon thereafter became interested in the scientific organization of society and never returned to mathematics. The term "Rodrigues' formula" is often applied by transference to similar expressions for other classical polynomials of which Rodrigues himself knew nothing.

is called the *generating function* of the Legendre polynomials. Assume that this relation is true, and use it

(a) to verify that $P_n(1)$ and $P_n(-1) = (-1)^n$;

(b) to show that $P_{2n+1}(0) = 0$ and $P_{2n}(0) = (-1)^n \dfrac{1 \cdot 3 \cdots (2n - 1)}{2^n n!}$.

2. Consider the generating relation in Problem 1,

$$\frac{1}{\sqrt{1 - 2xt + t^2}} = \sum_{n=0}^{\infty} P_n(x)t^n.$$

(a) By differentiating both sides with respect to t, show that

$$(x - t) \sum_{n=0}^{\infty} P_n(x)t^n = (1 - 2xt + t^2) \sum_{n=1}^{\infty} nP_n(x)t^{n-1}.$$

(b) Equate the coefficients of t^n in (a) to obtain the recursion formula

$$(n + 1)P_{n+1}(x) = (2n + 1)xP_n(x) - nP_{n-1}(x).$$

(c) Assume that $P_0(x) = 1$ and $P_1(x) = x$ are known, and use the recursion formula in (b) to calculate $P_2(x)$, $P_3(x)$, $P_4(x)$, and $P_5(x)$.

3. Establish the generating relation of Problems 1 and 2 by the following steps:

(a) Use the binomial series to write

$$[1 - t(2x - t)]^{-1/2} = 1 + \frac{1}{2}t(2x - t) + \frac{1 \cdot 3}{2^2 2!}t^2(2x - t)^2 + \cdots$$

$$+ \frac{1 \cdot 3 \cdots (2n - 3)}{2^{n-1}(n - 1)!}t^{n-1}(2x - t)^{n-1}$$

$$+ \frac{1 \cdot 3 \cdots (2n - 1)}{2^n n!}t^n(2x - t)^n + \cdots.$$

(b) It is clear that t^n can occur only in terms out to and including the last term written in (a). By expanding the various powers of $2x - t$, show that the total coefficient of t^n is

$$\frac{1 \cdot 3 \cdots (2n - 1)}{2^n n!}(2x)^n - \frac{1 \cdot 3 \cdots (2n - 3)}{2^{n-1}(n - 1)!} \frac{n - 1}{1!}(2x)^{n-2}$$

$$+ \frac{1 \cdot 3 \cdots (2n - 5)}{2^{n-2}(n - 2)!} \frac{(n - 2)(n - 3)}{2!}(2x)^{n-4} - \cdots.$$

(c) Show that the sum in (b) is $P_n(x)$ as given by (8).

4. This problem constitutes a direct verification that if $P_n(x)$ is *defined* by formula (9), then it satisfies Legendre's equation (1) and has the property that $P_n(1) = 1$. Consider the polynomials of degree n defined by

$$y(x) = \frac{d^n}{dx^n}(x^2 - 1)^n.$$

(a) If $w = (x^2 - 1)^n$, then $(x^2 - 1)w' - 2nxw = 0$. By differentiating this equation $k + 1$ times, show that

$$(x^2 - 1)w^{(k+2)} + 2(k + 1)xw^{(k+1)} + (k + 1)kw^{(k)}$$
$$- 2nxw^{(k+1)} - 2(k + 1)nw^{(k)} = 0,$$

and conclude that $y = w^{(n)}$ is a solution of equation (1).

(b) Put $u = (x - 1)^n$ and $v = (x + 1)^n$ and use the formula

$$y = (uv)^{(n)} = u^{(n)}v + nu^{(n-1)}v^{(1)} + \cdots + nu^{(1)}v^{(n-1)} + uv^{(n)}$$

to show that $y(1) = n!\, 2^n$.

45   PROPERTIES OF LEGENDRE POLYNOMIALS

In the previous section we defined the sequence of Legendre polynomials

$$P_0(x),\ P_1(x),\ P_2(x),\ \ldots,\ P_n(x),\ \ldots. \tag{1}$$

The reader is aware that these polynomials have a number of applications, which range from mathematical physics to the theory of approximation. We now discuss the fundamental ideas on which some of these applications depend.

Orthogonality. The most important property of the Legendre polynomials is the fact that

$$\int_{-1}^{1} P_m(x)P_n(x)\, dx = \begin{cases} 0 & \text{if } m \neq n, \\ \dfrac{2}{2n + 1} & \text{if } m = n. \end{cases} \tag{2}$$

This is often expressed by saying that (1) is a sequence of *orthogonal functions* on the interval $-1 \le x \le 1$. We shall explain the significance of this property after we prove it.

Let $f(x)$ be any function with at least n continuous derivatives on the interval $-1 \le x \le 1$, and consider the integral

$$I = \int_{-1}^{1} f(x)P_n(x)\, dx.$$

Rodrigues' formula enables us to write this as

$$I = \frac{1}{2^n n!} \int_{-1}^{1} f(x) \frac{d^n}{dx^n}(x^2 - 1)^n\, dx,$$

and an integration by parts gives

$$I = \frac{1}{2^n n!}\left[f(x) \frac{d^{n-1}}{dx^{n-1}}(x^2 - 1)^n \right]_{-1}^{1} - \frac{1}{2^n n!}\int_{-1}^{1} f'(x) \frac{d^{n-1}}{dx^{n-1}}(x^2 - 1)^n\, dx.$$

The expression in brackets vanishes at both limits, so

$$I = -\frac{1}{2^n n!}\int_{-1}^{1} f'(x) \frac{d^{n-1}}{dx^{n-1}}(x^2 - 1)^n\, dx;$$

and by continuing to integrate by parts, we obtain

$$I = \frac{(-1)^n}{2^n n!} \int_{-1}^{1} f^{(n)}(x)(x^2 - 1)^n \, dx.$$

If $f(x) = P_m(x)$ with $m < n$, then $f^{(n)}(x) = 0$ and consequently $I = 0$, which proves the first part of (2). To establish the second part, we put $f(x) = P_n(x)$. Since $P_n^{(n)}(x) = (2n)!/2^n n!$, it follows that

$$I = \frac{(2n)!}{2^{2n}(n!)^2} \int_{-1}^{1} (1 - x^2)^n \, dx$$

$$= \frac{2(2n)!}{2^{2n}(n!)^2} \int_{0}^{1} (1 - x^2)^n \, dx. \qquad (3)$$

If we change the variable by writing $x = \sin \theta$, and recall the formula (proved by an integration by parts)

$$\int \cos^{2n+1} \theta \, d\theta = \frac{1}{2n + 1} \cos^{2n} \theta \sin \theta + \frac{2n}{2n + 1} \int \cos^{2n-1} \theta \, d\theta, \quad (4)$$

then the definite integral in (3) becomes

$$\int_{0}^{\pi/2} \cos^{2n+1} \theta \, d\theta = \frac{2n}{2n + 1} \int_{0}^{\pi/2} \cos^{2n-1} \theta \, d\theta$$

$$= \frac{2n}{2n + 1} \frac{2n - 2}{2n - 1} \cdots \frac{2}{3} \int_{0}^{\pi/2} \cos \theta \, d\theta$$

$$= \frac{2^n n!}{1 \cdot 3 \cdots (2n - 1)(2n + 1)} = \frac{2^{2n}(n!)^2}{(2n)! \, (2n + 1)}.$$

We conclude that in this case $I = 2/(2n + 1)$, and the proof of (2) is complete.

Legendre series. As we illustrate in Appendix A, many problems of potential theory depend on the possibility of expanding a given function in a series of Legendre polynomials. It is easy to see that this can always be done when the given function is itself a polynomial. For example, formulas 44-(10) tell us that

$$1 = P_0(x), \qquad x = P_1(x), \qquad x^2 = \frac{1}{3} + \frac{2}{3} P_2(x) = \frac{1}{3} P_0(x) + \frac{2}{3} P_2(x),$$

$$x^3 = \frac{3}{5} x + \frac{2}{5} P_3(x) = \frac{3}{5} P_1(x) + \frac{2}{5} P_3(x);$$

and it follows that any third-degree polynomial $p(x) = b_0 + b_1 x +$

$b_2 x^2 + b_3 x^3$ can be written as

$$p(x) = b_0 P_0(x) + b_1 P_1(x) + b_2 \left[\frac{1}{3} P_0(x) + \frac{2}{3} P_2(x) \right]$$

$$+ b_3 \left[\frac{3}{5} P_1(x) + \frac{2}{5} P_3(x) \right]$$

$$= \left(b_0 + \frac{b_2}{3} \right) P_0(x) + \left(b_1 + \frac{3 b_3}{5} \right) P_1(x) + \frac{2 b_2}{3} P_2(x) + \frac{2 b_3}{5} P_3(x)$$

$$= \sum_{n=0}^{3} a_n P_n(x).$$

More generally, since $P_n(x)$ is a polynomial of degree n for every positive integer n, a simple extension of this procedure shows that x^n can always be expressed as a linear combination of $P_0(x), P_1(x), \ldots, P_n(x)$, so any polynomial $p(x)$ of degree k has an expansion of the form

$$p(x) = \sum_{n=0}^{k} a_n P_n(x).$$

An obvious problem that arises from these remarks—and also from the demands of the applications—is that of expanding an "arbitrary" function $f(x)$ in a so-called *Legendre series*:

$$f(x) = \sum_{n=0}^{\infty} a_n P_n(x). \tag{5}$$

It is clear that a new procedure is needed for calculating the coefficients a_n in (5), and the key lies in formulas (2).

If we throw mathematical caution to the winds, and multiply (5) by $P_m(x)$ and integrate term by term from -1 to 1, then the result is

$$\int_{-1}^{1} f(x) P_m(x)\, dx = \sum_{n=0}^{\infty} a_n \int_{-1}^{1} P_m(x) P_n(x)\, dx;$$

and in view of (2), this collapses to

$$\int_{-1}^{1} f(x) P_m(x)\, dx = \frac{2 a_m}{2m + 1}.$$

We therefore have the following formula for the a_n in (5):

$$a_n = \left(n + \frac{1}{2} \right) \int_{-1}^{1} f(x) P_n(x)\, dx. \tag{6}$$

These manipulations are easy to justify if $f(x)$ is known in advance to have a series expansion of the form (5) and this series is integrable term by term on the interval $-1 \le x \le 1$. Both conditions are obviously satisfied when $f(x)$ is a polynomial; but in the case of other types of functions we have no way of knowing this, and our conclusion that the coefficients a_n in (5) are given by (6) is of doubtful validity. Nevertheless, these formal procedures are highly suggestive, and can lead to legitimate mathematics if we ask the following question. If the a_n are *defined* by formula (6) and then used to form the series (5), for what kinds of functions $f(x)$ will these a_n exist and the expansion (5) be valid? This question has an answer, but this is not the place to go into details.[3]

The possibility of expansions of the form (5) obviously depends in a crucial way on the orthogonality property (2) of the Legendre polynomials. This is an instance of the following general phenomenon, which is often encountered in the theory of special functions. If a sequence of functions $\phi_1(x), \phi_2(x), \ldots, \phi_n(x), \ldots$ defined on an interval $a \leq x \leq b$ has the property that

$$\int_a^b \phi_m(x)\phi_n(x)\,dx = \begin{cases} 0 & \text{if } m \neq n, \\ \alpha_n \neq 0 & \text{if } m = n, \end{cases} \tag{7}$$

then the ϕ_n are said to be *orthogonal functions* on this interval. Just as above, the general problem that arises in connection with a sequence of this kind is that of representing "arbitrary" functions $f(x)$ by expansions of the form

$$f(x) = \sum_{n=1}^{\infty} a_n\phi_n(x),$$

and a formal use of (7) suggests that the coefficients a_n ought to be given by

$$a_n = \frac{1}{\alpha_n} \int_a^b f(x)\phi_n(x)\,dx.$$

Additional examples occur in Appendices B and D of Chapter 5, where

---

[3] The answer we refer to—often called the *Legendre expansion theorem*—is easy to understand, but its proof depends on many properties of the Legendre polynomials that we have not mentioned. This theorem makes the following statement: If both $f(x)$ and $f'(x)$ have at most a finite number of jump discontinuities on the interval $-1 \leq x \leq 1$, and if $f(x-)$ and $f(x+)$ denote the limits of $f(x)$ from the left and from the right at a point x, then the a_n exist and the Legendre series converges to

$$\frac{1}{2}[f(x-) + f(x+)]$$

for $-1 < x < 1$, to $f(-1+)$ at $x = -1$, and to $f(1-)$ at $x = 1$—and in particular, it converges to $f(x)$ at every point of continuity. See N. N. Lebedev, *Special Functions and Their Applications*, pp. 53–58, Prentice-Hall, Englewood Cliffs, N.J., 1965.

the orthogonality (with respect to suitable weight functions) of the Hermite polynomials and Chebyshev polynomials is briefly mentioned. The satisfactory solution of this group of problems was one of the main achievements of pure mathematics in the nineteenth and early twentieth centuries. Also, Chapter 6 contains a fairly full treatment of the classical problem that underlies all of these ideas—that of expanding suitably restricted functions in Fourier series.

Least squares approximation. Let $f(x)$ be a function defined on the interval $-1 \le x \le 1$, and consider the problem of approximating $f(x)$ as closely as possible in the sense of least squares by polynomials $p(x)$ of degree $\le n$. If we think of the integral

$$I = \int_{-1}^{1} [f(x) - p(x)]^2 \, dx \tag{8}$$

as representing the sum of the squares of the deviations of $p(x)$ from $f(x)$, then the problem is to minimize the value of this integral by a suitable choice of $p(x)$. It turns out that the minimizing polynomial is precisely the sum of the first $n + 1$ terms of the Legendre series (5),

$$p(x) = a_0 P_0(x) + \cdots + a_n P_n(x),$$

where the coefficients are given by (6).

To prove this, we use the fact that all polynomials of degree $\le n$ are expressible in the form $b_0 P_0(x) + \cdots + b_n P_n(x)$. The integral (8) can therefore be written as

$$
\begin{aligned}
I &= \int_{-1}^{1} \left[f(x) - \sum_{k=0}^{n} b_k P_k(x) \right]^2 dx \\
&= \int_{-1}^{1} f(x)^2 dx + \sum_{k=0}^{n} \frac{2}{2k+1} b_k^2 - 2 \sum_{k=0}^{n} b_k \left[\int_{-1}^{1} f(x) P_k(x) \, dx \right] \\
&= \int_{-1}^{1} f(x)^2 \, dx + \sum_{k=0}^{n} \frac{2}{2k+1} b_k^2 - 2 \sum_{k=0}^{n} b_k \frac{2 a_k}{2k+1} \\
&= \int_{-1}^{1} f(x)^2 \, dx + \sum_{k=0}^{n} \frac{2}{2k+1} (b_k - a_k)^2 - \sum_{k=0}^{n} \frac{2}{2k+1} a_k^2.
\end{aligned}
$$

Since the a_k are fixed and the b_k are at our disposal, it is clear that I assumes its minimum value when $b_k = a_k$ for $k = 0, \ldots, n$. The only hypothesis required by this argument is that $f(x)$ and $f(x)^2$ must be integrable. If the function $f(x)$ is sufficiently well behaved to have a power series expansion on the interval $-1 \le x \le 1$, then most students assume that the "best" polynomial approximations to $f(x)$ are given by the partial sums of this power series. The result we have established here shows that this is false if our criterion is approximation in the sense of least squares.

PROBLEMS

1. Verify formula (4).

2. Legendre's equation can also be written in the form

$$\frac{d}{dx}[(1 - x^2)y'] + n(n + 1)y = 0,$$

so that

$$\frac{d}{dx}[(1 - x^2)P_m'] + m(m + 1)P_m = 0$$

and

$$\frac{d}{dx}[(1 - x^2)P_n'] + n(n + 1)P_n = 0.$$

Use these two equations to give a proof of the first part of formula (2) that does not depend on the specific form of the Legendre polynomials *Hint*: Multiply the first equation by P_n and the second by P_m, subtract, and integrate from -1 to 1.

3. If the generating relation given in Problems 1 and 2 of Section 44 is squared and integrated from $x = -1$ to $x = 1$, then the first part of (2) implies that

$$\int_{-1}^{1} \frac{dx}{1 - 2xt + t^2} = \sum_{n=0}^{\infty} \left(\int_{-1}^{1} P_n(x)^2 \, dx \right) t^{2n}.$$

Establish the second part of (2) by showing that the integral on the left has the value

$$\sum_{n=0}^{\infty} \frac{2}{2n + 1} t^{2n}.$$

4. Find the first three terms of the Legendre series of

(a) $f(x) = \begin{cases} 0 & \text{if } -1 \le x < 0, \\ x & \text{if } 0 \le x \le 1; \end{cases}$

(b) $f(x) = e^x$.

5. If $p(x)$ is a polynomial of degree $n \ge 1$ such that

$$\int_{-1}^{1} x^k p(x) \, dx = 0 \qquad \text{for } k = 0, 1, \dots, n - 1,$$

show that $p(x) = cP_n(x)$ for some constant c.

6. If $P_n(x)$ is multiplied by the reciprocal r of the coefficient of x^n, then the resulting polynomial $rP_n(x)$ has leading coefficient 1. Show that this polynomial has the following minimum property: Among all polynomials of degree n with leading coefficient 1, $rP_n(x)$ deviates least from zero on the interval $-1 \le x \le 1$ in the sense of least squares.

46 BESSEL FUNCTIONS. THE GAMMA FUNCTION

The differential equation

$$x^2y'' + xy' + (x^2 - p^2)y = 0, \tag{1}$$

where p is a non-negative constant, is called *Bessel's equation,* and its solutions are known as *Bessel functions.* These functions first arose in Daniel Bernoulli's investigation of the oscillations of a hanging chain (Problem 40-4), and appeared again in Euler's theory of the vibrations of a circular membrane and Bessel's studies of planetary motion.[4] More recently, Bessel functions have turned out to have very diverse applications in physics and engineering, in connection with the propagation of waves, elasticity, fluid motion, and especially in many problems of potential theory and diffusion involving cylindrical symmetry. They even occur in some interesting problems of pure mathematics. We present a few applications in Appendix B, but first it is necessary to define the more important Bessel functions and obtain some of their simpler properties.[5]

The definition of the function $J_p(x)$. We begin our study of the solutions of (1) by noticing that after division by x^2 the coefficients of y' and y are $P(x) = 1/x$ and $Q(x) = (x^2 - p^2)/x^2$, so $xP(x) = 1$ and $x^2Q(x) = -p^2 + x^2$. The origin is therefore a regular singular point, the indicial equation 30-(5) is $m^2 - p^2 = 0$, and the exponents are $m_1 = p$ and $m_2 = -p$. It follows from Theorem 30-A that equation (1) has a solution

---

[4] Friedrich Wilhelm Bessel (1784–1846) was a famous German astronomer and an intimate friend of Gauss, with whom he corresponded for many years. He was the first man to determine accurately the distance of a fixed star: his parallax measurement of 1838 yielded a distance for the star 61 Cygni of 11 light-years or about 360,000 times the diameter of the earth's orbit. In 1844 he discovered that Sirius, the brightest star in the sky, has a traveling companion and is therefore what is now known as a binary star. This Companion of Sirius, with the size of a planet but the mass of a star, and consequently a density many thousands of times the density of water, is one of the most interesting objects in the universe. It was the first dead star to be discovered, and occupies a special place in modern theories of stellar evolution.

[5] The entire subject is treated on a vast scale in G. N. Watson, *A Treatise on the Theory of Bessel Functions,* 2d ed., Cambridge University Press, London, 1944. This is a gargantuan work of 752 pages, with a 36-page bibliography of 791 items. What we shall discuss amounts to little more than the froth on a heaving ocean of scientific effort extending over nearly three centuries.

of the form

$$y = x^p \sum a_n x^n = \sum a_n x^{n+p}, \tag{2}$$

where $a_0 \neq 0$ and the power series $\sum a_n x^n$ converges for all x. To find this solution, we write

$$y' = \sum (n + p)a_n x^{n+p-1}$$

and

$$y'' = \sum (n + p - 1)(n + p)a_n x^{n+p-2}.$$

These formulas enable us to express the terms on the left side of equation (1) in the form

$$x^2 y'' = \sum (n + p - 1)(n + p)a_n x^{n+p},$$

$$xy' = \sum (n + p)a_n x^{n+p},$$

$$x^2 y = \sum a_{n-2} x^{n+p},$$

$$-p^2 y = \sum -p^2 a_n x^{n+p}.$$

If we add these series and equate to zero the coefficient of x^{n+p}, then after a little simplification we obtain the following recursion formula for the a_n:

$$n(2p + n)a_n + a_{n-2} = 0 \tag{3}$$

or

$$a_n = -\frac{a_{n-2}}{n(2p + n)}. \tag{4}$$

We know that a_0 is nonzero and arbitrary. Since $a_{-1} = 0$, (4) tells us that $a_1 = 0$; and repeated application of (4) yields the fact that $a_n = 0$ for every odd subscript n. The nonzero coefficients of our solution (2) are therefore

$$a_0, \qquad a_2 = -\frac{a_0}{2(2p + 2)},$$

$$a_4 = -\frac{a_2}{4(2p + 4)} = \frac{a_0}{2 \cdot 4(2p + 2)(2p + 4)},$$

$$a_6 = -\frac{a_4}{6(2p + 6)} = -\frac{a_0}{2 \cdot 4 \cdot 6(2p + 2)(2p + 4))(2p + 6)}, \ldots,$$

and the solution itself is

$$y = a_0 x^p \left[1 - \frac{x^2}{2^2(p+1)} + \frac{x^4}{2^4 2! \, (p+1)(p+2)} \right.$$
$$\left. - \frac{x^6}{2^6 3! \, (p+1)(p+2)(p+3)} + \cdots \right]$$
$$= a_0 x^p \sum_{n=0}^{\infty} (-1)^n \frac{x^{2n}}{2^{2n} n! \, (p+1) \cdots (p+n)}. \tag{5}$$

The *Bessel function of the first kind of order p*, denoted by $J_p(x)$, is defined by putting $a_0 = 1/2^p p!$ in (5), so that

$$J_p(x) = \frac{x^p}{2^p p!} \sum_{n=0}^{\infty} (-1)^n \frac{x^{2n}}{2^{2n} n! \, (p+1) \cdots (p+n)}$$
$$= \sum_{n=0}^{\infty} (-1)^n \frac{(x/2)^{2n+p}}{n! \, (p+n)!}. \tag{6}$$

The most useful Bessel functions are those of order 0 and 1, which are

$$J_0(x) = \sum_{n=0}^{\infty} (-1)^n \frac{1}{(n!)^2} \left(\frac{x}{2}\right)^{2n}$$
$$= 1 - \frac{x^2}{2^2} + \frac{x^4}{2^2 \cdot 4^2} - \frac{x^6}{2^2 \cdot 4^2 \cdot 6^2} + \cdots \tag{7}$$

and

$$J_1(x) = \sum_{n=0}^{\infty} (-1)^n \frac{1}{n! \, (n+1)!} \left(\frac{x}{2}\right)^{2n+1}$$
$$= \frac{x}{2} - \frac{1}{1! \, 2!} \left(\frac{x}{2}\right)^3 + \frac{1}{2! \, 3!} \left(\frac{x}{2}\right)^5 - \cdots. \tag{8}$$

Their graphs are shown in Fig. 53. These graphs display several interesting properties of the functions $J_0(x)$ and $J_1(x)$: each has a damped

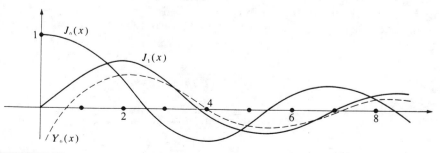

FIGURE 53

oscillatory behavior producing an infinite number of positive zeros; and these zeros occur alternately, in a manner suggesting the functions $\cos x$ and $\sin x$. This loose analogy is strengthened by the relation $J_0'(x) = -J_1(x)$, which we ask the reader to prove and apply in Problems 1 and 2.

We hope the reader has noticed the following flaw in this discussion—that $J_p(x)$ as defined by (6) is meaningless unless the non-negative real number p is an integer, since only in this case has any meaning been assigned to the factors $(p + n)!$ in the denominators. We next turn our attention to the problem of overcoming this difficulty.

The gamma function. The purpose of this digression is to give a reasonable and useful meaning to $p!$ [and more generally to $(p + n)!$ for $n = 0, 1, 2, \ldots$] when the non-negative real number p is not an integer. We accomplish this by introducing the *gamma function* $\Gamma(p)$, defined by

$$\Gamma(p) = \int_0^\infty t^{p-1} e^{-t} \, dt, \qquad p > 0. \tag{9}$$

The factor $e^{-t} \to 0$ so rapidly as $t \to \infty$ that this improper integral converges at the upper limit regardless of the value of p. However, at the lower limit we have $e^{-t} \to 1$, and the factor $t^{p-1} \to \infty$ whenever $p < 1$. The restriction that p must be positive is necessary in order to guarantee convergence at the lower limit.

It is easy to see that

$$\Gamma(p + 1) = p\Gamma(p); \tag{10}$$

for integration by parts yields

$$\Gamma(p + 1) = \lim_{b \to \infty} \int_0^b t^p e^{-t} \, dt$$

$$= \lim_{b \to \infty} \left(-t^p e^{-t} \Big|_0^b + p \int_0^b t^{p-1} e^{-t} \, dt \right)$$

$$= p \left(\lim_{b \to \infty} \int_0^b t^{p-1} e^{-t} \, dt \right) = p\Gamma(p),$$

since $b^p / e^b \to 0$ as $b \to \infty$. If we use the fact that

$$\Gamma(1) = \int_0^\infty e^{-t} \, dt = 1, \tag{11}$$

then (10) yields $\Gamma(2) = 1\Gamma(1) = 1$, $\Gamma(3) = 2\Gamma(2) = 2 \cdot 1$, $\Gamma(4) = 3\Gamma(3) = 3 \cdot 2 \cdot 1$, and in general

$$\Gamma(n + 1) = n! \tag{12}$$

for any integer $n \geq 0$.

We began our discussion of the gamma function under the assumption that p is non-negative, and we mentioned at the outset that the integral (9) does not exist if $p = 0$. However, we can define $\Gamma(p)$ for many negative p's without the aid of this integral if we write (10) in the form

$$\Gamma(p) = \frac{\Gamma(p + 1)}{p}. \tag{13}$$

This extension of the definition is necessary for the applications, and it begins as follows: If $-1 < p < 0$, then $0 < p + 1 < 1$, so the right side of equation (13) has a value and the left side of (13) is defined to have the value given by the right side. The next step is to notice that if $-2 < p < -1$, then $-1 < p + 1 < 0$, so we can use (13) again to define $\Gamma(p)$ on the interval $-2 < p < -1$ in terms of the values of $\Gamma(p + 1)$ already defined in the previous step. It is clear that this process can be continued indefinitely. Furthermore, it is easy to see from (11) that

$$\lim_{p \to 0} \Gamma(p) = \lim_{p \to 0} \frac{\Gamma(p + 1)}{p} = \pm\infty$$

according as $p \to 0$ from the right or left. The function $\Gamma(p)$ behaves in a similar way near all negative integers, and therefore its graph has the general appearance shown in Fig. 54. We will also need to know the curious fact that

$$\Gamma\left(\frac{1}{2}\right) = \sqrt{\pi}. \tag{14}$$

This is indicated in the figure, and its proof is left to the reader (in Problem 3). Since $\Gamma(p)$ never vanishes, the function $1/\Gamma(p)$ will be defined and well behaved for all values of p if we agree that $1/\Gamma(p) = 0$ for $p = 0, -1, -2, \ldots$.

These ideas enable us to define $p!$ by

$$p! = \Gamma(p + 1)$$

for all values of p except negative integers, and by formula (12) this function has its usual meaning when p is a non-negative integer. Its reciprocal, $1/p! = 1/\Gamma(p + 1)$, is defined for all p's and has the value 0 whenever p is a negative integer.

The gamma function is an extremely interesting function in its own right. However, our purpose in introducing it here is solely to guarantee that the function $J_p(x)$ as defined by formula (6) has a meaning for every $p \geq 0$. We point out that even more has been achieved—since $1/(p + n)!$ now has a meaning for every $p + n$, (6) defines a perfectly respectable function of x for all values of p, without exception.

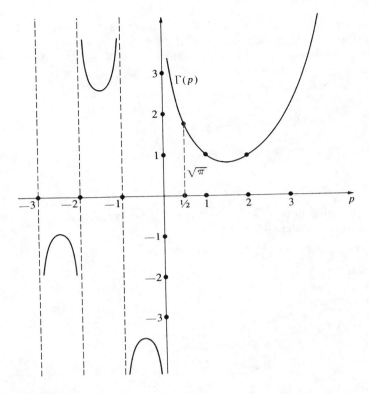

FIGURE 54

The general solution of Bessel's equation. Our present position is this: we have found a particular solution of (1) corresponding to the exponent $m_1 = p$, namely, $J_p(x)$. In order to find the general solution, we must now construct a second independent solution—that is, one that is not a constant multiple of $J_p(x)$. Any such solution is called a *Bessel function of the second kind*. The natural procedure is to try the other exponent, $m_2 = -p$. But in doing so, we expect to encounter difficulties whenever the difference $m_1 - m_2 = 2p$ is zero or a positive integer, that is, whenever the non-negative constant p is an integer or half an odd integer. It turns out that the expected difficulties are serious only in the first case.

We therefore begin by assuming that p is not an integer. In this case we replace p by $-p$ in our previous treatment, and it is easy to see that the discussion goes through almost without change. The only exception is that (3) becomes

$$n(-2p + n)a_n + a_{n-2} = 0;$$

and if it happens that $p = 1/2$, then by letting $n = 1$ we see that there is no compulsion to choose $a_1 = 0$. However, since all we want is a particular solution, it is certainly permissible to put $a_1 = 0$. The same problem arises when $p = 3/2$ and $n = 3$, and so on; and we solve it by putting $a_1 = a_3 = \cdots = 0$ in all cases. Everything else goes as before, and we obtain a second solution

$$J_{-p}(x) = \sum_{n=0}^{\infty} (-1)^n \frac{(x/2)^{2n-p}}{n!\,(-p+n)!}. \tag{15}$$

The first term of this series is

$$\frac{1}{(-p)!} \left(\frac{x}{2}\right)^{-p},$$

so $J_{-p}(x)$ is unbounded near $x = 0$. Since $J_p(x)$ is bounded near $x = 0$, these two solutions are independent and

$$y = c_1 J_p(x) + c_2 J_{-p}(x), \qquad p \text{ not an integer,} \tag{16}$$

is the general solution of (1).

The solution is entirely different when p is an integer $m \geq 0$. Formula (15) now becomes

$$J_{-m}(x) = \sum_{n=0}^{\infty} (-1)^n \frac{(x/2)^{2n-m}}{n!\,(-m+n)!}$$

$$= \sum_{n=m}^{\infty} (-1)^n \frac{(x/2)^{2n-n}}{n!\,(-m+n)!}$$

since the factors $1/(-m+n)!$ are zero when $n = 0, 1, \ldots, m-1$. On replacing the dummy variable n by $n+m$ and compensating by beginning the summation at $n = 0$, we obtain

$$J_{-m}(x) = \sum_{n=0}^{\infty} (-1)^{n+m} \frac{(x/2)^{2(n+m)-m}}{(n+m)!\,n!}$$

$$= (-1)^m \sum_{n=0}^{\infty} (-1)^n \frac{(x/2)^{2n+m}}{n!\,(m+n)!}$$

$$= (-1)^m J_m(x).$$

This show that $J_{-m}(x)$ is not independent of $J_m(x)$, so in this case

$$y = c_1 J_m(x) + c_2 J_{-m}(x)$$

is not the general solution of (1), and the search continues.

At this point the story becomes rather complicated, and we sketch it very briefly. One possible approach is to use the method of Section 16, which is easily seen to yield

$$J_m(x) \int \frac{dx}{x J_m(x)^2}$$

as a second solution independent of $J_m(x)$. It is customary, however, to proceed somewhat differently, as follows. When p is not an integer, any function of the form (16) with $c_2 \neq 0$ is a Bessel function of the second kind, including $J_{-p}(x)$ itself. The standard Bessel function of the second kind is defined by

$$Y_p(x) = \frac{J_p(x) \cos p\pi - J_{-p}(x)}{\sin p\pi}. \tag{17}$$

This seemingly eccentric choice is made for good reasons, which we describe in a moment. First, however, the reader should notice that (16) can certainly be written in the equivalent form

$$y = c_1 J_p(x) + c_2 Y_p(x), \qquad p \text{ not an integer.} \tag{18}$$

We still have the problem of what to do when p is an integer m, for (17) is meaningless in this case. It turns out after detailed analysis that the function defined by

$$Y_m(x) = \lim_{p \to m} Y_p(x) \tag{19}$$

exists and is a Bessel function of the second kind; and it follows that

$$y = c_1 J_p(x) + c_2 Y_p(x) \tag{20}$$

is the general solution of Bessel's equation in all cases, whether p is an integer or not. The graph of $Y_0(x)$ is shown by the dashed curve in Fig. 53. This graph illustrates the important fact that for every $p \geq 0$, the function $Y_p(x)$ is unbounded near the origin. Accordingly, if we are interested only in solutions of Bessel's equation that are bounded near $x = 0$, and this is often the case in the applications, then we must take $c_2 = 0$ in (20).

Now for the promised explanation of the surprising form of (17). We have pointed out that there are many ways of defining Bessel functions of the second kind. The definitions (17) and (19) are particularly convenient for two reasons. First, the form of (17) makes it fairly easy to show that the limit (19) exists (see Problem 9). And second, these definitions imply that the behavior of $Y_p(x)$, for large values of x, is matched in a natural way to the behavior of $J_p(x)$. To understand what is meant by this statement, we recall from Problem 24-3 that introducing a new dependent variable $u(x) = \sqrt{x}\, y(x)$ transforms Bessel's equation (1) into

$$u'' + \left(1 + \frac{1 - 4p^2}{4x^2}\right)u = 0. \tag{21}$$

When x is very large, equation (21) closely approximates the familiar differential equation $u'' + u = 0$, which has independent solutions

$u_1(x) = \cos x$ and $u_2(x) = \sin x$. We therefore expect that for large values of x, any Bessel function $y(x)$ will behave like some linear combination of

$$\frac{1}{\sqrt{x}}\cos x \quad \text{and} \quad \frac{1}{\sqrt{x}}\sin x.$$

This expectation is supported by the fact that

$$J_p(x) = \sqrt{\frac{2}{\pi x}}\cos\left(x - \frac{\pi}{4} - \frac{p\pi}{2}\right) + \frac{r_1(x)}{x^{3/2}}$$

and

$$Y_p(x) = \sqrt{\frac{2}{\pi x}}\sin\left(x - \frac{\pi}{4} - \frac{p\pi}{2}\right) + \frac{r_2(x)}{x^{3/2}},$$

where $r_1(x)$ and $r_2(x)$ are bounded as $x \to \infty.$[6]

PROBLEMS

1. Use (7) and (8) to show that

(a) $\dfrac{d}{dx}J_0(x) = -J_1(x)$;

(b) $\dfrac{d}{dx}[xJ_1(x)] = xJ_0(x)$.

2. Use Problem 1 and Rolle's theorem to show that:
   (a) Between any two positive zeros of $J_0(x)$ there is a zero of $J_1(x)$.
   (b) Between any two positive zeros of $J_1(x)$ there is a zero of $J_0(x)$.

3. According to the definition (9),

$$\Gamma\left(\frac{1}{2}\right) = \int_0^\infty t^{-1/2}e^{-t}\,dt.$$

(a) Show that the change of variable $t = s^2$ leads to

$$\Gamma\left(\frac{1}{2}\right) = 2\int_0^\infty e^{-s^2}\,ds.$$

(b) Since s in (a) is a dummy variable, we can write

$$\Gamma\left(\frac{1}{2}\right)^2 = 4\left(\int_0^\infty e^{-x^2}\,dx\right)\left(\int_0^\infty e^{-y^2}\,dy\right)$$

$$= 4\int_0^\infty \int_0^\infty e^{-(x^2+y^2)}\,dx\,dy.$$

---

[6] See Watson, *op. cit.*, chap. VII (footnote 5); or R. Courant and D. Hilbert, *Methods of Mathematical Physics*, vol. 1, pp. 331–334, 526, Interscience-Wiley, New York, 1953.

By changing this double integral to polar coordinates, show that

$$\Gamma\left(\frac{1}{2}\right)^2 = 4 \int_0^{\pi/2} \int_0^\infty e^{-r^2} r \, dr \, d\theta = \pi,$$

so $\Gamma\left(\frac{1}{2}\right) = \sqrt{\pi}.$

4. Since $p! = \Gamma(p + 1)$ whenever p is not a negative integer, (14) says that $(-\frac{1}{2})! = \sqrt{\pi}$. Calculate $(\frac{1}{2})!$ and $(\frac{3}{2})!$. More generally, show that

$$\left(n + \frac{1}{2}\right)! = \frac{(2n + 1)!}{2^{2n+1}n!} \sqrt{\pi}$$

and

$$\left(n - \frac{1}{2}\right)! = \frac{(2n)!}{2^{2n}n!} \sqrt{\pi}$$

for any non-negative integer n.

5. When $p = 1/2$, equation (21) shows that the general solution of Bessel's equation is expressible in either of the equivalent forms

$$y = \frac{1}{\sqrt{x}}(c_1 \cos x + c_2 \sin x)$$

and

$$y = c_1 J_{1/2}(x) + c_2 J_{-1/2}(x).$$

It therefore must be true that

$$\sqrt{x} J_{1/2}(x) = a \cos x + b \sin x$$

and

$$\sqrt{x} J_{-1/2}(x) = c \cos x + d \sin x$$

for certain constants a, b, c, and d. By evaluating these constants, show that

$$J_{1/2}(x) = \sqrt{\frac{2}{\pi x}} \sin x \quad \text{and} \quad J_{-1/2}(x) = \sqrt{\frac{2}{\pi x}} \cos x.$$

6. Establish the formulas in Problem 5 by direct manipulation of the series expansions of $J_{1/2}$ and $J_{-1/2}(x)$.

7. Many differential equations are really Bessel's equation in disguised form, and are therefore solvable by means of Bessel functions. For example, let Bessel's equation be written as

$$z^2 \frac{d^2 w}{dz^2} + z \frac{dw}{dz} + (z^2 - p^2)w = 0,$$

and show that the change of variables defined by $z = ax^b$ and $w = yx^c$ (where a, b, and c are constants) transforms it into

$$x^2 \frac{d^2 y}{dx^2} + (2c + 1)x \frac{dy}{dx} + [a^2 b^2 x^{2b} + (c^2 - p^2 b^2)]y = 0.$$

Write the general solution of this equation in terms of Bessel functions.

8. Use the result of Problem 7 to show that the general solution of Airy's equation $y'' + xy = 0$ (see Problem 28-5) is

$$y = x^{1/2}\left[c_1 J_{1/3}\left(\frac{2}{3}x^{3/2}\right) + c_2 J_{-1/3}\left(\frac{2}{3}x^{3/2}\right) \right].$$

9. Apply l'Hospital's rule to the limit (19) to show that

$$Y_m(x) = \frac{1}{\pi}\left[\frac{\partial}{\partial p}J_p(x) - (-1)^m \frac{\partial}{\partial p}J_{-p}(x) \right]_{p=m}.$$

47 PROPERTIES OF BESSEL FUNCTIONS

The Bessel function $J_p(x)$ has been defined for any real number p by

$$J_p(x) = \sum_{n=0}^{\infty} (-1)^n \frac{(x/2)^{2n+p}}{n!\,(p+n)!}. \tag{1}$$

In this section we develop several properties of these functions that are useful in their applications.

Identities and the functions $J_{m+1/2}(x)$. We begin by considering the formulas

$$\frac{d}{dx}[x^p J_p(x)] = x^p J_{p-1}(x) \tag{2}$$

and

$$\frac{d}{dx}[x^{-p}J_p(x)] = -x^{-p}J_{p+1}(x). \tag{3}$$

To establish (2), we simply multiply the series (1) by x^p and differentiate:

$$\frac{d}{dx}[x^p J_p(x)] = \frac{d}{dx}\sum_{n=0}^{\infty} \frac{(-1)^n x^{2n+2p}}{2^{2n+p}n!\,(p+n)!}$$

$$= \sum_{n=0}^{\infty} \frac{(-1)^n x^{2n+2p-1}}{2^{2n+p-1}n!\,(p+n-1)!}$$

$$= x^p \sum_{n=0}^{\infty} (-1)^n \frac{(x/2)^{2n+p-1}}{n!\,(p-1+n)!} = x^p J_{p-1}(x).$$

The verification of (3) is similar, and we leave the details to the reader in Problem 1 below. If the differentiations in (2) and (3) are carried out, and the results are divided by $x^{\pm p}$, then the formulas become

$$J_p'(x) + \frac{p}{x}J_p(x) = J_{p-1}(x) \tag{4}$$

and

$$J_p'(x) - \frac{p}{x}J_p(x) = J_{p+1}(x). \tag{5}$$

If (4) and (5) are first added and then subtracted, the results are

$$2J_p'(x) = J_{p-1}(x) - J_{p+1}(x) \tag{6}$$

and

$$\frac{2p}{x} J_p(x) = J_{p-1}(x) + J_{p+1}(x). \tag{7}$$

These formulas enable us to express Bessel functions and their derivatives in terms of other Bessel functions.

An interesting application of (7) begins with the formulas

$$J_{1/2}(x) = \sqrt{\frac{2}{\pi x}} \sin x \quad \text{and} \quad J_{-1/2}(x) = \sqrt{\frac{2}{\pi x}} \cos x,$$

which were established in Problem 46-5. It now follows from (7) that

$$J_{3/2}(x) = \frac{1}{x} J_{1/2}(x) - J_{-1/2}(x) = \sqrt{\frac{2}{\pi x}} \left(\frac{\sin x}{x} - \cos x \right)$$

and

$$J_{5/2}(x) = \frac{3}{x} J_{3/2}(x) - J_{1/2}(x) = \sqrt{\frac{2}{\pi x}} \left(\frac{3 \sin x}{x^2} - \frac{3 \cos x}{x} - \sin x \right).$$

Also,

$$J_{-3/2}(x) = -\frac{1}{x} J_{-1/2}(x) - J_{1/2}(x) = \sqrt{\frac{2}{\pi x}} \left(-\frac{\cos x}{x} - \sin x \right)$$

and

$$J_{-5/2}(x) = -\frac{3}{x} J_{-3/2}(x) - J_{-1/2}(x) = \sqrt{\frac{2}{\pi x}} \left(\frac{3 \cos x}{x^2} + \frac{3 \sin x}{x} - \cos x \right).$$

It is clear that calculations of this kind can be continued indefinitely, and therefore every Bessel function $J_{m+1/2}(x)$ (where m is an integer) is elementary. It has been proved by Liouville that these are the only cases in which $J_p(x)$ is elementary.[7]

Another application of formula (7) is given at the end of Appendix C, where we show how it yields Lambert's continued fraction for $\tan x$. This continued fraction is of great historical interest, for it led to the first proof of the fact that π is not a rational number.

---

[7] The details of this remarkable achievement can be found in Watson, *op. cit.*, chap. IV, and in J. F. Ritt, *Integration in Finite Terms*, Columbia University Press, New York, 1948. The functions $J_{m+1/2}(x)$ are often called *spherical Bessel functions* because they arise in solving the wave equation in spherical coordinates.

When the differentiation formulas (2) and (3) are written in the form

$$\int x^p J_{p-1}(x)\, dx = x^p J_p(x) + c \tag{8}$$

and

$$\int x^{-p} J_{p+1}(x)\, dx = -x^{-p} J_p(x) + c, \tag{9}$$

then they serve for the integration of many simple expressions containing Bessel functions. For example, when $p = 1$, (8) yields

$$\int x J_0(x)\, dx = x J_1(x) + c. \tag{10}$$

In the case of more complicated integrals, where the exponent does not match the order of the Bessel function as it does in (8) and (9), integration by parts is usually necessary as a supplementary tool.

Zeros and Bessel series. It follows from Problem 24-3 that for every value of p, the function $J_p(x)$ has an infinite number of positive zeros. This is true in particular of $J_0(x)$. The zeros of this function are known to a high degree of accuracy, and their values are given in many volumes of mathematical tables. The first five are approximately 2.4048, 5.5201, 8.6537, 11.7915, and 14.9309; their successive differences are 3.1153, 3.1336, 3.1378, and 3.1394. The corresponding positive zeros and differences for $J_1(x)$ are 3.8317, 7.0156, 10.1735, 13.3237, and 16.4706; and 3.1839, 3.1579, 3.1502, and 3.1469. Notice how these differences confirm the guarantees given in Problem 25-1.

What is the purpose of this concern with the zeros of $J_p(x)$? It is often necessary in mathematical physics to expand a given function in terms of Bessel functions, where the particular type of expansion depends on the problem at hand. The simplest and most useful expansions of this kind are series of the form

$$f(x) = \sum_{n=1}^{\infty} a_n J_p(\lambda_n x) = a_1 J_p(\lambda_1 x) + a_2 J_p(\lambda_2 x) + \cdots, \tag{11}$$

where $f(x)$ is defined on the interval $0 \le x \le 1$ and the λ_n are the positive zeros of some fixed Bessel function $J_p(x)$ with $p \ge 0$. We have chosen the interval $0 \le x \le 1$ only for the sake of simplicity, and all the formulas given below can be adapted by a simple change of variable to the case of a function defined on an interval of the form $0 \le x \le a$. The role of such expansions in physical problems is similar to that of Legendre series as illustrated in Appendix A, where the problem considered involves temperatures in a sphere. In Appendix B we demonstrate the use of (11) in solving the two-dimensional wave equation for a vibrating circular membrane.

In the light of our previous experience with Legendre series, we expect the determination of the coefficients in (11) to depend on certain integral properties of the functions $J_p(\lambda_n x)$. What we need here is the fact that

$$\int_0^1 x J_p(\lambda_m x) J_p(\lambda_n x)\, dx = \begin{cases} 0 & \text{if } m \neq n, \\ \dfrac{1}{2} J_{p+1}(\lambda_n)^2 & \text{if } m = n. \end{cases} \tag{12}$$

In terms of the ideas introduced in Section 43, these formulas say that the functions $J_p(\lambda_n x)$ are orthogonal with respect to the weight function x on the interval $0 \leq x \leq 1$. We shall prove them at the end of this section, but first we demonstrate their use.

If an expansion of the form (11) is assumed to be possible, then multiplying through by $x J_p(\lambda_m x)$, formally integrating term by term from 0 to 1, and using (12) yields

$$\int_0^1 x f(x) J_p(\lambda_m x)\, dx = \frac{a_m}{2} J_{p+1}(\lambda_m)^2;$$

and on replacing m by n we obtain the following formula for a_n:

$$a_n = \frac{2}{J_{p+1}(\lambda_n)^2} \int_0^1 x f(x) J_p(\lambda_n x)\, dx. \tag{13}$$

The series (11), with its coefficients calculated by (13), is called the *Bessel series*—or sometimes the *Fourier–Bessel series*—of the function $f(x)$. As usual, we state without proof a rather deep theorem that gives conditions under which this series actually converges and has the sum $f(x)$.[8]

> **Theorem A. (Bessel expansion theorem).** *Assume that $f(x)$ and $f'(x)$ have at most a finite number of jump discontinuities on the interval $0 \leq x \leq 1$. If $0 < x < 1$, then the Bessel series (11) converges to $f(x)$ when x is a point of continuity of this function, and converges to $\frac{1}{2}[f(x-) + f(x+)]$ when x is a point of discontinuity.*

It is natural to wonder what happens at the endpoints of the interval. At $x = 1$, the series converges to zero regardless of the nature of the function because every $J_p(\lambda_n)$ is zero. The series also converges at $x = 0$, to zero if $p > 0$ and to $f(0+)$ if $p = 0$.

As an illustration, we compute the Bessel series of the function $f(x) = 1$ for the interval $0 \leq x \leq 1$ in terms of the functions $J_0(\lambda_n x)$, where it is understood that the λ_n are the positive zeros of $J_0(x)$. In this

---

[8] For the proof, see Watson, *op. cit.*, chap. XVIII.

case, (13) is

$$a_n = \frac{2}{J_1(\lambda_n)^2} \int_0^1 x J_0(\lambda_n x) \, dx.$$

By (10), we see that

$$\int_0^1 x J_0(\lambda_n x) \, dx = \left[\frac{1}{\lambda_n} x J_1(\lambda_n x) \right]_0^1 = \frac{J_1(\lambda_n)}{\lambda_n},$$

so

$$a_n = \frac{2}{\lambda_n J_1(\lambda_n)}.$$

It follows that

$$1 = \sum_{n=1}^{\infty} \frac{2}{\lambda_n J_1(\lambda_n)} J_0(\lambda_n x) \qquad (0 \le x < 1)$$

is the desired Bessel series.

Proofs of the orthogonality properties. To establish (12), we begin with the fact that $y = J_p(x)$ is a solution of

$$y'' + \frac{1}{x} y' + \left(1 - \frac{p^2}{x^2} \right) y = 0.$$

If a and b are distinct positive constants, it follows that the functions $u(x) = J_p(ax)$ and $v(x) = J_p(bx)$ satisfy the equations

$$u'' + \frac{1}{x} u' + \left(a^2 - \frac{p^2}{x^2} \right) u = 0 \tag{14}$$

and

$$v'' + \frac{1}{x} v' + \left(b^2 - \frac{p^2}{x^2} \right) v = 0. \tag{15}$$

We now multiply these equations by v and u, the subtract the results, to obtain

$$\frac{d}{dx}(u'v - v'u) + \frac{1}{x}(u'v - v'u) = (b^2 - a^2)uv;$$

and after multiplication by x, this becomes

$$\frac{d}{dx}[x(u'v - v'u)] = (b^2 - a^2)xuv. \tag{16}$$

When (16) is integrated from $x = 0$ to $x = 1$, we get

$$(b^2 - a^2) \int_0^1 xuv \, dx = [x(u'v - v'u)]_0^1.$$

The expression in brackets clearly vanishes at $x = 0$, and at the other end of the interval we have $u(1) = J_p(a)$ and $v(1) = J_p(b)$. It therefore follows that the integral on the left is zero if a and b are distinct positive zeros λ_m and λ_n of $J_p(x)$; that is, we have

$$\int_0^1 xJ_p(\lambda_m x)J_p(\lambda_n x)\, dx = 0, \tag{17}$$

which is the first part of (12).

Our final task is to evaluate the integral in (17) when $m = n$. If (14) is multiplied by $2x^2u'$, it becomes

$$2x^2u'u'' + 2xu'^2 + 2a^2x^2uu' - 2p^2uu' = 0$$

or

$$\frac{d}{dx}(x^2u'^2) + \frac{d}{dx}(a^2x^2u^2) - 2a^2xu^2 - \frac{d}{dx}(p^2u^2) = 0,$$

so on integrating from $x = 0$ to $x = 1$, we obtain

$$2a^2 \int_0^1 xu^2\, dx = [x^2u'^2 + (a^2x^2 - p^2)u^2]_0^1. \tag{18}$$

When $x = 0$, the expression in brackets vanishes; and since $u'(1) = aJ_p'(a)$, (18) yields

$$\int_0^1 xJ_p(ax)^2\, dx = \frac{1}{2}J_p'(a)^2 + \frac{1}{2}\left(1 - \frac{p^2}{a^2}\right)J_p(a)^2.$$

We now put $a = \lambda_n$ and get

$$\int_0^1 xJ_p(\lambda_n x)^2\, dx = \frac{1}{2}J_p'(\lambda_n)^2 = \frac{1}{2}J_{p+1}(\lambda_n)^2,$$

where the last step makes use of (5), and the proof of (12) is complete.

PROBLEMS

1. Verify formula (3).
2. Prove that the positive zeros of $J_p(x)$ and $J_{p+1}(x)$ occur alternately, in the sense that between each pair of consecutive positive zeros of either there is exactly one zero of the other.
3. Express $J_2(x)$, $J_3(x)$, and $J_4(x)$ in terms of $J_0(x)$ and $J_1(x)$.
4. If $f(x)$ is defined by

$$f(x) = \begin{cases} 1 & 0 \le x < \frac{1}{2}, \\ \frac{1}{2} & x = \frac{1}{2}, \\ 0 & \frac{1}{2} < x \le 1, \end{cases}$$

show that

$$f(x) = \sum_{n=1}^{\infty} \frac{J_1(\lambda_n/2)}{\lambda_n J_1(\lambda_n)^2} J_0(\lambda_n x),$$

where the λ_n are the positive zeros of $J_0(x)$.

5. If $f(x) = x^p$ for the interval $0 \le x < 1$, show that its Bessel series in the functions $J_p(\lambda_n x)$, where the λ_n are the positive zeros of $J_p(x)$, is

$$x^p = \sum_{n=1}^{\infty} \frac{2}{\lambda_n J_{p+1}(\lambda_n)} J_p(\lambda_n x).$$

6. Use the notation of Problem 5 to show formally that if $g(x)$ is a well-behaved function on the interval $0 \le x \le 1$, then

$$\frac{1}{2} \int_0^1 x^{p+1} g(x) \, dx = \sum_{n=1}^{\infty} \frac{1}{\lambda_n J_{p+1}(\lambda_n)} \int_0^1 xg(x) J_p(\lambda_n x) \, dx.$$

By taking $g(x) = x^p$ and x^{p+2}, deduce that

$$\sum_{n=1}^{\infty} \frac{1}{\lambda_n^2} = \frac{1}{4(p+1)} \quad \text{and} \quad \sum_{n=1}^{\infty} \frac{1}{\lambda_n^4} = \frac{1}{16(p+1)^2(p+2)}.$$

7. The positive zeros of $\sin x$ are $\pi, 2\pi, 3\pi, \ldots$. Use the result of Problem 6 (and Problem 46-5) to show that

$$\sum_{n=1}^{\infty} \frac{1}{n^2} = 1 + \frac{1}{4} + \frac{1}{9} + \cdots = \frac{\pi^2}{6}$$

and

$$\sum_{n=1}^{\infty} \frac{1}{n^4} = 1 + \frac{1}{16} + \frac{1}{81} + \cdots = \frac{\pi^4}{90}.$$

8. Show that the change of dependent variable defined by

$$By = \frac{1}{u} \frac{du}{dx}$$

transforms the special Riccati equation

$$\frac{dy}{dx} + By^2 = Cx^m$$

into

$$\frac{d^2u}{dx^2} - BCx^m u = 0.$$

If $m \ne -2$, use Problem 46-7 to show that this equation is solvable in terms of elementary functions if and only if $m = -4k/(2k+1)$ for some integer k. (When $m = -2$, the substitution $y = v/x$ transforms Riccati's equation into an equation with separable variables that has an elementary solution.)

9. Show that the general solution of

$$\frac{dy}{dx} = x^2 + y^2$$

can be written as

$$y = x\,\frac{J_{-3/4}(\tfrac{1}{2}x^2) + cJ_{3/4}(\tfrac{1}{2}x^2)}{cJ_{-1/4}(\tfrac{1}{2}x^2) - J_{1/4}(\tfrac{1}{2}x^2)}.$$

APPENDIX A. LEGENDRE POLYNOMIALS AND POTENTIAL THEORY

If a number of particles of masses $m_1, m_2, \ldots m_n$, attracting according to the inverse square law of gravitation, are placed at points $P_1, P_2, \ldots, P_n$, then the *potential* due to these particles at any point P (that is, the work done against their attractive forces in moving a unit mass from P to an infinite distance) is

$$U = \frac{Gm_1}{PP_1} + \frac{Gm_2}{PP_2} + \cdots + \frac{Gm_n}{PP_n}, \tag{1}$$

where G is the gravitational constant.[9] If the points $P, P_1, P_2, \ldots, P_n$ have rectangular coordinates $(x,y,z), (x_1,y_1,z_1), (x_2,y_2,z_2), \ldots, (x_n,y_n,z_n)$, so that

$$PP_1 = \sqrt{(x - x_1)^2 + (y - y_1)^2 + (z - z_1)^2},$$

with similar expressions for the other distances, then it is easy to verify by partial differentiation that the potential U satisfies *Laplace's equation*:

$$\frac{\partial^2 U}{\partial x^2} + \frac{\partial^2 U}{\partial y^2} + \frac{\partial^2 U}{\partial x^2} = 0. \tag{2}$$

This partial differential equation does not involve either the particular masses or the coordinates of the points at which they are located, so it is satisfied by the potential produced in empty space by an arbitrary discrete or continuous distribution of particles. It is often written in the form

$$\nabla^2 U = 0, \tag{3}$$

where the symbol ∇^2 (del squared) is simply a concise notation for the differential operator

$$\frac{\partial^2}{\partial x^2} + \frac{\partial^2}{\partial y^2} + \frac{\partial^2}{\partial z^2}.$$

---

[9] See equation 21-(17).

The function U is called a *gravitational potential*. If we work instead with charged particles of charges $q_1, q_2, \ldots, q_n$, then their *electrostatic potential* has the same form as (1) with the m's replaced by q's and G by Coulomb's constant, so it also satisfies Laplace's equation. This equation has such a wide variety of applications that its study is a branch of analysis in its own right, known as *potential theory*. the related equation

$$a^2 \nabla^2 U = \frac{\partial U}{\partial t}, \qquad (4)$$

called the *heat equation*, occurs in problems of heat conduction, where U is now a function of the time t as well as the space coordinates. The *wave equation*

$$a^2 \nabla^2 U = \frac{\partial^2 U}{\partial t^2} \qquad (5)$$

is connected with vibratory phenomena.

We add a few brief comments on the physical meaning of equations (3) and (4). [Equation (5) is simply the three-dimensional counterpart of the one-dimensional wave equation 40-(8), which we have already discussed quite fully.] First, Laplace's equation (3) makes the same sort of statement about the function U as the one-dimensional equation $d^2y/dx^2 = 0$ makes about a function $y(x)$ of the single variable x. But the latter equation implies that $y(x)$ has the linear form $y = mx + b$; and every such function has the property that its value at the center of an interval equals the average of its values at the endpoints. It is clear from (1) that solutions of Laplace's equation need not be linear functions of x, y, and z; in fact, they can be very complicated indeed. Nevertheless, it can be proved (and was discovered by Gauss) that any solution of (3) has the very remarkable property that its value at the center of a sphere equals the average of its values on the surface of that sphere.[10] More generally, the function $\nabla^2 U$ can be thought of as a rough measure of the difference between the average value of U on the surface of a small sphere and its exact value at the center. Thus, for example, if U represents the temperature at an arbitrary point P in a solid body, and $\nabla^2 U$ is positive at a certain point P_0, then the value of U at P_0 is in general lower than its values at nearby points. We therefore expect heat to flow toward P_0, raising the temperature there; and since the temperature U is rising, $\partial U/\partial t$ is positive at P_0. This is essentially what the heat equation (4) says: that $\partial U/\partial t$ is proportional to $\nabla^2 U$ and has the same sign. If the temperature U reaches a steady state throughout the

---

[10] The two-dimensional version of this property is given in Problem 42-4.

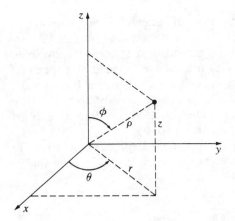

FIGURE 55

body, so that $\partial U/\partial t = 0$ at all points, then $\nabla^2 U = 0$ and we are back to the case of Laplace's equation.

We shall have occasion to use the formulas for $\nabla^2 U$ in cylindrical coordinates (r, θ, z) and spherical coordinates (ρ, θ, ϕ), as shown in Fig. 55. These coordinates are related to rectangular coordinates by the equations

$$x = r \cos \theta, \qquad y = r \sin \theta, \qquad z = z,$$

and

$$x = \rho \sin \phi \cos \theta, \qquad y = \rho \sin \phi \sin \theta, \qquad z = \rho \cos \phi.$$

By tedious but straightforward calculations one can show that in cylindrical coordinates,

$$\nabla^2 U = \frac{\partial^2 U}{\partial r^2} + \frac{1}{r} \frac{\partial U}{\partial r} + \frac{1}{r^2} \frac{\partial^2 U}{\partial \theta^2} + \frac{\partial^2 U}{\partial z^2}, \tag{6}$$

and in spherical coordinates,

$$\nabla^2 U = \frac{1}{\rho^2} \frac{\partial}{\partial \rho} \left(\rho^2 \frac{\partial U}{\partial \rho} \right) + \frac{1}{\rho^2 \sin \phi} \frac{\partial}{\partial \phi} \left(\sin \phi \frac{\partial U}{\partial \phi} \right) + \frac{1}{\rho^2 \sin^2 \phi} \frac{\partial^2 U}{\partial \theta^2}. \tag{7}$$

All students of mathematics or physics should carry out the necessary calculations at least once in their lives, but perhaps once is enough!

Steady-state temperatures in a sphere. Our purpose in this example is to illustrate as simply as possible the role of Legendre polynomials in solving certain boundary value problems of mathematical physics.[11]

Let a solid sphere of radius 1 be placed in a spherical coordinate system with its center at the origin. Let the surface be held at a specified temperature $f(\phi)$, which is assumed to be independent of θ for the sake of simplicity, until the flow of heat produces a steady state for the temperature $T(\rho, \phi)$ within the sphere. The problem is to find an explicit representation for the temperature function $T(\rho, \phi)$.

The steady-state temperature T satisfies Laplace's equation in spherical coordinates; and since T does not depend on θ, (7) allows us to write this equation in the form

$$\frac{\partial}{\partial \rho}\left(\rho^2 \frac{\partial T}{\partial \rho}\right) + \frac{1}{\sin \phi} \frac{\partial}{\partial \phi}\left(\sin \phi \frac{\partial T}{\partial \phi}\right) = 0. \tag{8}$$

To solve (8) subject to the given boundary condition

$$T(1, \phi) = f(\phi), \tag{9}$$

we use the method of *separation of variables*; that is, we seek a solution of (8) of the form $T(\rho, \phi) = u(\rho)v(\phi)$. When this expression is inserted in (8) and the variables are separated, we obtain

$$\frac{1}{u} \frac{d}{d\rho}\left(\rho^2 \frac{du}{d\rho}\right) = -\frac{1}{v \sin \phi} \frac{d}{d\phi}\left(\sin \phi \frac{dv}{d\phi}\right). \tag{10}$$

The crucial step in the method is the following observation: since the left side of equation (10) is a function of ρ alone and the right side is a function of ϕ alone, each side must be constant. If this constant—called the *separation constant*—is denoted by λ, then (10) splits into the two ordinary differential equations

$$\rho^2 \frac{d^2 u}{d\rho^2} + 2\rho \frac{du}{d\rho} - \lambda u = 0 \tag{11}$$

and

$$\frac{1}{\sin \phi} \frac{d}{d\phi}\left(\sin \phi \frac{dv}{d\phi}\right) + \lambda v = 0. \tag{12}$$

Equation (11) is an Euler equation with $p = 2$ and $q = -\lambda$, so its indicial equation is

$$m(m - 1) + 2m - \lambda = 0 \quad \text{or} \quad m^2 + m - \lambda = 0.$$

The exponents are therefore $\frac{1}{2}(-1 \pm \sqrt{1 + 4\lambda})$, and the general solution

---

[11] Many problems of greater complexity are discussed in Lebedev, *op. cit.*, chap. 8.

of (11) is

$$u = c_1 \rho^{-1/2 + \sqrt{\lambda + 1/4}} + c_2 \rho^{-1/2 - \sqrt{\lambda + 1/4}} \tag{13}$$

or

$$u = c_3 \rho^{-1/2} + c_4 \rho^{-1/2} \log \rho.$$

To guarantee that u is single-valued and bounded near $\rho = 0$, we discard the second possibility altogether, and in (13) put $c_2 = 0$ and $-\frac{1}{2} + \sqrt{\lambda + \frac{1}{4}} = n$ where n is a non-negative integer. It follows that $\lambda = n(n + 1)$, so (13) reduces to

$$u = c_1 \rho^n \tag{14}$$

and (12) becomes

$$\frac{d^2 v}{d\phi^2} + \frac{\cos \phi}{\sin \phi} \frac{dv}{d\phi} + n(n + 1)v = 0.$$

If the independent variable is changed from ϕ to $x = \cos \phi$, then this equation is transformed into

$$(1 - x^2) \frac{d^2 v}{dx^2} - 2x \frac{dv}{dx} + n(n + 1)v = 0, \tag{15}$$

which is precisely Legendre's equation. By the physics of the problem, the function v must be bounded for $0 \le \phi \le \pi$, or equivalently for $-1 \le x \le 1$; and we know from Section 44 that the only solutions of (15) with this property are constant multiples of the Legendre polynomials $P_n(x)$. If this result is combined with (14), then it follows that for each $n = 0, 1, 2, \ldots$, we have particular solutions of (8) of the form

$$a_n \rho^n P_n(\cos \phi), \tag{16}$$

where the a_n are arbitrary constants. We cannot hope to satisfy the boundary condition (9) by using these solutions individually. However, Laplace's equation is linear and sums of solutions are also solutions, so it is natural to put the particular solutions (16) together into an infinite series and hope that $T(\rho, \phi)$ can be expressed in the form

$$T(\rho, \phi) = \sum_{n=0}^{\infty} a_n \rho^n P_n(\cos \phi). \tag{17}$$

The boundary condition (9) now requires that

$$f(\phi) = \sum_{n=0}^{\infty} a_n P_n(\cos \phi),$$

or equivalently that

$$f(\cos^{-1} x) = \sum_{n=0}^{\infty} a_n P_n(x). \tag{18}$$

We know from Section 45 that if the function $f(\cos^{-1} x)$ is sufficiently well behaved, then it can be expanded into a Legendre series of the form (18) where the coefficients a_n are given by

$$a_n = \left(n + \frac{1}{2}\right) \int_{-1}^{1} f(\cos^{-1} x) P_n(x)\, dx. \tag{19}$$

With these coefficients, (17) is the desired solution of our problem.

We have found the solution (17) by rather formal procedures, and it should be pointed out that there are difficult questions of pure mathematics involved here that we have not touched on at all. To a physicist, it may seem obvious that a solid body whose surface temperature is specified will actually attain a definite and unique steady-state temperature at every interior point, but mathematicians are unhappily aware that the obvious is often false.[12] The so-called Dirichlet problem of potential theory requires a rigorous proof of the existence and uniqueness of a potential function throughout a region that assumes given values on the boundary. This problem was solved in the early twentieth century by the great German mathematician Hilbert, for very general but precisely defined types of boundaries and boundary functions.

The electrostatic dipole potential. The generating relation

$$(1 - 2xt + t^2)^{-1/2} = \sum_{n=0}^{\infty} P_n(x) t^n \tag{20}$$

for the Legendre polynomials is discussed in Problems 44-1, 44-2, and 44-3. As a direct physical illustration of its value, we use it to find the potential due to two point charges of equal magnitude q but opposite sign. If these charges are placed in a polar coordinate system (Fig. 56),

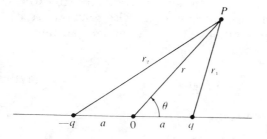

FIGURE 56

---

[12] Some fairly simple examples in which the statement just made is false are given in O. D. Kellogg, *Foundations of Potential Theory*, p. 285, Springer, New York, 1929. Einstein, a great maker of aphorisms, said: "The rarest and most valuable of all intellectual traits is the capacity to doubt the obvious."

then with suitable units of measurement the potential at P is

$$U = \frac{q}{r_1} - \frac{q}{r_2}, \tag{21}$$

where

$$r_1 = \sqrt{r^2 + a^2 - 2ar \cos \theta} \quad \text{and} \quad r_2 = \sqrt{r^2 + a^2 + 2ar \cos \theta}$$

by the law of cosines. When $r > a$, we can use (20) to write

$$\frac{1}{r_1} = \frac{1}{r} \frac{1}{\sqrt{1 - 2 \cos \theta (a/r) + (a/r)^2}} = \frac{1}{r} \sum_{n=0}^{\infty} P_n(\cos \theta) \left(\frac{a}{r}\right)^n,$$

and similarly

$$\frac{1}{r_2} = \frac{1}{r} \frac{1}{\sqrt{1 + 2 \cos \theta (a/r) + (a/r)^2}} = \frac{1}{r} \sum_{n=0}^{\infty} P_n(-\cos \theta) \left(\frac{a}{r}\right)^n.$$

Formula (21) can now be written

$$U = \frac{q}{r} \sum_{n=0}^{\infty} [P_n(\cos \theta) - P_n(-\cos \theta)] \left(\frac{a}{r}\right)^n. \tag{22}$$

We know that the nth Legendre polynomial $P_n(x)$ is even if n is even and odd if n is odd. The bracketed expression therefore equals 0 or $2P_n(\cos \theta)$ according as n is even or odd, and (22) becomes

$$U = \frac{2q}{r} \sum_{n=0}^{\infty} P_{2n+1}(\cos \theta) \left(\frac{a}{r}\right)^{2n+1}$$

$$= \frac{2q}{r} \left[P_1(\cos \theta) \left(\frac{a}{r}\right) + P_3(\cos \theta) \left(\frac{a}{r}\right)^3 + \cdots \right]. \tag{23}$$

If we now assume that all terms except the first can be neglected when r is large compared with a, and recall that $P_1(x) = x$, then (23) yields

$$U = 2aq \left(\frac{\cos \theta}{r^2} \right).$$

This is the approximation used by physicists for the dipole potential.

APPENDIX B. BESSEL FUNCTIONS AND THE VIBRATING MEMBRANE

One of the simplest physical applications of Bessel functions occurs in Euler's theory of the vibrations of a circular membrane. In this context a *membrane* is understood to be a uniform thin sheet of flexible material pulled taut into a state of uniform tension and clamped along a given closed curve in the xy-plane. When this membrane is slightly displaced from its equilibrium position and then released, the restoring forces due to the deformation cause it to vibrate. Our problem is to analyze this vibrational motion.

The equation of motion. Our discussion is similar to that given in Section 40 for the vibrating string; that is, we make several simplifying assumptions that enable us to formulate a partial differential equation, and we hope that this equation describes the motion with a reasonable degree of accuracy. These assumptions can be summarized in a single statement: we consider only *small* oscillations of a *freely vibrating* membrane. The various ways in which this is used will appear as we proceed.

First, we assume that the vibrations are so small that each point of the membrane moves only in the z direction, with displacement at time t given by some function $z = z(x,y,t)$. We consider a small piece of the membrane (Fig. 57) bounded by vertical planes through the following points in the xy-plane: (x,y), $(x + \Delta x, y)$, $(x + \Delta x, y + \Delta y)$, and $x, y + \Delta y)$. If m is the constant mass per unit area, then the mass of this piece is $m \, \Delta x \, \Delta y$, and by Newton's second law of motion we see that

$$F = m \, \Delta x \, \Delta y \frac{\partial^2 z}{\partial t^2} \tag{1}$$

is the force acting on it in the z direction.

When the membrane is in its equilibrium position, the constant tension T has the following physical meaning: Along any line segment of length Δs, the material on one side exerts a force, normal to the segment and of magnitude $T \, \Delta s$, on the material on the other side. In this case the forces on opposite edges of our small piece are parallel to the xy-plane and cancel one another. When the membrane is curved, as in the frozen instant of motion shown in Fig. 57, we assume that the deformation is so small that the tension is still T but now acts parallel to the tangent plane, and therefore has an appreciable vertical component. It is the curvature

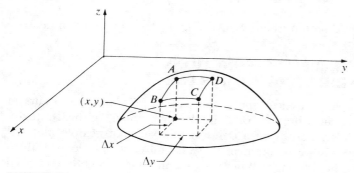

FIGURE 57

of our piece which produces different magnitudes for these vertical components on opposite edges, and this in turn is the source of the restoring forces that cause the motion.

We analyze these forces by assuming that the piece of the membrane denoted by $ABCD$ is only slightly tilted. This makes it possible to replace the sines of certain small angles by their tangents, as follows. Along the edges DC and AB, the forces are perpendicular to the x-axis and almost parallel to the y-axis, with small z components approximately equal to

$$T\,\Delta x\left(\frac{\partial z}{\partial y}\right)_{y+\Delta y} \quad \text{and} \quad -T\,\Delta x\left(\frac{\partial z}{\partial y}\right)_{y},$$

so their sum is approximately

$$T\,\Delta x\left[\left(\frac{\partial z}{\partial y}\right)_{y+\Delta y} - \left(\frac{\partial z}{\partial y}\right)_{y}\right].$$

The subscripts on these partial derivatives indicate their values at the points $(x, y + \Delta y)$ and (x,y). By working in the same way on the edges BC and AD, we find that the total force in the z direction (neglecting all external forces) is approximately

$$F = T\,\Delta y\left[\left(\frac{\partial z}{\partial x}\right)_{x+\Delta x} - \left(\frac{\partial z}{\partial x}\right)_{x}\right] + T\,\Delta x\left[\left(\frac{\partial z}{\partial y}\right)_{y+\Delta y} - \left(\frac{\partial z}{\partial y}\right)_{y}\right],$$

so (1) can be written

$$T\frac{(\partial z/\partial x)_{x+\Delta x} - (\partial z/\partial x)_{x}}{\Delta x} + T\frac{(\partial z/\partial y)_{y+\Delta y} - (\partial z/\partial y)_{y}}{\Delta y} = m\frac{\partial^2 z}{\partial t^2}.$$

If we now put $a^2 = T/m$ and let $\Delta x \to 0$ and $\Delta y \to 0$, this becomes

$$a^2\left(\frac{\partial^2 z}{\partial x^2} + \frac{\partial^2 z}{\partial y^2}\right) = \frac{\partial^2 z}{\partial t^2}, \tag{2}$$

which is the *two-dimensional wave equation*.

Students may be somewhat skeptical about the argument leading to equation (2). If so, they have plenty of company; for the question of what constitutes a satisfactory derivation of the differential equation describing a given physical system is never easy, and is particularly baffling in the case of the wave equation. To give a more refined treatment of the limits involved would get us nowhere, since the membrane is ultimately atomic and not continuous at all. Perhaps the most reasonable attitude is to accept our discussion as a plausibility argument that suggests the wave equation as a mathematical model. We can then adopt this equation as an axiom of rational mechanics describing an "ideal membrane" whose mathematical behavior may or may not match the actual behavior of real membranes.[13]

The circular membrane. We now specialize to the case of a circular membrane, in which it is natural to use polar coordinates with the origin located at the center. Formula (6) of Appendix A shows that in this case the wave equation (2) takes the form

$$a^2\left(\frac{\partial^2 z}{\partial r^2} + \frac{1}{r}\frac{\partial z}{\partial r} + \frac{1}{r^2}\frac{\partial^2 z}{\partial \theta^2}\right) = \frac{\partial^2 z}{\partial t^2}, \tag{3}$$

where $z = z(r, \theta, t)$ is a function of the polar coordinates and the time. For convenience we assume that the membrane has radius 1, and is therefore clamped to its plane of equilibrium along the circle $r = 1$. Accordingly, our boundary condition is

$$z(1, \theta, t) = 0. \tag{4}$$

The problem is to find a solution of (3) that satisfies this boundary condition and certain initial conditions to be specified later.

In applying the standard method of separation of variables, we begin with a search for particular solutions of the form

$$z(r, \theta, t) = u(r)v(\theta)w(t). \tag{5}$$

When (5) is inserted in (3) and the result is rearranged, we get

$$\frac{u''(r)}{u(r)} + \frac{1}{r}\frac{u'(r)}{u(r)} + \frac{1}{r^2}\frac{v''(\theta)}{v(\theta)} = \frac{1}{a^2}\frac{w''(t)}{w(t)}. \tag{6}$$

Since the left side of equation (6) is a function only of r and θ, and the right side is a function only of t, both sides must equal a constant. For the membrane to vibrate, $w(t)$ must be periodic; and the right side of (6) shows that in order to guarantee this, the separation constant must be negative. We therefore equate each side of (6) to $-\lambda^2$ with $\lambda > 0$, and obtain the two equations

$$w''(t) + \lambda^2 a^2 w(t) = 0 \tag{7}$$

and

$$\frac{u''(r)}{u(r)} + \frac{1}{r}\frac{u'(r)}{u(r)} + \frac{1}{r^2}\frac{v''(\theta)}{v(\theta)} = -\lambda^2. \tag{8}$$

---

[13] On the question, "What is rational mechanics?," we recommend the illuminating remarks of C. Truesdell, *Essays in the History of Mechanics*, pp. 334–340, Springer, New York, 1968.

It is clear that (7) has

$$w(t) = c_1 \cos \lambda at + c_2 \sin \lambda at \tag{9}$$

as its general solution, and (8) can be written as

$$r^2 \frac{u''(r)}{u(r)} + r \frac{u'(r)}{u(r)} + \lambda^2 r^2 = -\frac{v''(\theta)}{v(\theta)}. \tag{10}$$

In (10) we have a function of r on the left and a function of θ on the right, so again both sides must equal a constant. We now recall that the polar angle θ of a point in the plane is determined only up to an integral multiple of 2π; and by the nature of our problem, the value of v at any point must be independent of the value of θ used to describe that point. This requires that v must be either a constant or else nonconstant and periodic with period 2π. An inspection of the right side of equation (10) shows that these possibilities are covered by writing the separation constant in the form n^2 where $n = 0, 1, 2, \ldots$, and then (10) splits into

$$v''(\theta) + n^2 v(\theta) = 0 \tag{11}$$

and

$$r^2 u''(r) + r u'(r) + (\lambda^2 r^2 - n^2) u(r) = 0. \tag{12}$$

By recalling that v is either a constant or else nonconstant and periodic with period 2π, we see that (11) implies that

$$v(\theta) = d_1 \cos n\theta + d_2 \sin n\theta \tag{13}$$

for each n, regardless of the fact that (13) is not the general solution of (11) when $n = 0$. Next, it is clear from Problem 46-7 that (12) is a slightly disguised form of Bessel's equation of order n, with a bounded solution $J_n(\lambda r)$ and an independent unbounded solution $Y_n(\lambda r)$. Since $u(r)$ is necessarily bounded near $r = 0$, we discard the second solution and write

$$u(r) = k J_n(\lambda r). \tag{14}$$

The boundary condition (4) can now be satisfied by requiring that $u(1) = 0$ or

$$J_n(\lambda) = 0. \tag{15}$$

Thus the permissible values of λ are the positive zeros of the function $J_n(x)$, and we know from Section 47 that $J_n(x)$ has an infinite number of such zeros. We therefore conclude that the particular solutions (5) yielded by this analysis are constant multiples of the doubly infinite array of functions

$$J_n(\lambda r)(d_1 \cos n\theta + d_2 \sin n\theta)(c_1 \cos \lambda at + c_2 \sin \lambda at), \tag{16}$$

where $n = 0, 1, 2, \ldots$, and for each n the corresponding λ's are the positive roots of (15).

Special initial conditions. The above discussion is intended to show how Bessel functions of integral order arise in physical problems. It also demonstrates the significance of the positive zeros of these functions. For the sake of simplicity, we confine our further treatment to the following special case: the membrane is displaced into a shape $z = f(r)$ independent of the variable θ, and then released from rest at the instant $t = 0$. This means that we impose the initial conditions

$$z(r,\theta,0) = f(r) \tag{17}$$

and

$$\frac{\partial z}{\partial t}\bigg|_{t=0} = 0. \tag{18}$$

The problem is to determine the shape $z(r,\theta,t)$ at any subsequent time $t > 0$.

Our strategy is to adapt the particular solutions already found to the given initial conditions. First, the part of (17) that says that the initial shape is independent of θ implies that $v(\theta)$ is constant, so (13) tells us that $n = 0$. If the positive zeros of $J_0(x)$ are denoted by $\lambda_1, \lambda_2, \ldots, \lambda_n, \ldots$, then this remark reduces the array of functions (16) to

$$J_0(\lambda_n r)(c_1 \cos \lambda_n at + c_2 \sin \lambda_n at), \qquad n = 1, 2, \ldots.$$

Next, (18) implies that $c_2 = 0$, and this leaves us with constant multiples of the functions

$$J_0(\lambda_n r) \cos \lambda_n at, \qquad n = 1, 2, \ldots.$$

Up to this point we have not used the fact that sums of solutions of (3) are also solutions. Accordingly, the most general formal solutions now available to us are the infinite series

$$z = \sum_{n=1}^{\infty} a_n J_0(\lambda_n r) \cos \lambda_n at. \tag{19}$$

Our final step is to try to satisfy (17) by putting $t = 0$ in (19) and equating the result to $f(r)$:

$$f(r) = \sum_{n=1}^{\infty} a_n J_0(\lambda_n r).$$

The Bessel expansion theorem of Section 47 guarantees that this representation is valid whenever $f(r)$ is sufficiently well behaved, if the coefficients are defined by

$$a_n = \frac{2}{J_1(\lambda_n)^2} \int_0^1 r f(r) J_0(\lambda_n r) \, dr.$$

With these coefficients, (19) is a formal solution of (3) that satisfies the given boundary condition and initial conditions, and this concludes our discussion.[14]

APPENDIX C. ADDITIONAL PROPERTIES OF BESSEL FUNCTIONS

In Sections 46 and 47 we had no space for several remarkable properties of Bessel functions that should not go unmentioned, so we present them here. Unfortunately, a full justification of our procedures requires several theorems from more advanced parts of analysis, but this does not detract from the validity of the results themselves.

The generating function. The Bessel functions $J_n(x)$ of integral order are linked together by the fact that

$$e^{(x/2)(t-1/t)} = J_0(x) + \sum_{n=1}^{\infty} J_n(x)[t^n + (-1)^n t^{-n}]. \tag{1}$$

Since $J_{-n}(x) = (-1)^n J_n(x)$, this is often written in the form

$$e^{(x/2)(t-1/t)} = \sum_{n=-\infty}^{\infty} J_n(x) t^n. \tag{2}$$

To establish (1), we formally multiply the two series

$$e^{xt/2} = \sum_{j=0}^{\infty} \frac{1}{j!} \frac{x^j}{2^j} t^j \quad \text{and} \quad e^{-xt^{-1}/2} = \sum_{k=0}^{\infty} \frac{(-1)^k}{k!} \frac{x^k}{2^k} t^{-k}. \tag{3}$$

The result is a so-called double series, whose terms are all possible products of a term from the first series and a term from the second. The fact that each of the series (3) is absolutely convergent permits us to conclude that this double series converges to the proper sum regardless of the order of its terms. For each fixed integer $n \geq 0$, we obtain a term of the double series containing t^n precisely when $j = n + k$; and when all possible values of k are accounted for, the total coefficient of t^n is

$$\sum_{k=0}^{\infty} \frac{1}{(n+k)!} \frac{x^{n+k}}{2^{n+k}} \frac{(-1)^k}{k!} \frac{x^k}{2^k} = \sum_{k=0}^{\infty} (-1)^k \frac{(x/2)^{2k+n}}{k!\,(n+k)!} = J_n(x).$$

Similarly, a term containing t^{-n} ($n \geq 1$) arises precisely when $k = n + j$,

---

[14] Many additional applications of Bessel functions can be found in Lebedev, *op. cit.*, chap. 6. See also A. Gray and G. B. Mathews, *A Treatise on Bessel Functions and Their Applications to Physics*, Macmillan, New York, 1952.

so the total coefficient of t^{-n} is

$$\sum_{j=0}^{\infty} \frac{1}{j!} \frac{x^j}{2^j} \frac{(-1)^{n+j}}{(n+j)!} \frac{x^{n+j}}{2^{n+j}} = (-1)^n \sum_{j=0}^{\infty} (-1)^j \frac{(x/2)^{2j+n}}{j!\,(n+j)!}$$

$$= (-1)^n J_n(x),$$

and the proof of (1) is complete.

A simple consequence of (2) is the *addition formula*

$$J_n(x+y) = \sum_{k=-\infty}^{\infty} J_{n-k}(x)J_k(y). \tag{4}$$

To prove this, we notice first that

$$e^{(x/2)(t-1/t)}e^{(y/2)(t-1/t)} = e^{[(x+y)/2](t-1/t)} = \sum_{n=-\infty}^{\infty} J_n(x+y)t^n.$$

However, the product of the two exponentials on the left is also

$$\left[\sum_{j=-\infty}^{\infty} J_j(x)t^j\right]\left[\sum_{k=-\infty}^{\infty} J_k(y)t^k\right] = \sum_{n=-\infty}^{\infty} \left[\sum_{k=-\infty}^{\infty} J_{n-k}(x)J_k(y)\right]t^n,$$

and (4) follows at once on equating the coefficients of t^n in these expressions. When $n = 0$, (4) can be written as

$$J_0(x+y) = \sum_{k=-\infty}^{\infty} J_{-k}(x)J_k(y)$$

$$= J_0(x)J_0(y) + \sum_{k=1}^{\infty} J_{-k}(x)J_k(y) + \sum_{k=1}^{\infty} J_k(x)J_{-k}(y)$$

$$= J_0(x)J_0(y) + \sum_{k=1}^{\infty} (-1)^k[J_k(x)J_k(y) + J_k(x)J_k(y)]$$

$$= J_0(x)J_0(y) + \sum_{k=1}^{\infty} (-1)^k 2J_k(x)J_k(y)$$

or

$$J_0(x+y) = J_0(x)J_0(y) - 2J_1(x)J_1(y) + 2J_2(x)J_2(y) - \cdots. \tag{5}$$

If we replace y by $-x$ and use the fact that $J_n(x)$ is even or odd according as n is even or odd, then (5) yields the remarkable identity

$$1 = J_0(x)^2 + 2J_1(x)^2 + 2J_2(x)^2 + \cdots, \tag{6}$$

which shows that $|J_0(x)| \le 1$ and $|J_n(x)| \le 1/\sqrt{2}$ for $n = 1, 2, \ldots$.

Bessel's integral formula. When $t = e^{i\theta}$, the exponent on the left side of (2) becomes

$$x\frac{e^{i\theta} - e^{-i\theta}}{2} = ix\sin\theta,$$

and (2) itself assumes the form

$$e^{ix \sin \theta} = \sum_{n=-\infty}^{\infty} J_n(x)e^{in\theta}. \tag{7}$$

Since $e^{ix \sin \theta} = \cos (x \sin \theta) + i \sin (x \sin \theta)$ and $e^{in\theta} = \cos n\theta + i \sin n\theta$, equating real and imaginary parts in (7) yields

$$\cos (x \sin \theta) = \sum_{n=-\infty}^{\infty} J_n(x) \cos n\theta \tag{8}$$

and

$$\sin (x \sin \theta) = \sum_{n=-\infty}^{\infty} J_n(x) \sin n\theta. \tag{9}$$

If we now use the relations $J_{-n}(x) = (-1)^n J_n(x)$, $\cos (-n\theta) = \cos n\theta$, and $\sin (-n\theta) = -\sin n\theta$, then (8) and (9) become

$$\cos (x \sin \theta) = J_0(x) + 2 \sum_{n=1}^{\infty} J_{2n}(x) \cos 2n\theta \tag{10}$$

and

$$\sin (x \sin \theta) = 2 \sum_{n=1}^{\infty} J_{2n-1}(x) \sin (2n - 1)\theta. \tag{11}$$

As a special case of (10), we note that $\theta = 0$ yields the interesting series

$$1 = J_0(x) + 2J_2(x) + 2J_4(x) + \cdots.$$

Also, on putting $\theta = \pi/2$ in (10) and (11), we obtain the formulas

$$\cos x = J_0(x) - 2J_2(x) + 2J_4(x) - \cdots$$

and

$$\sin x = 2J_1(x) - 2J_3(x) + 2J_5(x) - \cdots,$$

which demonstrate once again the close ties between the Bessel functions and the trigonometric functions.

The most important application of (8) and (9) is to the proof of *Bessel's integral formula*

$$J_n(x) = \frac{1}{\pi} \int_0^\pi \cos (n\theta - x \sin \theta) \, d\theta. \tag{12}$$

To establish this, we multiply (8) by $\cos m\theta$, (9) by $\sin m\theta$, and add:

$$\cos (m\theta - x \sin \theta) = \sum_{n=-\infty}^{\infty} J_n(x) \cos (m - n)\theta.$$

When both sides of this are integrated from $\theta = 0$ to $\theta = \pi$, the right side reduces to $\pi J_m(x)$, and replacing m by n yields formula (12). In his astronomical work, Bessel encountered the functions $J_n(x)$ in the form of these integrals, and on this basis developed many of their properties.[15]

Some continued fractions. If we write the identity 47-(7) in the form

$$J_{p-1}(x) = \frac{2p}{x} J_p(x) - J_{p+1}(x),$$

then dividing by $J_p(x)$ yields

$$\frac{J_{p-1}(x)}{J_p(x)} = \frac{2p}{x} - \frac{1}{J_p(x)/J_{p+1}(x)}.$$

When this formula is itself applied to the second denominator on the right, with p replaced by $p + 1$, and this process is continued indefinitely, we obtain

$$\frac{J_{p-1}(x)}{J_p(x)} = \frac{2p}{x} - \cfrac{1}{2p+2 \over x} - \cfrac{1}{2p+4 \over x} - \cdots.$$

This is an infinite continued fraction expansion of the ratio $J_{p-1}(x)/J_p(x)$. We cannot investigate the theory of such expansions here. Nevertheless, it may be of interest to point out that when $p = 1/2$, it follows from Problem 46-5 that $J_{-1/2}(x)/J_{1/2}(x) = \cot x$, so

$$\tan x = \cfrac{1}{\cfrac{1}{x} - \cfrac{1}{\cfrac{3}{x} - \cfrac{1}{\cfrac{5}{x} - \cdots}}}.$$

This continued fraction was discovered in 1761 by Lambert, who used it to prove that π is irrational. He reasoned as follows: If x is a nonzero rational number, then the form of this continued fraction implies that $\tan x$ cannot be rational; but $\tan \pi/4 = 1$, so neither $\pi/4$ nor π is rational. Several minor flaws in Lambert's argument were patched up by Legendre about 30 years later.

---

[15] For a description of Bessel's original problem, see Gray and Mathews, *op. cit.*, pp. 4–7.

CHAPTER
9

LAPLACE
TRANSFORMS

48 INTRODUCTION

In recent years there has been a considerable growth of interest in the use of Laplace transforms as an efficient method for solving certain types of differential and integral equations. In addition to such applications, Laplace transforms also have a number of close connections with important parts of pure mathematics. We shall try to give the reader an adequate idea of some of these matters without dwelling too much on the analytic fine points and computational techniques that would be appropriate in a more extensive treatment.

Before entering into the details, we offer a few general remarks aimed at placing the ideas of this chapter in their proper context. We begin by noting that the operation of differentiation transforms a function $f(x)$ into another function, its derivative $f'(x)$. If the letter D is used to denote differentiation, then this transformation can be written

$$D[f(x)] = f'(x). \tag{1}$$

Another important transformation of functions is that of integration:

$$I[f(x)] = \int_0^x f(t)\, dt. \tag{2}$$

An even simpler transformation is the operation of multiplying all functions by a specific function $g(x)$:

$$M_g[f(x)] = g(x)f(x). \tag{3}$$

The basic feature these examples have in common is that *each transformation operates on functions to produce other functions*. It is clear that in most cases some restriction must be placed on the functions $f(x)$ to which a given transformation is applied. Thus, in (1) $f(x)$ must be differentiable, and in (2) it must be integrable. In each of our examples, the function on the right is called the *transform* of $f(x)$ under the corresponding transformation.

A general transformation T of functions is said to be *linear* if the relation

$$T[\alpha f(x) + \beta g(x)] = \alpha T[f(x)] + \beta T[g(x)] \tag{4}$$

holds for all admissible functions $f(x)$ and $g(x)$ and all constants α and β. Verbally, equation (4) says that the transform of any linear combination of two functions is the same linear combination of their transforms. It is worth observing that (4) reduces to

$$T[f(x) + g(x)] = T[f(x)] + T[g(x)]$$

and

$$T[\alpha f(x)] = \alpha T[f(x)]$$

when $\alpha = \beta = 1$ and when $\beta = 0$. It is easy to see that the transformations defined by (1), (2), and (3) are all linear.

A class of linear transformations of particular importance is that of the *integral transformations*. To get an idea of what these are, we consider functions $f(x)$ defined on a finite or infinite interval $a \leq x \leq b$, and we choose a fixed function $K(p,x)$ of the variable x and a parameter p. Then the general integral transformation is given by

$$T[f(x)] = \int_a^b K(p,x)f(x)\,dx = F(p). \tag{5}$$

The function $K(p,x)$ is called the *kernel* of the transformation T, and it is clear that T is linear regardless of the nature of K. The concept of a linear integral transformation, in generalized form, has been the source of some of the most fruitful ideas in modern analysis. Also, in classical analysis, various special cases of (5) have been minutely studied, and have led to specific transformations useful in handling particular types of problems.

When $a = 0$, $b = \infty$, and $K(p,x) = e^{-px}$, we obtain the special case of (5) that concerns us—the *Laplace transformation L*, defined by

$$L[f(x)] = \int_0^\infty e^{-px}f(x)\,dx = F(p). \tag{6}$$

Thus, the Laplace transformation L acts on any function $f(x)$ for which this integral exists, and produces its *Laplace transform* $L[f(x)] = F(p)$, a function of the parameter p.[1] We remind the reader that the improper integral in (6) is defined to be the following limit, and exists only when this limit exists:

$$\int_0^\infty e^{-px}f(x)\,dx = \lim_{b \to \infty} \int_0^b e^{-px}f(x)\,dx. \tag{7}$$

When the limit on the right exists, the improper integral on the left is said to *converge*.

The following Laplace transforms are quite easy to compute:

$$f(x) = 1, \qquad F(p) = \int_0^\infty e^{-px}\,dx = \frac{1}{p}; \tag{8}$$

$$f(x) = x, \qquad F(p) = \int_0^\infty e^{-px}x\,dx = \frac{1}{p^2}; \tag{9}$$

$$f(x) = x^n, \qquad F(p) = \int_0^\infty e^{-px}x^n\,dx = \frac{n!}{p^{n+1}}; \tag{10}$$

$$f(x) = e^{ax}, \qquad F(p) = \int_0^\infty e^{-px}e^{ax}\,dx = \frac{1}{p - a}; \tag{11}$$

$$f(x) = \sin ax, \qquad F(p) = \int_0^\infty e^{-px} \sin ax\,dx = \frac{a}{p^2 + a^2}; \tag{12}$$

$$f(x) = \cos ax, \qquad F(p) = \int_0^\infty e^{-px} \cos ax\,dx = \frac{p}{p^2 + a^2}. \tag{13}$$

The integral in (11) converges for $p > a$, and all the others converge for $p > 0$. Students should perform the necessary calculations themselves, so that the source of these restrictions on p is perfectly clear (see Problem 1). As an illustration, we provide the details for (10), in which n is assumed to be a positive integer:

$$L[x^n] = \int_0^\infty e^{-px}x^n\,dx = -\frac{x^n e^{-px}}{p}\bigg]_0^\infty + \frac{n}{p}\int_0^\infty e^{-px}x^{n-1}\,dx$$

$$= \frac{n}{p}L[x^{n-1}] = \frac{n}{p}\left(\frac{n-1}{p}\right)L[x^{n-2}]$$

$$= \cdots = \frac{n!}{p^n}L[1] = \frac{n!}{p^{n+1}}.$$

---

[1] As this remark suggests, we shall consistently use small letters to denote functions of x and the corresponding capital letters to denote the transforms of these functions.

It will be noted that we have made essential use here of the fact that

$$\lim_{x \to \infty} \frac{x^n}{e^{px}} = 0 \quad \text{for } p > 0.$$

The above formulas will be found in Table 1 in Section 50. Additional simple transforms can readily be determined without integration by using the linearity of L, as in

$$L[2x + 3] = 2L[x] + 3L[1] = \frac{2}{p^2} + \frac{3}{p}.$$

In later sections we shall develop methods for finding Laplace transforms of more complicated functions.

As we stated above, the Laplace transformation L can be regarded as the special case of the general integral transformation (5) obtained by taking $a = 0$, $b = \infty$, and $K(p,x) = e^{-px}$. Why do we choose these limits and this particular kernel? In order to see why this might be a fruitful choice, it is useful to consider a suggestive analogy with power series.

If we write a power series in the form

$$\sum_{n=0}^{\infty} a(n)x^n,$$

then its natural analog is the improper integral

$$\int_0^\infty a(t)x^t \, dt.$$

We now change the notation slightly by writing $x = e^{-p}$, and this integral becomes

$$\int_0^\infty e^{-pt}a(t) \, dt,$$

which is precisely the Laplace transform of the function $a(t)$. Laplace transforms are therefore the continuous analogs of power series; and since power series are important in analysis, we have reasonable grounds for expecting that Laplace transforms will also be important.

A short account of Laplace is given in Appendix A.

PROBLEMS

1. Evaluate the integrals in (8), (9), (11), (12), and (13).

2. Without integrating, show that

(a) $L[\sinh ax] = \dfrac{a}{p^2 - a^2}$, $\quad p > |a|$;

(b) $L[\cosh ax] = \dfrac{p}{p^2 - a^2}$, $\quad p > |a|$.

3. Find $L[\sin^2 ax]$ and $L[\cos^2 ax]$ without integrating. How are these two transforms related to one another?

4. Use the formulas given in the text to find the transform of each of the following functions:

(a) 10;

(b) $x^5 + \cos 2x$;

(c) $2e^{3x} - \sin 5x$;

(d) $4 \sin x \cos x + 2e^{-x}$;

(e) $x^6 \sin^2 3x + x^6 \cos^2 3x$.

5. Find a function $f(x)$ whose transform is

(a) $\dfrac{30}{p^4}$;

(b) $\dfrac{2}{p+3}$;

(c) $\dfrac{4}{p^3} + \dfrac{6}{p^2+4}$;

(d) $\dfrac{1}{p^2+p}$;

(e) $\dfrac{1}{p^4+p^2}$.

6. Give a reasonable definition of $\frac{1}{2}!$.

49 A FEW REMARKS ON THE THEORY

Before proceeding to the applications, it is desirable to consider more carefully the circumstances under which a function has a Laplace transform. A detailed and rigorous treatment of this problem would require familiarity with the general theory of improper integrals, which we do not assume. On the other hand, it is customary to give a brief introduction to this subject in elementary calculus, and a grasp of the following simple statements will suffice for our purposes.

First, the integral

$$\int_0^\infty f(x)\,dx \tag{1}$$

is said to *converge* if the limit

$$\lim_{b \to \infty} \int_0^b f(x)\,dx$$

exists, and in this case the value of (1) is by definition the value of this limit:

$$\int_0^\infty f(x)\,dx = \lim_{b \to \infty} \int_0^b f(x)\,dx.$$

Next, (1) converges whenever the integral

$$\int_0^\infty |f(x)|\,dx$$

converges, and in this case (1) is said to *converge absolutely*. And finally, (1) converges absolutely—and therefore converges—if there exists a function $g(x)$ such that $|f(x)| \le g(x)$ and

$$\int_0^\infty g(x)\, dx$$

converges (this is known as the *comparison test*).

Accordingly, if $f(x)$ is a given function defined for $x \ge 0$, the convergence of (1) requires first of all that the integral $\int_0^b f(x)\, dx$ must exist for each finite $b > 0$. To guarantee this, it suffices to assume that $f(x)$ is continuous, or at least is *piecewise continuous*. By the latter we mean that $f(x)$ is continuous over every finite interval $0 \le x \le b$, except possibly at a finite number of points where there are jump discontinuities, at which the function approaches different limits from the left and right. Figure 58 illustrates the appearance of a typical piecewise continuous function; its integral from 0 to b is the sum of the integrals of its continuous parts over the corresponding subintervals. This class of functions contains virtually all that are likely to arise in practice. In particular, it includes the discontinuous step functions and sawtooth functions expressing the sudden application or removal of forces and voltages in problems of physics and engineering.

If $f(x)$ is piecewise continuous for $x \ge 0$, then the only remaining threat to the existence of its Laplace transform

$$F(p) = \int_0^\infty e^{-px} f(x)\, dx$$

is the behavior of the integrand $e^{-px} f(x)$ for large x. In order to make sure that this integrand diminishes rapidly enough for convergence—or that $f(x)$ does not grow too rapidly—we shall further assume that $f(x)$ is

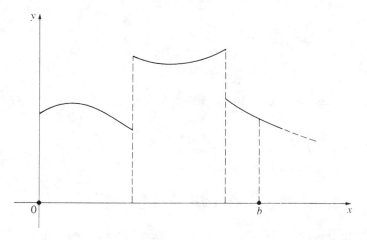

FIGURE 58

of *exponential order*. This means that there exist constants M and c such that

$$|f(x)| \le Me^{cx}. \tag{2}$$

Thus, although $f(x)$ may become infinitely large as $x \to \infty$, it must grow less rapidly than a multiple of some exponential function e^{cx}. It is clear that any bounded function is of exponential order with $c = 0$. As further examples, we mention e^{ax} (with $c = a$) and x^n (with c any positive number). On the other hand, e^{x^2} is not of exponential order. If $f(x)$ satisfies (2), then we have

$$|e^{-px}f(x)| \le Me^{-(p-c)x};$$

and since the integral of the function on the right converges for $p > c$, the Laplace transform of $f(x)$ converges absolutely for $p > c$. In addition, we note that

$$|F(p)| = \left| \int_0^\infty e^{-px}f(x)\, dx \right| \le \int_0^\infty |e^{-px}f(x)|\, dx$$

$$\le M \int_0^\infty e^{-(p-c)x}\, dx = \frac{M}{p-c}, \qquad p > c,$$

so

$$F(p) \to 0 \qquad \text{as} \qquad p \to \infty. \tag{3}$$

Actually, it can be shown that (3) is true whenever $F(p)$ exists, regardless of whether or not $f(x)$ is piecewise continuous and of exponential order. Thus, if $\phi(p)$ is a function of p with the property that its limit as $p \to \infty$ does not exist or is not equal to zero, then it cannot be the Laplace transform of *any* $f(x)$. In particular, polynomials in p, $\sin p$, $\cos p$, e^p, and $\log p$ cannot be Laplace transforms. On the other hand, a rational function is a Laplace transform if the degree of the numerator is less than that of the denominator.

The above remarks show that any piecewise continuous function of exponential order has a Laplace transform, so these conditions are sufficient for the existence of $L[f(x)]$. However, they are not necessary, as the example $f(x) = x^{-1/2}$ shows. This function has an infinite discontinuity at $x = 0$, so it is not piecewise continuous, but nevertheless its integral from 0 to b exists; and since it is bounded for large x, its Laplace transform exists. Indeed, for $p > 0$ we have

$$L[x^{-1/2}] = \int_0^\infty e^{-px}x^{-1/2}\, dx,$$

and the change of variable $px = t$ gives

$$L[x^{-1/2}] = p^{-1/2} \int_0^\infty e^{-t}t^{-1/2}\, dt.$$

Another change of variable, $t = s^2$, leads to

$$L[x^{-1/2}] = 2p^{-1/2} \int_0^\infty e^{-s^2} \, ds. \tag{4}$$

In most treatments of elementary calculus it is shown that the last-written integral has the value $\sqrt{\pi}/2$ (see Problem 1), so we have

$$L[x^{-1/2}] = \sqrt{\frac{\pi}{p}}. \tag{5}$$

This result will be useful in a later section.

In the remainder of this chapter we shall concentrate on the uses of Laplace transforms, and will not attempt to study the purely mathematical theory behind our procedures. Naturally these procedures need justification, and readers who are impatient with formalism can find what they want in more extensive discussions of the subject.

PROBLEMS

1. If I denotes the integral in (4), then (s being a dummy variable) we can write

$$I^2 = \left(\int_0^\infty e^{-x^2} \, dx \right) \left(\int_0^\infty e^{-y^2} \, dy \right) = \int_0^\infty \int_0^\infty e^{-(x^2+y^2)} \, dx \, dy.$$

Evaluate this double integral by changing to polar coordinates, and thereby show that $I = \sqrt{\pi}/2$.

2. In each of the following cases, graph the function and find its Laplace transform:
   (a) $f(x) = u(x - a)$ where a is a positive number and $u(x)$ is the unit step function defined by

$$u(x) = \begin{cases} 0 & \text{if } x < 0 \\ 1 & \text{if } x \geq 0; \end{cases}$$

   (b) $f(x) = [x]$ where $[x]$ denotes the greatest integer $\leq x$;
   (c) $f(x) = x - [x]$;
   (d) $f(x) = \begin{cases} \sin x & \text{if } 0 \leq x \leq \pi \\ 0 & \text{if } x > \pi. \end{cases}$

3. Show explicitly that $L[e^{x^2}]$ does not exist. *Hint:* $x^2 - px = (x - p/2)^2 - p^2/4$.
4. Show explicitly that $L[x^{-1}]$ does not exist.
5. Let ϵ be a positive number and consider the function $f_\epsilon(x)$ defined by

$$f_\epsilon(x) = \begin{cases} 1/\epsilon & \text{if } 0 \leq x \leq \epsilon \\ 0 & \text{if } x > \epsilon. \end{cases}$$

The graph of this function is shown in Fig. 59. It is clear that for every $\epsilon > 0$

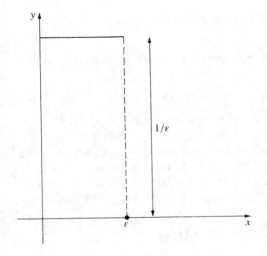

FIGURE 59

we have $\int_0^\infty f_\epsilon(x)\,dx = 1$. Show that

$$L[f_\epsilon(x)] = \frac{1 - e^{-p\epsilon}}{p\epsilon}$$

and

$$\lim_{\epsilon \to 0} L[f_\epsilon(x)] = 1.$$

Strictly speaking, $\lim_{\epsilon \to 0} f_\epsilon(x)$ does not exist as a function, so $L[\lim_{\epsilon \to 0} f_\epsilon(x)]$ is not defined; but if we throw caution to the winds, then

$$\delta(x) = \lim_{\epsilon \to 0} f_\epsilon(x)$$

is seen to be some kind of quasi-function that is infinite at $x = 0$ and zero for $x > 0$, and has the properties

$$\int_0^\infty \delta(x)\,dx = 1 \qquad \text{and} \qquad L[\delta(x)] = 1.$$

This quasi-function $\delta(x)$ is called the *Dirac delta function* or *unit impulse function*.[2]

---

[2] P.A.M. Dirac (1902–1984) was an English theoretical physicist who won the Nobel Prize at the age of thirty-one for his work in quantum theory. There are several ways of making good mathematical sense out of his delta function. See, for example, I. Halperin, *Introduction to the Theory of Distributions*, University of Toronto Press, Toronto, 1952; or A. Erdélyi, *Operational Calculus and Generalized Functions*, Holt, New York, 1962. Dirac's own discussion of his function is interesting and easy to read; see pp. 58–61 of his treatise, *The Principles of Quantum Mechanics*, Oxford University Press, 4th ed., 1958.

50   APPLICATIONS TO DIFFERENTIAL EQUATIONS

Suppose we wish to find the particular solution of the differential equation

$$y'' + ay' + by = f(x) \tag{1}$$

that satisfies the initial conditions $y(0) = y_0$ and $y'(0) = y_0'$. It is clear that we could try to apply the methods of Chapter 3 to find the general solution and then evaluate the arbitrary constants in accordance with the given initial conditions. However, the use of Laplace transforms provides an alternate way of attacking this problem that has several advantages.

To see how this method works, let us apply the Laplace transformation L to both sides of (1):

$$L[y'' + ay' + by] = L[f(x)].$$

By the linearity of L, this can be written as

$$L[y''] + aL[y'] + bL[y] = L[f(x)]. \tag{2}$$

Our next step is to express $L[y']$ and $L[y'']$ in terms of $L[y]$. First, an integration by parts gives

$$L[y'] = \int_0^\infty e^{-px} y' \, dx$$

$$= ye^{-px} \Big]_0^\infty + p \int_0^\infty e^{-px} y \, dx$$

$$= -y(0) + pL[y],$$

so

$$L[y'] = pL[y] - y(0). \tag{3}$$

Next,

$$L[y''] = L[(y')'] = pL[y'] + y'(0),$$

so

$$L[y''] = p^2 L[y] - py(0) - y'(0). \tag{4}$$

If we now insert the given initial conditions in (3) and (4), and substitute these expressions in (2), we obtain an algebraic equation for $L[y]$,

$$p^2 L[y] - py_0 - y_0' + apL[y] - ay_0 + bL[y] = L[f(x)];$$

and solving for $L[y]$ yields

$$L[y] = \frac{L[f(x)] + (p + a)y_0 + y_0'}{p^2 + ap + b}. \tag{5}$$

The function $f(x)$ is known, so its Laplace transform $L[f(x)]$ is a specific function of p; and since a, b, y_0, and y_0' are known constants, $L[y]$ is completely known as a function of p. If we can now find which function

$y(x)$ has the right side of equation (5) as its Laplace transform, then this function will be the solution of our problem—initial conditions and all. These procedures are particularly suited to solving equations of the form (1) in which the function $f(x)$ is discontinuous, for in this case the methods of Chapter 3 may be difficult to apply.

There is an obvious flaw in this discussion: in order for (2) to have any meaning, the functions $f(x)$, y, y', and y'' must have Laplace transforms. This difficulty can be remedied by simply assuming that $f(x)$ is piecewise continuous and of exponential order. Once this assumption is made, then it can be shown (we omit the proof) that y, y', and y'' necessarily have the same properties, so they also have Laplace transforms. Another difficulty is that in obtaining (3) and (4) we took it for granted that

$$\lim_{x \to \infty} ye^{-px} = 0 \quad \text{and} \quad \lim_{x \to \infty} y'e^{-px} = 0.$$

However, since y and y' are automatically of exponential order, these statements are valid for all sufficiently large values of p.

Example 1. Find the solution of

$$y'' + 4y = 4x \tag{6}$$

that satisfies the initial conditions $y(0) = 1$ and $y'(0) = 5$.

When L is applied to both sides of (6), we get

$$L[y''] + 4L[y] = 4L[x]. \tag{7}$$

If we recall that $L(x) = 1/p^2$, and use (4) and the initial conditions, then (7) becomes

$$p^2L[y] - p - 5 + 4L[y] = \frac{4}{p^2}$$

or

$$(p^2 + 4)L[y] = p + 5 + \frac{4}{p^2},$$

so

$$L[y] = \frac{p}{p^2 + 4} + \frac{5}{p^2 + 4} + \frac{4}{p^2(p^2 + 4)}$$

$$= \frac{p}{p^2 + 4} + \frac{5}{p^2 + 4} + \frac{1}{p^2} - \frac{1}{p^2 + 4}$$

$$= \frac{p}{p^2 + 4} + \frac{4}{p^2 + 4} + \frac{1}{p^2}. \tag{8}$$

On referring to the transforms obtained in Section 48, we see that (8) can be written

$$L[y] = L[\cos 2x] + L[2 \sin 2x] + L[x]$$
$$= L[\cos 2x + 2 \sin 2x + x],$$

SO

$$y = \cos 2x + 2 \sin 2x + x$$

is the desired solution. We can easily check this result, for the general solution of (6) is seen by inspection to be

$$y = c_1 \cos 2x + c_2 \sin 2x + x,$$

and the initial conditions imply at once that $c_1 = 1$ and $c_2 = 2$.

The validity of this procedure clearly rests on the assumption that only one function $y(x)$ has the right side of equation (8) as its Laplace transform. This is true if we restrict outselves to continuous $y(x)$'s—and any solution of a differential equation is necessarily continuous. When $f(x)$ is assumed to be continuous, the equation $L[f(x)] = F(p)$ is often written in the form

$$L^{-1}[F(p)] = f(x).$$

It is customary to call L^{-1} the *inverse Laplace transformation*, and to refer to $f(x)$ as the *inverse Laplace transform* of $F(p)$. Since L is linear, it is evident that L^{-1} is also linear. In Example 1 we made use of the following inverse transforms:

$$L^{-1}\left[\frac{p}{p^2 + 4}\right] = \cos 2x, \qquad L^{-1}\left[\frac{2}{p^2 + 4}\right] = \sin 2x, \qquad L^{-1}\left[\frac{1}{p^2}\right] = x.$$

This example also illustrates the value of decomposition into partial fractions as a method of finding inverse transforms.

For the convenience of the reader, we give a short list of useful transform pairs in Table 1. Much more extensive tables are available for the use of those who find it desirable to apply Laplace transforms frequently in their work.

We shall consider a number of general properties of Laplace transforms that greatly increase the flexibility of Table 1. The first of these is the *shifting formula*:

$$L[e^{ax}f(x)] = F(p - a). \qquad (9)$$

To establish this, it suffices to observe that

$$L[e^{ax}f(x)] = \int_0^\infty e^{-px}e^{ax}f(x)\, dx$$

$$= \int_0^\infty e^{-(p-a)x}f(x)\, dx$$

$$= F(p - a).$$

Formula (9) can be used to find transforms of products of the form $e^{ax}f(x)$ when $F(p)$ is known, and also to find inverse transforms of functions of the form $F(p - a)$ when $f(x)$ is known.

TABLE 1
Simple transform pairs

$f(x)$	$F(p) = L[f(x)]$
1	$\dfrac{1}{p}$
x	$\dfrac{1}{p^2}$
x^n	$\dfrac{n!}{p^{n+1}}$
e^{ax}	$\dfrac{1}{p - a}$
$\sin ax$	$\dfrac{a}{p^2 + a^2}$
$\cos ax$	$\dfrac{p}{p^2 + a^2}$
$\sinh ax$	$\dfrac{a}{p^2 - a^2}$
$\cosh ax$	$\dfrac{p}{p^2 - a^2}$

Example 2.

$$L[\sin bx] = \frac{b}{p^2 + b^2},$$

so

$$L[e^{ax} \sin bx] = \frac{b}{(p - a)^2 + b^2}.$$

Example 3.

$$L^{-1}\left[\frac{1}{p^2}\right] = x,$$

so

$$L^{-1}\left[\frac{1}{(p - a)^2}\right] = e^{ax}x.$$

The methods of this section can be applied to systems of linear differential equations with constant coefficients, and also to certain types of partial differential equations. Discussions of these further applications can be found in more extended works on Laplace transforms.[3]

---

[3] For example, see R. V. Churchill, *Operational Mathematics*, 2d ed., McGraw-Hill, New York, 1958.

PROBLEMS

1. Find the Laplace transforms of
   (a) $x^5 e^{-2x}$;
   (b) $(1 - x^2)e^{-x}$;
   (c) $e^{3x} \cos 2x$.

2. Find the inverse Laplace transforms of

   (a) $\dfrac{6}{(p + 2)^2 + 9}$;

   (b) $\dfrac{12}{(p + 3)^4}$;

   (c) $\dfrac{p + 3}{p^2 + 2p + 5}$.

3. Solve each of the following differential equations by the method of Laplace transforms:
   (a) $y' + y = 3e^{2x}$, $\quad y(0) = 0$;
   (b) $y'' - 4y' + 4y = 0$, $\quad y(0) = 0$ and $y'(0) = 3$;
   (c) $y'' + 2y' + 2y = 2$, $\quad y(0) = 0$ and $y'(0) = 1$;
   (d) $y'' + y' = 3x^2$, $\quad y(0) = 0$ and $y'(0) = 1$;
   (e) $y'' + 2y' + 5y = 3e^{-x} \sin x$, $\quad y(0) = 0$ and $y'(0) = 3$.

4. Find the solution of $y'' - 2ay' + a^2 y = 0$ in which the initial conditions $y(0) = y_0$ and $y'(0) = y_0'$ are left unrestricted. (This provides an additional derivation of our earlier solution, in Section 17, for the case in which the auxiliary equation has a double root.)

5. Apply (3) to establish the formula for the Laplace transform of an integral,

$$L\left[\int_0^x f(x)\, dx\right] = \frac{F(p)}{p},$$

   and verify this by finding

$$L^{-1}\left[\frac{1}{p(p + 1)}\right]$$

   in two ways.

6. Solve $y' + 4y + 5\int_0^x y\, dx = e^{-x}$, $\quad y(0) = 0$.

51 DERIVATIVES AND INTEGRALS OF LAPLACE TRANSFORMS

Consider the general Laplace transform formula

$$F(p) = \int_0^\infty e^{-px} f(x)\, dx.$$

The differentiation of this with respect to p under the integral sign can be justified, and yields

$$F'(p) = \int_0^\infty e^{-px}(-x)f(x)\, dx \tag{1}$$

or

$$L[-xf(x)] = F'(p).$$ (2)

By differentiating (1), we find that

$$L[x^2f(x)] = F''(p),$$ (3)

and, more generally, that

$$L[(-1)^n x^n f(x)] = F^{(n)}(p)$$ (4)

for any positive integer n. These formulas can be used to find transforms of functions of the form $x^n f(x)$ when $F(p)$ is known.

Example 1. Since $L[\sin ax] = a/(p^2 + a^2)$, we have

$$L[x \sin ax] = -\frac{d}{dp}\left(\frac{a}{p^2 + a^2}\right) = \frac{2ap}{(p^2 + a^2)^2}.$$

Example 2. We know from Section 49 that $L[x^{-1/2}] = \sqrt{\pi/p}$, so

$$L[x^{1/2}] = L[x(x^{-1/2})] = -\frac{d}{dp}\left(\sqrt{\frac{\pi}{p}}\right) = \frac{1}{2p}\sqrt{\frac{\pi}{p}}.$$

If we apply (2) to a function $y(x)$ and its derivatives—and remember formulas 50-(3) and 50-(4)—then we get

$$L[xy] = -\frac{d}{dp}L[y] = -\frac{dY}{dp},$$ (5)

$$L[xy'] = -\frac{d}{dp}L[y'] = -\frac{d}{dp}[pY - y(0)] = -\frac{d}{dp}[pY],$$ (6)

and

$$L[xy''] = -\frac{d}{dp}L[y''] = -\frac{d}{dp}[p^2Y - py(0) - y'(0)]$$

$$= -\frac{d}{dp}[p^2Y - py(0)].$$ (7)

These formulas can sometimes be used to solve linear differential equations whose coefficients are first degree polynomials in the independent variable.

Example 3. Bessel's equation of order zero is

$$xy'' + y' + xy = 0.$$ (8)

It is known to have a single solution $y(x)$ with the property that $y(0) = 1$. To find this solution, we apply L to (8) and use (5) and (7), which gives

$$-\frac{d}{dp}[p^2Y - p] + pY - 1 - \frac{dY}{dp} = 0$$

or

$$(p^2 + 1)\frac{dY}{dp} = -pY.$$ (9)

If we separate the variables in (9) and integrate, we get

$$Y = \frac{c}{\sqrt{p^2 + 1}} = c(p^2 + 1)^{-1/2}$$

$$= \frac{c}{p}\left(1 + \frac{1}{p^2}\right)^{-1/2}. \tag{10}$$

On expanding the last factor by the binomial series

$$(1 + z)^a = 1 + az + \frac{a(a - 1)}{2!}z^2 + \frac{a(a - 1)(a - 2)}{3!}z^3 + \cdots$$

$$+ \frac{a(a - 1)\cdots(a - n + 1)}{n!}z^n + \cdots,$$

(10) becomes

$$Y = \frac{c}{p}\left[1 - \frac{1}{2}\cdot\frac{1}{p^2} + \frac{1}{2!}\cdot\frac{1}{2}\cdot\frac{3}{2}\cdot\frac{1}{p^4} - \frac{1}{3!}\cdot\frac{1}{2}\cdot\frac{3}{2}\cdot\frac{5}{2}\cdot\frac{1}{p^6} + \cdots\right.$$

$$\left. + \frac{1\cdot 3\cdot\cdot 5\cdots(2n - 1)}{2^n n!}\frac{(-1)^n}{p^{2n}} + \cdots\right]$$

$$= c\sum_{n=0}^{\infty}\frac{(2n)!}{2^{2n}(n!)^2}\frac{(-1)^n}{p^{2n+1}}.$$

If we now proceed formally, and compute the inverse transform of this series term by term, then we find that

$$y(x) = c\sum_{n=0}^{\infty}\frac{(-1)^n}{2^{2n}(n!)^2}x^{2n}$$

$$= c\left(1 - \frac{x^2}{2^2} + \frac{x^4}{2^2\cdot 4^2} - \frac{x^6}{2^2\cdot 4^2\cdot 6^2} + \cdots\right).$$

Since $y(0) = 1$, it follows that $c = 1$, and our solution is

$$y(x) = 1 - \frac{x^2}{2^2} + \frac{x^4}{2^2\cdot 4^2} - \frac{x^6}{2^2\cdot 4^2\cdot 6^2} + \cdots.$$

This series defines the important Bessel function $J_0(x)$, whose Laplace transform we have found to be $1/\sqrt{p^2 + 1}$. We obtained this series in Chapter 8 in a totally different way, and it is interesting to see how easily it can be derived by Laplace transform methods.

We now turn to the problem of integrating transforms, and our main result is

$$L\left[\frac{f(x)}{x}\right] = \int_{p}^{\infty} F(p)\,dp. \tag{11}$$

To establish this, we put $L[f(x)/x] = G(p)$. An application of (2) yields

$$\frac{dG}{dp} = L\left[(-x)\frac{f(x)}{x}\right] = -L[f(x)] = -F(p),$$

so

$$G(p) = -\int_a^p F(p)\,dp$$

for some a. Since we want to make $G(p) \to 0$ as $p \to \infty$, we put $a = \infty$ and get

$$G(p) = \int_p^\infty F(p)\,dp,$$

which is (11). This formula is useful in finding transforms of functions of the form $f(x)/x$ when $F(p)$ is known. Furthermore, if we write (11) as

$$\int_0^\infty e^{-px}\frac{f(x)}{x}\,dx = \int_p^\infty F(p)\,dp$$

and let $p \to 0$, we obtain

$$\int_0^\infty \frac{f(x)}{x}\,dx = \int_0^\infty F(p)\,dp, \tag{12}$$

which is valid whenever the integral on the left exists. This formula can sometimes be used to evaluate integrals that are difficult to handle by other methods.

Example 4. Since $L[\sin x] = 1/(p^2 + 1)$, (12) gives

$$\int_0^\infty \frac{\sin x}{x}\,dx = \int_0^\infty \frac{dp}{p^2 + 1} = \tan^{-1} p\,\Big]_0^\infty = \frac{\pi}{2}.$$

For easy reference, we list the main general properties of Laplace transforms in Table 2. It will be noted that the last item in the list is new. We shall discuss this formula and its applications in the next section.

PROBLEMS

1. Show that

$$L[x\cos ax] = \frac{p^2 - a^2}{(p^2 + a^2)^2},$$

and use this result to find

$$L^{-1}\left[\frac{1}{(p^2 + a^2)^2}\right].$$

TABLE 2
General properties of
$$L[f(x)] = F(p)$$

$L[\alpha f(x) + \beta g(x)] = \alpha F(p) + \beta G(p)$

$L[e^{ax}f(x)] = F(p - a)$

$L[f'(x)] = pF(p) - f(0);$
$\quad L[f''(x)] = p^2F(p) - pf(0) - f'(0)$

$L\left[\int_0^x f(x)\,dx\right] = \dfrac{f(p)}{p}$

$L[-xf(x)] = F'(p);$
$\quad L[(-1)^n x^n f(x)] = F^{(n)}(p)$

$L\left[\dfrac{f(x)}{x}\right] = \displaystyle\int_p^\infty F(p)\,dp$

$L\left[\displaystyle\int_0^x f(x - t)g(t)\,dt\right] = F(p)G(p)$

---

2. Find each of the following transforms:
   (a) $L[x^2 \sin ax]$;
   (b) $L[x^{3/2}]$.

3. Solve each of the following differential equations:
   (a) $xy'' + (3x - 1)y' - (4x + 9)y = 0$, $\quad y(0) = 0$;
   (b) $xy'' + (2x + 3)y' + (x + 3)y = 3e^{-x}$, $\quad y(0 = 0$.

4. If $y(x)$ satisfies the differential equation

$$y'' + x^2 y = 0,$$

where $y(0) = y_0$ and $y'(0) = y_0'$, show that its transform $Y(p)$ satisfies the equation

$$Y'' + p^2 Y = py_0 + y_0'.$$

Observe that the second equation is of the same type as the first, so that no progress has been made. The method of Example 3 is advantageous only when the coefficients are first degree polynomials.

5. If a and b are positive constants, evaluate the following integrals:

   (a) $\displaystyle\int_0^\infty \dfrac{e^{-ax} - e^{-bx}}{x}\,dx$;

   (b) $\displaystyle\int_0^\infty \dfrac{e^{-ax} \sin bx}{x}\,dx$.

6. Show formally that

   (a) $\displaystyle\int_0^\infty J_0(x)\,dx = 1$;

   (b) $J_0(x) = \dfrac{1}{\pi}\displaystyle\int_0^\pi \cos(x \cos t)\,dt.$

7. If $x > 0$, show formally that

(a) $f(x) = \displaystyle\int_0^\infty \frac{\sin xt}{t} dt = \frac{\pi}{2}$;

(b) $f(x) = \displaystyle\int_0^\infty \frac{\cos xt}{1 + t^2} dt = \frac{\pi}{2} e^{-x}$.

8. (a) If $f(x)$ is periodic with period a, so that $f(x + a) = f(x)$, show that

$$F(p) = \frac{1}{1 - e^{-ap}} \int_0^a e^{-px} f(x) \, dx.$$

(b) Find $F(p)$ if $f(x) = 1$ in the intervals from 0 to 1, 2 to 3, 4 to 5, etc., and $f(x) = 0$ in the remaining intervals.

52 CONVOLUTIONS AND ABEL'S MECHANICAL PROBLEM

If $L[f(x)] = F(p)$ and $L[g(x)] = G(p)$, what is the inverse transform of $F(p)G(p)$?

To answer this question formally, we use dummy variables s and t in the integrals defining the transforms and write

$$F(p)G(p) = \left[\int_0^\infty e^{-ps} f(s) \, ds \right]\left[\int_0^\infty e^{-pt} g(t) \, dt \right]$$

$$= \int_0^\infty \int_0^\infty e^{-p(s+t)} f(s)g(t) \, ds \, dt$$

$$= \int_0^\infty \left[\int_0^\infty e^{-p(s+t)} f(s) \, ds \right] g(t) \, dt,$$

where the integration is extended over the first quadrant ($s \geq 0$, $t \geq 0$) in the st-plane. We now introduce a new variable x in the inner integral of the last expression by putting $s + t = x$, so that $s = x - t$ and (t being fixed during this integration) $ds = dx$. This enables us to write

$$F(p)G(p) = \int_0^\infty \left[\int_t^\infty e^{-px} f(x - t) \, dx \right] g(t) \, dt$$

$$= \int_0^\infty \int_t^\infty e^{-px} f(x - t) g(t) \, dx \, dt.$$

This integration is extended over the first half of the first quadrant ($x - t \geq 0$) in the xt-plane, and reversing the order as suggested in

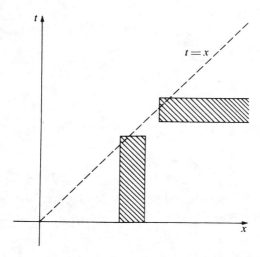

FIGURE 60

Fig. 60, we get

$$F(p)G(p) = \int_0^\infty \left[\int_0^x e^{-px} f(x - t)g(t)\, dt \right] dx$$

$$= \int_0^\infty e^{-px} \left[\int_0^x f(x - t)g(t)\, dt \right] dx$$

$$= L\left[\int_0^x f(x - t)g(t)\, dt \right]. \tag{1}$$

The integral in the last expression is a function of the upper limit x, and provides the answer to our question. This integral is called the *convolution* of the functions $f(x)$ and $g(x)$. It can be regarded as a "generalized product" of these functions. The fact stated in equation (1)—namely, that the product of the Laplace transforms of two functions is the transform of their convolution—is called the *convolution theorem*.

The convolution theorem can be used to find inverse transforms. For instance, since $L[x] = 1/p^2$ and $L[\sin x] = 1/(p^2 + 1)$, we have

$$L^{-1}\left[\frac{1}{p^2(p^2 + 1)} \right] = L^{-1}\left[\frac{1}{p^2} \left(\frac{1}{p^2 + 1} \right) \right]$$

$$= \int_0^x (x - t) \sin t\, dt$$

$$= x - \sin x,$$

as can easily be verified by partial fractions. A more interesting class of applications arises as follows. If $f(x)$ and $k(x)$ are given functions, then the equation

$$f(x) = y(x) + \int_0^x k(x - t)y(t)\,dt, \tag{2}$$

in which the unknown function $y(x)$ appears under the integral sign, is called an *integral equation*. Because of its special form, in which the integral is the convolution of the two functions $k(x)$ and $y(x)$, this equation lends itself to solution by means of Laplace transforms. In fact, if we apply L to both sides of equation (2), we get

$$L[f(x)] = L[y(x)] + L[k(x)]L[y(x)],$$

so

$$L[y(x)] = \frac{L[f(x)]}{1 + L[k(x)]}. \tag{3}$$

The right side of (3) is presumably known as a function of p; and if this function is a recognizable transform, then we have our solution $y(x)$.

Example 1. The integral equation

$$y(x) = x^3 + \int_0^x \sin(x - t)y(t)\,dt \tag{4}$$

is of this type, and by applying L we get

$$L[y(x)] = L[x^3] + L[\sin x]L[y(x)].$$

Solving for $L[y(x)]$ yields

$$L[y(x)] = \frac{L[x^3]}{1 - L[\sin x]} = \frac{3!/p^4}{1 - 1/(p^2 + 1)}$$

$$= \frac{3!}{p^4}\left(\frac{p^2 + 1}{p^2}\right) = \frac{3!}{p^4} + \frac{3!}{p^6},$$

so

$$y(x) = x^3 + \frac{1}{20}x^5$$

is the solution of (4).

As a further illustration of this technique, we analyze a classical problem in mechanics that leads to an integral equation of the above type. Consider a wire bent into a smooth curve (Fig. 61) and let a bead of mass m start from rest and slide without friction down the wire to the origin under the action of its own weight. Suppose that (x, y) is the starting point and (u, v) is any intermediate point. If the shape of the wire is specified by a given function $y = y(x)$, then the total time of descent will be a definite function $T(y)$ of the initial height y. Abel's *mechanical problem* is the converse: specify the function $T(y)$ in advance and then find the shape of the wire that yields this $T(y)$ as the total time of descent.

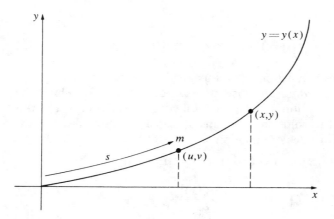

FIGURE 61

To formulate this problem mathematically, we start with the principle of conservation of energy:

$$\frac{1}{2}m\left(\frac{ds}{dt}\right)^2 = mg(y - v) \qquad \text{or} \qquad -\frac{ds}{dt} = \sqrt{2g(y - v)},$$

which can be written as

$$dt = -\frac{ds}{\sqrt{2g(y - v)}}.$$

On integrating this from $v = y$ to $v = 0$, we get

$$T(y) = \int_{v=y}^{v=0} dt = \int_{v=0}^{v=y} \frac{ds}{\sqrt{2g(y - v)}} = \frac{1}{\sqrt{2g}} \int_0^y \frac{s'(v)\, dv}{\sqrt{y - v}}. \tag{5}$$

Now

$$s = s(y) = \int_0^y \sqrt{1 + \left(\frac{dx}{dy}\right)^2}\, dy$$

is known whenever the curve $y = y(x)$ is known, so its derivative

$$f(y) = s'(y) = \sqrt{1 + \left(\frac{dx}{dy}\right)^2} \tag{6}$$

is also known. If we insert (6) in (5), then we see that

$$T(y) = \frac{1}{\sqrt{2g}} \int_0^y \frac{f(v)\, dv}{\sqrt{y - v}}, \tag{7}$$

and this enables us to calculate $T(y)$ whenever the curve is given. In Abel's problem we want to find the curve when $T(y)$ is given; and from this point of view, the function $f(y)$ in equation (7) is the unknown and (7) itself is called *Abel's integral equation*. Note that the integral in (7) is the convolution of the functions $y^{-1/2}$ and $f(y)$, so on applying the Laplace transformation L we get

$$L[T(y)] = \frac{1}{\sqrt{2g}} L[y^{-1/2}]L[f(y)].$$

If we now recall that $L[y^{-1/2}] = \sqrt{\pi/p}$, then this yields

$$L[f(y)] = \sqrt{2g} \frac{L[T(y)]}{\sqrt{\pi/p}}$$

$$= \sqrt{\frac{2g}{\pi}} p^{1/2}L[T(y)]. \tag{8}$$

When $T(y)$ is given, the right side of equation (8) is known as a function of p, so hopefully we can find $f(y)$ by taking the inverse transform. Once $f(y)$ is known, the curve itself can be found by solving the differential equation (6).

As a concrete example, we now specialize our discussion to the case in which $T(y)$ is a constant T_0. This assumption means that the time of descent is to be independent of the starting point. The curve defined by this property is called the *tautochrone*, so our problem is that of finding the tautochrone. In this case, (8) becomes

$$L[f(y)] = \sqrt{\frac{2g}{\pi}} p^{1/2}L[T_0] = \sqrt{\frac{2g}{\pi}} p^{1/2} \frac{T_0}{p} = b^{1/2}\sqrt{\frac{\pi}{p}},$$

where $b = 2gT_0^2/\pi^2$. The inverse transform of $\sqrt{\pi/p}$ is $y^{-1/2}$, so

$$f(y) = \sqrt{\frac{b}{y}}. \tag{9}$$

With this $f(y)$, (6) now yields

$$1 + \left(\frac{dx}{dy}\right)^2 = \frac{b}{y}$$

as the differential equation of the curve, so

$$x = \int \sqrt{\frac{b-y}{y}} \, dy.$$

On substituting $y = b \sin^2 \phi$, this becomes

$$x = 2b \int \cos^2 \phi \, d\phi = b \int (1 + \cos 2\phi) \, d\phi$$

$$= \frac{b}{2} (2\phi + \sin 2\phi) + c,$$

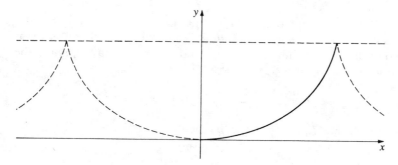

FIGURE 62

SO

$$x = \frac{b}{2}(2\phi + \sin 2\phi) + c \quad \text{and} \quad y = \frac{b}{2}(1 - \cos 2\phi). \quad (10)$$

The curve must pass through the origin $(0,0)$, so $c = 0$; and if we put $a = b/2$ and $\theta = 2\phi$, then (10) take the simpler form

$$x = a(\theta + \sin \theta) \quad \text{and} \quad y = a(1 - \cos \theta).$$

These are the parametric equations of the cycloid shown in Fig. 62, which is generated by a fixed point on a circle of radius a rolling under the horizontal dashed line $y = 2a$. Since $2a = b = 2gT_0^2/\pi^2$, the diameter of the generating circle is determined by the constant time of descent.

Accordingly, the tautochrone is a cycloid. In Problems 6-5 and 11-5 we verified this property of cycloids by other methods. Our present discussion has the advantage of enabling us to find the tautochrone without knowing in advance what the answer will be.

PROBLEMS

1. Find $L^{-1}[1/(p^2 + a^2)^2]$ by convolution. (See Problem 51-1.)

2. Solve each of the following integral equations:

(a) $y(x) = 1 - \int_0^x (x - t)y(t)\, dt$;

(b) $y(x) = e^x\left[1 + \int_0^x e^{-t}y(t)\, dt\right]$;

(c) $e^{-x} = y(x) + 2\int_0^x \cos(x - t)y(t)\, dt$;

(d) $3\sin 2x = y(x) + \int_0^x (x - t)y(t)\, dt$.

3. Deduce

$$f(y) = \frac{\sqrt{2g}}{\pi} \frac{d}{dy} \int_0^y \frac{T(t)\, dt}{\sqrt{y - t}}$$

from equation (8), and use this to verify (9) when $T(y)$ is a constant T_0.

4. Find the equation of the curve of descent if $T(y) = k\sqrt{y}$ for some constant k.

5. Show that the differential equation

$$y'' + a^2 y = f(x), \qquad y(0) = y'(0) = 0,$$

has

$$y(x) = \frac{1}{a} \int_0^x f(t) \sin a(x - t)\, dt$$

as its solution.

53 MORE ABOUT CONVOLUTIONS. THE UNIT STEP AND IMPULSE FUNCTIONS

In the preceding section we found that the product of the Laplace transforms of two functions is the transform of a certain combination of these functions called their *convolution*. If we use the time t as the independent variable and if the two functions are $f(t)$ and $g(t)$, then this *convolution theorem* [equation 52-(1)] can be expressed as follows:

$$L[f(t)]L[g(t)] = L\left[\int_0^t f(t - \tau)g(\tau)\, d\tau \right]. \tag{1}$$

It is customary to denote the convolution of $f(t)$ and $g(t)$ by $f(t) * g(t)$, so that

$$f(t) * g(t) = \int_0^t f(t - \tau)g(\tau)\, d\tau. \tag{2}$$

The convolution theorem (1) can then be written in the form

$$L[f(t) * g(t)] = L[f(t)]L[g(t)]. \tag{3}$$

Our purpose in this section is to discuss an application of this theorem that makes it possible to determine the response of a mechanical or electrical system to a general stimulus if its response to the unit step function is known. These ideas have important uses in electrical engineering and other areas of applied science.

Any physical system capable of responding to a stimulus can be thought of as a device that transforms an *input function* (the stimulus) into an *output function* (the response). If we assume that all initial conditions are zero at the moment $t = 0$ when the input $f(t)$ begins to act, then by setting up the differential equation that describes the system, operating on this equation with the Laplace transformation L, and solving for the transform of the output $y(t)$, we obtain an equation of the form

$$L[y(t)] = \frac{L[f(t)]}{z(p)}, \tag{4}$$

where $z(p)$ is a polynomial whose coefficients depend only on the parameters of the system itself. This equation is the main source of the explicit formulas for $y(t)$ that we obtain below with the aid of the convolution theorem.

Let us be more specific. We seek solutions $y(t)$ of the linear differential equation

$$y'' + ay' + by = f(t) \tag{5}$$

that satisfy the initial conditions

$$y(0) = y'(0) = 0 \tag{6}$$

describing a mechanical or electrical system at rest in its equilibrium position. The input $f(t)$ can be thought of as an impressed external force F or electromotive force E that begins to act at time $t = 0$, as discussed in Section 20. When this input is the unit step function $u(t)$ defined in Problem 49-2(a), the solution (or output) $y(t)$ is denoted by $A(t)$ and called the *indicial response*; that is,

$$A'' + aA' + bA = u(t).$$

By applying the Laplace transformation L and using formulas (3) and (4) in Section 50, we obtain

$$p^2 L[A] + apL[A] + bL[A] = L[u(t)] = \frac{1}{p},$$

so

$$L[A] = \frac{1}{p} \frac{1}{p^2 + ap + b} = \frac{1}{p} \frac{1}{z(p)}, \tag{7}$$

where $z(p)$ is defined by the last equality. We now apply L in the same way to the general equation (5), which yields (4); and dividing both sides of this by p and using (7) gives

$$\frac{1}{p} L[y] = \frac{1}{pz(p)} L[f] = L[A]L[f]. \tag{8}$$

The convolution theorem now enables us to write (8) in the form

$$\frac{1}{p} L[y] = L[A(t) * f(t)] = L\left[\int_0^t A(t - \tau)f(\tau)\, d\tau\right].$$

By using formula 50-(3) once more we get

$$L[y] = pL\left[\int_0^t A(t - \tau)f(\tau)\,d\tau\right]$$

$$= L\left[\frac{d}{dt}\int_0^t A(t - \tau)f(\tau)\,d\tau\right],$$

so

$$y(t) = \frac{d}{dt}\int_0^t A(t - \tau)f(\tau)\,d\tau. \tag{9}$$

By applying Leibniz's rule for differentiating integrals[4] to (9), we now have

$$y(t) = \int_0^t A'(t - \tau)f(\tau)\,d\tau + A(0)f(t). \tag{10}$$

Next, since $L[A]L[f] = L[f]L[A]$, (8) also enables us to write

$$\frac{1}{p}L[y] = L[f(t)*A(t)] = L\left[\int_0^t f(t - \sigma)A(\sigma)\,d\sigma\right],$$

and by following the same reasoning as before, we obtain

$$y(t) = \int_0^t f'(t - \sigma)A(\sigma)\,d\sigma + f(0)A(t). \tag{11}$$

In formula (10) we notice that $A(0) = 0$ because of the initial conditions (6); and (11) takes a more convenient form under the change of variable $\tau = t - \sigma$. Our two formulas (10) and (11) for $y(t)$ therefore become

$$y(t) = \int_0^t A'(t - \tau)f(\tau)\,d\tau \tag{12}$$

and

$$y(t) = \int_0^t A(t - \tau)f'(\tau)\,d\tau + f(0)A(t). \tag{13}$$

Each of these formulas provides a solution of (5) for a general input $f(t)$

---

[4] *Leibniz's rule* states that if $F(t) = \int_u^v G(t, x)\,dx$, where u and v are functions of t and x is a dummy variable, then

$$\frac{d}{dt}F(t) = \int_u^v \frac{\partial}{\partial t}G(t, x)\,dx + G(t, v)\frac{dv}{dt} - G(t, u)\frac{du}{dt}.$$

See p. 613 of George F. Simmons, *Calculus With Analytic Geometry*, McGraw-Hill, New York, 1985.

in terms of the indicial response $A(t)$ to the unit step function. Formula (13) is sometimes called the *principle of superposition*; it has been variously attributed to the famous nineteenth century physicists James Clerk Maxwell and Ludwig Boltzmann, and also to the English applied mathematician Oliver Heaviside.

Example 1. Use formula (13) to solve $y'' + y' - 6y = 2e^{3t}$, where $y(0) = y'(0) = 0$.

Here we have

$$L[A(t)] = \frac{1}{p(p^2 + p - 6)},$$

so by partial fractions and inversion we find that

$$A(t) = -\frac{1}{6} + \frac{1}{15} e^{-3t} + \frac{1}{10} e^{2t}.$$

Since $f(t) = 2e^{3t}$, $f'(t) = 6e^{3t}$ and $f(0) = 2$, (13) gives

$$y(t) = \int_0^t \left[-\frac{1}{6} + \frac{1}{15} e^{-3(t-\tau)} + \frac{1}{10} e^{2(t-\tau)} \right] 6e^{3\tau} \, d\tau$$

$$+ 2\left[-\frac{1}{6} + \frac{1}{15} e^{-3t} + \frac{1}{10} e^{2t} \right]$$

$$= \frac{1}{3} e^{3t} + \frac{1}{15} e^{-3t} - \frac{2}{5} e^{2t}.$$

This solution can be verified by substituting directly in the given equation, and also by solving the equation by the method already studied in Section 50.

We can also use formula (12) to solve the equation in this example, but before doing this, it is desirable to express (12) in a simpler form. We accomplish this by using the unit impulse function $\delta(t)$ described in Problem 49-5. In physics, the *impulse* due to a constant force F acting over a time interval Δt is defined to be $F \Delta t$. The "function" $\delta(t)$ can be thought of as a limit of constant functions of unit impulse acting over shorter and shorter intervals of time; it is used to describe forces and voltages that act very suddenly, as in the case of a hammer blow on a mechanical system or a lightning stroke on a transmission line.

For us, the essential property of $\delta(t)$ is that expressed by the equation

$$L[\delta(t)] = 1,$$

obtained in Problem 49-5. When the input $f(t)$ in the differential equation (5) is the unit impulse function $\delta(t)$, the output $y(t)$ is denoted by $h(t)$ and called the *impulsive response*. Applying L in this case yields

$$L[h(t)] = \frac{1}{z(p)}, \tag{14}$$

so

$$h(t) = L^{-1}\left[\frac{1}{z(p)}\right].$$

By (7) and (14),

$$L[A(t)] = \frac{1}{p}\frac{1}{z(p)} = \frac{L[h(t)]}{p},$$

and it follows from Problem 50-5 that

$$A(t) = \int_0^t h(t)\, dt.$$

This shows that $A'(t) = h(t)$, so formula (12) becomes

$$y(t) = \int_0^t h(t - \tau)f(\tau)\, d\tau. \tag{15}$$

Thus, the solution of (5) with a general input $f(t)$ can be written as the convolution of the impulsive response $h(t)$ with $f(t)$.

Example 2. Consider again the equation $y'' + y' - 6y = 2e^{3t}$ solved in Example 1. We have

$$h(t) = L^{-1}\left[\frac{1}{(p + 3)(p - 2)}\right] = \frac{1}{5}(e^{2t} - e^{-3t}),$$

so that

$$y(t) = \int_0^t \frac{1}{5}[e^{2(t-\tau)} - e^{-3(t-\tau)}]2e^{3\tau}\, d\tau$$

$$= \frac{1}{3}e^{3t} + \frac{1}{15}e^{-3t} - \frac{2}{5}e^{2t},$$

as before.

Remark 1. In complicated practical situations electrical engineers are sometimes compelled to work with indicial or impulsive responses $A(t)$ or $h(t)$ that are only accessible experimentally, by means of oscilloscope pictures responding to generator-produced step functions or impulse functions. In such a case the output must be calculated from (13) or (15) by methods of graphical integration that permit the plotting of individual points on the output curve. For a discussion of these topics see Chapter 9 of W. D. Day, *Introduction to Laplace Transforms for Radio and Electronic Engineers*, Interscience, New York, 1960.

Remark 2. To form a more general view of the meaning of convolution, let us consider a linear physical system in which the effect at the present time t of a small stimulus $g(\tau)\,d\tau$ at any past time τ is proportional to the size of the stimulus. We further assume that the proportionality factor depends only on the elapsed time $t - \tau$, and thus has the form $f(t - \tau)$. The effect at the present time t is therefore

$$f(t - \tau)g(\tau)\,d\tau.$$

Since the system is linear, the total effect at the present time t due to the stimulus acting throughout the entire past history of the system is obtained by adding these separate effects, and this leads to the convolution integral

$$\int_0^t f(t - \tau)g(\tau)\,d\tau.$$

The lower limit here is 0 because we assume that the stimulus started acting at time $t = 0$, that is, that $g(\tau) = 0$ for $\tau < 0$. The importance of convolution is difficult to exaggerate: it provides a reasonable way of taking account of the past in the study of wave motion, heat conduction, diffusion, and other areas of mathematical physics.

PROBLEMS

1. Show that $f(t) * g(t) = g(t) * f(t)$ directly from the definition (2), by introducing a new dummy variable $\sigma = t - \tau$. This shows that the operation of forming convolutions is commutative. It is also associative and distributive:

$$f(t) * [g(t) * h(t)] = [f(t) * g(t)] * h(t)$$

and

$$f(t) * [g(t) + h(t)] = f(t) * g(t) + f(t) * h(t),$$
$$[f(t) + g(t)] * h(t) = f(t) * h(t) + g(t) * h(t).$$

An interesting discussion of the abstract properties of convolution is given by Mark Kac and Stanislaw Ulam on pp. 140–142 of *Mathematics and Logic*, New American Library, New York, 1969.

2. Find the convolution of each of the following pairs of functions:
   (a) $1,\ \sin at$;
   (b) $e^{at},\ e^{bt}$, where $a \neq b$;
   (c) $t,\ e^{at}$;
   (d) $\sin at,\ \sin bt$, where $a \neq b$.

3. Verify the convolution theorem for each of the pairs of functions considered in Problem 2.

4. Use the methods of both Examples 1 and 2 to solve each of the following differential equations:
   (a) $y'' + 5y' + 6y = 5e^{3t}$, $\quad y(0) = y'(0) = 0$;
   (b) $y'' + y' - 6y = t$, $\quad y(0) = y'(0) = 0$;
   (c) $y'' - y' = t^2$, $\quad y(0) = y'(0) = 0$.

5. When the polynomial $z(p)$ has distinct real zeros a and b, so that

$$\frac{1}{z(p)} = \frac{1}{(p-a)(p-b)} = \frac{A}{p-a} + \frac{B}{p-b}$$

for suitable constants A and B, then

$$h(t) = Ae^{at} + Be^{bt}$$

and (15) takes the form

$$y(t) = \int_0^t f(\tau)[Ae^{a(t-\tau)} + Be^{b(t-\tau)}]\, d\tau.$$

This sometimes called the *Heaviside expansion theorem.*
   (a) Use this theorem to write the solution of $y'' + 3y' + 2y = f(t)$, $y(0) = y'(0) = 0$.
   (b) Give an explicit evaluation of the solution in (a) for the cases $f(t) = e^{3t}$ and $f(t) = t$.
   (c) Find the solutions in (b) by using the superposition principle (13).
6. Formula (13) can also be derived from (4) as follows, without the use of Leibniz's rule for differentiating integrals:

$$L[y(t)] = \frac{L[f(t)]}{z(p)} = \frac{1}{pz(p)} \cdot pL[f(t)]$$

$$= L[A(t)] \cdot pL[f(t)]$$

$$= L[A(t)] \cdot \{L[f'(t)] + f(0)\}$$

$$= L[A(t) * f'(t)] + f(0)L[A(t)]$$

$$= L\left[\int_0^t A(t-\tau)f'(\tau)\, d\tau + f(0)A(t)\right].$$

Check the steps.
7. As we know from Section 20, the forced vibrations of an undamped spring–mass system are described by the differential equation

$$Mx'' + kx = f(t),$$

where $x(t)$ is the displacement and $f(t)$ is the impressed external force or "forcing function." If $x(0) = x'(0) = 0$, find the functions $A(t)$ and $h(t)$ and write down the solution $x(t)$ for any $f(t)$.
8. The current $I(t)$ in an electric circuit with inductance L and resistance R is given by equation (4) in Section 13:

$$L\frac{dI}{dt} + RI = E(t),$$

where $E(t)$ is the impressed electromotive force. If $I(0) = 0$, use the methods of this section to find $I(t)$ in each of the following cases:
   (a) $E(t) = E_0 u(t)$;
   (b) $E(t) = E_0 \delta(t)$;
   (c) $E(t) = E_0 \sin \omega t$.

APPENDIX A. LAPLACE

Pierre Simon de Laplace (1749–1827) was a French mathematician and theoretical astronomer who was so famous in his own time that he was known as the Newton of France. His main interests throughout his life were celestial mechanics, the theory of probability, and personal advancement.

At the age of twenty-four he was already deeply engaged in the detailed application of Newton's law of gravitation to the solar system as a whole, in which the planets and their satellites are not governed by the sun alone but interact with one another in a bewildering variety of ways. Even Newton had been of the opinion that divine intervention would occasionally be needed to prevent this complex mechanism from degenerating into chaos. Laplace decided to seek reassurance elsewhere, and succeeded in proving that the ideal solar system of mathematics is a stable dynamical system that will endure unchanged for all time. This achievement was only one of the long series of triumphs recorded in his monumental treatise *Mécanique Céleste* (published in five volumes from 1799 to 1825), which summed up the work on gravitation of several generations of illustrious mathematicians. Unfortunately for his later reputation, he omitted all reference to the discoveries of his predecessors and contemporaries, and left it to be inferred that the ideas were entirely his own. Many anecdotes are associated with this work. One of the best known describes the occasion on which Napoleon tried to get a rise out of Laplace by protesting that he had written a huge book on the system of the world without once mentioning God as the author of the universe. Laplace is supposed to have replied, "Sire, I had no need of that hypothesis." The principal legacy of the *Mécanique Céleste* to later generations lay in Laplace's wholesale development of potential theory, with its far-reaching implications for a dozen different branches of physical science ranging from gravitation and fluid mechanics to electromagnetism and atomic physics. Even though he lifted the idea of the potential from Lagrange without acknowledgment, he exploited it so extensively that ever since his time the fundamental differential equation of potential theory has been known as Laplace's equation.

His other masterpiece was the treatise *Théorie Analytique des Probabilités* (1812), in which he incorporated his own discoveries in probability from the preceding 40 years. Again he failed to acknowledge the many ideas of others he mixed in with his own; but even discounting this, his book is generally agreed to be the greatest contribution to this part of mathematics by any one man. In the introduction he says: "At bottom, the theory of probability is only common sense reduced to calculation." This may be so, but the following 700 pages of intricate analysis—in which he freely used Laplace transforms, generating functions, and many other highly nontrivial tools—has been said by some to surpass in complexity even the *Mécanique Céleste*.

After the French Revolution Laplace's political talents and greed for position came to full flower. His countrymen speak ironically of his "suppleness" and "versatility" as a politician. What this really means is that each time there was a change of regime (and there were many), Laplace smoothly adapted himself by changing his principles—back and forth between fervent republicanism and fawning royalism—and each time he emerged with a better job and grander titles. He has been aptly compared with the apocryphal Vicar of Bray in English literature, who was twice a Catholic and twice a Protestant. The Vicar is said to have replied as follows to the charge of being a turncoat: "Not so, neither, for if I changed my religion, I am sure I kept true to my principle, which is to live and die the Vicar of Bray."

To balance his faults, Laplace was always generous in giving assistance and encouragement to younger scientists. From time to time he helped forward in their careers such men as the chemist Gay-Lussac, the traveler and naturalist Humboldt, the physicist Poisson, and—appropriately—the young Cauchy, who was destined to become one of the chief architects of nineteenth century mathematics.

APPENDIX B. ABEL

Niels Henrik Abel (1802–1829) was one of the foremost mathematicians of the nineteenth century and probably the greatest genius produced by the Scandinavian countries. Along with his contemporaries Gauss and Cauchy, Abel was one of the pioneers in the development of modern mathematics, which is characterized by its insistence on rigorous proof. His career was a poignant blend of good-humored optimism under the strains of poverty and neglect, modest satisfaction in the many towering achievements of his brief maturity, and patient resignation in the face of an early death.

Abel was one of six children in the family of a poor Norwegian country minister. His great abilities were recognized and encouraged by one of his teachers when he was only sixteen, and soon he was reading and digesting the works of Newton, Euler, and Lagrange. As a comment on this experience, he inserted the following marginal remark in one of his later mathematical notebooks: "It appears to me that if one wants to make progress in mathematics, one should study the masters and not the pupils." When Abel was only eighteen his father died and left the family destitute. They subsisted by the aid of friends and neighbors, and somehow the boy, helped by contributions from several professors, managed to enter the University of Oslo in 1821. His earliest researches were published in 1823, and included his solution of the classic tautochrone problem by means of the integral equation discussed in Section 52. This was the first solution of an equation of this kind, and foreshadowed the extensive development of integral equations in the late nineteenth and early twentieth centuries. He also proved that the general fifth degree equation $ax^5 + bx^4 + cx^3 + dx^2 + ex + f = 0$ cannot be solved in terms of radicals, as is possible for equations of lower degree, and thus disposed of a problem that had baffled mathematicians for 300 years. He published his proof in a small pamphlet at his own expense.

In his scientific development Abel soon outgrew Norway, and longed to visit France and Germany. With the backing of his friends and professors he applied to the government, and after the usual red tape and delays, he received a fellowship for a mathematical grand tour of the Continent. He spent most of his first year abroad in Berlin. Here he had the great good fortune to make the acquaintance of August Leopold Crelle, an enthusiastic mathematical amateur who became his close friend, advisor, and protector. In turn, Abel inspired Crelle to launch his famous *Journal für die Reine und Angewandte Mathematik,* which was the world's first periodical devoted wholly to mathematical research. The first three volumes contained 22 contributions by Abel.

Abel's early mathematical training had been exclusively in the older formal tradition of the eighteenth century, as typified by Euler. In Berlin he came under the influence of the new school of thought led by Gauss and Cauchy, which emphasized rigorous deduction as opposed to formal calculation. Except for Gauss's great work on the hypergeometric series, there were hardly any proofs in analysis that would be accepted as valid today. As Abel expressed it in a letter to a friend: "If you disregard the very simplest cases, there is in all of mathematics not a single infinite series whose sum has been rigorously determined. In other words, the most important parts of mathematics stand without a foundation." In this period he wrote his classic study of the binomial series, in which he founded the general theory of convergence and gave the first satisfactory proof of the validity of this series expansion.

Abel had sent to Gauss in Göttingen his pamphlet on the fifth degree equation, hoping that it would serve as a kind of scientific passport. However, for some reason Gauss put it aside without looking at it, for it was found uncut among his papers after his death 30 years later. Unfortunately for both men, Abel felt that he had been snubbed, and decided to go on to Paris without visiting Gauss.

In Paris he met Cauchy, Legendre, Dirichlet, and others, but these meetings were perfunctory and he was not recognized for what he was. He had already published a number of important articles in Crelle's *Journal,* but the French were hardly aware yet of the existence of this new periodical and Abel was much too shy to speak of his own work to people he scarcely knew. Soon after his arrival he finished his great *Mémoire sur une Propriété Générale d'une Classe Très Étendue des Fonctions Transcendantes*, which he regarded as his masterpiece. This work contains the discovery about integrals of algebraic functions now known as Abel's theorem, and is the foundation for the later theory of Abelian integrals, Abelian functions, and much of algebraic geometry. Decades later, Hermite is said to have remarked of this *Mémoire*: "Abel has left mathematicians enough to keep them busy for 500 years." Jacobi described Abel's theorem as the greatest discovery in integral calculus of the nineteenth century. Abel submitted his manuscript to the French Academy. He hoped that it would bring him to the notice of the French mathematicians, but he waited in vain until his purse was empty and he was forced to return to Berlin. What happened was this: the manuscript was given to Cauchy and Legendre for examination; Cauchy took it home, mislaid it, and forgot all about it; and it was not published until 1841, when again the manuscript was lost before the proof sheets were read. The original finally turned up in Florence in 1952.[5] In Berlin, Abel finished his first revolutionary article on elliptic functions, a subject he had been working on for several years, and then went back to Norway, deeply in debt.

He had expected on his return to be appointed to a professorship at the university, but once again his hopes were dashed. He lived by tutoring, and for a brief time held a substitute teaching positon. During this period he worked incessantly, mainly on the theory of the elliptic functions that he had discovered as the inverses of elliptic integrals. This theory quickly took its place as one of the major fields of nineteenth century analysis, with many applications to number theory, mathematical physics, and algebraic geometry. Meanwhile, Abel's fame had spread to all the mathematical centers of Europe and he stood among the elite of the world's mathematicians, but in his isolation he was unaware of it. By early 1829 the tuberculosis he contracted on his journey had progressed to the point where he was unable to work, and in the spring of that year he died, at the age of twenty-six. As an ironic postscript, shortly after his death Crelle wrote that his efforts had been successful, and that Abel would be appointed to the chair of mathematics in Berlin.

Crelle eulogized Abel in his *Journal* as follows: "All of Abel's works carry the imprint of an ingenuity and force of thought which is amazing. One may say that he was able to penetrate all obstacles down to the very foundation of the problem, with a force which appeared

---

[5] For the details of this astonishing story, see the fine book by O. Ore, *Niels Henrik Abel: Mathematician Extraordinary*, University of Minnesota Press, Minneapolis, 1957.

irresistible . . . He distinguished himself equally by the purity and nobility of his character and by a rare modesty which made his person cherished to the same unusual degree as was his genius." Mathematicians, however, have their own ways of remembering their great men, and so we speak of Abel's integral equation, Abelian integrals and functions, Abelian groups, Abel's series, Abel's partial summation formula, Abel's limit theorem in the theory of power series, and Abel summability. Few have had their names linked to so many concepts and theorems in modern mathematics, and what he might have accomplished in a normal lifetime is beyond conjecture.

CHAPTER 10

SYSTEMS OF FIRST ORDER EQUATIONS

54 GENERAL REMARKS ON SYSTEMS

One of the fundamental concepts of analysis is that of a system of n simultaneous first order differential equations. If $y_1(x)$, $y_2(x)$, . . . , $y_n(x)$ are unknown functions of a single independent variable x, then the most general system of interest to us is one in which their derivatives y_1', y_2', . . . , y_n' are explicitly given as functions of x and y_1, y_2, . . . , y_n:

$$y_1' = f_1(x, y_1, y_2, \ldots, y_n)$$
$$y_2' = f_2(x, y_1, y_2, \ldots, y_n)$$
$$\cdots \tag{1}$$
$$y_n' = f_n(x, y_1, y_2, \ldots, y_n).$$

Systems of differential equations arise quite naturally in many scientific problems. In Section 22 we used a system of two second order linear equations to describe the motion of coupled harmonic oscillators; in the example below we shall see how they occur in connection with dynamical systems having several degrees of freedom; and in Section 57 we will use them to analyze a simple biological community composed of different species of animals interacting with one another.

An important mathematical reason for studying systems is that the single nth order equation

$$y^{(n)} = f(x,y,y', \ldots, y^{(n-1)}) \tag{2}$$

can always be regarded as a special case of (1). To see this, we put

$$y_1 = y, \qquad y_2 = y', \qquad \ldots, \qquad y_n = y^{(n-1)} \tag{3}$$

and observe that (2) is equivalent to the system

$$y_1' = y_2$$
$$y_2' = y_3 \tag{4}$$
$$\ldots$$
$$y_n' = f(x,y_1,y_2, \ldots, y_n),$$

which is clearly a special case of (1). The statement that (2) and (4) are equivalent is understood to mean the following: if $y(x)$ is a solution of equation (2), then the functions $y_1(x), y_2(x), \ldots, y_n(x)$ defined by (3) satisfy (4); and conversely, if $y_1(x), y_2(x), \ldots, y_n(x)$ satisfy (4), then $y(x) = y_1(x)$ is a solution of (2).

This reduction of an nth order equation to a system of n first order equations has several advantages. We illustrate by considering the relation between the basic existence and uniqueness theorems for the system (1) and for equation (2).

If a fixed point $x = x_0$ is chosen and the values of the unknown functions

$$y_1(x_0) = a_1, \qquad y_2(x_0) = a_2, \qquad \ldots, \qquad y_n(x_0) = a_n \tag{5}$$

are assigned arbitrarily in such a way that the functions $f_1, f_2, \ldots, f_n$ are defined, then (1) gives the values of the derivatives $y_1'(x_0)$, $y_2'(x_0), \ldots, y_n'(x_0)$. The similarity between this situation and that discussed in Section 2 suggests the following analog of Picard's theorem.

> **Theorem A.** *Let the functions* $f_1, f_2, \ldots, f_n$ *and the partial derivatives* $\partial f_1/\partial y_1, \ldots, \partial f_1/\partial y_n, \ldots, \partial f_n/\partial y_1, \ldots, \partial f_n/\partial y_n$ *be continuous in a region R of $(x,y_1,y_2, \ldots, y_n)$ space. If $(x_0,a_1,a_2, \ldots, a_n)$ is an interior point of R, then the system* (1) *has a unique solution $y_1(x), y_2(x), \ldots, y_n(x)$ that satisfies the initial conditions* (5).

We will not prove this theorem, but instead remark that when the ground has been properly prepared, its proof is identical with that of Picard's theorem as given in Chapter 13. Furthermore, by virtue of the above reduction, Theorem A includes as a special case the following corresponding theorem for equation (2).

Theorem B. *Let the function f and the partial derivatives $\partial f / \partial y$, $\partial f / \partial y'$, ..., $\partial f / \partial y^{(n-1)}$ be continuous in a region R of $(x, y, y', ..., y^{(n-1)})$ space. If $(x_0, a_1, a_2, ..., a_n)$ is an interior point of R, then equation (2) has a unique solution $y(x)$ that satisfies the initial conditions $y(x_0) = a_1$, $y'(x_0) = a_2$, ..., $y^{(n-1)}(x_0) = a_n$.*

As a further illustration of the value of reducing higher order equations to systems of first order equations, we consider the famous n-body problem of classical mechanics.

Let n particles with masses m_i be located at points (x_i, y_i, z_i) and assume that they attract one another according to Newton's law of gravitation. If r_{ij} is the distance between m_i and m_j, and if θ is the angle from the positive x-axis to the segment joining them (Fig. 63), then the x component of the force exerted on m_i by m_j is

$$\frac{Gm_im_j}{r_{ij}^2} \cos \theta = \frac{Gm_im_j(x_j - x_i)}{r_{ij}^3},$$

where G is the gravitational constant. Since the sum of these components for all $j \neq i$ equals $m_i(d^2x_i/dt^2)$, we have n second order differential equations

$$\frac{d^2x_i}{dt^2} = G \sum_{j \neq i} \frac{m_j(x_j - x_i)}{r_{ij}^3},$$

and similarly

$$\frac{d^2y_i}{dt^2} = G \sum_{j \neq i} \frac{m_j(y_j - y_i)}{r_{ij}^3}$$

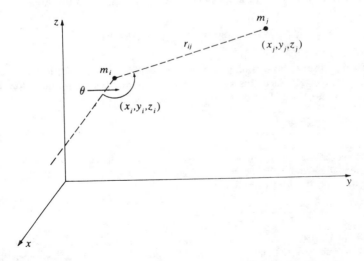

FIGURE 63

and

$$\frac{d^2 z_i}{dt^2} = G \sum_{j \neq i} \frac{m_j(z_j - z_i)}{r_{ij}^3}.$$

If we put $v_{x_i} = dx_i/dt$, $v_{y_i} = dy_i/dt$, and $v_{z_i} = dz_i/dt$, and apply the above reduction, then we obtain a system of $6n$ equations of the form (1) in the unknown functions x_1, v_{x_1}, ..., x_n, v_{x_n}, y_1, v_{y_1}, ..., y_n, v_{y_n}, z_1, v_{z_1}, ..., z_n, v_{z_n}. If we now make use of the fact that

$$r_{ij}^3 = [(x_i - x_j)^2 + (y_i - y_j)^2 + (z_i - z_j)^2]^{3/2},$$

then Theorem A yields the following conclusion: if the initial positions and initial velocities of the particles, i.e., the values of the unknown functions at a certain instant $t = t_0$, are given, and if the particles do not collide in the sense that the r_{ij} do not vanish, then their subsequent positions and velocities are uniquely determined. This conclusion underlies the once popular philosophy of mechanistic determinism, according to which the universe is nothing more than a gigantic machine whose future is inexorably fixed by its state at any given moment.[1]

PROBLEMS

1. Replace each of the following differential equations by an equivalent system of first order equations:
   (a) $y'' - x^2 y' - xy = 0$;
   (b) $y''' = y'' - x^2(y')^2$.

2. If a particle of mass m moves in the xy-plane, its equations of motion are

$$m\frac{d^2 x}{dt^2} = f(t,x,y) \quad \text{and} \quad m\frac{d^2 y}{dt^2} = g(t,x,y),$$

where f and g represent the x and y components, respectively, of the force acting on the particle. Replace this system of two second order equations by an equivalent system of four first order equations of the form (1).

---

[1] It also led Sir James Jeans to define the universe as "a self-solving system of $6N$ simultaneous differential equations, where N is Eddington's number." Sir Arthur Eddington asserted (with more poetry than truth) that

$$N = \frac{3}{2} \times 136 \times 2^{256}$$

is the total number of particles of matter in the universe. See Jeans, *The Astronomical Horizon*, Oxford University Press, London, 1945; or Eddington, *The Expanding Universe*, Cambridge University Press, London, 1952.

55 LINEAR SYSTEMS

For the sake of convenience and clarity, we restrict our attention through the rest of this chapter to systems of only two first order equations in two unknown functions, of the form

$$\begin{cases} \dfrac{dx}{dt} = F(t,x,y) \\[2mm] \dfrac{dy}{dt} = G(t,x,y). \end{cases} \tag{1}$$

The brace notation is used to emphasize the fact that the equations are linked together, and the choice of the letter t for the independent variable and x and y for the dependent variables is customary in this case for reasons that will appear later.

In this and the next section we specialize even further, to *linear systems,* of the form

$$\begin{cases} \dfrac{dx}{dt} = a_1(t)x + b_1(t)y + f_1(t) \\[2mm] \dfrac{dy}{dt} = a_2(t)x + b_2(t)y + f_2(t). \end{cases} \tag{2}$$

We shall assume in the present discussion, and in the theorems stated below, that the functions $a_i(t)$, $b_i(t)$, and $f_i(t)$, $i = 1, 2$, are continuous on a certain closed interval $[a,b]$ of the t-axis. If $f_1(t)$ and $f_2(t)$ are identically zero, then the system (2) is called *homogeneous*; otherwise it is said to be *nonhomogeneous*. A solution of (2) on $[a,b]$ is of course a pair of functions $x(t)$ and $y(t)$ that satisfy both equations of (2) throughout this interval. We shall write such a solution in the form

$$\begin{cases} x = x(t) \\ y = y(t). \end{cases}$$

Thus, it is easy to verify that the homogeneous linear system (with constant coefficients)

$$\begin{cases} \dfrac{dx}{dt} = 4x - y \\[2mm] \dfrac{dy}{dt} = 2x + y \end{cases} \tag{3}$$

has both

$$\begin{cases} x = e^{3t} \\ y = e^{3t} \end{cases} \quad \text{and} \quad \begin{cases} x = e^{2t} \\ y = 2e^{2t} \end{cases} \tag{4}$$

as solutions on any closed interval.

We now give a brief sketch of the general theory of the linear system (2). It will be observed that this theory is very similar to that of the second order linear equation as described in Sections 14 and 15. We begin by stating the following fundamental existence and uniqueness theorem, whose proof is given in Chapter 13.

Theorem A. *If t_0 is any point of the interval $[a,b]$, and if x_0 and y_0 are any numbers whatever, then (2) has one and only one solution*

$$\begin{cases} x = x(t) \\ y = y(t), \end{cases}$$

valid throughout $[a,b]$, such that $x(t_0) = x_0$ and $y(t_0) = y_0$.

Our next step is to study the structure of the solutions of the homogeneous system obtained from (2) by removing the terms $f_1(t)$ and $f_2(t)$:

$$\begin{cases} \dfrac{dx}{dt} = a_1(t)x + b_1(t)y \\[2mm] \dfrac{dy}{dt} = a_2(t)x + b_2(t)y. \end{cases} \tag{5}$$

It is obvious that (5) is satisfied by the so-called *trivial solution*, in which $x(t)$ and $y(t)$ are both identically zero. Our main tool in constructing more useful solutions is the next theorem.

Theorem B. *If the homogeneous system (5) has two solutions*

$$\begin{cases} x = x_1(t) \\ y = y_1(t) \end{cases} \quad and \quad \begin{cases} x = x_2(t) \\ y = y_2(t) \end{cases} \tag{6}$$

on $[a,b]$, then

$$\begin{cases} x = c_1 x_1(t) + c_2 x_2(t) \\ y = c_1 y_1(t) + c_2 y_2(t) \end{cases} \tag{7}$$

is also a solution on $[a,b]$ for any constants c_1 and c_2.

Proof. The proof is a routine verification, and is left to the reader.

The solution (7) is obtained from the pair of solutions (6) by multiplying the first by c_1, the second by c_2, and adding; (7) is therefore called a *linear combination* of the solutions (6). With this terminology, we can restate Theorem B as follows: any linear combination of two solutions of the homogeneous system (5) is also a solution. Accordingly, (3) has

$$\begin{cases} x = c_1 e^{3t} + c_2 e^{2t} \\ y = c_1 e^{3t} + 2c_2 e^{2t} \end{cases} \tag{8}$$

as a solution for every choice of the constants c_1 and c_2.

The next question we must settle is that of whether (7) contains *all* solutions of (5) on $[a,b]$, that is, whether it is the *general solution* of (5) on $[a,b]$. By Theorem A, (7) will be the general solution if the constants c_1 and c_2 can be chosen so as to satisfy arbitrary conditions $x(t_0) = x_0$ and $y(t_0) = y_0$ at an arbitrary point t_0 in $[a,b]$, or equivalently, if the system of linear algebraic equations

$$c_1 x_1(t_0) + c_2 x_2(t_0) = x_0$$
$$c_1 y_1(t_0) + c_2 y_2(t_0) = y_0$$

in the unknowns c_1 and c_2 can be solved for each t_0 in $[a,b]$ and every pair of numbers x_0 and y_0. By the elementary theory of determinants, this is possible whenever the determinant of the coefficients,

$$W(t) = \begin{vmatrix} x_1(t) & x_2(t) \\ y_1(t) & y_2(t) \end{vmatrix},$$

does not vanish on the interval $[a,b]$. This determinant is called the *Wronskian* of the two solutions (6) (see Problem 4), and the above remarks prove the next theorem.

> **Theorem C.** *If the two solutions* (6) *of the homogeneous system* (5) *have a Wronskian* $W(t)$ *that does not vanish on* $[a,b]$, *then* (7) *is the general solution of* (5) *on this interval.*

It follows from this theorem that (8) is the general solution of (3) on any closed interval, for the Wronskian of the two solutions (4) is

$$W(t) = \begin{vmatrix} e^{3t} & e^{2t} \\ e^{3t} & 2e^{2t} \end{vmatrix} = e^{5t},$$

which never vanishes. It is useful to know, as this example suggests, that the vanishing or nonvanishing of the Wronskian $W(t)$ of two solutions does not depend on the choice of t. To state it formally, we have

> **Theorem D.** *If* $W(t)$ *is the Wronskian of the two solutions* (6) *of the homogeneous system* (5), *then* $W(t)$ *is either identically zero or nowhere zero on* $[a,b]$.

Proof. A simple calculation shows that $W(t)$ satisfies the first order differential equation

$$\frac{dW}{dt} = [a_1(t) + b_2(t)]W, \tag{9}$$

from which it follows that

$$W(t) = ce^{\int [a_1(t) + b_2(t)]dt} \tag{10}$$

for some constant c. The conclusion of the theorem is now evident from the fact that the exponential factor in (10) never vanishes on $[a,b]$.

Theorem C provides an adequate means of verifying that (7) is the general solution of (5): show that the Wronskian $W(t)$ of the two solutions (6) does not vanish. We now develop an equivalent test that is often more direct and convenient.

The two solutions (6) are called *linearly dependent* on $[a,b]$ if one is a constant multiple of the other in the sense that

$$x_1(t) = kx_2(t) \qquad x_2(t) = kx_1(t)$$
$$\text{or}$$
$$y_1(t) = ky_2(t) \qquad y_2(t) = ky_1(t)$$

for some constant k and all t in $[a,b]$, and *linearly independent* if neither is a constant multiple of the other. It is clear that linear dependence is equivalent to the condition that there exist two constants c_1 and c_2, at least one of which is not zero, such that

$$c_1 x_1(t) + c_2 x_2(t) = 0$$
$$c_1 y_1(t) + c_2 y_2(t) = 0 \tag{11}$$

for all t in $[a,b]$. We now have the next theorem.

Theorem E. *If the two solutions (6) of the homogeneous system (5) are linearly independent on $[a,b]$, then (7) is the general solution of (5) on this interval.*

Proof. In view of Theorems C and D, it suffices to show that the solutions (6) are linearly dependent if and only if their Wronskian $W(t)$ is identically zero. We begin by assuming that they are linearly dependent, so that, say,

$$x_1(t) = kx_2(t)$$
$$y_1(t) = ky_2(t). \tag{12}$$

Then

$$W(t) = \begin{vmatrix} x_1(t) & x_2(t) \\ y_1(t) & y_2(t) \end{vmatrix} = \begin{vmatrix} kx_2(t) & x_2(t) \\ ky_2(t) & y_2(t) \end{vmatrix}$$
$$= kx_2(t)y_2(t) - kx_2(t)y_2(t) = 0$$

for all t in $[a,b]$. The same argument works equally well if the constant k is on the other side of equations (12). We now assume that $W(t)$ is identically zero, and show that the solutions (6) are linearly dependent in the sense of equations (11). Let t_0 be a fixed point in $[a,b]$. Since $W(t_0) = 0$, the system of linear algebraic equations

$$c_1 x_1(t_0) + c_2 x_2(t_0) = 0$$
$$c_1 y_1(t_0) + c_2 y_2(t_0) = 0$$

has a solution c_1, c_2 in which these numbers are not both zero. Thus, the solution of (5) given by

$$\begin{cases} x = c_1 x_1(t) + c_2 x_2(t) \\ y = c_1 y_1(t) + c_2 y_2(t) \end{cases} \tag{13}$$

equals the trivial solution at t_0. It now follows from the uniqueness part of Theorem A that (13) must equal the trivial solution throughout the interval $[a,b]$, so (11) holds and the proof is complete.

The value of this test is that in specific problems it is usually a simple matter of inspection to decide whether two solutions of (5) are linearly independent or not.

We now return to the nonhomogeneous system (2) and conclude our discussion with

Theorem F. *If the two solutions* (6) *of the homogeneous system* (5) *are linearly independent on* $[a,b]$, *and if*

$$\begin{cases} x = x_p(t) \\ y = y_p(t) \end{cases}$$

is any particular solution of (2) *on this interval, then*

$$\begin{cases} x = c_1 x_1(t) + c_1 x_2(t) + x_p(t) \\ y = c_1 y_1(t) + c_2 y_2(t) + y_p(t) \end{cases} \tag{14}$$

is the general solution of (2) *on* $[a,b]$.

Proof. It suffices to show that if

$$\begin{cases} x = x(t) \\ y = y(t) \end{cases}$$

is an arbitrary solution of (2), then

$$\begin{cases} x = x(t) - x_p(t) \\ y = y(t) - y_p(t) \end{cases}$$

is a solution of (5), and this we leave to the reader.

The above treatment of the linear system (2) shows how its general solution (14) can be built up out of simpler pieces. But how do we find these pieces? Unfortunately—as in the case of second order linear equations—there does not exist any general method that always works. In the next section we discuss an important special case in which this problem can be solved: that in which the coefficients $a_i(t)$ and $b_i(t)$, $i = 1, 2$, are constants.

PROBLEMS

1. Prove Theorem B.
2. Finish the proof of Theorem F.
3. Verify equation (9).

4. Let the second order linear equation

$$\frac{d^2x}{dt^2} + P(t)\frac{dx}{dt} + Q(t)x = 0 \qquad (*)$$

be reduced to the system

$$\begin{cases} \dfrac{dx}{dt} = y \\[2mm] \dfrac{dy}{dt} = -Q(t)x - P(t)y. \end{cases} \qquad (**)$$

If $x_1(t)$ and $x_2(t)$ are solutions of equation (*), and if

$$\begin{cases} x = x_1(t) \\ y = y_1(t) \end{cases} \quad \text{and} \quad \begin{cases} x = x_2(t) \\ y = y_2(t) \end{cases}$$

are the corresponding solutions of (**), show that the Wronskian of the former in the sense of Section 15 is precisely the Wronskian of the latter in the sense of this section.

5. (a) Show that

$$\begin{cases} x = e^{4t} \\ y = e^{4t} \end{cases} \quad \text{and} \quad \begin{cases} x = e^{-2t} \\ y = -e^{-2t} \end{cases}$$

are solutions of the homogeneous system

$$\begin{cases} \dfrac{dx}{dt} = x + 3y \\[2mm] \dfrac{dy}{dt} = 3x + y. \end{cases}$$

(b) Show in two ways that the given solutions of the system in (a) are linearly independent on every closed interval, and write the general solution of this system.

(c) Find the particular solution

$$\begin{cases} x = x(t) \\ y = y(t) \end{cases}$$

of this system for which $x(0) = 5$ and $y(0) = 1$.

6. (a) Show that

$$\begin{cases} x = 2e^{4t} \\ y = 3e^{4t} \end{cases} \quad \text{and} \quad \begin{cases} x = e^{-t} \\ y = -e^{-t} \end{cases}$$

are solutions of the homogeneous system

$$\begin{cases} \dfrac{dx}{dt} = x + 2y \\[2mm] \dfrac{dy}{dt} = 3x + 2y. \end{cases}$$

(b) Show in two ways that the given solutions of the system in (a) are linearly independent on every closed interval, and write the general solution of this system.

(c) Show that

$$\begin{cases} x = 3t - 2 \\ y = -2t + 3 \end{cases}$$

is a particular solution of the nonhomogeneous system

$$\begin{cases} \dfrac{dx}{dt} = x + 2y + t - 1 \\ \dfrac{dy}{dt} = 3x + 2y - 5t - 2, \end{cases}$$

and write the general solution of this system.

7. Obtain the given solutions of the homogeneous system in Problem 6
   (a) by differentiating the first equation with respect to t and eliminating y;
   (b) by differentiating the second equation with respect to t and eliminating x.

8. Use a method suggested by Problem 7 to find the general solution of the system

$$\begin{cases} \dfrac{dx}{dt} = x + y \\ \dfrac{dy}{dt} = y. \end{cases}$$

9. (a) Find the general solution of the system

$$\begin{cases} \dfrac{dx}{dt} = x \\ \dfrac{dy}{dt} = y. \end{cases}$$

(b) Show that any second order equation obtained from the system in (a) is not equivalent to this system, in the sense that it has solutions that are not part of any solution of the system. Thus, although higher order equations are equivalent to systems, the reverse is not true, and systems are more general.

56 HOMOGENEOUS LINEAR SYSTEMS WITH CONSTANT COEFFICIENTS

We are now in a position to give a complete explicit solution of the simple system

$$\begin{cases} \dfrac{dx}{dt} = a_1 x + b_1 y \\ \dfrac{dy}{dt} = a_2 x + b_2 y, \end{cases} \tag{1}$$

where a_1, b_1, a_2, and b_2 are given constants. Some of the problems at the end of the previous section illustrate a procedure that can often be applied to this case: differentiate one equation, eliminate one of the dependent variables, and solve the resulting second order linear equation. The method we now describe is based instead on constructing a pair of linearly independent solutions directly from the given system.

If we recall that the exponential function has the property that its derivatives are constant multiples of the function itself, then (just as in Section 17) it is natural to seek solutions of (1) having the form

$$\begin{cases} x = Ae^{mt} \\ y = Be^{mt}. \end{cases} \tag{2}$$

If we substitute (2) into (1) we get

$$Ame^{mt} = a_1 Ae^{mt} + b_1 Be^{mt}$$
$$Bme^{mt} = a_2 Ae^{mt} + b_2 Be^{mt};$$

and dividing by e^{mt} yields the linear algebraic system

$$\begin{aligned} (a_1 - m)A + b_1 B &= 0 \\ a_2 A + (b_2 - m)B &= 0 \end{aligned} \tag{3}$$

in the unknowns A and B. It is clear that (3) has the trivial solution $A = B = 0$, which makes (2) the trivial solution of (1). Since we are looking for nontrivial solutions of (1), this is no help at all. However, we know that (3) has nontrivial solutions whenever the determinant of the coefficients vanishes, i.e., whenever

$$\begin{vmatrix} a_1 - m & b_1 \\ a_2 & b_2 - m \end{vmatrix} = 0.$$

When this determinant is expanded, we get the quadratic equation

$$m^2 - (a_1 + b_2)m + (a_1 b_2 - a_2 b_1) = 0 \tag{4}$$

for the unknown m. By analogy with our previous work, we call this the *auxiliary equation* of the system (1). Let m_1 and m_2 be the roots of (4). If we replace m in (3) by m_1, then we know that the resulting equations have a nontrivial solution A_1, B_1, so

$$\begin{cases} x = A_1 e^{m_1 t} \\ y = B_1 e^{m_1 t} \end{cases} \tag{5}$$

is a nontrivial solution of the system (1). By proceeding similarly with m_2, we find another nontrivial solution

$$\begin{cases} x = A_2 e^{m_2 t} \\ y = B_2 e^{m_2 t}. \end{cases} \tag{6}$$

In order to make sure that we obtain two linearly independent solutions—and hence the general solution—it is necessary to examine in detail each of the three possibilities for m_1 and m_2.

Distinct real roots. When m_1 and m_2 are distinct real numbers, then (5) and (6) are easily seen to be linearly independent (why?) and

$$\begin{cases} x = c_1 A_1 e^{m_1 t} + c_2 A_2 e^{m_2 t} \\ y = c_1 B_1 e^{m_1 t} + c_2 B_2 e^{m_2 t} \end{cases} \tag{7}$$

is the general solution of (1).

Example 1. In the case of the system

$$\begin{cases} \dfrac{dx}{dt} = x + y \\ \dfrac{dy}{dt} = 4x - 2y, \end{cases} \tag{8}$$

(3) is

$$(1 - m)A + B = 0$$
$$4A + (-2 - m)B = 0. \tag{9}$$

The auxiliary equation here is

$$m^2 + m - 6 = 0 \quad \text{or} \quad (m + 3)(m - 2) = 0,$$

so m_1 and m_2 are -3 and 2. With $m = -3$, (9) becomes

$$4A + B = 0$$
$$4A + B = 0.$$

A simple nontrivial solution of this system is $A = 1$, $B = -4$, so we have

$$\begin{cases} x = e^{-3t} \\ y = -4e^{-3t} \end{cases} \tag{10}$$

as a nontrivial solution of (8). With $m = 2$, (9) becomes

$$-A + B = 0$$
$$4A - 4B = 0,$$

and a simple nontrivial solution is $A = 1$, $B = 1$. This yields

$$\begin{cases} x = e^{2t} \\ y = e^{2t} \end{cases} \tag{11}$$

as another solution of (8); and since it is clear that (10) and (11) are linearly independent,

$$\begin{cases} x = c_1 e^{-3t} + c_2 e^{2t} \\ y = -4c_1 e^{-3t} + c_2 e^{2t} \end{cases} \tag{12}$$

is the general solution of (8).

Distinct complex roots. If m_1 and m_2 are distinct complex numbers, then they can be written in the form $a \pm ib$ where a and b are real numbers and $b \neq 0$. In this case we expect the A's and B's obtained from (3) to be complex numbers, and we have two linearly independent solutions

$$\begin{cases} x = A_1^* e^{(a+ib)t} \\ y = B_1^* e^{(a+ib)t} \end{cases} \quad \text{and} \quad \begin{cases} x = A_2^* e^{(a-ib)t} \\ y = B_2^* e^{(a-ib)t}. \end{cases} \tag{13}$$

However, these are complex-valued solutions, and to extract real-valued solutions we proceed as follows. If we express the numbers A_1^* and B_1^* in the standard form $A_1^* = A_1 + iA_2$ and $B_1^* = B_1 + iB_2$, and use Euler's formula 17-(7), then the first of the solutions (13) can be written as

$$\begin{cases} x = (A_1 + iA_2)e^{at}(\cos bt + i \sin bt) \\ y = (B_1 + iB_2)e^{at}(\cos bt + i \sin bt) \end{cases}$$

or

$$\begin{cases} x = e^{at}[(A_1 \cos bt - A_2 \sin bt) + i(A_1 \sin bt + A_2 \cos bt)] \\ y = e^{at}[(B_1 \cos bt - B_2 \sin bt) + i(B_1 \sin bt + B_2 \cos bt)]. \end{cases} \tag{14}$$

It is easy to see that if a pair of complex-valued functions is a solution of (1), in which the coefficients are *real* constants, then their two real parts and their two imaginary parts are real-valued solutions. It follows from this that (14) yields the two real-valued solutions

$$\begin{cases} x = e^{at}(A_1 \cos bt - A_2 \sin bt) \\ y = e^{at}(B_1 \cos bt - B_2 \sin bt) \end{cases} \tag{15}$$

and

$$\begin{cases} x = e^{at}(A_1 \sin bt + A_2 \cos bt) \\ y = e^{at}(B_1 \sin bt + B_2 \cos bt). \end{cases} \tag{16}$$

It can be shown that these solutions are linearly independent (we ask the reader to prove this in Problem 3), so the general solution in this case is

$$\begin{cases} x = e^{at}[c_1(A_1 \cos bt - A_2 \sin bt) + c_2(A_1 \sin bt + A_2 \cos bt)] \\ y = e^{at}[c_1(B_1 \cos bt - B_2 \sin bt) + c_2(B_1 \sin bt + B_2 \cos bt)]. \end{cases} \tag{17}$$

Since we have already found the general solution, it is not necessary to consider the second of the two solutions (13).

Equal real roots. When m_1 and m_2 have the same value m, then (5) and (6) are not linearly independent and we essentially have only one solution

$$\begin{cases} x = Ae^{mt} \\ y = Be^{mt}. \end{cases} \tag{18}$$

Our experience in Section 17 would lead us to expect a second linearly independent solution of the form

$$\begin{cases} x = Ate^{mt} \\ y = Bte^{mt}. \end{cases}$$

Unfortunately the matter is not quite as simple as this, and we must actually look for a second solution of the form

$$\begin{cases} x = (A_1 + A_2 t)e^{mt} \\ y = (B_1 + B_2 t)e^{mt}, \end{cases} \tag{19}$$

so that the general solution is

$$\begin{cases} x = c_1 A e^{mt} + c_2(A_1 + A_2 t)e^{mt} \\ y = c_1 B e^{mt} + c_2(B_1 + B_2 t)e^{mt}.^2 \end{cases} \tag{20}$$

The constants A_1, A_2, B_1, and B_2 are found by substituting (19) into the system (1). Instead of trying to carry this through in the general case, we illustrate the method by showing how it works in a simple example.

Example 2. In the case of the system

$$\begin{cases} \dfrac{dx}{dt} = 3x - 4y \\ \dfrac{dy}{dt} = x - y, \end{cases} \tag{21}$$

(3) is

$$(3 - m)A - 4B = 0 \tag{22}$$

$$A + (-1 - m)B = 0.$$

The auxiliary equation is

$$m^2 - 2m + 1 = 0 \quad \text{or} \quad (m - 1)^2 = 0,$$

---

[2] The only exception to this statement occurs when $a_1 = b_2 = a$ and $a_2 = b_1 = 0$, so that the auxiliary equation is $m^2 - 2am + a^2 = 0$, $m = a$, and the constants A and B in (18) are completely unrestricted. In this case the general solution of (1) is obviously

$$\begin{cases} x = c_1 e^{mt} \\ y = c_2 e^{mt}, \end{cases}$$

and the system is said to be *uncoupled* (since each equation can be solved independently of the other).

which has equal real roots 1 and 1. With $m = 1$, (22) becomes

$$2A - 4B = 0$$

$$A - 2B = 0.$$

A simple nontrivial solution of this system is $A = 2$, $B = 1$, so

$$\begin{cases} x = 2e^t \\ y = e^t \end{cases} \tag{23}$$

is a nontrivial solution of (21). We now seek a second linearly independent solution of the form

$$\begin{cases} x = (A_1 + A_2 t)e^t \\ y = (B_1 + B_2 t)e^t. \end{cases} \tag{24}$$

When this is substituted into (21), we obtain

$$(A_1 + A_2 t + A_2)e^t = 3(A_1 + A_2 t)e^t - 4(B_1 + B_2 t)e^t$$

$$(B_1 + B_2 t + B_2)e^t = (A_1 + A_2 t)e^t - (B_1 + B_2 t)e^t,$$

which reduces at once to

$$(2A_2 - 4B_2)t + (2A_1 - A_2 - 4B_1) = 0$$

$$(A_2 - 2B_2)t + (A_1 - 2B_1 - B_2) = 0.$$

Since these are to be identities in the variable t, we must have

$$2A_2 - 4B_2 = 0 \qquad 2A_1 - A_2 - 4B_1 = 0$$

$$A_2 - 2B_2 = 0, \qquad A_1 - 2B_1 - B_2 = 0.$$

The two equations on the left have $A_2 = 2$, $B_2 = 1$ as a simple nontrivial solution. With this, the two equations on the right become

$$2A_1 - 4B_1 = 2$$

$$A_1 - 2B_1 = 1,$$

so we may take $A_1 = 1$, $B_1 = 0$. We now insert these numbers into (24) and obtain

$$\begin{cases} x = (1 + 2t)e^t \\ y = te^t \end{cases} \tag{25}$$

as our second solution. It is obvious that (23) and (25) are linearly independent, so

$$\begin{cases} x = 2c_1 e^t + c_2(1 + 2t)e^t \\ y = c_1 e^t + c_2 te^t \end{cases} \tag{26}$$

is the general solution of the system (21).

PROBLEMS

1. Use the methods described in this section to find the general solution of each of the following systems:

(a) $\begin{cases} \dfrac{dx}{dt} = -3x + 4y \\[2mm] \dfrac{dy}{dt} = -2x + 3y; \end{cases}$

(e) $\begin{cases} \dfrac{dx}{dt} = 2x \\[2mm] \dfrac{dy}{dt} = 3y; \end{cases}$

(b) $\begin{cases} \dfrac{dx}{dt} = 4x - 2y \\[2mm] \dfrac{dy}{dt} = 5x + 2y; \end{cases}$

(f) $\begin{cases} \dfrac{dx}{dt} = -4x - y \\[2mm] \dfrac{dy}{dt} = x - 2y; \end{cases}$

(c) $\begin{cases} \dfrac{dx}{dt} = 5x + 4y \\[2mm] \dfrac{dy}{dt} = -x + y; \end{cases}$

(g) $\begin{cases} \dfrac{dx}{dt} = 7x + 6y \\[2mm] \dfrac{dy}{dt} = 2x + 6y; \end{cases}$

(d) $\begin{cases} \dfrac{dx}{dt} = 4x - 3y \\[2mm] \dfrac{dy}{dt} = 8x - 6y; \end{cases}$

(h) $\begin{cases} \dfrac{dx}{dt} = x - 2y \\[2mm] \dfrac{dy}{dt} = 4x + 5y. \end{cases}$

2. Show that the condition $a_2 b_1 > 0$ is sufficient, but not necessary, for the system (1) to have two real-valued linearly independent solutions of the form (2).

3. Show that the Wronskian of the two solutions (15) and (16) is given by

$$W(t) = (A_1 B_2 - A_2 B_1)e^{2at},$$

and prove that $A_1 B_2 - A_2 B_1 \neq 0$.

4. Show that in formula (20) the constants A_2 and B_2 satisfy the same linear algebraic system as the constants A and B, and that consequently we may put $A_2 = A$ and $B_2 = B$ without any loss of generality.

5. Consider the nonhomogeneous linear system

$$\begin{cases} \dfrac{dx}{dt} = a_1(t)x + b_1(t)y + f_1(t) \\[2mm] \dfrac{dy}{dt} = a_2(t)x + b_2(t)y + f_2(t) \end{cases} \qquad (*)$$

and the corresponding homogeneous system

$$\begin{cases} \dfrac{dx}{dt} = a_1(t)x + b_1(t)y \\[2mm] \dfrac{dy}{dt} = a_2(t)x + b_2(t)y. \end{cases} \qquad (**)$$

(a) If

$$\begin{cases} x = x_1(t) \\ y = y_1(t) \end{cases} \quad \text{and} \quad \begin{cases} x = x_2(t) \\ y = y_2(t) \end{cases}$$

are linearly independent solutions of (**), so that

$$\begin{cases} x = c_1 x_1(t) + c_2 x_2(t) \\ y = c_1 y_1(t) + c_2 y_2(t) \end{cases}$$

is its general solution, show that

$$\begin{cases} x = v_1(t)x_1(t) + v_2(t)x_2(t) \\ y = v_1(t)y_1(t) + v_2(t)y_2(t) \end{cases}$$

will be a particular solution of (*) if the functions $v_1(t)$ and $v_2(t)$ satisfy the system

$$v_1' x_1 + v_2' x_2 = f_1$$

$$v_1' y_1 + v_2' y_2 = f_2.$$

This technique for finding particular solutions of nonhomogeneous linear systems is called the *method of variation of parameters*.

(b) Apply the method outlined in (a) to find a particular solution of the nonhomogeneous system

$$\begin{cases} \dfrac{dx}{dt} = x + y - 5t + 2 \\ \dfrac{dy}{dt} = 4x - 2y - 8t - 8, \end{cases}$$

whose corresponding homogeneous system is solved in Example 1.

57 NONLINEAR SYSTEMS. VOLTERRA'S PREY–PREDATOR EQUATIONS

Everyone knows that there is a constant struggle for survival among different species of animals living in the same environment. One kind of animal survives by eating another; a second, by developing methods of evasion to avoid being eaten; and so on.

As a simple example of this universal conflict between the predator and its prey, let us imagine an island inhabited by foxes and rabbits. The foxes eat rabbits, and the rabbits eat clover. We assume that there is so much clover that the rabbits always have an ample supply of food. When the rabbits are abundant, then the foxes flourish and their population grows. When the foxes become too numerous and eat too many rabbits, they enter a period of famine and their population begins to decline. As the foxes decrease, the rabbits become relatively safe and their population starts to increase again. This triggers a new increase in the fox population, and as time goes on we see an endlessly repeated cycle of interrelated increases and decreases in the populations of the two species. These fluctuations are represented graphically in Fig. 64, where the sizes of the populations are plotted against time.

Problems of this kind have been studied by both mathematicians and biologists, and it is quite interesting to see how the mathematical conclusions we shall develop confirm and extend the intuitive ideas arrived at in the preceding paragraph. In discussing the interaction between the foxes and the rabbits, we shall follow the approach of Volterra, who initiated the quantitative treatment of such problems.[3]

If x is the number of rabbits at time t, then we should have

$$\frac{dx}{dt} = ax, \qquad a > 0,$$

as a consequence of the unlimited supply of clover, if the number y of foxes is zero. It is natural to assume that the number of encounters per

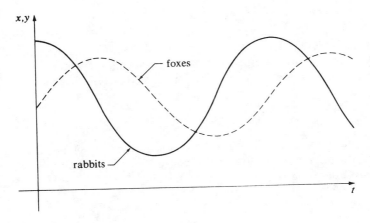

FIGURE 64

[3] Vito Volterra (1860–1940) was an eminent Italian mathematician. His early work on integral equations (together with that of Fredholm and Hilbert) began the full-scale development of linear analysis that dominated so much of mathematics during the first half of the twentieth century. His vigorous excursions in later life into mathematical biology enriched both mathematics and biology. For further details, see his *Lecons sur la théorie mathématique de la lutte pour la vie,* Gauthier-Villars, Paris, 1931; or A. J. Lotka, *Elements of Mathematical Biology,* pp. 88–94, Dover, New York, 1956. A modern discussion, with the Hudson's Bay Company data on the numbers of lynx and hares in Canada from 1847 to 1903, can be found in E. R. Leigh, "The Ecological Role of Volterra's Equations," in *Some Mathematical Problems in Biology,* American Mathematical Society, Providence, R.I., 1968.

unit time between rabbits and foxes is jointly proportional to x and y. If we further assume that a certain proportion of these encounters result in a rabbit being eaten, then we have

$$\frac{dx}{dt} = ax - bxy, \qquad a \text{ and } b > 0.$$

In the same way

$$\frac{dy}{dt} = -cy + dxy, \qquad c \text{ and } d > 0;$$

for in the absence of rabbits the foxes die out, and their increase depends on the number of their encounters with rabbits. We therefore have the following nonlinear system describing the interaction of these two species:

$$\begin{cases} \dfrac{dx}{dt} = x(a - by) \\ \dfrac{dy}{dt} = -y(c - dx). \end{cases} \tag{1}$$

Equations (1) are called *Volterra's prey-predator equations*. Unfortunately this system cannot be solved in terms of elementary functions. On the other hand, if we think of its unknown solution

$$\begin{cases} x = x(t) \\ y = y(t) \end{cases}$$

as constituting the parametric equations of a curve in the xy-plane, then we can find the rectangular equation of this curve. On eliminating t in (1) by division, and separating the variables, we obtain

$$\frac{(a - by)\,dy}{y} = -\frac{(c - dx)\,dx}{x}.$$

Integration now yields

$$a \log y - by = -c \log x + dx + \log K$$

or

$$y^a e^{-by} = K x^{-c} e^{dx}, \tag{2}$$

where the constant K is given by

$$K = x_0^c y_0^a e^{-dx_0 - by_0}$$

in terms of the initial values of x and y.

Although we cannot solve (2) for either x or y, we can determine points on the curve by an ingenious method due to Volterra. To do this, we equate the left and right sides of (2) to new variables z and w, and then plot the graphs C_1 and C_2 of the functions

$$z = y^a e^{-by} \quad \text{and} \quad w = Kx^{-c}e^{dx} \tag{3}$$

as shown in Fig. 65. Since $z = w$, we are confined in the third quadrant to the dotted line L. To the maximum value of z given by the point A on C_1, there corresponds one y and—via M on L and the corresponding points A' and A'' on C_2—two x's, and these determine the bounds between which x may vary. Similarly, the minimum value of w given by B on C_2 leads to N on L and hence to B' and B'' on C_1, and these points determine the bounds for y. In this way we find the points P_1, P_2 and Q_1, Q_2 on the desired curve C_3. Additional points are easily found by starting on L at a point R anywhere between M and N and projecting up to C_1 and over to C_3, and then over to C_2 and up to C_3, as indicated in Fig. 65. It is clear that changing the value of K raises or lowers the point B, and this expands or contracts the curve C_3. Accordingly, when K is given various values, we obtain a family of ovals about the point S, which is all there is of C_3 when the minimum value of w equals the maximum value of z.

We next show that as t increases, the corresponding point (x,y) on C_3 moves around the curve in a counterclockwise direction. To see this,

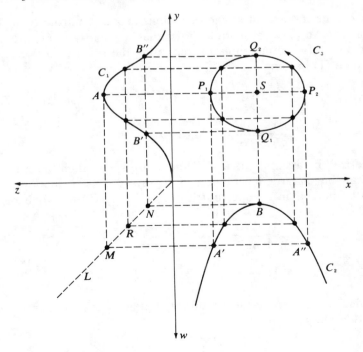

FIGURE 65

we begin by noting that equations (1) give the horizontal and vertical components of the velocity of this point. A simple calculation based on formulas (3) shows that the point S has coordinates $x = c/d$, $y = a/b$. When $x < c/d$, it follows from the second equation of (1) that dy/dt is negative, so our point on C_3 moves down as it traverses the arc $Q_2P_1Q_1$. Similarly, it moves up along the arc $Q_1P_2Q_2$, so the assertion is proved.

Finally, we use the fox–rabbit problem to illustrate the important *method of linearization*. First, we observe that if the rabbit and fox populations are

$$x = \frac{c}{d} \quad \text{and} \quad y = \frac{a}{b}, \tag{4}$$

then the system (1) is satisfied and we have $dx/dt = 0$ and $dy/dt = 0$, so there are no increases or decreases in x or y. The populations (4) are called *equilibrium populations*, for x and y can maintain themselves indefinitely at these constant levels. It is obvious that this is the special case in which the minimum of w equals the maximum of z, so that the oval C_3 reduces to the point S. If we now return to the general case and put

$$x = \frac{c}{d} + X \quad \text{and} \quad y = \frac{a}{b} + Y,$$

then X and Y can be thought of as the deviations of x and y from their equilibrium values. An easy calculation shows that if x and y in (1) are replaced by X and Y [which amounts to translating the point $(c/d, a/b)$ to the origin] then (1) becomes

$$\begin{cases} \dfrac{dX}{dt} = -\dfrac{bc}{d} Y - bXY \\ \dfrac{dY}{dt} = \dfrac{ad}{b} X + dXY. \end{cases} \tag{5}$$

We now "linearize" by assuming that if X and Y are small, then the XY terms in (5) can be discarded without serious error. This assumption amounts to little more than a hope, but it does simplify (5) to a linear system

$$\begin{cases} \dfrac{dX}{dt} = -\dfrac{bc}{d} Y \\ \dfrac{dY}{dt} = \dfrac{ad}{b} X. \end{cases} \tag{6}$$

It is easy to find the general solution of (6), but it is even easier to eliminate t by division and obtain

$$\frac{dY}{dX} = -\frac{ad^2}{b^2c} \frac{X}{Y},$$

whose solution is immediately seen to be

$$ad^2X^2 + b^2cY^2 = C^2.$$

This is a family of ellipses surrounding the origin in the XY-plane. Since ellipses are qualitatively similar to the ovals of Fig. 65, we have reasonable grounds for hoping that (6) is an acceptable approximation to (5).

We trust that the reader agrees that the fox–rabbit problem is interesting for its own sake. Beyond this, however, we have come to appreciate the fact that nonlinear systems present us with problems of a different nature from those we have considered before. In studying a system like (1), we have learned to direct our attention to the behavior of solutions near points in the xy-plane at which the right sides both vanish; we have seen why periodic solutions (i.e., those that yield simple closed curves like C_3 in Fig. 65) are important and desirable; and we have a hint of a method for studying nonlinear systems by means of linear systems that approximate them. In the next chapter we shall study nonlinear systems more fully, and each of these themes will be worked out in greater detail and generality.

PROBLEMS

1. Eliminate y from the system (1) and obtain the nonlinear second order equation satisfied by the function $x(t)$.
2. Show that $d^2y/dt^2 > 0$ whenever $dx/dt > 0$. What is the meaning of this result in terms of Fig. 64?

CHAPTER
11

NONLINEAR
EQUATIONS

58 AUTONOMOUS SYSTEMS. THE
PHASE PLANE AND ITS PHENOMENA

There have been two major trends in the historical development of differential equations. The first and oldest is characterized by attempts to find explicit solutions, either in closed form—which is rarely possible—or in terms of power series. In the second, one abandons all hope of solving equations in any traditional sense, and instead concentrates on a search for qualitative information about the general behavior of solutions. We applied this point of view to linear equations in Chapter 4. The qualitative theory of nonlinear equations is totally different. It was founded by Poincaré around 1880, in connection with his work in celestial mechanics, and since that time has been the object of steadily increasing interest on the part of both pure and applied mathematicians.[1]

The theory of linear differential equations has been studied deeply and extensively for the past 200 years, and is a fairly complete and well-rounded body of knowledge. However, very little of a general

---

[1] See Appendix A for a general account of Poincaré's work in mathematics and science.

nature is known about nonlinear equations. Our purpose in this chapter is to survey some of the central ideas and methods of this subject, and also to demonstrate that it presents a wide variety of interesting and distinctive new phenomena that do not appear in the linear theory. The reader will be surprised to find that most of these phenomena can be treated quite easily without the aid of sophisticated mathematical machinery, and in fact require little more than elementary differential equations and two-dimensional vector algebra.

Why should one be interested in nonlinear differential equations? The basic reason is that many physical systems—and the equations that describe them—are simply nonlinear from the outset. The usual linearizations are approximating devices that are partly confessions of defeat in the face of the original nonlinear problems and partly expressions of the practical view that half a loaf is better than none. It should be added at once that there are many physical situations in which a linear approximation is valuable and adequate for most purposes. This does not alter the fact that in many other situations linearization is unjustified.[2]

It is quite easy to give simple examples of problems that are essentially nonlinear. For instance, if x is the angle of deviation of an undamped pendulum of length a whose bob has mass m, then we saw in Section 5 that its equation of motion is

$$\frac{d^2x}{dt^2} + \frac{g}{a}\sin x = 0;$$ (1)

and if there is present a damping force proportional to the velocity of the bob, then the equation becomes

$$\frac{d^2x}{dt^2} + \frac{c}{m}\frac{dx}{dt} + \frac{g}{a}\sin x = 0.$$ (2)

In the usual linearization we replace $\sin x$ by x, which is reasonable for small oscillations but amounts to a gross distortion when x is large. An example of a different type can be found in the theory of the vacuum tube, which leads to the important *van der Pol equation*

$$\frac{d^2x}{dt^2} + \mu(x^2 - 1)\frac{dx}{dt} + x = 0.$$ (3)

---

[2] It has even been suggested by Einstein that since the basic equations of physics are nonlinear, all of mathematical physics will have to be done over again. If his crystal ball was clear on the day he said this, the mathematics of the future will certainly be very different from that of the past and present.

It will be seen later that each of these nonlinear equations has interesting properties not shared by the others.

Throughout this chapter we shall be concerned with second order nonlinear equations of the form

$$\frac{d^2x}{dt^2} = f\left(x, \frac{dx}{dt}\right), \tag{4}$$

which includes equations (1), (2), and (3) as special cases. If we imagine a simple dynamical system consisting of a particle of unit mass moving on the x-axis, and if $f(x, dx/dt)$ is the force acting on it, then (4) is the equation of motion. The values of x (position) and dx/dt (velocity), which at each instant characterize the state of the system, are called its *phases*, and the plane of the variables x and dx/dt is called the *phase plane*. If we introduce the variable $y = dx/dt$, then (4) can be replaced by the equivalent system

$$\begin{cases} \dfrac{dx}{dt} = y \\ \dfrac{dy}{dt} = f(x,y). \end{cases} \tag{5}$$

We shall see that a good deal can be learned about the solutions of (4) by studying the solutions of (5). When t is regarded as a parameter, then in general a solution of (5) is a pair of functions $x(t)$ and $y(t)$ defining a curve in the xy-plane, which is simply the phase plane mentioned above. We shall be interested in the total picture formed by these curves in the phase plane.

More generally, we study systems of the form

$$\begin{cases} \dfrac{dx}{dt} = F(x,y) \\ \dfrac{dy}{dt} = G(x,y), \end{cases} \tag{6}$$

where F and G are continuous and have continuous first partial derivatives throughout the plane. A system of this kind, in which the independent variable t does not appear in the functions F and G on the right, is said to be *autonomous*. We now turn to a closer examination of the solutions of such a system.

It follows from our assumptions and Theorem 54-A that if t_0 is any number and (x_0, y_0) is any point in the phase plane, then there exists a unique solution

$$\begin{cases} x = x(t) \\ y = y(t) \end{cases} \tag{7}$$

of (6) such that $x(t_0) = x_0$ and $y(t_0) = y_0$. If $x(t)$ and $y(t)$ are not both constant functions, then (7) defines a curve in the phase plane called a *path* of the system.[3] It is clear that if (7) is a solution of (6), then

$$\begin{cases} x = x(t + c) \\ y = y(t + c) \end{cases} \tag{8}$$

is also a solution for any constant c. Thus each path is represented by many solutions, which differ from one another only by a translation of the parameter. Also, it is quite easy to prove (see Problem 2) that any path through the point (x_0, y_0) must correspond to a solution of the form (8). It follows from this that at most one path passes through each point of the phase plane. Furthermore, the direction of increasing t along a given path is the same for all solutions representing the path. A path is therefore a *directed curve,* and in our figures we shall use arrows to indicate the direction in which the path is traced out as t increases.

The above remarks show that in general the paths of (6) cover the entire phase plane and do not intersect one another. The only exceptions to this statement occur at points (x_0, y_0) where both F and G vanish:

$$F(x_0, y_0) = 0 \quad \text{and} \quad G(x_0, y_0) = 0.$$

These points are called *critical points,* and at such a point the unique solution guaranteed by Theorem 54-A is the constant solution $x = x_0$ and $y = y_0$. A constant solution does not define a path, and therefore no path goes through a critical point. In our work we will always assume that each critical point (x_0, y_0) is *isolated,* in the sense that there exists a circle centered on (x_0, y_0) that contains no other critical point.

In order to obtain a physical interpretation of critical points, let us consider the special autonomous system (5) arising from the dynamical equation (4). In this case a critical point is a point $(x_0, 0)$ at which $y = 0$ and $f(x_0, 0) = 0$; that is, it corresponds to a state of the particle's motion in which both the velocity dx/dt and the acceleration $dy/dt = d^2x/dt^2$ vanish. This means that the particle is at rest with no force acting on it, and is therefore in a state of equilibrium.[4] It is obvious that the states of equilibrium of a physical system are among its most important features, and this accounts in part for our interest in critical points.

The general autonomous system (6) does not necessarily arise from any dynamical equation of the form (4). What sort of physical meaning can be attached to the paths and critical points in this case? Here it is convenient to consider Fig. 66 and the two-dimensional vector field

---

[3] The terms *trajectory* and *characteristic* are used by some writers.

[4] For this reason, some writers use the term *equilibrium point* instead of critical point.

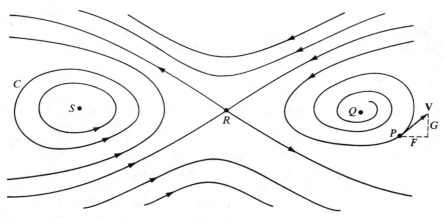

FIGURE 66

defined by

$$\mathbf{V}(x,y) = F(x,y)\mathbf{i} + G(x,y)\mathbf{j},$$

which at a typical point $P = (x,y)$ has horizontal component $F(x,y)$ and vertical component $G(x,y)$. Since $dx/dt = F$ and $dy/dt = G$, this vector is tangent to the path at P and points in the direction of increasing t. If we think of t as time, then $\mathbf{V}$ can be interpreted as the velocity vector of a particle moving along the path. We can also imagine that the entire phase plane is filled with particles, and that each path is the trail of a moving particle preceded and followed by many others on the same path and accompanied by yet others on nearby paths. This situation can be described as a two-dimensional *fluid motion*; and since the system (6) is autonomous, which means that the vector $\mathbf{V}(x,y)$ at a fixed point (x,y) does not change with time, the fluid motion is *stationary*. The paths are the trajectories of the moving particles, and the critical points Q, R, and S are points of zero velocity where the particles are at rest (i.e., stagnation points of the fluid motion).

The most striking features of the fluid motion illustrated in Fig. 66 are:

(a) the critical points;
(b) the arrangement of the paths near critical points;
(c) the stability or instability of critical points, that is, whether a particle near such a point remains near or wanders off into another part of the plane;
(d) closed paths (like C in the figure), which correspond to periodic solutions.

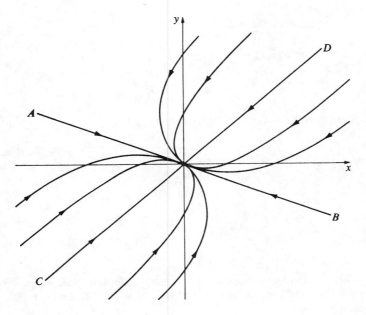

FIGURE 67

These features constitute a major part of the *phase portrait* (or overall picture of the paths) of the system (6). Since in general nonlinear equations and systems cannot be solved explicitly, the purpose of the qualitative theory discussed in this chapter is to discover as much as possible about the phase portrait directly from the functions F and G. To gain some insight into the sort of information we might hope to obtain, observe that if $x(t)$ is a periodic solution of the dynamical equation (4), then its derivative $y(t) = dx/dt$ is also periodic and the corresponding path of the system (5) is therefore closed. Conversely, if any path of (5) is closed, then (4) has a periodic solution. As a concrete example of the application of this idea, we point out that the van der Pol equation—which cannot be solved—can nevertheless be shown to have a unique periodic solution (if $\mu > 0$) by showing that its equivalent autonomous system has a unique closed path.

PROBLEMS

1. Derive equation (2) by applying Newton's second law of motion to the bob of the pendulum.
2. Let (x_0, y_0) be a point in the phase plane. If $x_1(t)$, $y_1(t)$ and $x_2(t)$, $y_2(t)$ are solutions of (6) such that $x_1(t_1) = x_0$, $y_1(t_1) = y_0$ and $x_2(t_2) = x_0$, $y_2(t_2) = y_0$ for suitable t_1 and t_2, show that there exists a constant c such that

$$x_1(t + c) = x_2(t) \quad \text{and} \quad y_1(t + c) = y_2(t).$$

3. Describe the relation between the phase portraits of the systems

$$\begin{cases} \dfrac{dx}{dt} = F(x,y) \\[2mm] \dfrac{dy}{dt} = G(x,y) \end{cases} \quad \text{and} \quad \begin{cases} \dfrac{dx}{dt} = -F(x,y) \\[2mm] \dfrac{dy}{dt} = -G(x,y). \end{cases}$$

4. Describe the phase portrait of each of the following systems:

(a) $\begin{cases} \dfrac{dx}{dt} = 0 \\[2mm] \dfrac{dy}{dt} = 0; \end{cases}$
$\qquad$
(c) $\begin{cases} \dfrac{dx}{dt} = 1 \\[2mm] \dfrac{dy}{dt} = 2; \end{cases}$

(b) $\begin{cases} \dfrac{dx}{dt} = x \\[2mm] \dfrac{dy}{dt} = 0; \end{cases}$
$\qquad$
(d) $\begin{cases} \dfrac{dx}{dt} = -x \\[2mm] \dfrac{dy}{dt} = -y. \end{cases}$

5. The critical points and paths of equation (4) are by definition those of the equivalent system (5). Find the critical points of equations (1), (2), and (3).

6. Find the critical points of

(a) $\dfrac{d^2x}{dt^2} + \dfrac{dx}{dt} - (x^3 + x^2 - 2x) = 0;$

(b) $\begin{cases} \dfrac{dx}{dt} = y^2 - 5x + 6 \\[2mm] \dfrac{dy}{dt} = x - y. \end{cases}$

7. Find all solutions of the nonautonomous system

$$\begin{cases} \dfrac{dx}{dt} = x \\[2mm] \dfrac{dy}{dt} = x + e^t, \end{cases}$$

and sketch (in the xy-plane) some of the curves defined by these solutions.

59 TYPES OF CRITICAL POINTS. STABILITY

Consider an autonomous system

$$\begin{cases} \dfrac{dx}{dt} = F(x,y) \\[2mm] \dfrac{dy}{dt} = G(x,y). \end{cases} \tag{1}$$

We assume, as usual, that the functions F and G are continuous and have continuous first partial derivatives throughout the xy-plane. The critical points of (1) can be found, at least in principle, by solving the simultaneous equations $F(x,y) = 0$ and $G(x,y) = 0$. There are four simple types of critical points that occur quite frequently, and our purpose in this section is to describe them in terms of the configurations of nearby paths. First, however, we need two definitions.

Let (x_0,y_0) be an isolated critical point of (1). If $C = [x(t),y(t)]$ is a path of (1), then we say that C *approaches* (x_0,y_0) as $t \to \infty$ if

$$\lim_{t\to\infty} x(t) = x_0 \quad \text{and} \quad \lim_{t\to\infty} y(t) = y_0.^5 \tag{2}$$

Geometrically, this means that if $P = (x,y)$ is a point that traces out C in accordance with the equations $x = x(t)$ and $y = y(t)$, then $P \to (x_0,y_0)$ as $t \to \infty$. If it is also true that

$$\lim_{t\to\infty} \frac{y(t) - y_0}{x(t) - x_0} \tag{3}$$

exists, or if the quotient in (3) becomes either positively or negatively infinite as $t \to \infty$, then we say that C *enters* the critical point (x_0,y_0) as $t \to \infty$. The quotient in (3) is the slope of the line joining (x_0,y_0) and the point P with coordinates $x(t)$ and $y(t)$, so the additional requirement means that this line approaches a definite direction as $t \to \infty$. In the above definitions, we may also consider limits as $t \to -\infty$. It is clear that these properties are properties of the path C, and do not depend on which solution is used to represent this path.

It is sometimes possible to find explicit solutions of the system (1), and these solutions can then be used to determine the paths. In most cases, however, to find the paths it is necessary to eliminate t between the two equations of the system, which yields

$$\frac{dy}{dx} = \frac{G(x,y)}{F(x,y)}. \tag{4}$$

This first order equation gives the slope of the tangent to the path of (1) that passes through the point (x,y), provided that the functions F and G are not both zero at this point. In this case, of course, the point is a critical point and no path passes through it. The paths of (1) therefore coincide with the one-parameter family of integral curves of (4), and this

---

[5] It can be proved that if (2) is true for some solution $x(t)$, $y(t)$, then (x_0, y_0) is necessarily a critical point. See F. G. Tricomi, *Differential Equations*, p. 47, Blackie, Glasgow, 1961.

family can often be obtained by the methods of Chapter 2. It should be noted, however, that while the paths of (1) are directed curves, the integral curves of (4) have no direction associated with them. Each of these techniques for determining the paths will be illustrated in the examples below.

We now give geometric descriptions of the four main types of critical points. In each case we assume that the critical point under discussion is the origin $O = (0,0)$.

Nodes. A critical point like that in Fig. 67 is called a *node*. Such a point is approached and also entered by each path as $t \to \infty$ (or as $t \to -\infty$). For the node shown in Fig. 67, there are four half-line paths, AO, BO, CO, and DO, which together with the origin make up the lines AB and CD. All other paths resemble parts of parabolas, and as each of these paths approaches O its slope approaches that of the line AB.

Example 1. Consider the system

$$\begin{cases} \dfrac{dx}{dt} = x \\[2mm] \dfrac{dy}{dt} = -x + 2y. \end{cases} \tag{5}$$

It is clear that the origin is the only critical point, and the general solution can be found quite easily by the methods of Section 56:

$$\begin{cases} x = c_1 e^t \\ y = c_1 e^t + c_2 e^{2t}. \end{cases} \tag{6}$$

When $c_1 = 0$, we have $x = 0$ and $y = c_2 e^{2t}$. In this case the path (Fig. 68) is the positive y-axis when $c_2 > 0$, and the negative y-axis when $c_2 < 0$, and each path approaches and enters the origin as $t \to -\infty$. When $c_2 = 0$, we have $x = c_1 e^t$ and $y = c_1 e^t$. This path is the half-line $y = x$, $x > 0$, when $c_1 > 0$, and the half-line $y = x$, $x < 0$, when $c_1 < 0$, and again both paths approach and enter the origin as $t \to -\infty$. When both c_1 and c_2 are $\neq 0$, the paths lie on the parabolas $y = x + (c_2/c_1^2)x^2$, which go through the origin with slope 1. It should be understood that each of these paths consists of only part of a parabola, the part with $x > 0$ if $c_1 > 0$, and the part with $x < 0$ if $c_1 < 0$. Each of these paths also approaches and enters the origin as $t \to -\infty$; this can be seen at once from (6). If we proceed directly from (5) to the differential equation

$$\frac{dy}{dx} = \frac{-x + 2y}{x}, \tag{7}$$

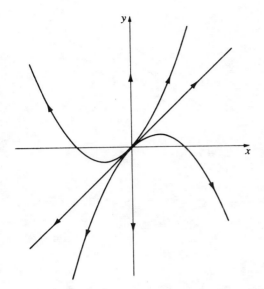

FIGURE 68

giving the slope of the tangent to the path through (x,y) [provided $(x,y) \neq (0,0)$], then on solving (7) as a homogeneous equation, we find that $y = x + cx^2$. This procedure yields the curves on which the paths lie (except those on the y axis), but gives no information about the manner in which the paths are traced out. It is clear from this discussion that the critical point $(0,0)$ of the system (5) is a node.

Saddle points. A critical point like that in Fig. 69 is called a *saddle point*. It is approached and entered by two half-line paths AO and BO as $t \to \infty$, and these two paths lie on a line AB. It is also approached and entered by two half-line paths CO and DO at $t \to -\infty$, and these two paths lie on another line CD. Between the four half-line paths there are four regions, and each contains a family of paths resembling hyperbolas. These paths do not approach O as $t \to \infty$ or as $t \to -\infty$, but instead are asymptotic to one or another of the half-line paths as $t \to \infty$ and as $t \to -\infty$.

Centers. A *center* (sometimes called a *vortex*) is a critical point that is surrounded by a family of closed paths. It is not approached by any path as $t \to \infty$ or as $t \to -\infty$.

Example 2. The system

$$\begin{cases} \dfrac{dx}{dt} = -y \\ \dfrac{dy}{dt} = x \end{cases} \tag{8}$$

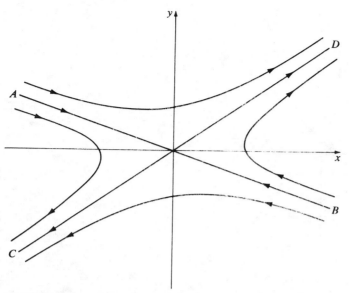

FIGURE 69

has the origin as its only critical point, and its general solution is

$$\begin{cases} x = -c_1 \sin t + c_2 \cos t \\ y = c_1 \cos t + c_2 \sin t. \end{cases} \tag{9}$$

The solution satisfying the conditions $x(0) = 1$ and $y(0) = 0$ is clearly

$$\begin{cases} x = \cos t \\ y = \sin t; \end{cases} \tag{10}$$

and the solution determined by $x(0) = 0$ and $y(0) = -1$ is

$$\begin{cases} x = \sin t = \cos\left(t - \dfrac{\pi}{2}\right) \\ y = -\cos t = \sin\left(t - \dfrac{\pi}{2}\right). \end{cases} \tag{11}$$

These two different solutions define the same path C (Fig. 70), which is evidently the circle $x^2 + y^2 = 1$. Both (10) and (11) show that this path is traced out in the counterclockwise direction. If we eliminate t between the equations of the system, we get

$$\frac{dy}{dx} = -\frac{x}{y},$$

whose general solution $x^2 + y^2 = c^2$ yields all the paths (but without their directions). It is obvious that the critical point $(0,0)$ of the system (8) is a center.

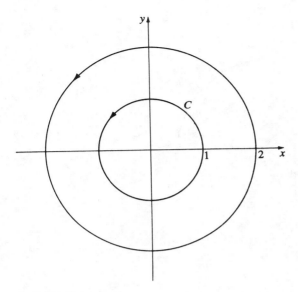

FIGURE 70

Spirals. A critical point like that in Fig. 71 is called a *spiral* (or sometimes a *focus*). Such a point is approached in a spiral-like manner by a family of paths that wind around it an infinite number of times as $t \to \infty$ (or as $t \to -\infty$). Note particularly that while the paths approach O, they do not enter it. That is, a point P moving along such a path approaches O as $t \to \infty$ (or as $t \to -\infty$), but the line OP does not approach any definite direction.

Example 3. If a is an arbitrary constant, then the system

$$\begin{cases} \dfrac{dx}{dt} = ax - y \\[2mm] \dfrac{dy}{dt} = x + ay \end{cases} \tag{12}$$

has the origin as its only critical point (why?). The differential equation of the paths,

$$\frac{dy}{dx} = \frac{x + ay}{ax - y}, \tag{13}$$

is most easily solved by introducing polar coordinates r and θ defined by $x = r \cos \theta$ and $y = r \sin \theta$. Since

$$r^2 = x^2 + y^2 \qquad \text{and} \qquad \theta = \tan^{-1} \frac{y}{x},$$

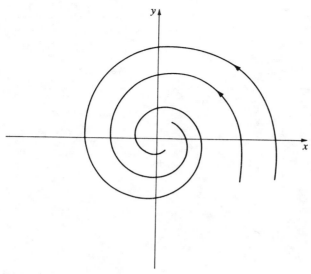

FIGURE 71

we see that

$$r\frac{dr}{dx} = x + y\frac{dy}{dx} \quad \text{and} \quad r^2\frac{d\theta}{dx} = x\frac{dy}{dx} - y.$$

With the aid of these equations, (13) can easily be written in the very simple form

$$\frac{dr}{d\theta} = ar,$$

so

$$r = ce^{a\theta} \tag{14}$$

is the polar equation of the paths. The two possible spiral configurations are shown in Fig. 72 and the direction in which these paths are traversed can be seen from the fact that $dx/dt = -y$ when $x = 0$. If $a = 0$, then (12) collapses to (8) and (14) becomes $r = c$, which is the polar equation of the family $x^2 + y^2 = c^2$ of all circles centered on the origin. This example therefore generalizes Example 2; and since the center shown in Fig. 70 stands on the borderline between the spirals of Fig. 72, a critical point that is a center is often called a *borderline case*. We will encounter other borderline cases in the next section.

We now introduce the concept of *stability* as it applies to the critical points of the system (1).

It was pointed out in the previous section that one of the most important questions in the study of a physical system is that of its steady states. However, a steady state has little physical significance unless it has a reasonable degree of permanence, i.e., unless it is stable. As a simple

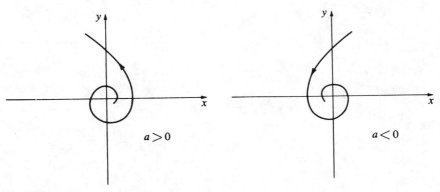

FIGURE 72

example, consider the pendulum of Fig. 73. There are two steady states possible here: when the bob is at rest at the highest point, and when the bob is at rest at the lowest point. The first state is clearly unstable, and the second is stable. We now recall that a steady state of a simple physical system corresponds to an equilibrium point (or critical point) in the phase plane. These considerations suggest in a general way that a small disturbance at an unstable equilibrium point leads to a larger and larger departure from this point, while the opposite is true at a stable equilibrium point.

We now formulate these intuitive ideas in a more precise way. Consider an isolated critical point of the system (1), and assume for the sake of convenience that this point is located at the origin $O = (0,0)$ of the phase plane. This critical point is said to be *stable* if for each positive number R there exists a positive number $r \leq R$ such that every path which is inside the circle $x^2 + y^2 = r^2$ for some $t = t_0$ remains inside the

m **FIGURE 73**

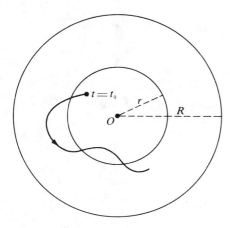

FIGURE 74

circle $x^2 + y^2 = R^2$ for all $t > t_0$ (Fig. 74). Loosely speaking, a critical point is stable if all paths that get sufficiently close to the point stay close to the point. Further, our critical point is said to be *asymptotically stable* if it is stable and there exists a circle $x^2 + y^2 = r_0^2$ such that every path which is inside this circle for some $t = t_0$ approaches the origin as $t \to \infty$. Finally, if our critical point is not stable, then it is called *unstable*.

As examples of these concepts, we point out that the node in Fig. 68, the saddle point in Fig. 69, and the spiral on the left in Fig. 72 are unstable, while the center in Fig. 70 is stable but not asymptotically stable. The node in Fig. 67, the spiral in Fig. 71, and the spiral on the right in Fig. 72 are asymptotically stable.

PROBLEMS

1. For each of the following nonlinear systems: (i) find the critical points; (ii) find the differential equation of the paths; (iii) solve this equation to find the paths; and (iv) sketch a few of the paths and show the direction of increasing t.

(a) $\begin{cases} \dfrac{dx}{dt} = y(x^2 + 1) \\ \dfrac{dy}{dt} = 2xy^2; \end{cases}$

(c) $\begin{cases} \dfrac{dx}{dt} = e^y \\ \dfrac{dy}{dt} = e^y \cos x; \end{cases}$

(b) $\begin{cases} \dfrac{dx}{dt} = y(x^2 + 1) \\ \dfrac{dy}{dt} = -x(x^2 + 1); \end{cases}$

(d) $\begin{cases} \dfrac{dx}{dt} = -x \\ \dfrac{dy}{dt} = 2x^2y^2. \end{cases}$

2. Each of the following linear systems has the origin as an isolated critical point. (i) Find the general solution. (ii) Find the differential equation of the paths.

(iii) Solve the equation found in (ii) and sketch a few of the paths, showing the direction of increasing t. (iv) Discuss the stability of the critical point.

(a) $\begin{cases} \dfrac{dx}{dt} = x \\[2mm] \dfrac{dy}{dt} = -y; \end{cases}$ (b) $\begin{cases} \dfrac{dx}{dt} = -x \\[2mm] \dfrac{dy}{dt} = -2y; \end{cases}$ (c) $\begin{cases} \dfrac{dx}{dt} = 4y \\[2mm] \dfrac{dy}{dt} = -x. \end{cases}$

3. Sketch the phase portrait of the equation $d^2x/dt^2 = 2x^3$, and show that it has an unstable isolated critical point at the origin.

60 CRITICAL POINTS AND STABILITY FOR LINEAR SYSTEMS

Our goal in this chapter is to learn as much as we can about nonlinear differential equations by studying the phase portraits of nonlinear autonomous systems of the form

$$\begin{cases} \dfrac{dx}{dt} = F(x,y) \\[3mm] \dfrac{dy}{dt} = G(x,y). \end{cases}$$

One aspect of this is the problem of classifying the critical points of such a system with respect to their nature and stability. It will be seen in Section 62 that under suitable conditions this problem can be solved for a given nonlinear system by studying a related linear system. We therefore devote this section to a complete analysis of the critical points of linear autonomous systems.

We consider the system

$$\begin{cases} \dfrac{dx}{dt} = a_1 x + b_1 y \\[3mm] \dfrac{dy}{dt} = a_2 x + b_2 y, \end{cases} \tag{1}$$

which has the origin $(0,0)$ as an obvious critical point. We assume throughout this section that

$$\begin{vmatrix} a_1 & b_1 \\ a_2 & b_2 \end{vmatrix} \neq 0, \tag{2}$$

so that $(0,0)$ is the only critical point. It was proved in Section 56 that (1) has a nontrivial solution of the form

$$\begin{cases} x = Ae^{mt} \\ y = Be^{mt} \end{cases}$$

whenever m is a root of the quadratic equation

$$m^2 - (a_1 + b_2)m + (a_1b_2 - a_2b_1) = 0, \tag{3}$$

which is called the *auxiliary equation* of the system. Observe that condition (2) implies that zero cannot be a root of (3).

Let m_1 and m_2 be the roots of (3). We shall prove that the nature of the critical point $(0,0)$ of the system (1) is determined by the nature of the numbers m_1 and m_2. It is reasonable to expect that three possibilities will occur, according as m_1 and m_2 are real and distinct, real and equal, or conjugate complex. Unfortunately the situation is a little more complicated than this, and it is necessary to consider five cases, subdivided as follows.

Major cases:

Case A.   The roots m_1 and m_2 are real, distinct, and of the same sign (node).

Case B.   The roots m_1 and m_2 are real, distinct, and of opposite signs (saddle point).

Case C.   The roots m_1 and m_2 are conjugate complex but not pure imaginary (spiral).

Borderline cases:

Case D.   The roots m_1 and m_2 are real and equal (node).

Case E.   The roots m_1 and m_2 are pure imaginary (center).

The reason for the distinction between the major cases and the borderline cases will become clear in Section 62. For the present it suffices to remark that while the borderline cases are of mathematical interest they have little significance for applications, because the circumstances defining them are unlikely to arise in physical problems. We now turn to the proofs of the assertions in parentheses.

Case A. If the roots m_1 and m_2 are real, distinct, and of the same sign, then the critical point $(0,0)$ is a node.

Proof. We begin by assuming that m_1 and m_2 are both negative, and we choose the notation so that $m_1 < m_2 < 0$. By Section 56, the general solution of (1) in this case is

$$\begin{cases} x = c_1A_1e^{m_1t} + c_2A_2e^{m_2t} \\ y = c_1B_1e^{m_1t} + c_2B_2e^{m_2t}, \end{cases} \tag{4}$$

where the A's and B's are definite constants such that $B_1/A_1 \neq B_2/A_2$, and where the c's are arbitrary constants. When $c_2 = 0$, we obtain the solutions

$$\begin{cases} x = c_1A_1e^{m_1t} \\ y = c_1B_1e^{m_1t}, \end{cases} \tag{5}$$

and when $c_1 = 0$, we obtain the solutions

$$\begin{cases} x = c_2 A_2 e^{m_2 t} \\ y = c_2 B_2 e^{m_2 t}. \end{cases} \tag{6}$$

For any $c_1 > 0$, the solution (5) represents a path consisting of half of the line $A_1 y = B_1 x$ with slope B_1/A_1; and for any $c_1 < 0$, it represents a path consisting of the other half of this line (the half on the other side of the origin). Since $m_1 < 0$, both of these half-line paths approach $(0,0)$ as $t \to \infty$; and since $y/x = B_1/A_1$, both enter $(0,0)$ with slope B_1/A_1 (Fig. 75). In exactly the same way, the solutions (6) represent two half-line paths lying on the line $A_2 y = B_2 x$ with slope B_2/A_2. These two paths also approach $(0,0)$ as $t \to \infty$, and enter it with slope B_2/A_2.

If $c_1 \neq 0$ and $c_2 \neq 0$, the general solution (4) represents curved paths. Since $m_1 < 0$ and $m_2 < 0$, these paths also approach $(0,0)$ as $t \to \infty$. Furthermore, since $m_1 - m_2 < 0$ and

$$\frac{y}{x} = \frac{c_1 B_1 e^{m_1 t} + c_2 B_2 e^{m_2 t}}{c_1 A_1 e^{m_1 t} + c_2 A_2 e^{m_2 t}} = \frac{(c_1 B_1/c_2) e^{(m_1 - m_2)t} + B_2}{(c_1 A_1/c_2) e^{(m_1 - m_2)t} + A_2},$$

it is clear that $y/x \to B_2/A_2$ as $t \to \infty$, so all of these paths enter $(0,0)$ with slope B_2/A_2. Figure 75 presents a qualitative picture of the situation. It is evident that our critical point is a node, and that it is asymptotically stable.

If m_1 and m_2 are both positive, and if we choose the notation so that $m_1 > m_2 > 0$, then the situation is exactly the same except that all the paths now approach and enter $(0,0)$ as $t \to -\infty$. The picture of the paths

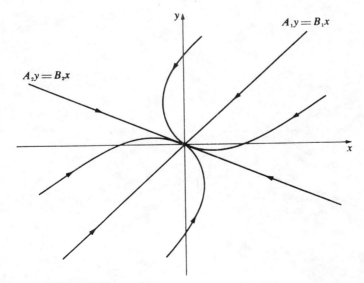

FIGURE 75

given in Fig. 75 is unchanged except that the arrows showing their directions are all reversed. We still have a node, but now it is unstable.

Case B. If the roots m_1 and m_2 are real, distinct, and of opposite signs, then the critical point $(0,0)$ is a saddle point.

Proof. We may choose the notation so that $m_1 < 0 < m_2$. The general solution of (1) can still be written in the form (4), and again we have particular solutions of the forms (5) and (6). The two half-line paths represented by (5) still approach and enter $(0,0)$ as $t \to \infty$, but this time the two half-line paths represented by (6) approach and enter $(0,0)$ as $t \to -\infty$. If $c_1 \neq 0$ and $c_2 \neq 0$, the general solution (4) still represents curved paths, but since $m_1 < 0 < m_2$, none of these paths approaches $(0,0)$ as $t \to \infty$ or $t \to -\infty$. Instead, as $t \to \infty$, each of these paths is asymptotic to one of the half-line paths represented by (6); and as $t \to -\infty$, each is asymptotic to one of the half-line paths represented by (5). Figure 76 gives a qualitative picture of this behavior. In this case the critical point is a saddle point, and it is obviously unstable.

Case C. If the roots m_1 and m_2 are conjugate complex but not pure imaginary, then the critical point $(0,0)$ is a spiral.

Proof. In this case we can write m_1 and m_2 in the form $a \pm ib$ where a and b are nonzero real numbers. Also, for later use, we observe that the

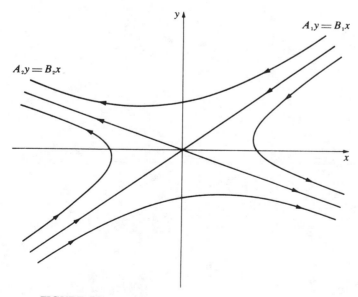

FIGURE 76

discriminant D of equation (3) is negative:

$$D = (a_1 + b_2)^2 - 4(a_1b_2 - a_2b_1)$$
$$= (a_1 - b_2)^2 + 4a_2b_1 < 0. \tag{7}$$

By Section 56, the general solution of (1) in this case is

$$\begin{cases} x = e^{at}[c_1(A_1 \cos bt - A_2 \sin bt) + c_2(A_1 \sin bt + A_2 \cos bt)] \\ y = e^{at}[c_1(B_1 \cos bt - B_2 \sin bt) + c_2(B_1 \sin bt + B_2 \cos bt)], \end{cases} \tag{8}$$

where the A's and B's are definite constants and the c's are arbitrary constants.

Let us first assume that $a < 0$. Then it is clear from formulas (8) that $x \to 0$ and $y \to 0$ as $t \to \infty$, so all the paths approach $(0,0)$ as $t \to \infty$. We now prove that the paths do not enter the point $(0,0)$ as $t \to \infty$, but instead wind around it in a spiral-like manner. To accomplish this we introduce the polar cordinate θ and show that, along any path, $d\theta/dt$ is either positive for all t or negative for all t. We begin with the fact that $\theta = \tan^{-1}(y/x)$, so

$$\frac{d\theta}{dt} = \frac{x \, dy/dt - y \, dx/dt}{x^2 + y^2};$$

and by using equations (1) we obtain

$$\frac{d\theta}{dt} = \frac{a_2x^2 + (b_2 - a_1)xy - b_1y^2}{x^2 + y^2}. \tag{9}$$

Since we are interested only in solutions that represent paths, we assume that $x^2 + y^2 \neq 0$. Now (7) implies that a_2 and b_1 have opposite signs. We consider the case in which $a_2 > 0$ and $b_1 < 0$. When $y = 0$, (9) yields $d\theta/dt = a_2 > 0$. If $y \neq 0$, $d\theta/dt$ cannot be 0; for if it were, then (9) would imply that

$$a_2x^2 + (b_2 - a_1)xy - b_1y^2 = 0$$

or

$$a_2\left(\frac{x}{y}\right)^2 + (b_2 - a_1)\frac{x}{y} - b_1 = 0 \tag{10}$$

for some real number x/y—and this cannot be true because the discriminant of the quadratic equation (10) is D, which is negative by (7). This shows that $d\theta/dt$ is always positive when $a_2 > 0$, and in the same way we see that it is always negative when $a_2 < 0$. Since by (8), x and y change sign infinitely often as $t \to \infty$, all paths must spiral in to the origin (counterclockwise or clockwise according as $a_2 > 0$ or $a_2 < 0$). The critical point in this case is therefore a spiral, and it is asymptotically stable.

If $a > 0$, the situation is the same except that the paths approach $(0,0)$ as $t \to -\infty$ and the critical point is unstable. Figure 72 illustrates the arrangement of the paths when $a_2 > 0$.

Case D. If the roots m_1 and m_2 are real and equal, then the critical point $(0,0)$ is a node.

Proof. We begin by assuming that $m_1 = m_2 = m < 0$. There are two subcases that require separate discussion: (i) $a_1 = b_2 \neq 0$ and $a_2 = b_1 = 0$; (ii) all other possibilities leading to a double root of equation (3).

We first consider the subcase (i), which is the situation described in the footnote in Section 56. If a denotes the common value of a_1 and b_2, then equation (3) becomes $m^2 - 2am + a^2 = 0$ and $m = a$. The system (1) is thus

$$\begin{cases} \dfrac{dx}{dt} = ax \\ \dfrac{dy}{dt} = ay, \end{cases}$$

and its general solution is

$$\begin{cases} x = c_1 e^{mt} \\ y = c_2 e^{mt}, \end{cases} \tag{11}$$

where c_1 and c_2 are arbitrary constants. The paths defined by (11) are half-lines of all possible slopes (Fig. 77), and since $m < 0$ we see that each path approaches and enters (0,0) as $t \to \infty$. The critical point is therefore a node, and it is asymptotically stable. If $m > 0$, we have the same situation except that the paths enter (0,0) as $t \to -\infty$, the arrows in Fig. 77 are reversed, and (0,0) is unstable.

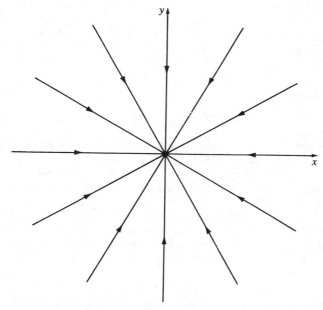

FIGURE 77

We now discuss subcase (ii). By formulas 56-(20) and Problem 56-(4), the general solution of (1) can be written in the form

$$\begin{cases} x = c_1 A e^{mt} + c_2(A_1 + At)e^{mt} \\ y = c_1 B e^{mt} + c_2(B_1 + Bt)e^{mt}, \end{cases} \tag{12}$$

where the A's and B's are definite constants and the c's are arbitrary constants. When $c_2 = 0$, we obtain the solutions

$$\begin{cases} x = c_1 A e^{mt} \\ y = c_1 B e^{mt}. \end{cases} \tag{13}$$

We know that these solutions represent two half-line paths lying on the line $Ay = Bx$ with slope B/A, and since $m < 0$ both paths approach $(0,0)$ as $t \to \infty$ (Fig. 78). Also, since $y/x = B/A$, both paths enter $(0,0)$ with slope B/A. If $c_2 \neq 0$, the solutions (12) represent curved paths, and since $m < 0$ it is clear from (12) that these paths approach $(0,0)$ as $t \to \infty$. Furthermore, it follows from

$$\frac{y}{x} = \frac{c_1 B e^{mt} + c_2(B_1 + Bt)e^{mt}}{c_1 A e^{mt} + c_2(A_1 + At)e^{mt}} = \frac{c_1 B/c_2 + B_1 + Bt}{c_1 A/c_2 + A_1 + At}$$

that $y/x \to B/A$ as $t \to \infty$, so these curved paths all enter $(0,0)$ with slope B/A. We also observe that $y/x \to B/A$ as $t \to -\infty$. Figure 78 gives a qualitative picture of the arrangement of these paths. It is clear that $(0,0)$ is a node that is asymptotically stable. If $m > 0$, the situation is unchanged

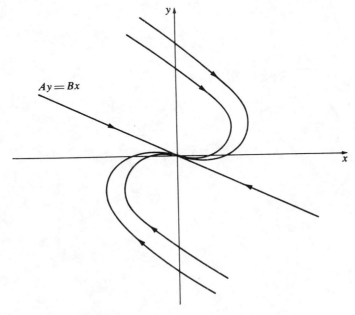

FIGURE 78

except that the directions of the paths are reversed and the critical point is unstable.

Case E. If the roots m_1 and m_2 are pure imaginary, then the critical point $(0,0)$ is a center.

Proof. It suffices here to refer back to the discussion of Case C, for now m_1 and m_2 are of the form $a \pm ib$ with $a = 0$ and $b \neq 0$. The general solution of (1) is therefore given by (8) with the exponential factor missing, so $x(t)$ and $y(t)$ are periodic and each path is a closed curve surrounding the origin. As Fig. 79 suggests, these curves are actually ellipses; this can be proved (see Problem 5) by solving the differential equation of the paths,

$$\frac{dy}{dx} = \frac{a_2 x + b_2 y}{a_1 x + b_1 y}. \tag{14}$$

Our critical point $(0,0)$ is evidently a center that is stable but not asymptotically stable.

In the above discussions we have made a number of statements about stability. It will be convenient to summarize this information as follows.

Theorem A. *The critical point $(0,0)$ of the linear system (1) is stable if and only if both roots of the auxiliary equation (3) have nonpositive real parts, and it is asymptotically stable if and only if both roots have negative real parts.*

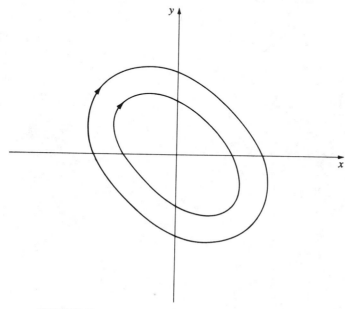

FIGURE 79

If we now write equation (3) in the form

$$(m - m_1)(m - m_2) = m^2 + pm + q = 0, \tag{15}$$

so that $p = -(m_1 + m_2)$ and $q = m_1 m_2$, then our five cases can be described just as readily in terms of the coefficients p and q as in terms of the roots m_1 and m_2. In fact, if we interpret these cases in the pq-plane, then we arrive at a striking diagram (Fig. 80) that displays at a glance the nature and stability properties of the critical point $(0,0)$. The first thing to notice is that the p-axis $q = 0$ is excluded, since by condition (2) we know that $m_1 m_2 \neq 0$. In the light of what we have learned about our five cases, all of the information contained in the diagram follows directly from the fact that

$$m_1, m_2 = \frac{-p \pm \sqrt{p^2 - 4q}}{2}.$$

Thus, above the parabola $p^2 - 4q = 0$, we have $p^2 - 4q < 0$, so m_1 and m_2 are conjugate complex numbers that are pure imaginary if and only if $p = 0$; these are Cases C and E comprising the spirals and centers. Below the p-axis we have $q < 0$, which means that m_1 and m_2 are real, distinct, and have opposite signs; this yields the saddle points of Case B. And finally, the zone between these two regions (including the parabola but excluding the p-axis) is characterized by the relations $p^2 - 4q \geq 0$ and $q > 0$, so m_1 and m_2 are real and of the same sign; here we have the nodes of Cases A and D. Furthermore, it is clear that there is precisely

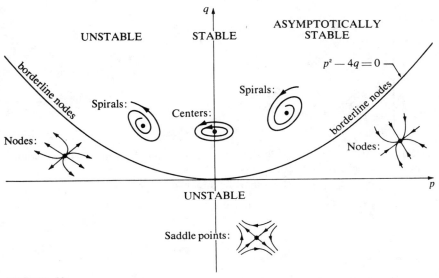

FIGURE 80

one region of asymptotic stability: the first quadrant. We state this formally as follows.

Theorem B. *The critical point $(0,0)$ of the linear system (1) is asymptotically stable if and only if the coefficients $p = -(a_1 + b_2)$ and $q = a_1 b_2 - a_2 b_1$ of the auxiliary equation (3) are both positive.*

Finally, it should be emphasized that we have studied the paths of our linear system near a critical point by analyzing explicit solutions of the system. In the next two sections we enter more fully into the spirit of the subject by investigating similar problems for nonlinear systems, which in general cannot be solved explicitly.

PROBLEMS

1. Determine the nature and stability properties of the critical point $(0,0)$ for each of the following linear autonomous systems:

(a) $\begin{cases} \dfrac{dx}{dt} = 2x \\ \dfrac{dy}{dt} = 3y; \end{cases}$
(e) $\begin{cases} \dfrac{dx}{dt} = -4x - y \\ \dfrac{dy}{dt} = x - 2y; \end{cases}$

(b) $\begin{cases} \dfrac{dx}{dt} = -x - 2y \\ \dfrac{dy}{dt} = 4x - 5y; \end{cases}$
(f) $\begin{cases} \dfrac{dx}{dt} = 4x - 3y \\ \dfrac{dy}{dt} = 8x - 6y; \end{cases}$

(c) $\begin{cases} \dfrac{dx}{dt} = -3x + 4y \\ \dfrac{dy}{dt} = -2x + 3y; \end{cases}$
(g) $\begin{cases} \dfrac{dx}{dt} = 4x - 2y \\ \dfrac{dy}{dt} = 5x + 2y. \end{cases}$

(d) $\begin{cases} \dfrac{dx}{dt} = 5x + 2y \\ \dfrac{dy}{dt} = -17x - 5y; \end{cases}$

2. If $a_1 b_2 - a_2 b_1 = 0$, show that the system (1) has infinitely many critical points, none of which are isolated.

3. (a) If $a_1 b_2 - a_2 b_1 \neq 0$, show that the system

$$\begin{cases} \dfrac{dx}{dt} = a_1 x + b_1 y + c_1 \\ \dfrac{dy}{dt} = a_2 x + b_2 y + c_2 \end{cases}$$

has a single isolated critical point (x_0, y_0).

(b) Show that the system in (a) can be written in the form of (1) by means of the change of variables $\bar{x} = x - x_0$ and $\bar{y} = y - y_0$.

(c) Find the critical point of the system

$$\begin{cases} \dfrac{dx}{dt} = 2x - 2y + 10 \\[2mm] \dfrac{dy}{dt} = 11x - 8y + 49, \end{cases}$$

write the system in the form of (1) by changing the variables, and determine the nature and stability properties of the critical point.

4. In Section 20 we studied the free vibrations of a mass attached to a spring by solving the equation

$$\frac{d^2x}{dt^2} + 2b\frac{dx}{dt} + a^2x = 0,$$

where $b \geq 0$ and $a > 0$ are constants representing the viscosity of the medium and the stiffness of the spring, respectively. Consider the equivalent autonomous system

$$\begin{cases} \dfrac{dx}{dt} = y \\[2mm] \dfrac{dy}{dt} = -a^2x - 2by, \end{cases} \tag{*}$$

which has $(0,0)$ as its only critical point.

(a) Find the auxiliary equation of (*). What are p and q?

(b) For each of the following four cases, describe the nature and stability properties of the critical point, and give a brief physical interpretation of the corresponding motion of the mass:

  (i) $b = 0$; (iii) $b = a$;
  (ii) $0 < b < a$; (iv) $b > a$.

5. Solve equation (14) under the hypotheses of Case E, and show that the result is a one-parameter family of ellipses surrounding the origin. *Hint:* Recall that if $Ax^2 + Bxy + Cy^2 = D$ is the equation of a real curve, then the curve is an ellipse if and only if the discriminant $B^2 - 4AC$ is negative.

61 STABILITY BY LIAPUNOV'S DIRECT METHOD

It is intuitively clear that if the total energy of a physical system has a local minimum at a certain equilibrium point, then that point is stable. This idea was generalized by Liapunov[6] into a simple but powerful

---

[6] Alexander Mikhailovich Liapunov (1857–1918) was a Russian mathematician and mechanical engineer. He had the very rare merit of producing a doctoral dissertation of lasting value. This classic work was originally published in 1892 in Russian, but is now available in an English translation, *Stability of Motion*, Academic Press, New York, 1966. Liapunov died by violence in Odessa, which cannot be considered a surprising fate for a middle-class intellectual in the chaotic aftermath of the Russian Revolution.

method for studying stability problems in a broader context. We shall discuss Liapunov's method and some of its applications in this and the next section.

Consider an autonomous system

$$\begin{cases} \dfrac{dx}{dt} = F(x,y) \\[2mm] \dfrac{dy}{dt} = G(x,y), \end{cases} \tag{1}$$

and assume that this system has an isolated critical point, which as usual we take to be the origin $(0,0)$.[7] Let $C = [x(t),y(t)]$ be a path of (1), and consider a function $E(x,y)$ that is continuous and has continuous first partial derivatives in a region containing this path. If a point (x,y) moves along the path in accordance with the equations $x = x(t)$ and $y = y(t)$, then $E(x,y)$ can be regarded as a function of t along C [we denote this function by $E(t)$] and its rate of change is

$$\frac{dE}{dt} = \frac{\partial E}{\partial x}\frac{dx}{dt} + \frac{\partial E}{\partial y}\frac{dy}{dt}$$

$$= \frac{\partial E}{\partial x}F + \frac{\partial E}{\partial y}G. \tag{2}$$

This formula is at the heart of Liapunov's ideas, and in order to exploit it we need several definitions that specify the kinds of functions we shall be interested in.

Suppose that $E(x,y)$ is continuous and has continuous first partial derivatives in some region containing the origin. If E vanishes at the origin, so that $E(0,0) = 0$, then it is said to be *positive definite* if $E(x,y) > 0$ for $(x,y) \neq (0,0)$, and *negative definite* if $E(x,y) < 0$ for $(x,y) \neq (0,0)$. Similarly, E is called *positive semidefinite* if $E(0,0) = 0$ and $E(x,y) \geq 0$ for $(x,y) \neq (0,0)$, and *negative semidefinite* if $E(0,0) = 0$ and $E(x,y) \leq 0$ for $(x,y) \neq (0,0)$. It is clear that functions of the form $ax^{2m} + by^{2n}$, where a and b are positive constants and m and n are positive integers, are positive definite. Since $E(x,y)$ is negative definite if and only if $-E(x,y)$ is positive definite, functions of the form $ax^{2m} + by^{2n}$ with $a < 0$ and $b < 0$ are negative definite. The functions x^{2m}, y^{2m}, and $(x - y)^{2m}$ are not positive definite, but are nevertheless positive

---

[7] A critical point (x_0,y_0) can always be moved to the origin by a simple translation of coordinates $\bar{x} = x - x_0$ and $\bar{y} = y - y_0$, so there is no loss of generality in assuming that it lies at the origin in the first place.

semidefinite. If $E(x,y)$ is positive definite, then $z = E(x,y)$ can be interpreted as the equation of a surface (Fig. 81) that resembles a paraboloid opening upward and tangent to the xy-plane at the origin.

A positive definite function $E(x,y)$ with the property that

$$\frac{\partial E}{\partial x} F + \frac{\partial E}{\partial y} G \tag{3}$$

is negative semidefinite is called a *Liapunov function* for the system (1). By formula (2), the requirement that (3) be negative semidefinite means that $dE/dt \leq 0$—and therefore E is nonincreasing—along the paths of (1) near the origin. These functions generalize the concept of the total energy of a physical system. Their relevance for stability problems is made clear in the following theorem, which is Liapunov's basic discovery.

Theorem A. *If there exists a Liapunov function $E(x,y)$ for the system* (1), *then the critical point* (0,0) *is stable. Furthermore, if this function has the additional property that the function* (3) *is negative definite, then the critical point* (0,0) *is asymptotically stable.*

Proof. Let C_1 be a circle of radius $R > 0$ centered on the origin (Fig. 82), and assume also that C_1 is small enough to lie entirely in the domain of definition of the function E. Since $E(x,y)$ is continuous and positive

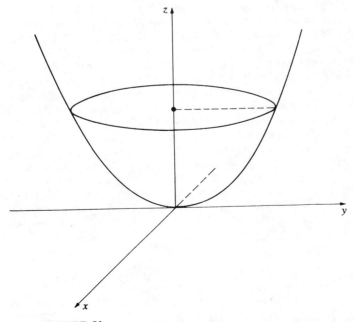

FIGURE 81

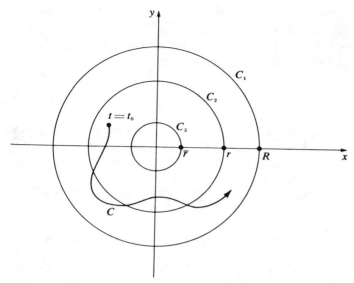

FIGURE 82

definite, it has a positive minimum m on C_1. Next, $E(x,y)$ is continuous at the origin and vanishes there, so we can find a positive number $r < R$ such that $E(x,y) < m$ whenever (x,y) is inside the circle C_2 of radius r. Now let $C = [x(t), y(t)]$ be any path which is inside C_2 for $t = t_0$. Then $E(t_0) < m$, and since (3) is negative semidefinite we have $dE/dt \leq 0$, which implies that $E(t) \leq E(t_0) < m$ for all $t > t_0$. It follows that the path C can never reach the circle C_1 for any $t > t_0$, so we have stability.

To prove the second part of the theorem, it suffices to show that under the additional hypothesis we also have $E(t) \to 0$, for since $E(x,y)$ is positive definite this will imply that the path C approaches the critical point $(0,0)$. We begin by observing that since $dE/dt < 0$, it follows that $E(t)$ is a decreasing function; and since by hypothesis $E(t)$ is bounded below by 0, we conclude that $E(t)$ approaches some limit $L \geq 0$ as $t \to \infty$. To prove that $E(t) \to 0$ it suffices to show that $L = 0$, so we assume that $L > 0$ and deduce a contradiction. Choose a positive number $\bar{r} < r$ with the property that $E(x,y) < L$ whenever (x,y) is inside the circle C_3 with radius $\bar{r}$. Since the function (3) is continuous and negative definite, it has a negative maximum $-k$ in the ring consisting of the circles C_1 and C_3 and the region between them. This ring contains the entire path C for $t \geq t_0$, so the equation

$$E(t) = E(t_0) + \int_{t_0}^{t} \frac{dE}{dt}\, dt$$

yields the inequality

$$E(t) \leq E(t_0) - k(t - t_0) \tag{4}$$

for all $t \geq t_0$. However, the right side of (4) becomes negatively infinite as

$t \to \infty$, so $E(t) \to -\infty$ as $t \to \infty$. This contradicts the fact that $E(x,y) \geq 0$, so we conclude that $L = 0$ and the proof is complete.

Example 1. Consider the equation of motion of a mass m attached to a spring:

$$m\frac{d^2x}{dt^2} + c\frac{dx}{dt} + kx = 0. \tag{5}$$

Here $c \geq 0$ is a constant representing the viscosity of the medium through which the mass moves, and $k > 0$ is the spring constant. The autonomous system equivalent to (5) is

$$\begin{cases} \dfrac{dx}{dt} = y \\ \dfrac{dy}{dt} = -\dfrac{k}{m}x - \dfrac{c}{m}y, \end{cases} \tag{6}$$

and its only critical point is $(0,0)$. The kinetic energy of the mass is $my^2/2$, and the potential energy (or the energy stored in the spring) is

$$\int_0^x kx\,dx = \frac{1}{2}kx^2.$$

Thus the total energy of the system is

$$E(x,y) = \frac{1}{2}my^2 + \frac{1}{2}kx^2. \tag{7}$$

It is easy to see that (7) is positive definite; and since

$$\frac{\partial E}{\partial x}F + \frac{\partial E}{\partial y}G = kxy + my\left(-\frac{k}{m}x - \frac{c}{m}y\right)$$

$$= -cy^2 \leq 0,$$

(7) is a Liapunov function for (6) and the critical point $(0,0)$ is stable. We know from Problem 60-4 that when $c > 0$ this critical point is asymptotically stable, but the particular Liapunov function discussed here is not capable of detecting this fact.[8]

---

[8] It is known that both stability and asymptotic stability can always be detected by suitable Liapunov functions, but knowing in principle that such a function exists is a very different matter from actually finding one. For references on this point, see L. Cesari, *Asymptotic Behavior and Stability Problems in Ordinary Differential Equations*, p. 111, Academic Press, New York, 1963; or G. Sansone and R. Conti, *Non-Linear Differential Equations*, p. 481, Macmillan, New York, 1964.

Example 2. The system

$$\begin{cases} \dfrac{dx}{dt} = -2xy \\[2ex] \dfrac{dy}{dt} = x^2 - y^3 \end{cases} \tag{8}$$

has (0,0) as an isolated critical point. Let us try to prove stability by constructing a Liapunov function of the form $E(x,y) = ax^{2m} + by^{2n}$. It is clear that

$$\frac{\partial E}{\partial x} F + \frac{\partial E}{\partial y} G = 2max^{2m-1}(-2xy) + 2nby^{2n-1}(x^2 - y^3)$$

$$= (-4max^{2m}y + 2nbx^2y^{2n-1}) - 2nby^{2n+2}.$$

We wish to make the expression in parentheses vanish, and inspection shows that this can be done by choosing $m = 1$, $n = 1$, $a = 1$, and $b = 2$. With these choices we have $E(x,y) = x^2 + 2y^2$ (which is positive definite) and $(\partial E/\partial x)F + (\partial E/\partial y)G = -4y^4$ (which is negative semidefinite). The critical point (0,0) of the system (8) is therefore stable.

It is clear from this example that in complicated situations it may be very difficult indeed to construct suitable Liapunov functions. The following result is sometimes helpful in this connection.

Theorem B *The function* $E(x,y) = ax^2 + bxy + cy^2$ *is positive definite if and only if* $a > 0$ *and* $b^2 - 4ac < 0$, *and is negative definite if and only if* $a < 0$ *and* $b^2 - 4ac < 0$.

Proof. If $y = 0$, we have $E(x,0) = ax^2$, so $E(x,0) > 0$ for $x \neq 0$ if and only if $a > 0$. If $y \neq 0$, we have

$$E(x,y) = y^2 \left[a\left(\frac{x}{y}\right)^2 + b\left(\frac{x}{y}\right) + c \right];$$

and when $a > 0$ the bracketed polynomial in x/y (which is positive for large x/y) is positive for all x/y if and only if $b^2 - 4ac < 0$. This proves the first part of the theorem, and the second part follows at once by considering the function $-E(x,y)$.

PROBLEMS

1. Determine whether each of the following functions is positive definite, negative definite, or neither:
   (a) $x^2 - xy - y^2$;
   (b) $2x^2 - 3xy + 3y^2$;
   (c) $-2x^2 + 3xy - y^2$;
   (d) $-x^2 - 4xy - 5y^2$.
2. Show that a function of the form $ax^3 + bx^2y + cxy^2 + dy^3$ cannot be either positive definite or negative definite.

3. Show that $(0,0)$ is an asymptotically stable critical point for each of the following systems:

(a) $\begin{cases} \dfrac{dx}{dt} = -3x^3 - y \\[2mm] \dfrac{dy}{dt} = x^5 - 2y^3; \end{cases}$

(b) $\begin{cases} \dfrac{dx}{dt} = -2x + xy^3 \\[2mm] \dfrac{dy}{dt} = -x^2y^2 - y^3. \end{cases}$

4. Prove that the critical point $(0,0)$ of the system (1) is unstable if there exists a function $E(x,y)$ with the following properties:
   (a) $E(x,y)$ is continuous and has continuous first partial derivatives in some region containing the origin;
   (b) $E(0,0) = 0$;
   (c) every circle centered on $(0,0)$ contains at least one point where $E(x,y)$ is positive;
   (d) $(\partial E/\partial x)F + (\partial E/\partial y)G$ is positive definite.

5. Show that $(0,0)$ is an unstable critical point for the system

$$\begin{cases} \dfrac{dx}{dt} = 2xy + x^3 \\[2mm] \dfrac{dy}{dt} = -x^2 + y^5. \end{cases}$$

6. Assume that $f(x)$ is a function such that $f(0) = 0$ and $xf(x) > 0$ for $x \neq 0$ [that is, $f(x) > 0$ when $x > 0$ and $f(x) < 0$ when $x < 0$].
   (a) Show that

$$E(x,y) = \frac{1}{2}y^2 + \int_0^x f(x)\,dx$$

   is positive definite.
   (b) Show that the equation

$$\frac{d^2x}{dt^2} + f(x) = 0$$

   has $x = 0$, $y = dx/dt = 0$ as a stable critical point.
   (c) If $g(x) \geq 0$ in some neighborhood of the origin, show that the equation

$$\frac{d^2x}{dt^2} + g(x)\frac{dx}{dt} + f(x) = 0$$

   has $x = 0$, $y = dx/dt = 0$ as a stable critical point.

62 SIMPLE CRITICAL POINTS OF NONLINEAR SYSTEMS

Consider an autonomous system

$$\begin{cases} \dfrac{dx}{dt} = F(x,y) \\[2mm] \dfrac{dy}{dt} = G(x,y) \end{cases} \tag{1}$$

with an isolated critical point at $(0,0)$. If $F(x,y)$ and $G(x,y)$ can be expanded in power series in x and y, then (1) takes the form

$$\begin{cases} \dfrac{dx}{dt} = a_1 x + b_1 y + c_1 x^2 + d_1 xy + e_1 y^2 + \cdots \\[2mm] \dfrac{dy}{dt} = a_2 x + b_2 y + c_2 x^2 + d_2 xy + e_2 y^2 + \cdots . \end{cases} \tag{2}$$

When $|x|$ and $|y|$ are small—that is, when (x,y) is close to the origin—the terms of second degree and higher are very small. It is therefore natural to discard these nonlinear terms and conjecture that the qualitative behavior of the paths of (2) near the critical point $(0,0)$ is similar to that of the paths of the related linear system

$$\begin{cases} \dfrac{dx}{dt} = a_1 x + b_1 y \\[2mm] \dfrac{dy}{dt} = a_2 x + b_2 y. \end{cases} \tag{3}$$

We shall see that in general this is actually the case. The process of replacing (2) by the linear system (3) is usually called *linearization*.

More generally, we shall consider systems of the form

$$\begin{cases} \dfrac{dx}{dt} = a_1 x + b_1 y + f(x,y) \\[2mm] \dfrac{dy}{dt} = a_2 x + b_2 y + g(x,y). \end{cases} \tag{4}$$

It will be assumed that

$$\begin{vmatrix} a_1 & b_1 \\ a_2 & b_2 \end{vmatrix} \neq 0, \tag{5}$$

so that the related linear system (3) has $(0,0)$ as an isolated critical point; that $f(x,y)$ and $g(x,y)$ are continuous and have continuous first partial derivatives for all (x,y); and that as $(x,y) \to (0,0)$ we have

$$\lim \frac{f(x,y)}{\sqrt{x^2 + y^2}} = 0 \quad \text{and} \quad \lim \frac{g(x,y)}{\sqrt{x^2 + y^2}} = 0. \tag{6}$$

Observe that conditions (6) imply that $f(0,0) = 0$ and $g(0,0) = 0$, so $(0,0)$ is a critical point of (4); also, it is not difficult to prove that this critical point is isolated (see Problem 1). With the restrictions listed above, $(0,0)$ is said to be a *simple critical point* of the system (4).

Example 1. In the case of the system

$$\begin{cases} \dfrac{dx}{dt} = -2x + 3y + xy \\ \dfrac{dy}{dt} = -x + y - 2xy^2 \end{cases} \tag{7}$$

we have

$$\begin{vmatrix} a_1 & b_1 \\ a_2 & b_2 \end{vmatrix} = \begin{vmatrix} -2 & 3 \\ -1 & 1 \end{vmatrix} = 1 \neq 0,$$

so (5) is satisfied. Furthermore, by using polar coordinates we see that

$$\frac{|f(x,y)|}{\sqrt{x^2 + y^2}} = \frac{|r^2 \sin \theta \cos \theta|}{r} \leq r$$

and

$$\frac{|g(x,y)|}{\sqrt{x^2 + y^2}} = \frac{|2r^3 \sin^2 \theta \cos \theta|}{r} \leq 2r^2,$$

so $f(x,y)/r$ and $g(x,y)/r \to 0$ as $(x,y) \to (0,0)$ (or as $r \to 0$). This shows that conditions (6) are also satisfied, so $(0,0)$ is a simple critical point of the system (7).

The main facts about the nature of simple critical points are given in the following theorem of Poincaré, which we state without proof.[9]

Theorem A. *Let $(0,0)$ be a simple critical point of the nonlinear system* (4), *and consider the related linear system* (3). *If the critical point $(0,0)$ of* (3) *falls under any one of the three major cases described in Section 60, then the critical point $(0,0)$ of* (4) *is of the same type.*

As an illustration, we examine the nonlinear system (7) of Example 1, whose related linear system is

$$\begin{cases} \dfrac{dx}{dt} = -2x + 3y \\ \dfrac{dy}{dt} = -x + y. \end{cases} \tag{8}$$

The auxiliary equation of (8) is $m^2 + m + 1 = 0$, with roots

$$m_1, m_2 = \frac{-1 \pm \sqrt{3}\, i}{2}.$$

---

[9] Detailed treatments can be found in W. Hurewicz, *Lectures on Ordinary Differential Equations*, pp. 86–98, MIT, Cambridge, Mass., 1958; L. Cesari, *Asymptotic Behavior and Stability Problems in Ordinary Differential Equations*, pp. 157–163, Academic Press, New York, 1963; or F. G. Tricomi, *Differential Equations*, pp. 53–72, Blackie, Glasgow, 1961.

Since these roots are conjugate complex but not pure imaginary, we have Case C and the critical point (0,0) of the linear system (8) is a spiral. By Theorem A, the critical point (0,0) of the nonlinear system (7) is also a spiral.

It should be understood that while the type of the critical point (0,0) is the same for (4) as it is for (3) in the cases covered by the theorem, the actual appearance of the paths may be somewhat different. For example, Fig. 76 shows a typical saddle point for a linear system, whereas Fig. 83 suggests how a nonlinear saddle point might look. A certain amount of distortion is clearly present in the latter, but nevertheless the qualitative features of the two configurations are the same.

It is natural to wonder about the two borderline cases, which are not mentioned in Theorem A. The facts are these: if the related linear system (3) has a borderline node at the origin (Case D), then the nonlinear system (4) can have either a node or a spiral; and if (3) has a center at the origin (Case E), then (4) can have either a center or a spiral. For example, (0,0) is a critical point for each of the nonlinear systems

$$\begin{cases} \dfrac{dx}{dt} = -y - x^2 \\ \dfrac{dy}{dt} = x \end{cases} \quad \text{and} \quad \begin{cases} \dfrac{dx}{dt} = -y - x^3 \\ \dfrac{dy}{dt} = x. \end{cases} \tag{9}$$

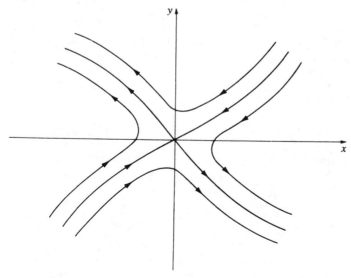

FIGURE 83

In each case the related linear system is

$$\begin{cases} \dfrac{dx}{dt} = -y \\[2mm] \dfrac{dy}{dt} = x. \end{cases} \tag{10}$$

It is easy to see that $(0,0)$ is a center for (10). However, it can be shown that while $(0,0)$ is a center for the first system of (9), it is a spiral for the second.[10]

We have already encountered a considerable variety of configurations at critical points of linear systems, and the above remarks show that no new phenomena appear at simple critical points of nonlinear systems. What about critical points that are not simple? The possibilities here can best be appreciated by examining a nonlinear system of the form (2). If the linear terms in (2) do not determine the pattern of the paths near the origin, then we must consider the second degree terms; if these fail to determine the pattern, then the third degree terms must be taken into account, and so on. This suggests that in addition to the linear configurations, a great many others can arise, of infinite variety and staggering complexity. Several are shown in Fig. 84. It is perhaps surprising to realize that such involved patterns as these can occur in connection with systems of rather simple appearance. For example, the three figures in the upper row show the arrangement of the paths of

$$\begin{cases} \dfrac{dx}{dt} = 2xy \\[2mm] \dfrac{dy}{dt} = y^2 - x^2, \end{cases} \qquad \begin{cases} \dfrac{dx}{dt} = x^3 - 2xy^2 \\[2mm] \dfrac{dy}{dt} = 2x^2y - y^3, \end{cases} \qquad \begin{cases} \dfrac{dx}{dt} = x - 4y\sqrt{|xy|} \\[2mm] \dfrac{dy}{dt} = -y + 4x\sqrt{|xy|}. \end{cases}$$

In the first case, this can be seen at once by looking at Fig. 3 and equation 3-(8).

We now discuss the question of stability for a simple critical point. The main result here is due to Liapunov: if (3) is asymptotically stable at the origin, then (4) is also. We state this formally as follows.

Theorem B. *Let* $(0,0)$ *be a simple critical point of the nonlinear system* (4), *and consider the related linear system* (3). *If the critical point* $(0,0)$ *of* (3) *is asymptotically stable, then the critical point* $(0,0)$ *of* (4) *is also asymptotically stable.*

---

[10] See Hurewicz, *op. cit.*, p. 99.

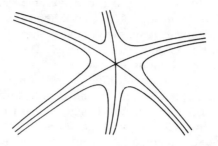

FIGURE 84

Proof. By Theorem 61-A, it suffices to construct a suitable Liapunov function for the system (4), and this is what we do.

Theorem 60–B tells us that the coefficients of the linear system (3) satisfy the conditions

$$p = -(a_1 + b_2) > 0 \quad \text{and} \quad q = a_1 b_2 - a_2 b_1 > 0. \quad (11)$$

Now define

$$E(x,y) = \frac{1}{2}(ax^2 + 2bxy + cy^2)$$

by putting

$$a = \frac{a_2^2 + b_2^2 + (a_1 b_2 - a_2 b_1)}{D},$$

$$b = -\frac{a_1 a_2 + b_1 b_2}{D},$$

and

$$c = \frac{a_1^2 + b_1^2 + (a_1 b_2 - a_2 b_1)}{D},$$

where

$$D = pq = -(a_1 + b_2)(a_1 b_2 - a_2 b_1).$$

By (11), we see that $D > 0$ and $a > 0$. Also, an easy calculation shows that

$$
\begin{aligned}
D^2(ac - b^2) &= (a_2^2 + b_2^2)(a_1^2 + b_1^2) \\
&\quad + (a_2^2 + b_2^2 + a_1^2 + b_1^2)(a_1b_2 - a_2b_1) \\
&\quad + (a_1b_2 - a_2b_1)^2 - (a_1a_2 + b_1b_2)^2 \\
&= (a_2^2 + b_2^2 + a_1^2 + b_1^2)(a_1b_2 - a_2b_1) \\
&\quad + 2(a_1b_2 - a_2b_1)^2 \\
&> 0,
\end{aligned}
$$

so $b^2 - ac < 0$. Thus, by Theorem 61-B, we know that the function $E(x,y)$ is positive definite. Furthermore, another calculation (whose details we leave to the reader) yields

$$
\frac{\partial E}{\partial x}(a_1x + b_1y) + \frac{\partial E}{\partial y}(a_2x + b_2y) = -(x^2 + y^2). \tag{12}
$$

This function is clearly negative definite, so $E(x,y)$ is a Liapunov function for the linear system (3).[11]

We next prove that $E(x,y)$ is also a Liapunov function for the nonlinear system (4). If F and G are defined by

$$
F(x,y) = a_1x + b_1y + f(x,y)
$$

and

$$
G(x,y) = a_2x + b_2y + g(x,y),
$$

then since E is known to be positive definite, it suffices to show that

$$
\frac{\partial E}{\partial x}F + \frac{\partial E}{\partial y}G \tag{13}
$$

is negative definite. If we use (12), then (13) becomes

$$
-(x^2 + y^2) + (ax + by)f(x,y) + (bx + cy)g(x,y);
$$

and by introducing polar coordinates we can write this as

$$
-r^2 + r[(a \cos \theta + b \sin \theta)f(x,y) + (b \cos \theta + c \sin \theta)g(x,y)].
$$

Denote the largest of the numbers $|a|$, $|b|$, $|c|$ by K. Our assumption (6) now implies that

$$
|f(x,y)| < \frac{r}{6K} \quad \text{and} \quad |g(x,y)| < \frac{r}{6K}
$$

for all sufficiently small $r > 0$, so

$$
\frac{\partial E}{\partial x}F + \frac{\partial E}{\partial y}G < -r^2 + \frac{4Kr^2}{6K} = -\frac{r^2}{3} < 0
$$

---

[11] The reason for the definitions of a, b, and c can now be understood: we want (12) to be true.

for these r's. Thus $E(x,y)$ is a positive definite function with the property that (13) is negative definite. Theorem 61-A now implies that $(0,0)$ is an asymptotically stable critical point of (4), and the proof is complete.

To illustrate this theorem, we again consider the nonlinear system (7) of Example 1, whose related linear system is (8). For (8) we have $p = 1 > 0$ and $q = 1 > 0$, so the critical point $(0,0)$ is asymptotically stable, both for the linear system (8) and for the nonlinear system (7).

Example 2. We know from Section 58 that the equation of motion for the damped vibrations of a pendulum is

$$\frac{d^2x}{dt^2} + \frac{c}{m}\frac{dx}{dt} + \frac{g}{a}\sin x = 0,$$

where c is a positive constant. The equivalent nonlinear system is

$$\begin{cases} \dfrac{dx}{dt} = y \\ \dfrac{dy}{dt} = -\dfrac{g}{a}\sin x - \dfrac{c}{m}y. \end{cases} \tag{14}$$

Let us now write (14) in the form

$$\begin{cases} \dfrac{dx}{dt} = y \\ \dfrac{dy}{dt} = -\dfrac{g}{a}x - \dfrac{c}{m}y + \dfrac{g}{a}(x - \sin x). \end{cases} \tag{15}$$

It is easy to see that

$$\frac{x - \sin x}{\sqrt{x^2 + y^2}} \to 0$$

as $(x,y) \to (0,0)$, for if $x \neq 0$, we have

$$\frac{|x - \sin x|}{\sqrt{x^2 + y^2}} \le \frac{|x - \sin x|}{|x|} = \left|1 - \frac{\sin x}{x}\right| \to 0;$$

and since $(0,0)$ is evidently an isolated critical point of the related linear system

$$\begin{cases} \dfrac{dx}{dt} = y \\ \dfrac{dy}{dt} = -\dfrac{g}{a}x - \dfrac{c}{m}y, \end{cases} \tag{16}$$

it follows that $(0,0)$ is a simple critical point of (15). Inspection shows ($p = c/m > 0$ and $q = g/a > 0$) that $(0,0)$ is an asymptotically stable critical point of (16), so by Theorem B it is also an asymptotically stable critical point of (15). This reflects the obvious physical fact that if the pendulum is slightly disturbed, then the resulting motion will die out with the passage of time.

PROBLEMS

1. Prove that if $(0,0)$ is a simple critical point of (4), then it is necessarily isolated. *Hint*: Write conditions (6) in the form $f(x,y)/r = \epsilon_1 \rightarrow 0$ and $g(x,y)/r = \epsilon_2 \rightarrow 0$, and in the light of (5) use polar coordinates to deduce a contradiction from the assumption that the right sides of (4) both vanish at points arbitrarily close to the origin but different from it.

2. Sketch the family of curves whose polar equation is $r = a \sin 2\theta$ (see Fig. 84), and express the differential equation of this family in the form $dy/dx = G(x,y)/F(x,y)$.

3. If $(0,0)$ is a simple critical point of (4) and $q = a_1 b_2 - a_2 b_1 < 0$, then Theorem A implies that $(0,0)$ is a saddle point of (4) and is therefore unstable. Prove that if $p = -(a_1 + b_2) < 0$ and $q = a_1 b_2 - a_2 b_1 > 0$, then $(0,0)$ is an unstable critical point of (4). *Hint*: Adapt the proof of Theorem B to show that there exists a positive definite function $E(x,y)$ such that

$$\frac{\partial E}{\partial x}(a_1 x + b_1 y) + \frac{\partial E}{\partial y}(a_2 x + b_2 y) = x^2 + y^2,$$

and apply Problem 61-4. (Observe that these facts together with Theorem B demonstrate that all the information in Fig. 80 about asymptotic stability and instability carries over directly to nonlinear systems with simple critical points from their related linear systems.)

4. Show that $(0,0)$ is an asymptotically stable critical point of

$$\begin{cases} \dfrac{dx}{dt} = -y - x^3 \\ \dfrac{dy}{dt} = x - y^3, \end{cases}$$

but is an unstable critical point of

$$\begin{cases} \dfrac{dx}{dt} = -y + x^3 \\ \dfrac{dy}{dt} = x + y^3. \end{cases}$$

How are these facts related to the parenthetical remark in Problem 3?

5. Verify that $(0,0)$ is a simple critical point for each of the following systems, and determine its nature and stability properties:

(a) $\begin{cases} \dfrac{dx}{dt} = x + y - 2xy \\ \dfrac{dy}{dt} = -2x + y + 3y^2; \end{cases}$ (b) $\begin{cases} \dfrac{dx}{dt} = -x - y - 3x^2 y \\ \dfrac{dy}{dt} = -2x - 4y + y \sin x. \end{cases}$

6. The van der Pol equation

$$\frac{d^2 x}{dt^2} + \mu(x^2 - 1)\frac{dx}{dt} + x = 0$$

is equivalent to the system

$$\begin{cases} \dfrac{dx}{dt} = y \\ \dfrac{dy}{dt} = -x - \mu(x^2 - 1)y. \end{cases}$$

Investigate the stability properties of the critical point (0,0) for the cases $\mu > 0$ and $\mu < 0$.

63   NONLINEAR MECHANICS. CONSERVATIVE SYSTEMS

It is well known that energy is dissipated in the action of any real dynamical system, usually through some form of friction. However, in certain situations this dissipation is so slow that it can be neglected over relatively short periods of time. In such cases we assume the law of conservation of energy, namely, that the sum of the kinetic energy and the potential energy is constant. A system of this kind is said to be *conservative*. Thus the rotating earth can be considered a conservative system over short intervals of time involving only a few centuries, but if we want to study its behavior throughout millions of years we must take into account the dissipation of energy by tidal friction.

The simplest conservative system consists of a mass m attached to a spring and moving in a straight line through a vacuum. If x denotes the displacement of m from its equilibrium position, and the restoring force exerted on m by the spring is $-kx$ where $k > 0$, then we know that the equation of motion is

$$m\frac{d^2x}{dt^2} + kx = 0.$$

A spring of this kind is called a *linear spring* because the restoring force is a linear function of x. If m moves through a resisting medium, and the resistance (or damping force) exerted on m is $-c(dx/dt)$ where $c > 0$, then the equation of motion of this nonconservative system is

$$m\frac{d^2x}{dt^2} + c\frac{dx}{dt} + kx = 0.$$

Here we have *linear damping* because the damping force is a linear function of dx/dt. By analogy, if f and g are arbitrary functions with the property that $f(0) = 0$ and $g(0) = 0$, then the more general equation

$$m\frac{d^2x}{dt^2} + g\left(\frac{dx}{dt}\right) + f(x) = 0 \tag{1}$$

can be interpreted as the equation of motion of a mass m under the action of a *restoring force* $-f(x)$ and a *damping force* $-g(dx/dt)$. In general these forces are nonlinear, and equation (1) can be regarded as the basic equation of nonlinear mechanics. In this section we shall briefly consider the special case of a nonlinear conservative system described by the equation

$$m \frac{d^2x}{dt^2} + f(x) = 0, \tag{2}$$

in which the damping force is zero and there is consequently no dissipation of energy.[12]

Equation (2) is equivalent to the autonomous system

$$\begin{cases} \dfrac{dx}{dt} = y \\[2mm] \dfrac{dy}{dt} = -\dfrac{f(x)}{m}. \end{cases} \tag{3}$$

If we eliminate dt, we obtain the differential equation of the paths of (3) in the phase plane,

$$\frac{dy}{dx} = -\frac{f(x)}{my}, \tag{4}$$

and this can be written in the form

$$my\, dy = -f(x)\, dx. \tag{5}$$

If $x = x_0$ and $y = y_0$ when $t = t_0$, then integrating (5) from t_0 to t yields

$$\frac{1}{2} my^2 - \frac{1}{2} my_0^2 = -\int_{x_0}^{x} f(x)\, dx$$

or

$$\frac{1}{2} my^2 + \int_{0}^{x} f(x)\, dx = \frac{1}{2} my_0^2 + \int_{0}^{x_0} f(x)\, dx. \tag{6}$$

To interpret this result, we observe that $\frac{1}{2} my^2 = \frac{1}{2} m(dx/dt)^2$ is the kinetic energy of the dynamical system and

$$V(x) = \int_{0}^{x} f(x)\, dx \tag{7}$$

---

[12] Extensive discussions of (1), with applications to a variety of physical problems, can be found in J. J. Stoker, *Nonlinear Vibrations*, Interscience-Wiley, New York, 1950; and in A. A. Andronow and C. E. Chaikin, *Theory of Oscillations*, Princeton University Press, Princeton, N.J., 1949.

is its potential energy. Equation (6) therefore expresses the law of conservation of energy,

$$\frac{1}{2}my^2 + V(x) = E, \tag{8}$$

where $E = \frac{1}{2}my_0^2 + V(x_0)$ is the constant total energy of the system. It is clear that (8) is the equation of the paths of (3), since we obtained it by solving (4). The particular path determined by specifying a value of E is a curve of constant energy in the phase plane. The critical points of the system (3) are the points $(x_c, 0)$ where the x_c are the roots of the equation $f(x) = 0$. As we pointed out in Section 58, these are the equilibrium points of the dynamical system described by (2). It is evident from (4) that the paths cross the x-axis at right angles and are horizontal when they cross the lines $x = x_c$. Equation (8) also shows that the paths are symmetric with respect to the x-axis.

If we write (8) in the form

$$y = \pm\sqrt{\frac{2}{m}[E - V(x)]}, \tag{9}$$

then the paths can be constructed by the following easy steps. First, establish an xz-plane with the z-axis on the same vertical line as the y-axis of the phase plane (Fig. 85). Next, draw the graph of $z = V(x)$ and several horizontal lines $z = E$ in the xz-plane (one such line is shown in the figure), and observe the geometric meaning of the difference $E - V(x)$. Finally, for each x, multiply $E - V(x)$ as obtained in the preceding step by $2/m$ and use formula (9) to plot the corresponding values of y in the phase plane directly below. Note that since $dx/dt = y$, the positive direction along any path is to the right above the x-axis and to the left below this axis.

Example 1. We saw in Section 58 that the equation of motion of an undamped pendulum is

$$\frac{d^2x}{dt^2} + k\sin x = 0, \tag{10}$$

where k is a positive constant. Since this equation is of the form (2), it can be interpreted as describing the undamped rectilinear motion of a unit mass under the influence of a nonlinear spring whose restoring force is $-k\sin x$. The autonomous system equivalent to (10) is

$$\begin{cases} \dfrac{dx}{dt} = y \\[2mm] \dfrac{dy}{dt} = -k\sin x, \end{cases} \tag{11}$$

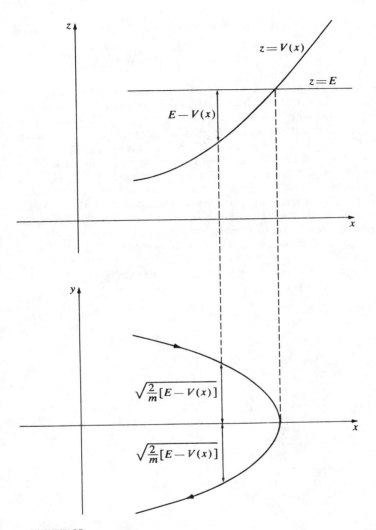

FIGURE 85

and its critical points are $(0,0)$, $(\pm\pi,0)$, $(\pm 2\pi,0)$, ... The differential equation of the paths is

$$\frac{dy}{dx} = -\frac{k \sin x}{y},$$

and by separating variables and integrating, we see that the equation of the family of paths is

$$\frac{1}{2}y^2 + (k - k \cos x) = E.$$

This is evidently of the form (8), where $m = 1$ and

$$V(x) = \int_0^x f(x)\,dx = k - k\cos x$$

is the potential energy. We now construct the paths by first drawing the graph of $z = V(x)$ and several lines $z = E$ in the xz-plane (Fig. 86, where $z = E = 2k$ is the only line shown). From this we read off the values $E - V(x)$ and sketch the paths in the phase plane directly below by using $y = \pm\sqrt{2[E - V(x)]}$. It is clear from this phase portrait that if the total energy E is between 0 and $2k$, then the corresponding paths are closed and equation (10) has periodic solutions. On the other hand, if $E > 2k$, then the path is not closed and the corresponding solution of (10) is not periodic. The value $E = 2k$ separates the two types of motion, and for this reason a path corresponding to $E = 2k$ is called a *separatrix*. The wavy paths outside the separatrices correspond to whirling motions of the pendulum, and the closed paths inside to oscillatory motions. It is evident that the critical points are alternately unstable saddle points and stable but not asymptotically stable centers. For the sake of contrast, it is interesting to consider the

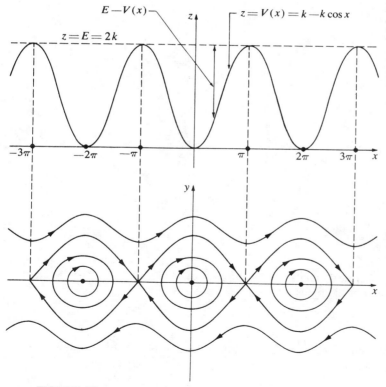

FIGURE 86

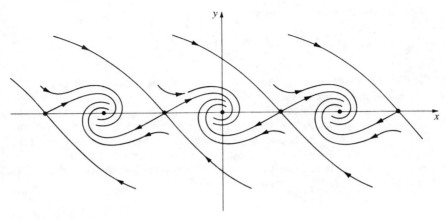

FIGURE 87

effect of transforming this conservative dynamical system into a nonconservative system by introducing a linear damping force. The equation of
motion then takes the form

$$\frac{d^2x}{dt^2} + c\frac{dx}{dt} + k \sin x = 0, \quad c > 0,$$

and the configuration of the paths is suggested in Fig. 87. We find that the
centers in Fig. 86 become asymptotically stable spirals, and also that every
path—except the separatrices entering the saddle points as $t \to \infty$—
ultimately winds into one of these spirals.

PROBLEMS

1. If $f(0) = 0$ and $xf(x) > 0$ for $x \neq 0$, show that the paths of

$$\frac{d^2x}{dt^2} + f(x) = 0$$

are closed curves surrounding the origin in the phase plane; that is, show that
the critical point $x = 0$, $y = dx/dt = 0$ is a stable but not asymptotically stable
center. Describe this critical point with respect to its nature and stability if
$f(0) = 0$ and $xf(x) < 0$ for $x \neq 0$.
2. Most actual springs are not linear. A nonlinear spring is called *hard* or *soft*
according as the magnitude of the restoring force increases more rapidly or
less rapidly than a linear function of the displacement. The equation

$$\frac{d^2x}{dt^2} + kx + \alpha x^3 = 0, \quad k > 0,$$

describes the motion of a hard spring if $\alpha > 0$ and a soft spring if $\alpha < 0$.
Sketch the paths in each case.

3. Find the equation of the paths of

$$\frac{d^2x}{dt^2} - x + 2x^3 = 0,$$

and sketch these paths in the phase plane. Locate the critical points and determine the nature of each.

4. Since by equation (7) we have $dV/dx = f(x)$, the critical points of (3) are the points on the x-axis in the phase plane at which $V'(x) = 0$. In terms of the curve $z = V(x)$—if this curve is smooth and well behaved—there are three possibilities: maxima, minima, and points of inflection. Sketch all three possibilities, and determine the type of critical point associated with each (a critical point of the third type is called a *cusp*).

64 PERIODIC SOLUTIONS. THE POINCARÉ–BENDIXSON THEOREM

Consider a nonlinear autonomous system

$$\begin{cases} \dfrac{dx}{dt} = F(x,y) \\[2mm] \dfrac{dy}{dt} = G(x,y) \end{cases} \tag{1}$$

in which the functions $F(x,y)$ and $G(x,y)$ are continuous and have continuous first partial derivatives throughout the phase plane. Our work so far has told us practically nothing about the paths of (1) except in the neighborhood of certain types of critical points. However, in many problems we are much more interested in the global properties of paths than we are in these local properties. *Global properties* of paths are those that describe their behavior over large regions of the phase plane, and in general they are very difficult to establish.

The central problem of the global theory is that of determining whether (1) has closed paths. As we remarked in Section 58, this problem is important because of its close connection with the issue of whether (1) has periodic solutions. A solution $x(t)$ and $y(t)$ of (1) is said to be *periodic* if neither function is constant, if both are defined for all t, and if there exists a number $T > 0$ such that $x(t + T) = x(t)$ and $y(t + T) = y(t)$ for all t. The smallest T with this property is called the *period* of the solution.[13] It is evident that each periodic solution of (1) defines a closed path that is traversed once as t increases from t_0 to $t_0 + T$ for any t_0.

---

[13] Every periodic solution has a period in this sense. Why?

Conversely, it is easy to see that if $C = [x(t), y(t)]$ is a closed path of (1), then $x(t)$, $y(t)$ is a periodic solution. Accordingly, the search for periodic solutions of (1) reduces to a search for closed paths.

We know from Section 60 that a linear system has closed paths if and only if the roots of the auxiliary equation are pure imaginary, and in this case every path is closed. Thus, for a linear system, either every path is closed or else no path is closed. On the other hand, a nonlinear system can perfectly well have a closed path that is isolated, in the sense that no other closed paths are near to it. The following is a well-known example of such a system:

$$\begin{cases} \dfrac{dx}{dt} = -y + x(1 - x^2 - y^2) \\[2mm] \dfrac{dy}{dt} = x + y(1 - x^2 - y^2). \end{cases} \tag{2}$$

To solve this system we introduce polar coordinates r and θ, where $x = r \cos \theta$ and $y = r \sin \theta$. If we differentiate the relations $x^2 + y^2 = r^2$ and $\theta = \tan^{-1}(y/x)$, we obtain the useful formulas

$$x\frac{dx}{dt} + y\frac{dy}{dt} = r\frac{dr}{dt} \quad \text{and} \quad x\frac{dy}{dt} - y\frac{dx}{dt} = r^2\frac{d\theta}{dt}. \tag{3}$$

On multiplying the first equation of (2) by x and the second by y, and adding, we find that

$$r\frac{dr}{dt} = r^2(1 - r^2). \tag{4}$$

Similarly, if we multiply the second by x and the first by y, and subtract, we get

$$r^2\frac{d\theta}{dt} = r^2. \tag{5}$$

The system (2) has a single critical point at $r = 0$. Since we are concerned only with finding the paths, we may assume that $r > 0$. In this case, (4) and (5) show that (2) becomes

$$\begin{cases} \dfrac{dr}{dt} = r(1 - r^2) \\[2mm] \dfrac{d\theta}{dt} = 1. \end{cases} \tag{6}$$

These equations are easy to solve separately, and the general solution of the system (6) is found to be

$$\begin{cases} r = \dfrac{1}{\sqrt{1 + ce^{-2t}}} \\ \theta = t + t_0. \end{cases} \tag{7}$$

The corresponding general solution of (2) is

$$\begin{cases} x = \dfrac{\cos(t + t_0)}{\sqrt{1 + ce^{-2t}}} \\ y = \dfrac{\sin(t + t_0)}{\sqrt{1 + ce^{-2t}}}. \end{cases} \tag{8}$$

Let us analyze (7) geometrically (Fig. 88). If $c = 0$, we have the solutions $r = 1$ and $\theta = t + t_0$, which trace out the closed circular path $x^2 + y^2 = 1$ in the counterclockwise direction. If $c < 0$, it is clear that $r > 1$ and that $r \to 1$ as $t \to \infty$. Also, if $c > 0$, we see that $r < 1$, and again $r \to 1$ as $t \to \infty$. These observations show that there exists a single closed path $(r = 1)$ which all other paths approach spirally from the outside or the inside as $t \to \infty$.

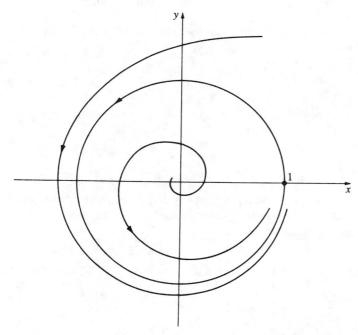

FIGURE 88

In the above discussion we have shown that the system (2) has a closed path by actually finding such a path. In general, of course, we cannot hope to be able to do this. What we need are tests that make it possible for us to conclude that certain regions of the phase plane do or do not contain closed paths. Our first test is given in the following theorem of Poincaré. A proof is sketched in Problem 1.

Theorem A. *A closed path of the system* (1) *necessarily surrounds at least one critical point of this system.*

This result gives a negative criterion of rather limited value: a system without critical points in a given region cannot have closed paths in that region.

Our next theorem provides another negative criterion, and is due to Bendixson.[14]

Theorem B. *If* $\partial F / \partial x + \partial G / \partial y$ *is always positive or always negative in a certain region of the phase plane, then the system* (1) *cannot have closed paths in that region.*

Proof. Assume that the region contains a closed path $C = [x(t), y(t)]$ with interior R. Then Green's theorem and our hypothesis yield

$$\int_C (F \, dy - G \, dx) = \iint_R \left(\frac{\partial F}{\partial x} + \frac{\partial G}{\partial y} \right) dx \, dy \neq 0.$$

However, along C we have $dx = F \, dt$ and $dy = G \, dt$, so

$$\int_C (F \, dy - G \, dx) = \int_0^T (FG - GF) \, dt = 0.$$

This contradiction shows that our initial assumption is false, so the region under consideration cannot contain any closed path.

These theorems are sometimes useful, but what we really want are positive criteria giving sufficient conditions for the existence of closed paths of (1). One of the few general theorems of this kind is the classical *Poincaré–Bendixson theorem*, which we now state without proof.[15]

---

[14] Ivar Otto Bendixson (1861–1935) was a Swedish mathematician who published one important memoir in 1901 supplementing some of Poincaré's earlier work. He served as professor (and later as president) at the University of Stockholm, and was an energetic long-time member of the Stockholm City Council.

[15] For details, see Hurewicz, *loc. cit.*, pp. 102–111, or Cesari, *loc. cit.*, pp. 163–167.

Theorem C. *Let R be a bounded region of the phase plane together with its boundary, and assume that R does not contain any critical points of the system* (1). *If C = [x(t),y(t)] is a path of* (1) *that lies in R for some t_0 and remains in R for all $t \geq t_0$, then C is either itself a closed path or it spirals toward a closed path as $t \to \infty$. Thus in either case the system* (1) *has a closed path in R.*

In order to understand this statement, let us consider the situation suggested in Fig. 89. Here R consists of the two dashed curves together with the ring-shaped region between them. Suppose that the vector

$$\mathbf{V}(x,y) = F(x,y)\mathbf{i} + G(x,y)\mathbf{j}$$

points *into* R at every boundary point. Then every path C through a boundary point (at $t = t_0$) must enter R and can never leave it, and under these circumstances the theorem asserts that C must spiral toward a closed path C_0. We have chosen a ring-shaped region R to illustrate the theorem because a closed path like C_0 must surround a critical point (P in the figure) and R must exclude all critical points.

The system (2) provides a simple application of these ideas. It is clear that (2) has a critical point at (0,0), and also that the region R between the circles $r = \frac{1}{2}$ and $r = 2$ contains no critical points. In our earlier analysis we found that

$$\frac{dr}{dt} = r(1 - r^2) \qquad \text{for } r > 0.$$

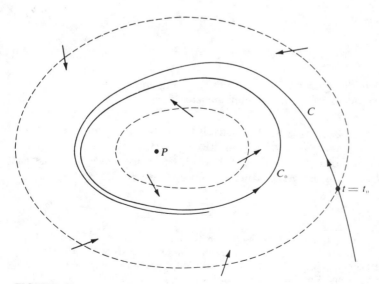

FIGURE 89

This shows that $dr/dt > 0$ on the inner circle and $dr/dt < 0$ on the outer circle, so the vector **V** points into R at all boundary points. Thus any path through a boundary point will enter R and remain in R as $t \to \infty$, and by the Poincaré–Bendixson theorem we know that R contains a closed path C_0. We have already seen that the circle $r = 1$ is the closed path whose existence is guaranteed in this way.

The Poincaré–Bendixson theorem is quite satisfying from a theoretical point of view, but in general it is rather difficult to apply. A more practical criterion has been developed that assures the existence of closed paths for equations of the form

$$\frac{d^2x}{dt^2} + f(x)\frac{dx}{dt} + g(x) = 0, \tag{9}$$

which is called *Liénard's equation*.[16] When we speak of a closed path for such an equation, we of course mean a closed path of the equivalent system

$$\begin{cases} \dfrac{dx}{dt} = y \\[2mm] \dfrac{dy}{dt} = -g(x) - f(x)y; \end{cases} \tag{10}$$

and as we know, a closed path of (10) corresponds to a periodic solution of (9). The fundamental statement about the closed paths of (9) is the following theorem.

Theorem D. (Liénard's Theorem.) *Let the functions $f(x)$ and $g(x)$ satisfy the following conditions: (i) both are continuous and have continuous derivatives for all x; (ii) $g(x)$ is an odd function such that $g(x) > 0$ for $x > 0$, and $f(x)$ is an even function; and (iii) the odd function $F(x) = \int_0^x f(x)\,dx$ has exactly one positive zero at $x = a$, is negative for $0 < x < a$, is positive and nondecreasing for $x > a$, and $F(x) \to \infty$ as $x \to \infty$. Then equation (9) has a unique closed path surrounding the origin in the phase plane, and this path is approached spirally by every other path as $t \to \infty$.*

For the benefit of the skeptical and tenacious reader who is rightly reluctant to accept unsupported assertions, a proof of this theorem is

---

[16] Alfred Liénard (1869–1958) was a French scientist who spent most of his career teaching applied physics at the School of Mines in Paris, of which he became director in 1929. His physical research was mainly in the areas of electricity and magnetism, elasticity, and hydrodynamics. From time to time he worked on mathematical problems arising from his other scientific investigations, and in 1933 was elected president of the French Mathematical Society. He was an unassuming bachelor whose life was devoted entirely to his work and his students.

given in Appendix B. An intuitive understanding of the role of the hypotheses can be gained by thinking of (9) in terms of the ideas of the previous section. From this point of view, equation (9) is the equation of motion of a unit mass attached to a spring and subject to the dual influence of a restoring force $-g(x)$ and a damping force $-f(x)\,dx/dt$. The assumption about $g(x)$ amounts to saying that the spring acts as we would expect, and tends to diminish the magnitude of any displacement. On the other hand, the assumptions about $f(x)$—roughly, that $f(x)$ is negative for small $|x|$ and positive for large $|x|$—mean that the motion is intensified for small $|x|$ and retarded for large $|x|$, and therefore tends to settle down into a steady oscillation. This rather peculiar behavior of $f(x)$ can also be expressed by saying that the physical system absorbs energy when $|x|$ is small and dissipates it when $|x|$ is large.

The main application of Liénard's theorem is to the van der Pol[17] equation

$$\frac{d^2x}{dt^2} + \mu(x^2 - 1)\frac{dx}{dt} + x = 0, \tag{11}$$

where μ is assumed to be a positive constant for physical reasons. Here $f(x) = \mu(x^2 - 1)$ and $g(x) = x$, so condition (i) is clearly satisfied. It is equally clear that condition (ii) is true. Since

$$F(x) = \mu\left(\frac{1}{3}x^3 - x\right) = \frac{1}{3}\mu x(x^2 - 3),$$

we see that $F(x)$ has a single positive zero at $x = \sqrt{3}$, is negative for $0 < x < \sqrt{3}$, is positive for $x > \sqrt{3}$, and that $F(x) \to \infty$ as $x \to \infty$. Finally, $F'(x) = \mu(x^2 - 1)$ is positive for $x > 1$, so $F(x)$ is certainly nondecreasing (in fact, increasing) for $x > \sqrt{3}$. Accordingly, all the conditions of the theorem are met, and we conclude that equation (11) has a unique closed path (periodic solution) that is approached spirally (asymptotically) by every other path (nontrivial solution).

PROBLEMS

1. A proof of Theorem A can be built on the following geometric ideas (Fig. 90). Let C be a simple closed curve (not necessarily a path) in the phase plane, and assume that C does not pass through any critical point of the system (1). If

---

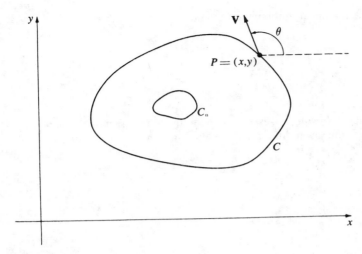

FIGURE 90

$P = (x,y)$ is a point on C, then

$$\mathbf{V}(x,y) = F(x,y)\mathbf{i} + G(x,y)\mathbf{j}$$

is a nonzero vector, and therefore has a definite direction given by the angle θ. If P moves once around C in the counterclockwise direction, the angle θ changes by an amount $\Delta\theta = 2\pi n$, where n is a positive integer, zero, or a negative integer. This integer n is called the *index* of C. If C shrinks continuously to a smaller simple closed curve C_0 without passing over any critical point, then its index varies continuously; and since the index is an integer, it cannot change.

(a) If C is a path of (1), show that its index is 1.

(b) If C is a path of (1) that contains no critical points, show that a small C_0 has index 0, and from this infer Theorem A.

2. Consider the nonlinear autonomous system

$$\begin{cases} \dfrac{dx}{dt} = 4x + 4y - x(x^2 + y^2) \\[2mm] \dfrac{dy}{dt} = -4x + 4y - y(x^2 + y^2). \end{cases}$$

(a) Transform the system into polar coordinate form.

(b) Apply the Poincaré–Bendixson theorem to show that there is a closed path between the circles $r = 1$ and $r = 3$.

(c) Find the general nonconstant solution $x = x(t)$ and $y = y(t)$ of the original system, and use this to find a periodic solution corresponding to the closed path whose existence was established in (b).

(d) Sketch the closed path and at least two other paths in the phase plane.

3. Show that the nonlinear autonomous system

$$\begin{cases} \dfrac{dx}{dt} = 3x - y - xe^{x^2+y^2} \\[2mm] \dfrac{dy}{dt} = x + 3y - ye^{x^2+y^2} \end{cases}$$

has a periodic solution.

4. In each of the following cases use a theorem of this section to determine whether or not the given differential equation has a periodic solution:

(a) $\dfrac{d^2x}{dt^2} + (5x^4 - 9x^2)\dfrac{dx}{dt} + x^5 = 0;$

(b) $\dfrac{d^2x}{dt^2} - (x^2 + 1)\dfrac{dx}{dt} + x^5 = 0;$

(c) $\dfrac{d^2x}{dt^2} - \left(\dfrac{dx}{dt}\right)^2 - (1 + x^2) = 0;$

(d) $\dfrac{d^2x}{dt^2} + \dfrac{dx}{dt} + \left(\dfrac{dx}{dt}\right)^5 - 3x^3 = 0;$

(e) $\dfrac{d^2x}{dt^2} + x^6\dfrac{dx}{dt} - x^2\dfrac{dx}{dt} + x = 0.$

5. Show that any differential equation of the form

$$a\dfrac{d^2x}{dt^2} + b(x^2 - 1)\dfrac{dx}{dt} + cx = 0 \quad (a, b, c > 0)$$

can be transformed into the van der Pol equation by a change of the independent variable.

APPENDIX A.   POINCARÉ

Jules Henri Poincaré (1854–1912) was universally recognized at the beginning of the twentieth century as the greatest mathematician of his generation. He began his academic career at Caen in 1879, but only two years later he was appointed to a professorship at the Sorbonne. He remained there for the rest of his life, lecturing on a different subject each year. In his lectures—which were edited and published by his students—he treated with great originality and mastery of technique virtually all known fields of pure and applied mathematics, and many that were not known until he discovered them. Altogether he produced more than 30 technical books on mathematical physics and celestial mechanics, half a dozen books of a more popular nature, and almost 500 research papers on mathematics. He was a quick, powerful, and restless thinker, not given to lingering over details, and was described by one of his contemporaries as "a conquerer, not a colonist." He also had the advantage of a prodigious memory, and habitually did his mathematics in his head as he paced back and forth in his study, writing it down only after it was complete in his mind. He was elected to the Academy of Sciences at the very early age of thirty-two. The academician who proposed him for membership said that "his work is above ordinary praise, and reminds us inevitably of what Jacobi wrote of Abel—that he had settled questions which, before him, were unimagined."

Poincaré's first great achievement in mathematics was in analysis. He generalized the idea of the periodicity of a function by creating his theory of automorphic functions. The elementary trigonometric and exponential functions are singly periodic, and the elliptic functions are doubly periodic. Poincaré's automorphic functions constitute a vast generalization of these, for they are invariant under a countably infinite group of linear fractional transformations and include the rich theory of elliptic functions as a detail. He applied them to solve linear differential equations with algebraic coefficients, and also showed how they can be used to uniformize algebraic curves, that is, to express the coordinates of any point on such a curve by means of single-valued functions $x(t)$ and $y(t)$ of a single parameter t. In the 1880s and 1890s automorphic functions developed into an extensive branch of mathematics, involving (in addition to analysis) group theory, number theory, algebraic geometry, and non-Euclidean geometry.

Another focal point of his thought can be found in his researches into celestial mechanics (*Les Méthodes Nouvelle de la Mécanique Céleste,* three volumes, 1892–1899). In the course of this work he developed his theory of asymptotic expansions (which kindled interest in divergent series), studied the stability of orbits, and initiated the qualitative theory of nonlinear differential equations. His celebrated investigations into the evolution of celestial bodies led him to study the equilibrium shapes of a rotating mass of fluid held together by gravitational attraction, and he discovered the pear-shaped figures that played an important role in the later work of Sir G. H. Darwin (Charles' son).[18] In Poincaré's summary of these discoveries, he writes: "Let us imagine a rotating fluid body contracting by cooling, but slowly enough to remain homogeneous and for the rotation to be the same in all its parts. At first very approximately a sphere, the figure of this mass will become an ellipsoid of revolution which will flatten more and more, then, at a certain moment, it will be transformed into an ellipsoid with three unequal axes. Later, the figure will cease to be an ellipsoid and will become pear-shaped until at last the mass, hollowing out more and more at its 'waist,' will separate into two distinct and unequal bodies." These ideas have gained additional interest

---

[18] See G. H. Darwin, *The Tides,* chap. XVIII, Houghton Mifflin, Boston, 1899.

in our own time; for with the aid of artificial satellites, geophysicists have recently found that the earth itself is slightly pear-shaped.

Many of the problems he encountered in this period were the seeds of new ways of thinking, which have grown and flourished in twentieth-century mathematics. We have already mentioned divergent series and nonlinear differential equations. In addition, his attempts to master the qualitative nature of curves and surfaces in higher dimensional spaces resulted in his famous memoir *Analysis situs* (1895), which most experts agree marks the beginning of the modern era in algebraic topology. Also, in his study of periodic orbits he founded the subject of topological (or qualitative) dynamics. The type of mathematical problem that arises here is illustrated by a theorem he conjectured in 1912 but did not live to prove: if a one-to-one continuous transformation carries the ring bounded by two concentric circles into itself in such a way as to preserve areas and to move the points of the inner circle clockwise and those of the outer circle counterclockwise, then at least two points must remain fixed. This theorem has important applications to the classical problem of three bodies (and also to the motion of a billiard ball on a convex billiard table). A proof was found in 1913 by Birkhoff, a young American mathematician.[19] Another remarkable discovery in this field, now known as the Poincaré recurrence theorem, relates to the long-range behavior of conservative dynamical systems. This result seemed to demonstrate the futility of contemporary efforts to deduce the second law of thermodynamics from classical mechanics, and the ensuing controversy was the historical source of modern ergodic theory.

One of the most striking of Poincaré's many contributions to mathematical physics was his famous paper of 1906 on the dynamics of the electron. He had been thinking about the foundations of physics for many years, and independently of Einstein had obtained many of the results of the special theory of relativity.[20] The main difference was that Einstein's treatment was based on elemental ideas relating to light signals, while Poincaré's was founded on the theory of electromagnetism and was therefore limited in its applicability to phenomena associated with this theory. Poincaré had a high regard for Einstein's abilities, and in 1911 recommended him for his first academic position.[21]

---

[19] See G. D. Birkhoff, *Dynamical Systems*, chap. VI, American Mathematical Society Colloquium Publications, vol. IX, Providence, R.I., 1927.

[20] A discussion of the historical background is given by Charles Scribner, Jr., "Henri Poincaré and the Principle of Relativity," *Am. J. Phys.*, vol. 32, p. 672 (1964).

[21] See M. Lincoln Schuster (ed.), *A Treasury of the World's Great Letters*, p. 453, Simon and Schuster, New York, 1940.

In 1902 he turned as a side interest to writing and lecturing for a wider public, in an effort to share with nonspecialists his enthusiasm for the meaning and human importance of mathematics and science. These lighter works have been collected in four books, *La Science et l'Hypothèse* (1903), *La Valeur de la Science* (1904), *Science et Méthode* (1908), and *Dernières Pensées* (1913).[22] They are clear, witty, profound, and altogether delightful, and show him to be a master of French prose at its best. In the most famous of these essays, the one on mathematical discovery, he looked into himself and analyzed his own mental processes, and in so doing provided the rest of us with some rare glimpses into the mind of a genius at work. As Jourdain wrote in his obituary, "One of the many reasons for which he will live is that he made it possible for us to understand him as well as to admire him."

At the present time mathematical knowledge is said to be doubling every 10 years or so, though some remain skeptical about the permanent value of this accumulation. It is generally believed to be impossible now for any human being to understand thoroughly more than one or two of the four main subdivisions of mathematics—analysis, algebra, geometry, and number theory—to say nothing of mathematical physics as well. Poincaré had creative command of the whole of mathematics as it existed in his day, and he was probably the last man who will ever be in this position.

APPENDIX B. PROOF OF LIÉNARD'S THEOREM

Consider Liénard's equation

$$\frac{d^2x}{dt^2} + f(x)\frac{dx}{dt} + g(x) = 0, \tag{1}$$

and assume that $f(x)$ and $g(x)$ satisfy the following conditions: (i) $f(x)$ and $g(x)$ are continuous and have continuous derivatives; (ii) $g(x)$ is an odd function such that $g(x) > 0$ for $x > 0$, and $f(x)$ is an even function; and (iii) the odd function $F(x) = \int_0^x f(x)\,dx$ has exactly one positive zero at $x = a$, is negative for $0 < x < a$, is positive and nondecreasing for $x > a$, and $F(x) \to \infty$ as $x \to \infty$. We shall prove that equation (1) has a unique closed path surrounding the origin in the phase plane, and that this path is approached spirally by every other path as $t \to \infty$.

---

[22] All have been published in English translation by Dover Publications, New York.

The system equivalent to (1) in the phase plane is

$$\begin{cases} \dfrac{dx}{dt} = y \\[2mm] \dfrac{dy}{dt} = -g(x) - f(x)y. \end{cases} \tag{2}$$

By condition (i), the basic theorem on the existence and uniqueness of solutions holds. It follows from condition (ii) that $g(0) = 0$ and $g(x) \neq 0$ for $x \neq 0$, so the origin is the only critical point. Also, we know that any closed path must surround the origin. The fact that

$$\frac{d^2x}{dt^2} + f(x)\frac{dx}{dt} = \frac{d}{dt}\left[\frac{dx}{dt} + \int_0^x f(x)\,dx\right]$$

$$= \frac{d}{dt}[y + F(x)]$$

suggests introducing a new variable,

$$z = y + F(x).$$

With this notation, equation (1) is equivalent to the system

$$\begin{cases} \dfrac{dx}{dt} = z - F(x) \\[2mm] \dfrac{dz}{dt} = -g(x) \end{cases} \tag{3}$$

in the xz-plane. Again we see that the existence and uniqueness theorem holds, that the origin is the only critical point, and that any closed path must surround the origin. The one-to-one correspondence $(x,y) \leftrightarrow (x,z)$ between the points of the two planes is continuous both ways, so closed paths correspond to closed paths and the configurations of the paths in the two planes are qualitatively similar. The differential equation of the paths of (3) is

$$\frac{dz}{dx} = \frac{-g(x)}{z - F(x)}. \tag{4}$$

These paths are easier to analyze than their corresponding paths in the phase plane, for the following reasons.

First, since both $g(x)$ and $F(x)$ are odd, equations (3) and (4) are unchanged when x and z are replaced by $-x$ and $-z$. This means that any curve symmetric to a path with respect to the origin is also a path. Thus if we know the paths in the right half-plane ($x > 0$), those in the left half-plane ($x < 0$) can be obtained at once by reflection through the origin.

Second, equation (4) shows that the paths become horizontal only as they cross the z-axis, and become vertical only as they cross the curve $z = F(x)$. Also, an inspection of the signs of the right sides of equations (3) shows that all paths are directed to the right above the curve $z = F(x)$ and to the left below this curve, and move downward or upward according as $x > 0$ or $x < 0$. These remarks mean that the curve $z = F(x)$, the z-axis, and the vertical line through any point Q on the right half of the curve $z = F(x)$ can be crossed only in the directions indicated by the arrows in Fig. 91. Suppose that the solution of (3) defining the path C through Q is so chosen that the point Q corresponds to the value $t = 0$ of the parameter. Then as t increases into positive values, a point on C with coordinates $x(t)$ and $y(t)$ moves down and to the left until it crosses the z-axis at a point R; and as t decreases into negative values, the point on C rises to the left until it crosses the z-axis at a point P. It will be convenient to let b be the abscissa of Q and to denote the path C by C_b.

It is easy to see from the symmetry property that when the path C_b is continued beyond P and R into the left half of the plane, the result will be a closed path if and only if the distances OP and OR are equal. To show that there is a unique closed path, it therefore suffices to show that there is a unique value of b with the property that $OP = OR$.

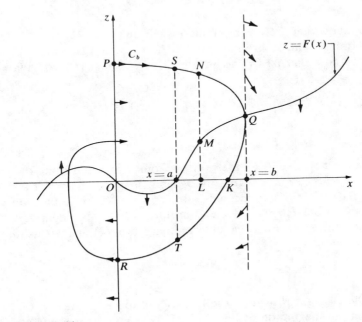

FIGURE 91

To prove this, we introduce

$$G(x) = \int_0^x g(x)\, dx$$

and consider the function

$$E(x,z) = \frac{1}{2}z^2 + G(x),$$

which reduces to $z^2/2$ on the z-axis. Along any path we have

$$\frac{dE}{dt} = g(x)\frac{dx}{dt} + z\frac{dz}{dt}$$

$$= -[z - F(x)]\frac{dz}{dt} + z\frac{dz}{dt}$$

$$= F(x)\frac{dz}{dt},$$

so

$$dE = F\, dz.$$

If we compute the line integral of $F\, dz$ along the path C_b from P to R, we obtain

$$I(b) = \int_{PR} F\, dz = \int_{PR} dE = E_R - E_P = \frac{1}{2}(OR^2 - OP^2),$$

so it suffices to show that there is a unique b such that $I(b) = 0$.

If $b \le a$, then F and dz are negative, so $I(b) > 0$ and C_b cannot be closed. Suppose now that $b > a$, as in Fig. 91. We split $I(b)$ into two parts,

$$I_1(b) = \int_{PS} F\, dz + \int_{TR} F\, dz \quad \text{and} \quad I_2(b) = \int_{ST} F\, dz,$$

so that

$$I(b) = I_1(b) + I_2(b).$$

Since F and dz are negative as C_b is traversed from P to S and from T to R, it is clear that $I_1(b) > 0$. On the other hand, if we go from S to T along C_b we have $F > 0$ and $dz < 0$, so $I_2(b) < 0$. Our immediate purpose is to show that $I(b)$ is a decreasing function of b by separately considering $I_1(b)$ and $I_2(b)$. First, we note that equation (4) enables us to write

$$F\, dz = F\frac{dz}{dx}dx = \frac{-g(x)F(x)}{z - F(x)}dx.$$

The effect of increasing b is to raise the arc PS and to lower the arc TR, which decreases the magnitude of $[-g(x)F(x)]/[z - F(x)]$ for a given x between 0 and a. Since the limits of integration for $I_1(b)$ are fixed, the result is a decrease in $I_1(b)$. Furthermore, since $F(x)$ is positive and nondecreasing to the right of a, we see that an increase in b gives rise to an increase in the positive number $-I_2(b)$, and hence to a decrease in $I_2(b)$. Thus $I(b) = I_1(b) + I_2(b)$ is a decreasing function for $b \geq a$. We now show that $I_2(b) \to -\infty$ as $b \to \infty$. If L in Fig. 91 is fixed and K is to the right of L, then

$$I_2(b) = \int_{ST} F \, dz < \int_{NK} f \, dz \leq -(LM) \cdot (LN);$$

and since $LN \to \infty$ as $b \to \infty$, we have $I_2(b) \to -\infty$.

Accordingly, $I(b)$ is a decreasing continuous function of b for $b \geq a$, $I(a) > 0$, and $I(b) \to -\infty$ as $b \to \infty$. It follows that $I(b) = 0$ for one and only one $b = b_0$, so there is one and only one closed path C_{b_0}.

Finally, we observe that $OR > OP$ for $b < b_0$; and from this and the symmetry we conclude that paths inside C_{b_0} spiral out to C_{b_0}. Similarly, the fact that $OR < OP$ for $b > b_0$ implies that paths outside C_{b_0} spiral in to C_{b_0}.

CHAPTER
12

THE CALCULUS OF VARIATIONS

65 INTRODUCTION. SOME TYPICAL PROBLEMS OF THE SUBJECT

The calculus of variations has been one of the major branches of analysis for more than two centuries. It is a tool of great power that can be applied to a wide variety of problems in pure mathematics. It can also be used to express the basic principles of mathematical physics in forms of the utmost simplicity and elegance.

The flavor of the subject is easy to grasp by considering a few of its typical problems. Suppose that two points P and Q are given in a plane (Fig. 92). There are infinitely many curves joining these points, and we can ask which of these curves is the shortest. The intuitive answer is of course a straight line. We can also ask which curve will generate the surface of revolution of smallest area when revolved about the x-axis, and in this case the answer is far from clear. If we think of a typical curve as a frictionless wire in a vertical plane, then another nontrivial problem is that of finding the curve down which a bead will slide from P to Q in the shortest time. This is the famous brachistochrone problem of John Bernoulli, which we discussed in Section 6. Intuitive answers to such questions are quite rare, and the calculus of variations provides a uniform analytical method for dealing with situations of this kind.

502

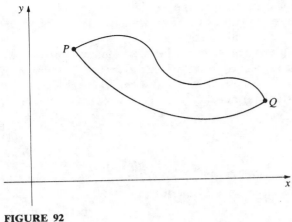

FIGURE 92

Every student of elementary calculus is familiar with the problem of finding points at which a function of a single variable has maximum or minimum values. The above problems show that in the calculus of variations we consider some quantity (arc length, surface area, time of descent) that depends on an entire curve, and we seek the curve that minimizes the quantity in question. The calculus of variations also deals with minimum problems depending on surfaces. For example, if a circular wire is bent in any manner and dipped into a soap solution, then the soap film spanning the wire will assume the shape of the surface of smallest area bounded by the wire. The mathematical problem is to find the surface from this minimum property and the known shape of the wire.

In addition, the calculus of variations has played an important role as a unifying influence in mechanics and as a guide in the mathematical interpretation of many physical phenomena. For instance, it has been found that if the configuration of a system of moving particles is governed by their mutual gravitational attractions, then their actual paths will be minimizing curves for the integral, with respect to time, of the difference between the kinetic and potential energies of the system. This far-reaching statement of classical mechanics is known as *Hamilton's principle* after its discoverer. Also, in modern physics, Einstein made extensive use of the calculus of variations in his work on general relativity, and Schrödinger used it to discover his famous wave equation, which is one of the cornerstones of quantum mechanics.

A few of the problems of the calculus of variations are very old, and were considered and partly solved by the ancient Greeks. The invention of ordinary calculus by Newton and Leibniz stimulated the study of a number of variational problems, and some of these were solved by ingenious special methods. However, the subject was launched as a coherent branch of analysis by Euler in 1744, with his discovery of the basic differential equation for a minimizing curve.

We shall discuss Euler's equation in the next section, but first we observe that each of the problems described in the second paragraph of this section is a special case of the following more general problem. Let P and Q have coordinates (x_1, y_1) and (x_2, y_2), and consider the family of functions

$$y = y(x) \tag{1}$$

that satisfy the boundary conditions $y(x_1) = y_1$ and $y(x_2) = y_2$—that is, the graph of (1) must join P and Q. Then we wish to find the function in this family that minimizes an integral of the form

$$I(y) = \int_{x_1}^{x_2} f(x, y, y')\, dx. \tag{2}$$

To see that this problem indeed contains the others, we note that the length of the curve (1) is

$$\int_{x_1}^{x_2} \sqrt{1 + (y')^2}\, dx, \tag{3}$$

and that the area of the surface of revolution obtained by revolving it about the x-axis is

$$\int_{x_1}^{x_2} 2\pi y \sqrt{1 + (y')^2}\, dx. \tag{4}$$

In the case of the curve of quickest descent, it is convenient to invert the coordinate system and take the point P at the origin, as in Fig. 93. Since the speed $v = ds/dt$ is given by $v = \sqrt{2gy}$, the total time of descent is

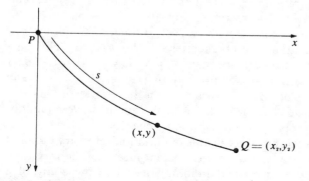

FIGURE 93

the integral of ds/v and the integral to be minimized is

$$\int_{x_1}^{x_2} \frac{\sqrt{1 + (y')^2}}{\sqrt{2gy}}\, dx. \tag{5}$$

Accordingly, the function $f(x,y,y')$ occurring in (2) has the respective forms $\sqrt{1 + (y')^2}$, $2\pi y\sqrt{1 + (y')^2}$ and $\sqrt{1 + (y')^2}/\sqrt{2gy}$ in our three problems.

It is necessary to be somewhat more precise in formulating the basic problem of minimizing the integral (2). First, we will always assume that the function $f(x,y,y')$ has continuous partial derivatives of the second order with respect to x, y, and y'. The next question is, What types of functions (1) are to be allowed? The integral (2) is a well-defined real number whenever the integrand is continuous as a function of x, and for this it suffices to assume that $y'(x)$ is continuous. However, in order to guarantee the validity of the operations we will want to perform, it is convenient to restrict outselves once and for all to considering only unknown functions $y(x)$ that have continuous second derivatives and satisfy the given boundary conditions $y(x_1) = y_1$ and $y(x_2) = y_2$. Functions of this kind will be called *admissible*. We can imagine a competition which only admissible functions are allowed to enter, and the problem is to select from this family the function or functions that yield the smallest value for I.

In spite of these remarks, we will not be seriously concerned with issues of mathematical rigor. Our point of view is deliberately naive, and our sole purpose is to reach the interesting applications as quickly and simply as possible. The reader who wishes to explore the very extensive theory of the subject can readily do so in the systematic treatises.[1]

66 EULER'S DIFFERENTIAL EQUATION FOR AN EXTREMAL

Assuming that there exists an admissible function $y(x)$ that minimizes the integral

$$I = \int_{x_1}^{x_2} f(x,y,y')\, dx, \tag{1}$$

how do we find this function? We shall obtain a differential equation for

---

[1] See, for example, I. M. Gelfand and S. V. Fomin, *Calculus of Variations*, Prentice-Hall, Englewood Cliffs, N.J., 1963; G. M. Ewing, *Calculus of Variations with Applications*, Norton, New York, 1969; or C. Carathéodory, *Calculus of Variations and Partial Differential Equations of the First Order, Part II: Calculus of Variations*, Holden-Day, San Francisco, 1967.

$y(x)$ by comparing the values of I that correspond to neighboring admissible functions. The central idea is that since $y(x)$ gives a minimum value to I, I will increase if we "disturb" $y(x)$ slightly. These disturbed functions are constructed as follows.

Let $\eta(x)$ be any function with the properties that $\eta''(x)$ is continuous and

$$\eta(x_1) = \eta(x_2) = 0. \tag{2}$$

If α is a small parameter, then

$$\bar{y}(x) = y(x) + \alpha\eta(x) \tag{3}$$

represents a one-parameter family of admissible functions. The vertical deviation of a curve in this family from the minimizing curve $y(x)$ is $\alpha\eta(x)$, as shown in Fig. 94.[2] The significance of (3) lies in the fact that for each family of this type, that is, for each choice of the function $\eta(x)$, the minimizing function $y(x)$ belongs to the family and corresponds to the value of the parameter $\alpha = 0$.

Now, with $\eta(x)$ fixed, we substitute $\bar{y}(x) = y(x) + \alpha\eta(x)$ and

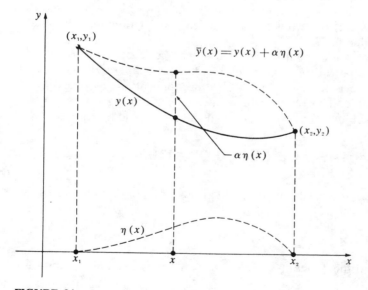

FIGURE 94

---

[2] The difference $\bar{y} - y = \alpha\eta$ is called the *variation* of the function y and is usually denoted by δy. This notation can be developed into a useful formalism (which we do not discuss) and is the source of the name *calculus of variations*.

$\bar{y}'(x) = y'(x) + \alpha\eta'(x)$ into the integral (1), and get a function of α,

$$I(\alpha) = \int_{x_1}^{x_2} f(x,\bar{y},\bar{y}') \, dx$$

$$= \int_{x_1}^{x_2} f[x,y(x) + \alpha\eta(x), y'(x) + \alpha\eta'(x)] \, dx. \tag{4}$$

When $\alpha = 0$, formula (3) yields $\bar{y}(x) = y(x)$; and since $y(x)$ minimizes the integral, we know that $I(\alpha)$ must have a minimum when $\alpha = 0$. By elementary calculus, a necessary condition for this is the vanishing of the derivative $I'(\alpha)$ when $\alpha = 0$: $I'(0) = 0$. The derivative $I'(\alpha)$ can be computed by differentiating (4) under the integral sign, that is,

$$I'(\alpha) = \int_{x_1}^{x_2} \frac{\partial}{\partial\alpha} f(x,\bar{y},\bar{y}') \, dx. \tag{5}$$

By the chain rule for differentiating functions of several variables, we have

$$\frac{\partial}{\partial\alpha} f(x,\bar{y},\bar{y}') = \frac{\partial f}{\partial x} \frac{\partial x}{\partial\alpha} + \frac{\partial f}{\partial\bar{y}} \frac{\partial\bar{y}}{\partial\alpha} + \frac{\partial f}{\partial\bar{y}'} \frac{\partial\bar{y}'}{\partial\alpha}$$

$$= \frac{\partial f}{\partial\bar{y}} \eta(x) + \frac{\partial f}{\partial\bar{y}'} \eta'(x),$$

so (5) can be written as

$$I'(\alpha) = \int_{x_1}^{x_2} \left[\frac{\partial f}{\partial\bar{y}} \eta(x) + \frac{\partial f}{\partial\bar{y}'} \eta'(x) \right] dx. \tag{6}$$

Now $I'(0) = 0$, so putting $\alpha = 0$ in (6) yields

$$\int_{x_1}^{x_2} \left[\frac{\partial f}{\partial y} \eta(x) + \frac{\partial f}{\partial y'} \eta'(x) \right] dx = 0. \tag{7}$$

In this equation the derivative $\eta'(x)$ appears along with the function $\eta(x)$. We can eliminate $\eta'(x)$ by integrating the second term by parts, which gives

$$\int_{x_1}^{x_2} \frac{\partial f}{\partial y'} \eta'(x) \, dx = \left[\eta(x) \frac{\partial f}{\partial y'} \right]_{x_1}^{x_2} - \int_{x_1}^{x_2} \eta(x) \frac{d}{dx} \left(\frac{\partial f}{\partial y'} \right) dx$$

$$= - \int_{x_1}^{x_2} \eta(x) \frac{d}{dx} \left(\frac{\partial f}{\partial y'} \right) dx$$

by virtue of (2). We can therefore write (7) in the form

$$\int_{x_1}^{x_2} \eta(x) \left[\frac{\partial f}{\partial y} - \frac{d}{dx} \left(\frac{\partial f}{\partial y'} \right) \right] dx = 0. \tag{8}$$

Our reasoning up to this point is based on a fixed choice of the function $\eta(x)$. However, since the integral in (8) must vanish for *every* such function, we at once conclude that the expression in brackets must also vanish. This yields

$$\frac{d}{dx}\left(\frac{\partial f}{\partial y'}\right) - \frac{\partial f}{\partial y} = 0, \tag{9}$$

which is *Euler's equation.*[3]

It is important to have a clear understanding of the exact nature of our conclusion: namely, if $y(x)$ is an admissible function that minimizes the integral (1), then y satisfies Euler's equation. Suppose an admissible function y can be found that satisfies this equation. Does this mean that y minimizes I? Not necessarily. The situation is similar to that in elementary calculus, where a function $g(x)$ whose derivative is zero at a point x_0 may have a maximum, a minimum, or a point of inflection at x_0. When no distinctions are made, these cases are often called *stationary values* of $g(x)$, and the points x_0 at which they occur are *stationary points*. In the same way, the condition $I'(0) = 0$ can perfectly well indicate a maximum or point of inflection for $I(\alpha)$ at $\alpha = 0$, instead of a minimum. Thus it is customary to call any admissible solution of Euler's equation a *stationary function* or *stationary curve,* and to refer to the corresponding value of the integral (1) as a *stationary value* of this integral—without committing ourselves as to which of the several possibilities actually occurs. Furthermore, solutions of Euler's equation which are unrestricted by the boundary conditions are called *extremals.*

In calculus we use the second derivative to give sufficient conditions distinguishing one type of stationary value from another. Similar sufficient conditions are available in the calculus of variations, but since these are quite complicated, we will not consider them here. In actual practice, the geometry or physics of the problem under discussion often makes it possible to determine whether a particular stationary function maximizes or minimizes the integral (or neither). The reader who is interested in sufficient conditions and other theoretical problems will find adequate discussions in the books mentioned in Section 65.

As it stands, Euler's equation (9) is not very illuminating. In order to interpret it and convert it into a useful tool, we begin by emphasizing

---

[3] In more detail, the indirect argument leading to (9) is as follows. Assume that the bracketed function in (8) is not zero (say, positive) at some point $x = a$ in the interval. Since this function is continuous, it will be positive throughout some subinterval about $x = a$. Choose an $\eta(x)$ that is positive inside the subinterval and zero outside. For this $\eta(x)$, the integral in (8) will be positive—which is a contradiction. When this argument is formalized, the resulting statement is known as the *fundamental lemma of the calculus of variations.*

that the partial derivatives $\partial f/\partial y$ and $\partial f/\partial y'$ are computed by treating x, y, and y' as independent variables. In general, however, $\partial f/\partial y'$ is a function of x explicitly, and also implicitly through y and y', so the first term in (9) can be written in the expanded form

$$\frac{\partial}{\partial x}\left(\frac{\partial f}{\partial y'}\right) + \frac{\partial}{\partial y}\left(\frac{\partial f}{\partial y'}\right)\frac{dy}{dx} + \frac{\partial}{\partial y'}\left(\frac{\partial f}{\partial y'}\right)\frac{dy'}{dx}.$$

Accordingly, Euler's equation is

$$f_{y'y'}\frac{d^2y}{dx^2} + f_{y'y}\frac{dy}{dx} + (f_{y'x} - f_y) = 0. \tag{10}$$

This equation is of the second order unless $f_{y'y'} = 0$, so in general the extremals—its solutions—constitute a two-parameter family of curves; and among these, the stationary functions are those in which the two parameters are chosen to fit the given boundary conditions. A second order nonlinear equation like (10) is usually impossible to solve, but fortunately many applications lead to special cases that can be solved.

CASE A. If x and y are missing from the function f, then Euler's equation reduces to

$$f_{y'y'}\frac{d^2y}{dx^2} = 0;$$

and if $f_{y'y'} \neq 0$, we have $d^2y/dx^2 = 0$ and $y = c_1x + c_2$, so the extremals are all straight lines.

CASE B. If y is missing from the function f, then Euler's equation becomes

$$\frac{d}{dx}\left(\frac{\partial f}{\partial y'}\right) = 0,$$

and this can be integrated at once to yield the first order equation

$$\frac{\partial f}{\partial y'} = c_1$$

for the extremals.

CASE C. If x is missing from the function f, then Euler's equation can be integrated to

$$\frac{\partial f}{\partial y'}y' - f = c_1.$$

This follows from the identity

$$\frac{d}{dx}\left(\frac{\partial f}{\partial y'}y' - f\right) = y'\left[\frac{d}{dx}\left(\frac{\partial f}{\partial y'}\right) - \frac{\partial f}{\partial y}\right] - \frac{\partial f}{\partial x},$$

since $\partial f/\partial x = 0$ and the expression in brackets on the right is zero by Euler's equation.

We now apply this machinery to the three problems formulated in Section 65.

Example 1. To find the shortest curve joining two points (x_1, y_1) and (x_2, y_2)—which we know intuitively to be a straight line—we must minimize the arc length integral

$$I = \int_{x_1}^{x_2} \sqrt{1 + (y')^2}\, dx.$$

The variables x and y are missing from $f(y') = \sqrt{1 + (y')^2}$, so this problem falls under Case A. Since

$$f_{y'y'} = \frac{\partial^2 f}{\partial y'^2} = \frac{1}{[1 + (y')^2]^{3/2}} \neq 0,$$

Case A tells us that the extremals are the two-parameter family of straight lines $y = c_1 x + c_2$. The boundary conditions yield

$$y - y_1 = \frac{y_2 - y_1}{x_2 - x_1}(x - x_1) \tag{11}$$

as the stationary curve, and this is of course the straight line joining the two points. It should be noted that this analysis shows only that if I has a stationary value, then the corresponding stationary curve must be the straight line (11). However, it is clear from the geometry that I has no maximizing curve but does have a minimizing curve, so we conclude in this way that (11) actually is the shortest curve joining our two points.

In this example we arrived at an obvious conclusion by analytical means. A much more difficult and interesting problem is that of finding the shortest curve joining two fixed points on a given surface and lying entirely on that surface. These curves are called *geodesics*, and the study of their properties is one of the focal points of the branch of mathematics known as differential geometry.

Example 2. To find the curve joining the points (x_1, y_1) and (x_2, y_2) that yields a surface of revolution of minimum area when revolved about the x-axis, we must minimize

$$I = \int_{x_1}^{x_2} 2\pi y \sqrt{1 + (y')^2}\, dx. \tag{12}$$

The variable x is missing from $f(y,y') = 2\pi y \sqrt{1 + (y')^2}$, so Case C tells us that Euler's equation becomes

$$\frac{y(y')^2}{\sqrt{1 + (y')^2}} - y\sqrt{1 + (y')^2} = c_1,$$

which simplifies to

$$c_1 y' = \sqrt{y^2 - c_1^2}.$$

On separating variables and integrating, we get

$$x = c_1 \int \frac{dy}{\sqrt{y^2 - c_1^2}} = c_1 \log\left(\frac{y + \sqrt{y^2 - c_1^2}}{c_1}\right) + c_2,$$

and solving for y gives

$$y = c_1 \cosh\left(\frac{x - c_2}{c_1}\right). \tag{13}$$

The extremals are therefore catenaries, and the required minimal surface—if it exists—must be obtained by revolving a catenary. The next problem is that of seeing whether the parameters c_1 and c_2 can indeed be chosen so that the curve (13) joins the points (x_1,y_1) and (x_2,y_2).

The choosing of these parameters turns out to be curiously complicated. If the curve (13) is made to pass through the first point (x_1,y_1), then one parameter is left free. Two members of this one-parameter family are shown in Fig. 95. It can be proved that all such curves are tangent to the

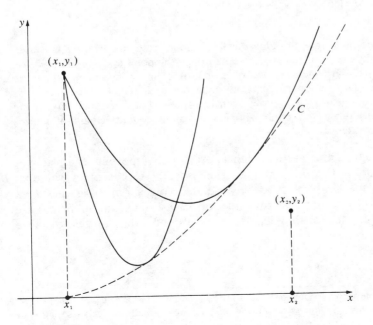

FIGURE 95

dashed curve C, so no curve in the family crosses C. Thus, when the second point (x_2,y_2) is below C, as in Fig. 95, there is no catenary through both points and no stationary function exists. In this case it is found that smaller and smaller surfaces are generated by curves that approach the dashed line from (x_1,y_1) to $(x_1,0)$ to $(x_2,0)$ to (x_2,y_2), so no admissible curve can generate a minimal surface. When the second point lies above C, there are two catenaries through the points, and hence two stationary functions, but only the upper catenary generates a minimal surface. Finally, when the second point is on C, there is only one stationary function but the surface it generates is not minimal.[4]

Example 3. To find the curve of quickest descent in Fig. 93, we must minimize

$$I = \int_{x_1}^{x_2} \frac{\sqrt{1 + (y')^2}}{\sqrt{2gy}} \, dx.$$

Again the variable x is missing from the function $f(y,y') = \sqrt{1 + (y')^2}/\sqrt{2gy}$, so by Case C, Euler's equation becomes

$$\frac{(y')^2}{\sqrt{y}\sqrt{1 + (y')^2}} - \frac{\sqrt{1 + (y')^2}}{\sqrt{y}} = c_1.$$

This reduces to

$$y[1 + (y')^2] = c,$$

which is precisely the differential equation 6-(4) arrived at in our earlier discussion of this famous problem. Its solution is given in Section 6. The resulting stationary curve is the cycloid

$$x = a(\theta - \sin \theta) \qquad \text{and} \qquad y = a(1 - \cos \theta) \tag{14}$$

generated by a circle of radius a rolling under the x axis, where a is chosen so that the first inverted arch passes through the point (x_2,y_2) in Fig. 93. As before, this argument shows only that if I has a minimum, then the corresponding stationary curve must be the cycloid (14). However, it is reasonably clear from physical considerations that I has no maximizing curve but does have a minimizing curve, so this cycloid actually minimizes the time of descent.

We conclude this section with an easy but important extension of our treatment of the integral (1). This integral represents variational problems of the simplest type because it involves only one unknown function. However, some of the situations we will encounter below are not quite so simple, for they lead to integrals depending on two or more unknown functions.

---

[4] A full discussion of these statements, with proofs, can be found in Chapter IV of G. A. Bliss's book *Calculus of Variations*, Carus Monograph no. 1, Mathematical Association of America, 1925.

For example, suppose we want to find conditions necessarily satisfied by two functions $y(x)$ and $z(x)$ that give a stationary value to the integral

$$I = \int_{x_1}^{x_2} f(x,y,z,y',z')\,dx, \tag{15}$$

where the boundary values $y(x_1)$, $z(x_1)$ and $y(x_2)$, $z(x_2)$ are specified in advance. Just as before, we introduce functions $\eta_1(x)$ and $\eta_2(x)$ that have continuous second derivatives and vanish at the endpoints. From these we form the neighboring functions $\bar{y}(x) = y(x) + \alpha\eta_1(x)$ and $\bar{z}(x) = z(x) + \alpha\eta_2(x)$, and then consider the function of α defined by

$$I(\alpha) = \int_{x_1}^{x_2} f(x,y + \alpha\eta_1, z + \alpha\eta_2, y' + \alpha\eta_1', z' + \alpha\eta_2')\,dx. \tag{16}$$

Again, if $y(x)$ and $z(x)$ are stationary functions we must have $I'(0) = 0$, so by computing the derivative of (16) and putting $\alpha = 0$ we get

$$\int_{x_1}^{x_2} \left(\frac{\partial f}{\partial y}\eta_1 + \frac{\partial f}{\partial z}\eta_2 + \frac{\partial f}{\partial y'}\eta_1' + \frac{\partial f}{\partial z'}\eta_2'\right) dx = 0,$$

or, if the terms involving η_1' and η_2' are integrated by parts,

$$\int_{x_1}^{x_2} \left\{\eta_1(x)\left[\frac{\partial f}{\partial y} - \frac{d}{dx}\left(\frac{\partial f}{\partial y'}\right)\right] + \eta_2(x)\left[\frac{\partial f}{\partial z} - \frac{d}{dx}\left(\frac{\partial f}{\partial z'}\right)\right]\right\} dx = 0. \tag{17}$$

Finally, since (17) must hold for all choices of the functions $\eta_1(x)$ and $\eta_2(x)$, we are led at once to Euler's equations

$$\frac{d}{dx}\left(\frac{\partial f}{\partial y'}\right) - \frac{\partial f}{\partial y} = 0 \quad \text{and} \quad \frac{d}{dx}\left(\frac{\partial f}{\partial z'}\right) - \frac{\partial f}{\partial z} = 0. \tag{18}$$

Thus, to find the extremals of our problem, we must solve the system (18). Needless to say, a system of intractable equations is harder to solve than only one; but if (18) can be solved, then the stationary functions are determined by fitting the resulting solutions to the given boundary conditions. Similar considerations apply without any essential change to integrals like (15) which involve more than two unknown functions.

PROBLEMS

1. Find the extremals for the integral (1) if the integrand is

(a) $\dfrac{\sqrt{1 + (y')^2}}{y}$;

(b) $y^2 - (y')^2$.

2. Find the stationary function of

$$\int_0^4 [xy' - (y')^2]\, dx$$

which is determined by the boundary conditions $y(0) = 0$ and $y(4) = 3$.

3. When the integrand in (1) is of the form

$$a(x)(y')^2 + 2b(x)yy' + c(x)y^2,$$

show that Euler's equation is a second order linear differential equation.

4. If P and Q are two points in a plane, then in terms of polar coordinates, the length of a curve from P to Q is

$$\int_P^Q ds = \int_P^Q \sqrt{dr^2 + r^2\, d\theta^2}\,.$$

Find the polar equation of a straight line by minimizing this integral
(a) with θ as the independent variable;
(b) with r as the independent variable.

5. Consider two points P and Q on the surface of the sphere $x^2 + y^2 + z^2 = a^2$, and coordinatize this surface by means of the spherical coordinates θ and ϕ, where $x = a \sin\phi \cos\theta$, $y = a \sin\phi \sin\theta$, and $z = a \cos\phi$. Let $\theta = F(\phi)$ be a curve lying on the surface and joining P and Q. Show that the shortest such curve (a geodesic) is an arc of a great circle, that is, that it lies on a plane through the center. *Hint:* Express the length of the curve in the form

$$\int_P^Q ds = \int_P^Q \sqrt{dx^2 + dy^2 + dz^2}$$

$$= a \int_P^Q \sqrt{1 + \left(\frac{d\theta}{d\phi}\right)^2 \sin^2\phi}\; d\phi,$$

solve the corresponding Euler equation for θ, and convert the result back into rectangular coordinates.

6. Prove that any geodesic on the right circular cone $z^2 = a^2(x^2 + y^2)$, $z \geq 0$, has the following property: If the cone is cut along a generator and flattened into a plane, then the geodesic becomes a straight line. *Hint:* Represent the cone parametrically by means of the equations

$$x = \frac{r \cos(\theta\sqrt{1 + a^2})}{\sqrt{1 + a^2}}, \qquad y = \frac{r \sin(\theta\sqrt{1 + a^2})}{\sqrt{1 + a^2}}, \qquad z = \frac{ar}{\sqrt{1 + a^2}};$$

show that the parameters r and θ represent ordinary polar coordinates on the flattened cone; and show that a geodesic $r = r(\theta)$ is a straight line in these polar coordinates.

7. If the curve $y = g(z)$ is revolved about the z-axis, then the resulting surface of revolution has $x^2 + y^2 = g(z)^2$ as its equation. A convenient parametric representation of this surface is given by

$$x = g(z) \cos\theta, \qquad y = g(z) \sin\theta, \qquad z = z,$$

where θ is the polar angle in the xy-plane. Show that a geodesic $\theta = \theta(z)$ on this surface has

$$\theta = c_1 \int \frac{\sqrt{1 + [g'(z)]^2}}{g(z)\sqrt{g(z)^2 - c_1^2}} \, dz + c_2$$

as its equation.

8. If the surface of revolution in Problem 7 is a right circular cylinder, show that every geodesic of the form $\theta = \theta(z)$ is a helix or a generator.

67 ISOPERIMETRIC PROBLEMS

The ancient Greeks proposed the problem of finding the closed plane curve of given length that encloses the largest area. They called this the *isoperimetric problem*, and were able to show in a more or less rigorous manner that the obvious answer—a circle—is correct.[5] If the curve is expressed parametrically by $x = x(t)$ and $y = y(t)$, and is traversed once counterclockwise as t increases from t_1 to t_2, then the enclosed area is known to be

$$A = \frac{1}{2} \int_{t_1}^{t_2} \left(x \frac{dy}{dt} - y \frac{dx}{dt} \right) dt, \tag{1}$$

which is an integral depending on two unknown functions.[6] Since the length of the curve is

$$L = \int_{t_1}^{t_2} \sqrt{\left(\frac{dx}{dt}\right)^2 + \left(\frac{dy}{dt}\right)^2} \, dt, \tag{2}$$

the problem is to maximize (1) subject to the side condition that (2) must have a constant value. The term *isoperimetric problem* is usually extended to include the general case of finding extremals for one integral subject to any constraint requiring a second integral to take on a prescribed value.

We will also consider finite side conditions, which do not involve integrals or derivatives. For example, if

$$G(x,y,z) = 0 \tag{3}$$

is a given surface, then a curve on this surface is determined parametrically by three functions $x = x(t)$, $y = y(t)$, and $z = z(t)$ that satisfy equation (3), and the problem of finding geodesics amounts to the

---

[5] See B. L. van der Waerden, *Science Awakening*, pp. 268–269, Oxford University Press, London, 1961; also, G. Polya, *Induction and Analogy in Mathematics*, Chapter 10, Princeton University Press, Princeton, N.J., 1954.

[6] Formula (1) is a special case of Green's theorem. Also, see Problem 1.

problem of minimizing the arc length integral

$$\int_{t_1}^{t_2} \sqrt{\left(\frac{dx}{dt}\right)^2 + \left(\frac{dy}{dt}\right)^2 + \left(\frac{dz}{dt}\right)^2} \, dt \tag{4}$$

subject to the side condition (3).

Lagrange multipliers. It is necessary to begin by considering some problems in elementary calculus that are quite similar to isoperimetric problems. For example, suppose we want to find the points (x,y) that yield stationary values for a function $z = f(x,y)$, where, however, the variables x and y are not independent but are constrained by a side condition

$$g(x,y) = 0. \tag{5}$$

The usual procedure is to arbitrarily designate one of the variables x and y in (5) as independent, say x, and the other as dependent on it, so that dy/dx can be computed from

$$\frac{\partial g}{\partial x} + \frac{\partial g}{\partial y}\frac{dy}{dx} = 0.$$

We next use the fact that since z is now a function of x alone, $dz/dx = 0$ is a necessary condition for z to have a stationary value, so

$$\frac{dz}{dx} = \frac{\partial f}{\partial x} + \frac{\partial f}{\partial y}\frac{dy}{dx} = 0$$

or

$$\frac{\partial f}{\partial x} - \frac{\partial f}{\partial y}\frac{\partial g/\partial x}{\partial g/\partial y} = 0. \tag{6}$$

On solving (5) and (6) simultaneously, we obtain the required points (x,y).[7]

One drawback to this approach is that the variables x and y occur symmetrically but are treated unsymmetrically. It is possible to solve the same problem by a different and more elegant method that also has many practical advantages. We form the function

$$F(x,y,\lambda) = f(x,y) + \lambda g(x,y)$$

and investigate its *unconstrained* stationary values by means of the

---

[7] In very simple cases, of course, we can solve (5) for y as a function of x and insert this in $z = f(x,y)$, which gives z as an explicit function of x; and all that remains is to compute dz/dx, solve the equation $dz/dx = 0$, and find the corresponding y's.

necessary conditions

$$\frac{\partial F}{\partial x} = \frac{\partial f}{\partial x} + \lambda \frac{\partial g}{\partial x} = 0,$$

$$\frac{\partial F}{\partial y} = \frac{\partial f}{\partial y} + \lambda \frac{\partial g}{\partial y} = 0, \tag{7}$$

$$\frac{\partial F}{\partial \lambda} = g(x,y) = 0.$$

If λ is eliminated from the first two of these equations, then the system clearly reduces to

$$\frac{\partial f}{\partial x} - \frac{\partial f}{\partial y} \frac{\partial g / \partial x}{\partial g / \partial y} = 0 \quad \text{and} \quad g(x,y) = 0,$$

and this is the system obtained in the above paragraph. It should be observed that this technique (solving the system (7) for x and y) solves the given problem in a way that has two major features important for theoretical work: it does not disturb the symmetry of the problem by making an arbitrary choice of the independent variable; and it removes the side condition at the small expense of introducing λ as another variable. The parameter λ is called a *Lagrange multiplier*, and this method is known as the method of Lagrange multipliers.[8] This discussion extends in an obvious manner to problems involving functions of more than two variables with several side conditions.

Integral side conditions. Here we want to find the differential equation that must be satisfied by a function $y(x)$ that gives a stationary value to the integral

$$I = \int_{x_1}^{x_2} f(x,y,y') \, dx, \tag{8}$$

where y is subject to the side condition

$$J = \int_{x_1}^{x_2} g(x,y,y') \, dx = c \tag{9}$$

and assumes prescribed values $y(x_1) = y_1$ and $y(x_2) = y_2$ at the endpoints. As before, we assume that $y(x)$ is the actual stationary function and disturb it slightly to find the desired analytic condition. However, this problem cannot be attacked by our earlier method of considering neighboring functions of the form $\bar{y}(x) = y(x) + \alpha \eta(x)$, for in general

---

[8] A brief account of Lagrange is given in Appendix A.

these will not maintain the second integral J at the constant value c. Instead, we consider a two-parameter family of neighboring functions

$$\bar{y}(x) = y(x) + \alpha_1 \eta_1(x) + \alpha_2 \eta_2(x), \tag{10}$$

where $\eta_1(x)$ and $\eta_2(x)$ have continuous second derivatives and vanish at the endpoints. The parameters α_1 and α_2 are not independent, but are related by the condition that

$$J(\alpha_1, \alpha_2) = \int_{x_1}^{x_2} g(x, \bar{y}, \bar{y}') \, dx = c. \tag{11}$$

Our problem is then reduced to that of finding necessary conditions for the function

$$I(\alpha_1, \alpha_2) = \int_{x_1}^{x_2} f(x, \bar{y}, \bar{y}') \, dx \tag{12}$$

to have a stationary value at $\alpha_1 = \alpha_2 = 0$, where α_1 and α_2 satisfy (11). This situation is made to order for the method of Lagrange multipliers. We therefore introduce the function

$$K(\alpha_1, \alpha_2, \lambda) = I(\alpha_1, \alpha_2) + \lambda J(\alpha_1, \alpha_2)$$

$$= \int_{x_1}^{x_2} F(x, \bar{y}, \bar{y}') \, dx, \tag{13}$$

where

$$F = f + \lambda g,$$

and investigate its unconstrained stationary value at $\alpha_1 = \alpha_2 = 0$ by means of the necessary conditions

$$\frac{\partial K}{\partial \alpha_1} = \frac{\partial K}{\partial \alpha_2} = 0 \qquad \text{when } \alpha_1 = \alpha_2 = 0. \tag{14}$$

If we differentiate (13) under the integral sign and use (10), we get

$$\frac{\partial K}{\partial \alpha_i} = \int_{x_1}^{x_2} \left[\frac{\partial F}{\partial \bar{y}} \eta_i(x) + \frac{\partial F}{\partial \bar{y}'} \eta_i'(x) \right] dx \qquad \text{for } i = 1, 2;$$

and setting $\alpha_1 = \alpha_2 = 0$ yields

$$\int_{x_1}^{x_2} \left[\frac{\partial F}{\partial y} \eta_i(x) + \frac{\partial F}{\partial y'} \eta_i'(x) \right] dx = 0$$

by virtue of (14). After the second term is integrated by parts, this becomes

$$\int_{x_1}^{x_2} \eta_i(x) \left[\frac{\partial F}{\partial y} - \frac{d}{dx} \left(\frac{\partial F}{\partial y'} \right) \right] dx = 0. \tag{15}$$

Since $\eta_1(x)$ and $\eta_2(x)$ are both arbitrary, the two conditions embodied in (15) amount to only one condition, and as usual we conclude that the stationary function $y(x)$ must satisfy Euler's equation

$$\frac{d}{dx}\left(\frac{\partial F}{\partial y'}\right) - \frac{\partial F}{\partial y} = 0. \tag{16}$$

The solutions of this equation (the extremals of our problem) involve three undetermined parameters: two constants of integration, and the Lagrange multiplier λ. The stationary function is then selected from these extremals by imposing the two boundary conditions and giving the integral J its prescribed value c.

In the case of integrals that depend on two or more functions, this result can be extended in the same way as in the previous section. For example, if

$$I = \int_{x_1}^{x_2} f(x,y,z,y',z')\, dx$$

has a stationary value subject to the side condition

$$J = \int_{x_1}^{x_2} g(x,y,z,y',z')\, dx = c,$$

then the stationary functions $y(x)$ and $z(x)$ must satisfy the system of equations

$$\frac{d}{dx}\left(\frac{\partial F}{\partial y'}\right) - \frac{\partial F}{\partial y} = 0 \quad \text{and} \quad \frac{d}{dx}\left(\frac{\partial F}{\partial z'}\right) - \frac{\partial F}{\partial z} = 0, \tag{17}$$

where $F = f + \lambda g$. The reasoning is similar to that already given, and we omit the details.

Example 1. We shall find the curve of fixed length L that joins the points $(0,0)$ and $(1,0)$, lies above the x-axis, and encloses the maximum area between itself and the x-axis. This is a restricted version of the original isoperimetric problem in which part of the curve surrounding the area to be maximized is required to be a line segment of length 1. Our problem is to maximize $\int_0^1 y\, dx$ subject to the side condition

$$\int_0^1 \sqrt{1 + (y')^2}\, dx = L$$

and the boundary conditions $y(0) = 0$ and $y(1) = 0$. Here we have $F = y + \lambda\sqrt{1 + (y')^2}$, so Euler's equation is

$$\frac{d}{dx}\left(\frac{\lambda y'}{\sqrt{1 + (y')^2}}\right) - 1 = 0, \tag{18}$$

or, after carrying out the differentiation,

$$\frac{y''}{[1 + (y')^2]^{3/2}} = \frac{1}{\lambda}. \tag{19}$$

In this case no integration is necessary, since (19) tells us at once that the curvature is constant and equals $1/\lambda$. It follows that the required maximizing curve is an arc of a circle (as might have been expected) with radius λ. As an alternate procedure, we can integrate (18) to get

$$\frac{y'}{\sqrt{1 + (y')^2}} = \frac{x - c_1}{\lambda}.$$

On solving this for y' and integrating again, we obtain

$$(x - c_1)^2 + (y - c_2)^2 = \lambda^2, \tag{20}$$

which of course is the equation of a circle with radius λ.

Example 2. In Example 1 it is clearly necessary to have $L > 1$. Also, if $L > \pi/2$, the circular arc determined by (20) will not define $y > 0$ as a single-valued function of x. We can avoid these artificial issues by considering curves in parametric form $x = x(t)$ and $y = y(t)$ and by turning our attention to the original isoperimetric problem of maximizing

$$\frac{1}{2} \int_{t_1}^{t_2} (x\dot{y} - y\dot{x})\, dt$$

(where $\dot{x} = dx/dt$ and $\dot{y} = dy/dt$) with the side condition

$$\int_{t_1}^{t_2} \sqrt{\dot{x}^2 + \dot{y}^2}\, dt = L.$$

Here we have

$$F = \frac{1}{2}(x\dot{y} + y\dot{x}) + \lambda\sqrt{\dot{x}^2 + \dot{y}^2},$$

so the Euler equations (17) are

$$\frac{d}{dt}\left(-\frac{1}{2}y + \frac{\lambda\dot{x}}{\sqrt{\dot{x}^2 + \dot{y}^2}}\right) - \frac{1}{2}\dot{y} = 0$$

and

$$\frac{d}{dt}\left(\frac{1}{2}x + \frac{\lambda\dot{y}}{\sqrt{\dot{x}^2 + \dot{y}^2}}\right) + \frac{1}{2}\dot{x} = 0.$$

These equations can be integrated directly, which yields

$$-y + \frac{\lambda\dot{x}}{\sqrt{\dot{x}^2 + \dot{y}^2}} = -c_1 \quad \text{and} \quad x + \frac{\lambda\dot{y}}{\sqrt{\dot{x}^2 + \dot{y}^2}} = c_2.$$

If we solve for $x - c_2$ and $y - c_1$, square, and add, then the result is

$$(x - c_2)^2 + (y - c_1)^2 = \lambda^2,$$

so the maximizing curve is a circle. This result can be expressed in the following way: if L is the length of a closed plane curve that encloses an area A, then $A \leq L^2/4\pi$, with equality if and only if the curve is a circle. A relation of this kind is called an *isoperimetric inequality*.[9]

---

[9] Students of physics may be interested in the ideas discussed in G. Polya and G. Szegö, *Isoperimetric Inequalities in Mathematical Physics*, Princeton University Press, Princeton, N.J., 1951.

Finite side conditions. At the beginning of this section we formulated the problem of finding geodesics on a given surface

$$G(x,y,z) = 0. \tag{21}$$

We now consider the slightly more general problem of finding a space curve $x = x(t)$, $y = y(t)$, $z = z(t)$ that gives a stationary value to an integral of the form

$$\int_{t_1}^{t_2} f(\dot{x}, \dot{y}, \dot{z}) \, dt, \tag{22}$$

where the curve is required to lie on the surface (21).

Our strategy is to eliminate the side condition (21), and to do this we proceed as follows. There is no loss of generality in assuming that the curve lies on a part of the surface where $G_z \neq 0$. On this part of the surface we can solve (21) for z, which gives $z = g(x,y)$ and

$$\dot{z} = \frac{\partial g}{\partial x} \dot{x} + \frac{\partial g}{\partial y} \dot{y}. \tag{23}$$

When (23) is inserted in (22), our problem is reduced to that of finding unconstrained stationary functions for the integral

$$\int_{t_1}^{t_2} f\left(\dot{x}, \dot{y}, \frac{\partial g}{\partial x} \dot{x} + \frac{\partial g}{\partial y} \dot{y}\right) dt.$$

We know from the previous section that the Euler equations 66-(18) for this problem are

$$\frac{d}{dt}\left(\frac{\partial f}{\partial \dot{x}} + \frac{\partial f}{\partial \dot{z}}\frac{\partial g}{\partial x}\right) - \frac{\partial f}{\partial \dot{z}}\frac{\partial \dot{z}}{\partial x} = 0,$$

and

$$\frac{d}{dt}\left(\frac{\partial f}{\partial \dot{y}} + \frac{\partial f}{\partial \dot{z}}\frac{\partial g}{\partial y}\right) - \frac{\partial f}{\partial \dot{z}}\frac{\partial \dot{z}}{\partial y} = 0.$$

It follows from (23) that

$$\frac{\partial \dot{z}}{\partial x} = \frac{d}{dt}\left(\frac{\partial g}{\partial x}\right) \qquad \text{and} \qquad \frac{\partial \dot{z}}{\partial y} = \frac{d}{dt}\left(\frac{\partial g}{\partial y}\right),$$

so the Euler equations can be written in the form

$$\frac{d}{dt}\left(\frac{\partial f}{\partial \dot{x}}\right) + \frac{\partial g}{\partial x}\frac{d}{dt}\left(\frac{\partial f}{\partial \dot{z}}\right) = 0 \qquad \text{and} \qquad \frac{d}{dt}\left(\frac{\partial f}{\partial \dot{y}}\right) + \frac{\partial g}{\partial y}\frac{d}{dt}\left(\frac{\partial f}{\partial \dot{z}}\right) = 0.$$

If we now define a function $\lambda(t)$ by

$$\frac{d}{dt}\left(\frac{\partial f}{\partial \dot{z}}\right) = \lambda(t)G_z, \tag{24}$$

and use the relations $\partial g/\partial x = -G_x/G_z$ and $\partial g/\partial y = -G_y/G_z$, then Euler's equations become

$$\frac{d}{dt}\left(\frac{\partial f}{\partial \dot{x}}\right) = \lambda(t)G_x, \tag{25}$$

and

$$\frac{d}{dt}\left(\frac{\partial f}{\partial \dot{y}}\right) = \lambda(t)G_y. \tag{26}$$

Thus a necessary condition for a stationary value is the existence of a function $\lambda(t)$ satisfying equations (24), (25), and (26). On eliminating $\lambda(t)$, we obtain the symmetric equations

$$\frac{(d/dt)(\partial f/\partial \dot{x})}{G_x} = \frac{(d/dt)(\partial f/\partial \dot{y})}{G_y} = \frac{(d/dt)(\partial f/\partial \dot{z})}{G_z}, \tag{27}$$

which together with (21) determine the extremals of the problem. It is worth remarking that equations (24), (25), and (26) can be regarded as the Euler equations for the problem of finding unconstrained stationary functions for the integral

$$\int_{t_1}^{t_2} [f(\dot{x},\dot{y},\dot{z}) + \lambda(t)G(x,y,z)]\,dt.$$

This is very similar to our conclusion for integral side conditions, except that here the multiplier is an undetermined function of t instead of an undetermined constant.

When we specialize this result to the problem of finding geodesics on the surface (21), we have

$$f = \sqrt{\dot{x}^2 + \dot{y}^2 + \dot{z}^2}.$$

The equations (27) become

$$\frac{(d/dt)(\dot{x}/f)}{G_x} = \frac{(d/dt)(\dot{y}/f)}{G_y} = \frac{(d/dt)(\dot{z}/f)}{G_z}, \tag{28}$$

and the problem is to extract information from this system.

Example 3. If we choose the surface (21) to be the sphere $x^2 + y^2 + z^2 = a^2$, then $G(x,y,z) = x^2 + y^2 + z^2 - a^2$ and (28) is

$$\frac{f\ddot{x} - \dot{x}\dot{f}}{2xf^2} = \frac{f\ddot{y} - \dot{y}\dot{f}}{2yf^2} = \frac{f\ddot{z} - \dot{z}\dot{f}}{2zf^2},$$

which can be rewritten in the form

$$\frac{x\ddot{y} - y\ddot{x}}{x\dot{y} - y\dot{x}} = \frac{\dot{f}}{f} = \frac{y\ddot{z} - z\ddot{y}}{y\dot{z} - z\dot{y}}.$$

If we ignore the middle term, this is

$$\frac{(d/dt)(x\dot{y} - y\dot{x})}{x\dot{y} - y\dot{x}} = \frac{(d/dt)(y\dot{z} - z\dot{y})}{y\dot{z} - z\dot{y}}.$$

One integration gives $x\dot{y} - y\dot{x} = c_1(y\dot{z} - z\dot{y})$ or

$$\frac{\dot{x} + c_1\dot{z}}{x + c_1 z} = \frac{\dot{y}}{y},$$

and a second yields $x + c_1 z = c_2 y$. This is the equation of a plane through the origin, so the geodesics on a sphere are arcs of great circles. A different method of arriving at this conclusion is given in Problem 66-5.

In this example we were able to solve equations (28) quite easily, but in general this task is extremely difficult. The main significance of these equations lies in their connection with the following very important result in mathematical physics: if a particle glides along a surface, free from the action of any external force, then its path is a geodesic. We shall prove this dynamical theorem in Appendix B. For the purpose of this argument it will be convenient to assume that the parameter t is the arc length s measured along the curve, so that $f = 1$ and equations (28) become

$$\frac{d^2x/ds^2}{G_x} = \frac{d^2y/ds^2}{G_y} = \frac{d^2z/ds^2}{G_z}. \tag{29}$$

PROBLEMS

1. Convince yourself of the validity of formula (1) for a closed convex curve like that shown in Fig. 96. *Hint:* What is the geometric meaning of

$$\int_P^Q y\,dx + \int_Q^P y\,dx,$$

where the first integral is taken from right to left along the upper part of the curve and the second from left to right along the lower part?
2. Verify formula (1) for the circle whose parametric equations are $x = a\cos t$ and $y = a\sin t$, $0 \le t \le 2\pi$.
3. Solve the following problems by the method of Lagrange multipliers.
   (a) Find the point on the plane $ax + by + cz = d$ that is nearest the origin. *Hint:* Minimize $w = x^2 + y^2 + z^2$ with the side condition $ax + by + cz - d = 0$.
   (b) Show that the triangle with greatest area A for a given perimeter is equilateral. *Hint:* If x, y, and z are the sides, then $A = \sqrt{s(s-x)(s-y)(s-z)}$ where $s = (x + y + z)/2$.
   (c) If the sum of n positive numbers $x_1, x_2, \ldots, x_n$ has a fixed value s, prove that their product $x_1 x_2 \cdots x_n$ has s^n/n^n as its maximum value, and

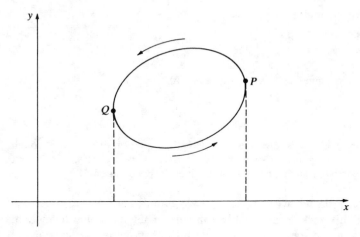

FIGURE 96

conclude from this that the geometric mean of n positive numbers can never exceed their arithmetic mean:

$$\sqrt[n]{x_1 x_2 \cdots x_n} \le \frac{x_1 + x_2 + \cdots + x_n}{n}.$$

4. A curve in the first quadrant joins $(0,0)$ and $(1,0)$ and has a given area beneath it. Show that the shortest such curve is an arc of a circle.

5. A uniform flexible chain of given length hangs between two points. Find its shape if it hangs in such a way as to minimize its potential energy.

6. Solve the original isoperimetric problem (Example 2) by using polar coordinates. *Hint:* Choose the origin to be any point on the curve and the polar axis to be the tangent line at that point; then maximize

$$\frac{1}{2} \int_0^\pi r^2 \, d\theta$$

with the side condition that

$$\int_0^\pi \sqrt{\left(\frac{dr}{d\theta}\right)^2 + r^2} \, d\theta$$

must be constant.

7. Show that the geodesics on any cylinder of the form $g(x,z) = 0$ make a constant angle with the y-axis.

APPENDIX A. LAGRANGE

Joseph Louis Lagrange (1736–1813) detested geometry but made outstanding discoveries in the calculus of variations and analytical mechanics. He also contributed to number theory and algebra, and fed the stream of thought that later nourished Gauss and Abel. His mathematical career can be viewed as a natural extension of the work of his older and greater contemporary, Euler, which in many respects he carried forward and refined.

Lagrange was born in Turin of mixed French–Italian ancestry. As a boy, his tastes were more classical than scientific; but his interest in mathematics was kindled while he was still in school by reading a paper by Edmund Halley on the uses of algebra in optics. He then began a course of independent study, and progressed so rapidly that at the age of nineteen he was appointed professor of mathematics at the Royal Artillery School in Turin.[10]

Lagrange's contributions to the calculus of variations were among his earliest and most important works. In 1755 he communicated to Euler his method of multipliers for solving isoperimetric problems. These problems had baffled Euler for years, since they lay beyond the reach of his own semigeometrical techniques. Euler was immediately able to answer many questions he had long contemplated; but he replied to Lagrange with admirable kindness and generosity, and withheld his own work from publication "so as not to deprive you of any part of the glory which is your due." Lagrange continued working for a number of years on his analytic version of the calculus of variations, and both he and Euler applied it to many new types of problems, especially in mechanics.

In 1766, when Euler left Berlin for St. Petersburg, he suggested to Frederick the Great that Lagrange be invited to take his place. Lagrange accepted and lived in Berlin for 20 years until Frederick's death in 1786. During this period he worked extensively in algebra and number theory and wrote his masterpiece, the treatise *Mécanique Analytique* (1788), in which he unified general mechanics and made of it, as Hamilton later said, "a kind of scientific poem." Among the enduring legacies of this work are Lagrange's equations of motion, generalized coordinates, and the concept of potential energy (which are all discussed in Appendix B).[11]

Men of science found the atmosphere of the Prussian court rather uncongenial after the death of Frederick, so Lagrange accepted an invitation from Louis XVI to move to Paris, where he was given

---

[10] See George Sarton's valuable essay, "Lagrange's Personality," *Proc. Am. Phil. Soc.*, vol. 88, pp. 457–496 (1944).

[11] For some interesting views on Lagrangian mechanics (and many other subjects), see S. Bochner, *The Role of Mathematics in the Rise of Science*, pp. 199–207, Princeton University Press, Princeton, N.J., 1966.

apartments in the Louvre. Lagrange was extremely modest and undogmatic for a man of his great gifts; and though he was a friend of aristocrats—and indeed an aristocrat himself—he was respected and held in affection by all parties throughout the turmoil of the French Revolution. His most important work during these years was his leading part in establishing the metric system of weights and measures. In mathematics, he tried to provide a satisfactory foundation for the basic processes of analysis, but these efforts were largely abortive. Toward the end of his life, Lagrange felt that mathematics had reached a dead end, and that chemistry, physics, biology, and other sciences would attract the ablest minds of the future. His pessimism might have been relieved if he had been able to forsee the coming of Gauss and his successors, who made the nineteenth century the richest in the long history of mathematics.

APPENDIX B. HAMILTON'S PRINCIPLE AND ITS IMPLICATIONS

One purpose of the mathematicians of the eighteenth century was to discover a general principle from which Newtonian mechanics could be deduced. In searching for clues, they noted a number of curious facts in elementary physics: for example, that a ray of light follows the quickest path through an optical medium; that the equilibrium shape of a hanging chain minimizes its potential energy; and that soap bubbles assume a shape having the least surface area for a given volume. These facts and others suggested to Euler that nature pursues its diverse ends by the most efficient and economical means, and that hidden simplicities underlie the apparent chaos of phenomena. It was this metaphysical idea that led him to create the calculus of variations as a tool for investigating such questions. Euler's dream was realized almost a century later by Hamilton.

Hamilton's principle. Consider a particle of mass m moving through space under the influence of a force

$$\mathbf{F} = F_1\mathbf{i} + F_2\mathbf{j} + F_3\mathbf{k},$$

and assume that this force is *conservative* in the sense that the work it does in moving the particle from one point to another is independent of the path. It is easy to show that there exists a scalar function $U(x,y,z)$ such that $\partial U/\partial x = F_1$, $\partial U/\partial y = F_2$, and $\partial U/\partial z = F_3$.[12] The function $V = -U$ is called the *potential energy* of the particle, since the change in

---

[12] In the language of vector analysis, $\mathbf{F}$ is the *gradient of U*.

its value from one point to another is the work done against **F** in moving the particle from the first point to the second. Furthermore, if $\mathbf{r}(t) = x(t)\mathbf{i} + y(t)\mathbf{j} + z(t)\mathbf{k}$ is the position vector of the particle, so that

$$\mathbf{v} = \frac{dx}{dt}\mathbf{i} + \frac{dy}{dt}\mathbf{j} + \frac{dz}{dt}\mathbf{k} \quad \text{and} \quad v = \sqrt{\left(\frac{dx}{dt}\right)^2 + \left(\frac{dy}{dt}\right)^2 + \left(\frac{dz}{dt}\right)^2}$$

are its velocity and speed, respectively, then $T = mv^2/2$ is its *kinetic energy*.

If the particle is at points P_1 and P_2 at times t_1 and t_2, then we are interested in the path it traverses in moving from P_1 to P_2. The *action* (or *Hamilton's integral*) is defined as

$$A = \int_{t_1}^{t_2} (T - V)\, dt,$$

and in general its value depends on the path along which the particle moves in passing from P_1 to P_2. We will show that the actual path of the particle is one that yields a stationary value for the action A.

The function $L = T - V$ is called the *Lagrangian*, and in the case under consideration it is given by

$$L = \frac{1}{2}m\left[\left(\frac{dx}{dt}\right)^2 + \left(\frac{dy}{dt}\right)^2 + \left(\frac{dz}{dt}\right)^2\right] - V(x,y,z).$$

The integrand of the action is therefore a function of the form $f(x,y,z,dx/dt,dy/dt,dz/dt)$, and if the action has a stationary value, then Euler's equations must be satisfied. These equations are

$$m\frac{d^2x}{dt^2} + \frac{\partial V}{\partial x} = 0, \quad m\frac{d^2y}{dt^2} + \frac{\partial V}{\partial y} = 0, \quad m\frac{d^2z}{dt^2} + \frac{\partial V}{\partial z} = 0,$$

and can be written in the form

$$m\frac{d^2\mathbf{r}}{dt^2} = -\frac{\partial V}{\partial x}\mathbf{i} - \frac{\partial V}{\partial y}\mathbf{j} - \frac{\partial V}{\partial z}\mathbf{k} = \mathbf{F}.$$

This is precisely Newton's second law of motion. Thus Newton's law is a necessary condition for the action of the particle to have a stationary value. Since Newton's law governs the motion of the particle, we have the following conclusion.

Hamilton's principle. *If a particle moves from a point P_1 to a point P_2 in a time interval $t_1 \le t \le t_2$, then the actual path it follows is one for which the action assumes a stationary value.*

It is quite easy to give simple examples in which the actual path of a particle maximizes the action. However, if the time interval is sufficiently short, then it can be shown that the action is necessarily a minimum. In this form, Hamilton's principle is sometimes called the *principle of least action,* and can be loosely interpreted as saying that nature tends to equalize the kinetic and potential energies throughout the motion.

In the above discussion we assumed Newton's law and deduced Hamilton's principle as a consequence. The same argument shows that Newton's law follows from Hamilton's principle, so these two approaches to the dynamics of a particle—the vectorial and the variational—are equivalent to one another. This result emphasizes the essential characteristic of variational principles in physics: they express the pertinent physical laws in terms of energy alone, without reference to any coordinate system.

The argument we have given extends at once to a system of n particles of masses m_i, with position vectors $\mathbf{r}_i(t) = x_i(t)\mathbf{i} + y_i(t)\mathbf{j} + z_i(t)\mathbf{k}$, which are moving under the influence of conservative forces $\mathbf{F}_i = F_{i1}\mathbf{i} + F_{i2}\mathbf{j} + F_{i3}\mathbf{k}$. Here the potential energy of the system is a function $V(x_1, y_1, z_1, \ldots, x_n, y_n, z_n)$ such that

$$\frac{\partial V}{\partial x_i} = -F_{i1}, \qquad \frac{\partial V}{\partial y_i} = -F_{i2}, \qquad \frac{\partial V}{\partial z_i} = -F_{i3},$$

the kinetic energy is

$$T = \frac{1}{2} \sum_{i=1}^{n} m_i \left[\left(\frac{dx_i}{dt} \right)^2 + \left(\frac{dy_i}{dt} \right)^2 + \left(\frac{dz_i}{dt} \right)^2 \right],$$

and the action over a time interval $t_1 \le t \le t_2$ is

$$A = \int_{t_1}^{t_2} (T - V) \, dt.$$

In just the same way as above, we see that Newton's equations of motion for the system,

$$m_i \frac{d^2 \mathbf{r}_i}{dt^2} = \mathbf{F}_i,$$

are a necessary condition for the action to have a stationary value. Hamilton's principle therefore holds for any finite system of particles in which the forces are conservative. It applies equally well to more general dynamical systems involving constraints and rigid bodies, and also to continuous media.

In addition, Hamilton's principle can be made to yield the basic laws of electricity and magnetism, quantum theory, and relativity. Its influence is so profound and far-reaching that many scientists regard it as the most powerful single principle in mathematical physics and place it at the pinnacle of physical science. Max Planck, the founder of quantum theory, expressed this view as follows: "The highest and most coveted aim of physical science is to condense all natural phenomena which have been observed and are still to be observed into one simple principle . . . . Amid the more or less general laws which mark the achievements of physical science during the course of the last centuries, the principle of least action is perhaps that which, as regards form and content, may claim to come nearest to this ideal final aim of theoretical research.''

Example 1. If a particle of mass m is constrained to move on a given surface $G(x,y,z) = 0$, and if no force acts on it, then it glides along a geodesic. To establish this, we begin by observing that since no force is present we have $V = 0$, so the Lagrangian $L = T - V$ reduces to T where

$$T = \frac{1}{2} m \left[\left(\frac{dx}{dt} \right)^2 + \left(\frac{dy}{dt} \right)^2 + \left(\frac{dz}{dt} \right)^2 \right].$$

We now apply Hamilton's principle, and require that the action

$$\int_{t_1}^{t_2} L \, dt = \int_{t_1}^{t_2} T \, dt$$

be stationary subject to the side condition $G(x,y,z) = 0$. By Section 67, this is equivalent to requiring that the integral

$$\int_{t_1}^{t_2} [T + \lambda(t)G(x,y,z)] \, dt$$

be stationary with no side condition, where $\lambda(t)$ is an undetermined function of t. Euler's equations for this unconstrained variational problem are

$$m \frac{d^2 x}{dt^2} - \lambda G_x = 0, \qquad m \frac{d^2 y}{dt^2} - \lambda G_y = 0, \qquad m \frac{d^2 z}{dt^2} - \lambda G_z = 0.$$

When m and λ are eliminated, these equations become

$$\frac{d^2 x/dt^2}{G_x} = \frac{d^2 y/dt^2}{G_y} = \frac{d^2 z/dt^2}{G_z}.$$

Now the total energy $T + V = T$ of the particle is constant (we prove this below), so its speed is also constant, and therefore $s = kt$ for some constant k if the arc length s is measured from a suitable point. This enables us to write our equations in the form

$$\frac{d^2 x/ds^2}{G_x} = \frac{d^2 y/ds^2}{G_y} = \frac{d^2 z/ds^2}{G_z}.$$

These are precisely equations 67-(29), so the path of the particle is a geodesic on the surface, as stated.

Lagrange's equations. In classical mechanics, Hamilton's principle can be viewed as the source of Lagrange's equations of motion, which occupy a dominant position in this subject. In order to trace the connection, we must first understand what is meant by degrees of freedom and generalized coordinates.

A single particle moving freely in three-dimensional space is said to have three *degrees of freedom,* since its position can be specified by three independent coordinates x, y, and z. By constraining it to move on a surface $G(x,y,z) = 0$, we reduce its degrees of freedom to two, since one of its coordinates can be expressed in terms of the other two. Similarly, an unconstrained system of n particles has $3n$ degrees of freedom, and the effect of introducing constraints is to reduce the number of independent coordinates needed to describe the configurations of the system. If the rectangular coordinates of the particles are x_i, y_i, and z_i ($i = 1, 2, \ldots, n$), and if the constraints are described by k consistent and independent equations of the form

$$G_j(x_1, y_1, z_1, \ldots, x_n, y_n, z_n) = 0, \qquad j = 1, 2, \ldots, k,$$

then the number of degrees of freedom is $m = 3n - k$. In principle these equations can be used to reduce the number of coordinates from $3n$ to m by expressing the $3n$ numbers x_i, y_i, and z_i ($i = 1, 2, \ldots, n$) in terms of m of these numbers. It is more convenient, however, to introduce Lagrange's *generalized coordinates* $q_1, q_2, \ldots, q_m$, which are any m independent coordinates whatever whose values determine the configurations of the system. This allows us full freedom to choose any coordinate system adapted to the problem at hand—rectangular, cylindrical, spherical, or any other—and renders our analysis independent of any particular coordinate system. We now express the rectangular coordinates of the particles in terms of these generalized coordinates and note that the resulting formulas automatically include the constraints: $x_i = x_i(q_1, \ldots, q_m)$, $y_i = y_i(q_1, \ldots, q_m)$, and $z_i = z_i(q_1, \ldots, q_m)$, where $i = 1, 2, \ldots, n$.

If m_i is the mass of the ith particle, then the kinetic energy of the system is

$$T = \frac{1}{2} \sum_{i=1}^{n} m_i \left[\left(\frac{dx_i}{dt} \right)^2 + \left(\frac{dy_i}{dt} \right)^2 + \left(\frac{dz_i}{dt} \right)^2 \right];$$

and in terms of the generalized coordinates this can be written as

$$T = \frac{1}{2} \sum_{i=1}^{n} m_i \left[\left(\sum_{j=1}^{m} \frac{\partial x_i}{\partial q_j} \dot{q}_j \right)^2 + \left(\sum_{j=1}^{m} \frac{\partial y_i}{\partial q_j} \dot{q}_j \right)^2 + \left(\sum_{j=1}^{m} \frac{\partial z_i}{\partial q_j} \dot{q}_j \right)^2 \right], \qquad (1)$$

where $\dot{q}_j = dq_j/dt$. For later use, we point out that T is a homogeneous function of degree 2 in the $\dot{q}_j$. The potential energy V of the system is assumed to be a function of the q_j alone, so the Lagrangian $L = T - V$ is a function of the form

$$L = L(q_1, q_2, \ldots, q_m, \dot{q}_1, \dot{q}_2, \ldots, \dot{q}_m).$$

Hamilton's principle tells us that the motion proceeds in such a way that the action $\int_{t_1}^{t_2} L \, dt$ is stationary over any interval of time $t_1 \le t \le t_2$, so Euler's equations must be satisfied. In this case these are

$$\frac{d}{dt}\left(\frac{\partial L}{\partial \dot{q}_j}\right) - \frac{\partial L}{\partial q_j} = 0, \qquad j = 1, 2, \ldots, m, \tag{2}$$

which are called *Lagrange's equations*. They constitute a system of m second order differential equations whose solution yields the q_j as functions of t.

We shall draw only one general deduction from Lagrange's equations, namely, the *law of conservation of energy*.

The first step in the reasoning is to note the following identity, which holds for any function L of the variables $t, q_1, q_2, \ldots, q_m, \dot{q}_1, \dot{q}_2, \ldots, \dot{q}_m$:

$$\frac{d}{dt}\left[\sum_{j=1}^{m} \dot{q}_j \frac{\partial L}{\partial \dot{q}_j} - L\right] = \sum_{j=1}^{m} \dot{q}_j\left[\frac{d}{dt}\left(\frac{\partial L}{\partial \dot{q}_j}\right) - \frac{\partial L}{\partial q_j}\right] - \frac{\partial L}{\partial t}. \tag{3}$$

Since the Lagrangian L of our system satisfies equations (2) and does not explicitly depend on t, the right side of (3) vanishes and we have

$$\sum_{j=1}^{m} \dot{q}_j \frac{\partial L}{\partial \dot{q}_j} - L = E \tag{4}$$

for some constant E. We next observe that $\partial V/\partial \dot{q}_j = 0$, so $\partial L/\partial \dot{q}_j = \partial T/\partial \dot{q}_j$. As we have already remarked, formula (1) shows that T is a homogeneous function of degree 2 in the $\dot{q}_j$, so

$$\sum_{j=1}^{m} \dot{q}_j \frac{\partial L}{\partial \dot{q}_j} = \sum_{j=1}^{m} \dot{q}_j \frac{\partial T}{\partial \dot{q}_j} = 2T$$

by Euler's theorem on homogeneous functions.[13] With this result, equation (4) becomes $2T - L = E$ or $2T - (T - V) = E$, so

$$T + V = E,$$

which states that during the motion, the sum of the kinetic and potential energies is constant.

---

[13] Recall that a function $f(x,y)$ is homogeneous of degree n in x and y if $f(kx,ky) = k^n f(x,y)$. If both sides of this are differentiated with respect to k and then k is set equal to 1, we obtain

$$x \frac{\partial f}{\partial x} + y \frac{\partial f}{\partial y} = nf(x,y),$$

which is Euler's theorem for this function. The same result holds for a homogeneous function of more than two variables.

In the following example we illustrate the way in which Lagrange's equations can be used in specific dynamical problems.

Example 2. If a particle of mass m moves in a plane under the influence of a gravitational force of magnitude km/r^2 directed toward the origin, then it is natural to choose polar coordinates as the generalized coordinates: $q_1 = r$ and $q_2 = \theta$. It is easy to see that $T = (m/2)(\dot{r}^2 + r^2\dot{\theta}^2)$ and $V = -km/r$, so the Lagrangian is

$$L = T - V = \frac{m}{2}(\dot{r}^2 + r^2\dot{\theta}^2) + \frac{km}{r}$$

and Lagrange's equations are

$$\frac{d}{dt}\left(\frac{\partial L}{\partial \dot{r}}\right) - \frac{\partial L}{\partial r} = 0, \tag{5}$$

$$\frac{d}{dt}\left(\frac{\partial L}{\partial \dot{\theta}}\right) - \frac{\partial L}{\partial \theta} = 0. \tag{6}$$

Since L does not depend explicitly on θ, equation (6) shows that $\partial L/\partial \dot{\theta} = mr^2\dot{\theta}$ is constant, so

$$r^2\frac{d\theta}{dt} = h \tag{7}$$

for some constant h assumed to be positive. We next observe that (5) can easily be written in the form

$$\frac{d^2r}{dt^2} - r\left(\frac{d\theta}{dt}\right)^2 = -\frac{k}{r^2}.$$

This is precisely equation 21-(12), which we solved in Section 21 to obtain the conclusion that the path of the particle is a conic section.

Variational problems for double integrals. Our general method of finding necessary conditions for an integral to be stationary can be applied equally well to multiple integrals. For example, consider a region R in the xy-plane bounded by a closed curve C (Fig. 97). Let $z = z(x,y)$ be a function that is defined in R and assumes prescribed boundary values on C, but is otherwise arbitrary (except for the usual differentiability conditions). This function can be thought of as defining a variable surface fixed along its boundary in space. An integral of the form

$$I(z) = \iint_R f(x,y,z,z_x,z_y)\, dx\, dy \tag{8}$$

will have values that depend on the choice of z, and we can pose the problem of finding a function z (a stationary function) that gives a stationary value to this integral.

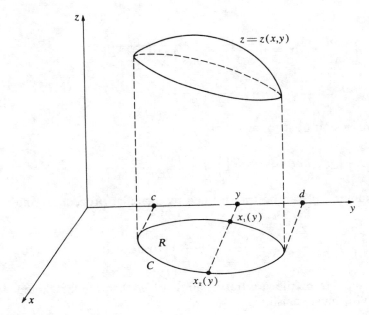

FIGURE 97

Our reasoning follows a familiar pattern. Assume that $z(x,y)$ is the desired stationary function and form the varied function $\bar{z}(x,y) = z(x,y) + \alpha\eta(x,y)$, where $\eta(x,y)$ vanishes on C. When $\bar{z}$ is substituted into the integral (8), we obtain a function $I(\alpha)$ of the parameter α, and just as before, the necessary condition $I'(0) = 0$ yields

$$\iint\limits_{R} \left(\frac{\partial f}{\partial z}\,\eta + \frac{\partial f}{\partial z_x}\,\eta_x + \frac{\partial f}{\partial z_y}\,\eta_y \right) dx\, dy = 0. \tag{9}$$

To simplify the task of eliminating η_x and η_y, we now assume that the curve C has the property that each line in the xy-plane parallel to an axis intersects C in at most two points.[14] Then, regarding the double integral of the second term in parentheses in (9) as a repeated integral (see Fig. 97), we get

$$\iint\limits_{R} \frac{\partial f}{\partial z_x}\,\eta_x\, dx\, dy = \int_{c}^{d} \int_{x_1(y)}^{x_2(y)} \frac{\partial f}{\partial z_x}\,\eta_x\, dx\, dy;$$

---

[14] This restriction is unnecessary, and can be avoided if we are willing to use Green's theorem.

and since

$$\int_{x_1}^{x_2} \frac{\partial f}{\partial z_x} \eta_x \, dx = \eta \frac{\partial f}{\partial z_x} \Big]_{x_1}^{x_2} - \int_{x_1}^{x_2} \eta \frac{\partial}{\partial x} \left(\frac{\partial f}{\partial z_x} \right) dx$$

$$= -\int_{x_1}^{x_2} \eta \frac{\partial}{\partial x} \left(\frac{\partial f}{\partial z_x} \right) dx$$

because η vanishes on C, it follows that

$$\iint_R \frac{\partial f}{\partial z_x} \eta_x \, dx \, dy = -\iint_R \eta \frac{\partial}{\partial x} \left(\frac{\partial f}{\partial z_x} \right) dx \, dy.$$

The term containing η_y can be transformed by a similar procedure, and (9) becomes

$$\iint_R \eta \left[\frac{\partial f}{\partial z} - \frac{\partial}{\partial x} \left(\frac{\partial f}{\partial z_x} \right) - \frac{\partial}{\partial y} \left(\frac{\partial f}{\partial z_y} \right) \right] dx \, dy = 0. \tag{10}$$

We now conclude from the arbitrary nature of η that the bracketed expression in (10) must vanish, so

$$\frac{\partial}{\partial x} \left(\frac{\partial f}{\partial z_x} \right) + \frac{\partial}{\partial y} \left(\frac{\partial f}{\partial z_y} \right) - \frac{\partial f}{\partial z} = 0 \tag{11}$$

is Euler's equation for an extremal in this case. As before, a stationary function (if one exists) is an extremal that satisfies the given boundary conditions.

Example 3. In its simplest form, the *problem of minimal surfaces* was first proposed by Euler as follows: to find the surface of smallest area bounded by a given closed curve in space. If we assume that this curve projects down to a closed curve C surrounding a region R in the xy-plane, and also that the surface is expressible in the form $z = z(x,y)$, then the problem is to minimize the surface area integral

$$\iint_R \sqrt{1 + z_x^2 + z_y^2} \, dx \, dy$$

subject to the boundary condition that $z(x,y)$ must assume prescribed values on C. Euler's equation (11) for this integral is

$$\frac{\partial}{\partial x} \left(\frac{z_x}{\sqrt{1 + z_x^2 + z_y^2}} \right) + \frac{\partial}{\partial y} \left(\frac{z_y}{\sqrt{1 + z_x^2 + z_y^2}} \right) = 0,$$

which can be written in the form

$$z_{xx}(1 + z_y^2) - 2z_x z_y z_{xy} + z_{yy}(1 + z_x^2) = 0. \tag{12}$$

This partial differential equation was discovered by Lagrange. Euler showed that every minimal surface not part of a plane must be saddle-shaped, and also that its mean curvature must be zero at every point.[15] The mathematical problem of proving that minimal surfaces exist, i.e., that (12) has a solution satisfying suitable boundary conditions, is extremely difficult. A complete solution was attained only in 1930 and 1931 by the independent work of T. Radó (Hungarian, 1895–1965) and J. Douglas (American, 1897–1965). An experimental method of finding minimal surfaces was devised by the blind Belgian physicist J. Plateau (1801–1883), who described it in his 1873 treatise on molecular forces in liquids. The essence of the matter is that if a piece of wire is bent into a closed curve and dipped in a soap solution, then the resulting soap film spanning the wire will assume the shape of a minimal surface in order to minimize the potential energy due to surface tension. Plateau performed many striking experiments of this kind, and since his time the problem of minimal surfaces has been known as *Plateau's problem*.[16]

Example 4. In Section 40 we obtained the one-dimensional wave equation from Newton's second law of motion. In this example we deduce it from Hamilton's principle with the aid of equation (11). Assume the following: a string of constant linear mass density m is stretched with a tension T and fastened to the x-axis at the points $x = 0$ and $x = \pi$; it is plucked and allowed to vibrate in the xy-plane; and its displacements $y(x,t)$ are relatively small, so that the tension remains essentially constant and powers of the slope higher than the second can be neglected. When the string is displaced, an element of length dx is stretched to a length ds, where

$$ds = \sqrt{1 + y_x^2}\, dx \cong \left(1 + \frac{1}{2} y_x^2\right) dx.$$

This approximation results from expanding $\sqrt{1 + y_x^2} = (1 + y_x^2)^{1/2}$ in the binomial series $1 + y_x^2/2 + \cdots$ and discarding all powers of y_x higher than the second. The work done on the element is $T(ds - dx) = \frac{1}{2}Ty_x^2\, dx$, so the potential energy of the whole string is

$$V = \frac{1}{2} T \int_0^\pi y_x^2\, dx.$$

The element has mass $m\, dx$ and velocity y_t, so its kinetic energy is $\frac{1}{2}my_t^2\, dx$,

---

[15] The *mean curvature* of a surface at a point is defined as follows. Consider the normal line to the surface at the point, and a plane containing this normal line. As this plane rotates about the line, the curvature of the curve in which it intersects the surface varies, and the mean curvature is one-half the sum of its maximum and minimum values.

[16] The standard mathematical work on this subject is R. Courant, *Dirichlet's Principle, Conformal Mapping, and Minimal Surfaces*, Interscience-Wiley, New York, 1950.

and for the whole string we have

$$T = \frac{1}{2} m \int_0^\pi y_t^2 \, dx.$$

The Lagrangian is therefore

$$L = T - V = \frac{1}{2} \int_0^\pi (my_t^2 - Ty_x^2) \, dx,$$

and the action, which must be stationary by Hamilton's principle, is

$$\frac{1}{2} \int_{t_1}^{t_2} \int_0^\pi (my_t^2 - Ty_x^2) \, dx \, dt.$$

In this case equation (11) becomes

$$\frac{T}{m} y_{xx} = y_{tt},$$

which we recognize as the wave equation 40-(8).

NOTE ON HAMILTON. The Irish mathematician and mathematical physicist William Rowan Hamilton (1805–1865) was a classic child prodigy. He was educated by an eccentric but learned clerical uncle. At the age of three he could read English; at four he began Greek, Latin, and Hebrew; at eight he added Italian and French; at ten he learned Sanskrit and Arabic; and at thirteen he is said to have mastered one language for each year he had lived. This forced flowering of linguistic futility was broken off at the age of fourteen, when he turned to mathematics, astronomy, and optics. At eighteen he published a paper correcting a mistake in Laplace's *Mécanique Céleste*; and while still an undergraduate at Trinity College in Dublin, he was appointed professor of astronomy at that institution and automatically became Astronomer Royal of Ireland.

His first important work was in geometrical optics. He became famous at twenty-seven as a result of his mathematical prediction of conical refraction. Even more significant was his demonstration that all optical problems can be solved by a single method that includes Fermat's principle of least time as a special case. He then extended this method to problems in mechanics, and by the age of thirty had arrived at a single principle (now called Hamilton's principle) that exhibits optics and mechanics as merely two aspects of the calculus of variations.

In 1835 he turned his attention to algebra, and constructed a rigorous theory of complex numbers based on the idea that a complex number is an ordered pair of real numbers. This work was done independently of Gauss, who had already published the same ideas in 1831, but with emphasis on the interpretation of complex numbers as points in the complex plane. Hamilton subsequently tried to extend the algebraic structure of the complex numbers, which can be thought of as vectors in a plane, to vectors in three-dimensional space. This project failed, but in 1843 his efforts led him to the discovery of quaternions. These are four-dimensional vectors that include the complex numbers as a subsystem; in modern terminology, they constitute the simplest noncommutative linear algebra in which division is possible.[17] The remainder of Hamilton's life was devoted to the detailed elaboration of the theory and applications of quaternions, and to the production of massive indigestible treatises on the subject. This work had little effect on physics and geometry, and was supplanted by the more practical vector analysis of Willard Gibbs and the multilinear algebra of Grassmann and E. Cartan. The significant residue of Hamilton's labors on quaternions was the demonstrated existence of a consistent number system in which the commutative law of multiplication does not hold. This liberated algebra from some of the preconceptions that had paralyzed it, and encouraged other mathematicians of the late nineteenth and twentieth centuries to undertake broad investigations of linear algebras of all types.

Hamilton was also a bad poet and friend of Wordsworth and Coleridge, with whom he corresponded voluminously on science, literature, and philosophy.

---

[17] Fortunately Hamilton never learned that Gauss had discovered quaternions in 1819 but kept his ideas to himself. See Gauss, *Werke,* vol. VIII, pp. 357–362.

CHAPTER
13

THE
EXISTENCE
AND UNIQUENESS
OF SOLUTIONS

68 THE METHOD OF SUCCESSIVE APPROXIMATIONS

One of the main recurring themes of this book has been the idea that only a few simple types of differential equations can be solved explicitly in terms of known elementary functions. Some of these types are described in the first three chapters, and Chapter 5 provides a detailed account of second order linear equations whose solutions are expressible in terms of power series. However, many differential equations fall outside these categories, and nothing we have done so far suggests a procedure that might work in such cases.

We begin by examining the initial value problem described in Section 2:

$$y' = f(x,y), \qquad y(x_0) = y_0, \tag{1}$$

where $f(x,y)$ is an arbitrary function defined and continuous in some neighborhood of the point (x_0,y_0). In geometric language, our purpose is to devise a method for constructing a function $y = y(x)$ whose graph passes through the point (x_0,y_0) and that satisfies the differential equation $y' = f(x,y)$ in some neighborhood of x_0 (Fig. 98). We are prepared for

538

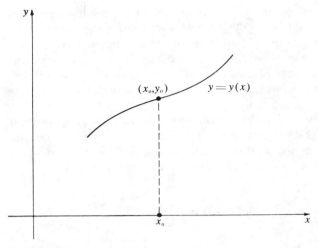

FIGURE 98

the idea that elementary procedures will not work and that in general some type of infinite process will be required.

The method we describe furnishes a line of attack for solving differential equations that is quite different from any the reader has encountered before. The key to this method lies in replacing the initial value problem (1) by the equivalent integral equation

$$y(x) = y_0 + \int_{x_0}^{x} f[t,y(t)] \, dt. \tag{2}$$

This is called an *integral equation* because the unknown function occurs under the integral sign. To see that (1) and (2) are indeed equivalent, suppose that $y(x)$ is a solution of (1). Then $y(x)$ is automatically continuous and the right side of

$$y'(x) = f[x,y(x)]$$

is a continuous function of x; and when we integrate this from x_0 to x and use $y(x_0) = y_0$, the result is (2). As usual, the dummy variable t is used in (2) to avoid confusion with the variable upper limit x on the integral. Thus any solution of (1) is a continuous solution of (2). Conversely, if $y(x)$ is a continuous solution of (2), then $y(x_0) = y_0$ because the integral vanishes when $x = x_0$, and by differentiation of (2) we recover the differential equation $y'(x) = f[x,y(x)]$. These simple arguments show that (1) and (2) are equivalent in the sense that the solutions of (1)—if any exist—are precisely the continuous solutions of (2). In particular, we automatically obtain a solution for (1) if we can construct a continuous solution for (2).

We now turn our attention to the problem of solving (2) by a process of iteration. That is, we begin with a crude approximation to a solution and improve it step by step by applying a repeatable operation which we hope will bring us as close as we please to an exact solution. The primary advantage that (2) has over (1) is that the integral equation provides a convenient mechanism for carrying out this process, as we now see.

A rough approximation to a solution is given by the constant function $y_0(x) = y_0$, which is simply a horizontal straight line through the point (x_0,y_0). We insert this approximation in the right side of equation (2) in order to obtain a new and perhaps better approximation $y_1(x)$ as follows:

$$y_1(x) = y_0 + \int_{x_0}^{x} f(t,y_0)\, dt.$$

The next step is to use $y_1(x)$ to generate another and perhaps even better approximation $y_2(x)$ in the same way:

$$y_2(x) = y_0 + \int_{x_0}^{x} f[t,y_1(t)]\, dt.$$

At the nth stage of the process we have

$$y_n(x) = y_0 + \int_{x_0}^{x} f[t,y_{n-1}(t)]\, dt. \tag{3}$$

This procedure is called *Picard's method of successive approximations.*[1] We show how it works by means of a few examples.

The simple initial value problem

$$y' = y, \qquad y(0) = 1$$

has the obvious solution $y(x) = e^x$. The equivalent integral equation is

$$y(x) = 1 + \int_{0}^{x} y(t)\, dt,$$

---

[1] Émile Picard (1856–1941), one of the most eminent French mathematicians of the past century, made two outstanding contributions to analysis: his method of successive approximations, which enabled him to perfect the theory of differential equations that Cauchy had initiated in the 1820s; and his famous theorem (called Picard's Great Theorem) about the values assumed by a complex analytic function near an essential singularity, which has stimulated much important research down to the present day. Like a true Frenchman, he was a connoisseur of fine food and was particularly fond of bouillabaisse.

and (3) becomes

$$y_n(x) = 1 + \int_0^x y_{n-1}(t) \, dt.$$

With $y_0(x) = 1$, it is easy to see that

$$y_1(x) = 1 + \int_0^x dt = 1 + x,$$

$$y_2(x) = 1 + \int_0^x (1 + t) \, dt = 1 + x + \frac{x^2}{2},$$

$$y_3(x) = 1 + \int_0^x \left(1 + t + \frac{t^2}{2}\right) dt = 1 + x + \frac{x^2}{2} + \frac{x^3}{2 \cdot 3},$$

and in general

$$y_n(x) = 1 + x + \frac{x^2}{2!} + \frac{x^3}{3!} + \cdots + \frac{x^n}{n!}.$$

In this case it is very clear that the successive approximations do in fact converge to the exact solution, for these approximations are the partial sums of the power series expansion of e^x.

Let us now consider the problem

$$y' = x + y, \qquad y(0) = 1. \tag{4}$$

This is a first order linear equation, and the solution satisfying the given initial condition is easily found to be $y(x) = 2e^x - x - 1$. The equivalent integral equation is

$$y(x) = 1 + \int_0^x [t + y(t)] \, dt,$$

and (3) is

$$y_n(x) = 1 + \int_0^x [t + y_{n-1}(t)] \, dt.$$

With $y_0(x) = 1$, Picard's method yields

$$y_1(x) = 1 + \int_0^x (t + 1) \, dt = 1 + x + \frac{x^2}{2!},$$

$$y_2(x) = 1 + \int_0^x \left(1 + 2t + \frac{t^2}{2!}\right) dt = 1 + x + x^2 + \frac{x^3}{3!},$$

$$y_3(x) = 1 + \int_0^x \left(1 + 2t + t^2 + \frac{t^3}{3!}\right) dt$$

$$= 1 + x + x^2 + \frac{x^3}{3} + \frac{x^4}{4!},$$

$$y_4(x) = 1 + \int_0^x \left(1 + 2t + t^2 + \frac{t^3}{3} + \frac{t^4}{4!}\right) dt$$

$$= 1 + x + x^2 + \frac{x^3}{3} + \frac{x^4}{3 \cdot 4} + \frac{x^5}{5!},$$

and in general

$$y_n(x) = 1 + x + 2\left(\frac{x^2}{2!} + \frac{x^3}{3!} + \cdots + \frac{x^n}{n!}\right) + \frac{x^{n+1}}{(n+1)!}.$$

This evidently converges to

$$1 + x + 2(e^x - x - 1) + 0 = 2e^x - x - 1,$$

so again we have the exact solution.

In spite of these examples, the reader may not be entirely convinced of the practical value of Picard's method. What are we to do, for instance, if the successive integrations are very complicated, or not possible at all except in principle? This skepticism is justified, for the real power of Picard's method lies mainly in the *theory* of differential equations—not in actually finding solutions, but in proving under very general conditions that an initial value problem has a solution and that this solution is unique. Theorems that make precise assertions of this kind are called *existence and uniqueness theorems*. We shall state and prove several of these theorems in the next two sections.

PROBLEMS

1. Find the exact solution of the initial value problem

$$y' = y^2, \qquad y(0) = 1.$$

   Starting with $y_0(x) = 1$, apply Picard's method to calculate $y_1(x)$, $y_2(x)$, $y_3(x)$, and compare these results with the exact solution.
2. Find the exact solution of the initial value problem

$$y' = 2x(1 + y), \qquad y(0) = 0.$$

   Starting with $y_0(x) = 0$, calculate $y_1(x)$, $y_2(x)$, $y_3(x)$, $y_4(x)$, and compare these results with the exact solution.
3. It is instructive to see how Picard's method works with a choice of the initial approximation other than the constant function $y_0(x) = y_0$. Apply the method to the initial value problem (4) with
   (a) $y_0(x) = e^x$;
   (b) $y_0(x) = 1 + x$;
   (c) $y_0(x) = \cos x$.

69 PICARD'S THEOREM

As we pointed out at the end of the last section, the principal value of Picard's method of successive approximations lies in the contribution it makes to the theory of differential equations. This contribution is most clearly illustrated in the proof of the following basic theorem.

Theorem A. (Picard's theorem.) *Let $f(x,y)$ and $\partial f / \partial y$ be continuous functions of x and y on a closed rectangle R with sides parallel to the axes (Fig. 99). If (x_0,y_0) is any interior point of R, then there exists a number $h > 0$ with the property that the initial value problem*

$$y' = f(x,y), \qquad y(x_0) = y_0 \tag{1}$$

has one and only one solution $y = y(x)$ on the interval $|x - x_0| \le h$.

Proof. The argument is fairly long and intricate, and is best absorbed in easy stages.

First, we know that every solution of (1) is also a continuous solution of the integral equation

$$y(x) = y_0 + \int_{x_0}^{x} f[t,y(t)] \, dt, \tag{2}$$

and conversely. This enables us to conclude that (1) has a unique solution on an interval $|x - x_0| \le h$ if and only if (2) has a unique continuous solution on the same interval. In Section 68 we presented some evidence

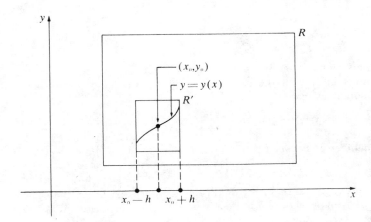

FIGURE 99

suggesting that the sequence of functions $y_n(x)$ defined by

$$y_0(x) = y_0,$$

$$y_1(x) = y_0 + \int_{x_0}^{x} f[t, y_0(t)]\, dt,$$

$$y_2(x) = y_0 + \int_{x_0}^{x} f[t, y_1(t)]\, dt, \tag{3}$$

$$\cdots$$

$$y_n(x) = y_0 + \int_{x_0}^{x} f[t, y_{n-1}(t)]\, dt,$$

$$\cdots$$

converges to a solution of (2). We next observe that $y_n(x)$ is the nth partial sum of the series of functions

$$y_0(x) + \sum_{n=1}^{\infty} [y_n(x) - y_{n-1}(x)] = y_0(x) + [y_1(x) - y_0(x)]$$

$$+ [y_2(x) - y_1(x)] + \cdots + [y_n(x) - y_{n-1}(x)] + \cdots, \tag{4}$$

so the convergence of the sequence (3) is equivalent to the convergence of this series. In order to complete the proof, we produce a number $h > 0$ that defines the interval $|x - x_0| \le h$, and then we show that on this interval the following statements are true: (i) the series (4) converges to a function $y(x)$; (ii) $y(x)$ is a continuous solution of (2); (iii) $y(x)$ is the only continuous solution of (2).

The hypotheses of the theorem are used to produce the positive number h, as follows. We have assumed that $f(x,y)$ and $\partial f/\partial y$ are continuous functions on the rectangle R. But R is closed (in the sense that it includes its boundary) and bounded, so each of these functions is necessarily bounded on R. This means that there exist constants M and K such that

$$|f(x,y)| \le M \tag{5}$$

and

$$\left| \frac{\partial}{\partial y} f(x,y) \right| \le K \tag{6}$$

for all points (x,y) in R. We next observe that if (x,y_1) and (x,y_2) are distinct points in R with the same x coordinate, then the mean value theorem guarantees that

$$|f(x,y_1) - f(x,y_2)| = \left| \frac{\partial}{\partial y} f(x,y^*) \right| |y_1 - y_2| \tag{7}$$

for some number y^* between y_1 and y_2. It is clear from (6) and (7) that

$$|f(x,y_1) - f(x,y_2)| \le K |y_1 - y_2| \tag{8}$$

for any points (x,y_1) and (x,y_2) in R (distinct or not) that lie on the same vertical line. We now choose h to be any positive number such that

$$Kh < 1 \tag{9}$$

and the rectangle R' defined by the inequalities $|x - x_0| \leq h$ and $|y - y_0| \leq Mh$ is contained in R. Since (x_0, y_0) is an interior point of R, there is no difficulty in seeing that such an h exists. The reasons for these apparently bizarre requirements will of course emerge as the proof continues.

From this point on, we confine our attention to the interval $|x - x_0| \leq h$. In order to prove (i), it suffices to show that the series

$$|y_0(x)| + |y_1(x) - y_0(x)| + |y_2(x) - y_1(x)|$$
$$+ \cdots + |y_n(x) - y_{n-1}(x)| + \cdots \quad (10)$$

converges; and to accomplish this, we estimate the terms $|y_n(x) - y_{n-1}(x)|$. It is first necessary to observe that each of the functions $y_n(x)$ has a graph that lies in R' and hence in R. This is obvious for $y_0(x) = y_0$, so the points $[t, y_0(t)]$ are in R', (5) yields $|f[t, y_0(t)]| \leq M$, and

$$|y_1(x) - y_0| = \left| \int_{x_0}^{x} f[t, y_0(t)] \, dt \right| \leq Mh,$$

which proves the statement for $y_1(x)$. It follows in turn from this inequality that the points $[t, y_1(t)]$ are in R', so $|f[t, y_1(t)]| \leq M$ and

$$|y_2(x) - y_0| = \left| \int_{x_0}^{x} f[t, y_1(t)] \, dt \right| \leq Mh.$$

Similarly,

$$|y_3(x) - y_0| = \left| \int_{x_0}^{x} f[t, y_2(t)] \, dt \right| \leq Mh,$$

and so on. Now for the estimates mentioned above. Since a continuous function on a closed interval has a maximum, and $y_1(x)$ is continuous, we can define a constant a by $a = \max |y_1(x) - y_0|$ and write

$$|y_1(x) - y_0(x)| \leq a.$$

Next, the points $[t, y_1(t)]$ and $[t, y_0(t)]$ lie in R', so (8) yields

$$|f[t, y_1(t)] - f[t, y_0(t)]| \leq K |y_1(t) - y_0(t)| \leq Ka$$

and we have

$$|y_2(x) - y_1(x)| = \left| \int_{x_0}^{x} (f[t, y_1(t)] - f[t, y_0(t)]) \, dt \right|$$
$$\leq Kah = a(Kh).$$

Similarly,

$$|f[t, y_2(t)] - f[t, y_1(t)]| \leq K |y_2(t) - y_1(t)| \leq K^2 ah,$$

so

$$|y_3(x) - y_2(x)| = \left| \int_{x_0}^{x} (f[t, y_2(t)] - f[t, y_1(t)]) \, dt \right|$$
$$\leq (K^2 ah)h = a(Kh)^2.$$

By continuing in this manner, we find that

$$|y_n(x) - y_{n-1}(x)| \le a(Kh)^{n-1}$$

for every $n = 1, 2, \ldots$. Each term of the series (10) is therefore less than or equal to the corresponding term of the series of constants

$$|y_0| + a + a(Kh) + a(Kh)^2 + \cdots + a(Kh)^{n-1} + \cdots.$$

But (9) guarantees that this series converges, so (10) converges by the comparison test, (4) converges to a sum which we denote by $y(x)$, and $y_n(x) \to y(x)$. Since the graph of each $y_n(x)$ lies in R', it is evident that the graph of $y(x)$ also has this property.

Now for the proof of (ii). The above argument shows not only that $y_n(x)$ converges to $y(x)$ in the interval, but also that this convergence is *uniform*. This means that by choosing n to be sufficiently large, we can make $y_n(x)$ as close as we please to $y(x)$ *for all x in the interval*; or more precisely, if $\epsilon > 0$ is given, then there exists a positive integer n_0 such that if $n \ge n_0$ we have $|y(x) - y_n(x)| < \epsilon$ for all x in the interval. Since each $y_n(x)$ is clearly continuous, this uniformity of the convergence implies that the limit function $y(x)$ is also continuous.[2] To prove that $y(x)$ is actually a solution of (2), we must show that

$$y(x) - y_0 - \int_{x_0}^{x} f[t, y(t)] \, dt = 0. \tag{11}$$

But we know that

$$y_n(x) - y_0 - \int_{x_0}^{x} f[t, y_{n-1}(t)] \, dt = 0, \tag{12}$$

so subtracting the left side of (12) from the left side of (11) gives

$$y(x) - y_0 - \int_{x_0}^{x} f[t, y(t)] \, dt = y(x) - y_n(x) + \int_{x_0}^{x} (f[t, y_{n-1}(t)] - f[t, y(t)]) \, dt,$$

and we obtain

$$\left| y(x) - y_0 - \int_{x_0}^{x} f[t, y(t)] \, dt \right|$$

$$\le |y(x) - y_n(x)| + \left| \int_{x_0}^{x} (f[t, y_{n-1}(t)] - f[t, y(t)]) \, dt \right|.$$

---

[2] We will not discuss this in detail, but the reasoning is quite simple and rests on the inequality

$$|y(x) - y(\bar{x})| = |[y(x) - y_n(x)] + [y_n(x) - y_n(\bar{x})] + [y_n(\bar{x}) - y(\bar{x})]|$$

$$\le |y(x) - y_n(x)| + |y_n(x) - y_n(\bar{x})| + |y_n(\bar{x}) - y(\bar{x})|.$$

Since the graph of $y(x)$ lies in R' and hence in R, (8) yields

$$\left| y(x) - y_0 - \int_{x_0}^{x} f[t,y(t)]\, dt \right|$$

$$\leq |y(x) - y_n(x)| + Kh \max |y_{n-1}(x) - y(x)|. \quad (13)$$

The uniformity of the convergence of $y_n(x)$ to $y(x)$ now implies that the right side of (13) can be made as small as we please by taking n large enough. The left side of (13) must therefore equal zero, and the proof of (11) is complete.

In order to prove (iii), we assume that $\bar{y}(x)$ is also a continuous solution of (2) on the interval $|x - x_0| \leq h$, and we show that $\bar{y}(x) = y(x)$ for every x in the interval. For the argument we give, it is necessary to know that the graph of $\bar{y}(x)$ lies in R' and hence in R, so our first step is to establish this fact. Let us suppose that the graph of $\bar{y}(x)$ leaves R' (Fig. 100). Then the properties of this function [continuity and the fact that $\bar{y}(x_0) = y_0$] imply that there exists an x_1 such that $|x_1 - x_0| < h$, $|\bar{y}(x_1) - y_0| = Mh$, and $|\bar{y}(x) - y_0| < Mh$ if $|x - x_0| < |x_1 - x_0|$. It follows that

$$\frac{|\bar{y}(x_1) - y_0|}{|x_1 - x_0|} = \frac{Mh}{|x_1 - x_0|} > \frac{Mh}{h} = M.$$

However, by the mean value theorem there exists a number x^* between x_0 and x_1 such that

$$\frac{|\bar{y}(x_1) - y_0|}{|x_1 - x_0|} = |\bar{y}'(x^*)| = |f[x^*, \bar{y}(x^*)]| \leq M,$$

since the point $[x^*, \bar{y}(x^*)]$ lies in R'. This contradiction shows that no point with the properties of x_1 can exist, so the graph of $\bar{y}(x)$ lies in R'. To complete the proof of (iii), we use the fact that $\bar{y}(x)$ and $y(x)$ are both

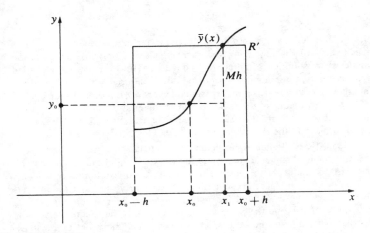

FIGURE 100

solutions of (2) to write

$$|\bar{y}(x) - y(x)| = \left| \int_{x_0}^{x} \{f[t,\bar{y}(t)] - f[t,y(t)]\} \, dt \right|.$$

Since the graphs of $\bar{y}(x)$ and $y(x)$ both lie in R', (8) yields

$$|\bar{y}(x) - y(x)| \le Kh \max |\bar{y}(x) - y(x)|,$$

so

$$\max |\bar{y}(x) - y(x)| \le Kh \max |\bar{y}(x) - y(x)|.$$

This implies that $\max |\bar{y}(x) - y(x)| = 0$, for otherwise we would have $1 \le Kh$ in contradiction to (9). It follows that $\bar{y}(x) = y(x)$ for every x in the interval $|x - x_0| \le h$, and Picard's theorem is fully proved.

Remark 1. This theorem can be strengthened in various ways by weakening its hypotheses. For instance, our assumption that $\partial f / \partial y$ is continuous on R is stronger than the proof requires, and is used only to obtain the inequality (8). We can therefore introduce this inequality into the theorem as an assumption that replaces the one about $\partial f / \partial y$. In this way we arrive at a stronger form of the theorem since there are many functions that lack a continuous partial derivative but nevertheless satisfy (8) for some constant K. This inequality, which says that the difference quotient

$$\frac{f(x,y_1) - f(x,y_2)}{y_1 - y_2}$$

is bounded on R, is called a *Lipschitz*[3] *condition* in the variable y.

Remark 2. If we drop the Lipschitz condition, and assume only that $f(x,y)$ is continuous on R, then it is still possible to prove that the initial value problem (1) has a solution. This result is known as *Peano's theorem*.[4] The only known proofs depend on more sophisticated argu-

---

[3] Rudolf Lipschitz (1832–1903) was a professor at Bonn for most of his life. He is remembered chiefly for his role in simplifying and clarifying Cauchy's original theory of the existence and uniqueness of solutions of differential equations. However, he also extended Dirichlet's theorem on the representability of a function by its Fourier series, obtained the formula for the number of ways a positive integer can be expressed as a sum of four squares as a consequence of his own theory of the factorization of integral quaternions, and made useful contributions to theoretical mechanics, the calculus of variations, Bessel functions, quadratic differential forms, and the theory of viscous fluids.

[4] Guiseppe Peano (1858–1932), Italian logician and mathematician, strongly influenced Hilbert's axiomatic treatment of plane geometry and the work of Whitehead and Russell on mathematical logic. His postulates for the positive integers have led generations of students to wonder whether all of modern algebra is some kind of conspiracy to render the obvious obscure (it is not!). In 1890 he astounded the mathematical world with his remarkable construction of a continuous curve in the plane that completely fills the square $0 \le x \le 1$, $0 \le y \le 1$. Unfortunately for a man who valued logic so highly, his 1886 proof of the above existence theorem for solutions of $y' = f(x,y)$ was inadequate, and a satisfactory proof was not found until many years later.

ments than those we have used above.[5] Furthermore, the solution whose existence this theorem guarantees is not necessarily unique. As an example, consider the problem

$$y' = 3y^{2/3}, \qquad y(0) = 0, \tag{14}$$

and let R be the rectangle $|x| \le 1$, $|y| \le 1$. Here $f(x,y) = 3y^{2/3}$ is plainly continuous on R. Also, $y_1(x) = x^3$ and $y_2(x) = 0$ are two different solutions valid for all x, so (14) certainly has a solution that is not unique. The explanation for this nonuniqueness lies in the fact that $f(x,y)$ does not satisfy a Lipschitz condition on the rectangle R, since the difference quotient

$$\frac{f(0,y) - f(0,0)}{y - 0} = \frac{3y^{2/3}}{y} = \frac{3}{y^{1/3}}$$

is unbounded in every neighborhood of the origin.

Remark 3. Theorem A is called a *local* existence and uniqueness theorem because it guarantees the existence of a unique solution only on some interval $|x - x_0| \le h$ where h may be very small. There are several important cases in which this restriction can be removed. Let us consider, for example, the first order linear equation

$$y' + P(x)y = Q(x),$$

where $P(x)$ and $Q(x)$ are defined and continuous on an interval $a \le x \le b$. Here we have

$$f(x,y) = -P(x)y + Q(x);$$

and if $K = \max |P(x)|$ for $a \le x \le b$, it is clear that

$$|f(x,y_1) - f(x,y_2)| = |-P(x)(y_1 - y_2)| \le K |y_1 - y_2|.$$

The function $f(x,y)$ is therefore continuous and satisfies a Lipschitz condition on the infinite vertical strip defined by $a \le x \le b$ and $-\infty < y < \infty$. Under these circumstances, the initial value problem

$$y' + P(x)y = Q(x), \qquad y(x_0) = y_0$$

has a unique solution on the entire interval $a \le x \le b$. Furthermore, the point (x_0,y_0) can be any point of the strip, interior or not. This statement is a special case of the next theorem.

---

[5] See, for example, A. N. Kolmogorov and S. V. Fomin, *Elements of the Theory of Functions and Functional Analysis*, vol, 1, p. 56, Graylock, Baltimore, 1957.

Theorem B. *Let $f(x,y)$ be a continuous function that satisfies a Lipschitz condition*

$$|f(x,y_1) - f(x,y_2)| \le K |y_1 - y_2|$$

on a strip defined by $a \le x \le b$ and $-\infty < y < \infty$. If (x_0,y_0) is any point of the strip, then the initial value problem

$$y' = f(x,y), \qquad y(x_0) = y_0 \tag{15}$$

has one and only one solution $y = y(x)$ on the interval $a \le x \le b$.

Proof. The argument is similar to that given for Theorem A, with certain simplifications permitted by the fact that the region under discussion is not bounded above or below. In particular, we start the proof in the same way and show that the series (4)— and therefore the sequence (3)—is uniformly convergent on the whole interval $a \le x \le b$. We accomplish this by using a somewhat different method of estimating the terms of the series (10).

First, we define M_0, M_1 and M by

$$M_0 = |y_0|, \qquad M_1 = \max |y_1(x)|, \qquad M = M_0 + M_1,$$

and we notice that $|y_0(x)| \le M$ and $|y_1(x) - y_0(x)| \le M$. Next, if $x_0 \le x \le b$, it follows that

$$
\begin{aligned}
|y_2(x) - y_1(x)| &= \left| \int_{x_0}^{x} \{f[t,y_1(t)] - f[t,y_0(t)]\} \, dt \right| \\
&\le \int_{x_0}^{x} |f[t,y_1(t)] - f[t,y_0(t)]| \, dt \\
&\le K \int_{x_0}^{x} |y_1(t) - y_0(t)| \, dt \\
&\le KM(x - x_0),
\end{aligned}
$$

$$
\begin{aligned}
|y_3(x) - y_2(x)| &= \left| \int_{x_0}^{x} \{f[t,y_2(t)] - f[t,y_1(t)]\} \, dt \right| \\
&\le K \int_{x_0}^{x} |y_2(t) - y_1(t)| \, dt \\
&\le K^2 M \int_{x_0}^{x} (t - x_0) \, dt = K^2 M \frac{(x - x_0)^2}{2},
\end{aligned}
$$

and in general

$$|y_n(x) - y_{n-1}(x)| \le K^{n-1} M \frac{(x - x_0)^{n-1}}{(n - 1)!}.$$

The same argument is also valid for $a \le x \le x_0$, provided only that $x - x_0$ is replaced by $|x - x_0|$, so we have

$$
\begin{aligned}
|y_n(x) - y_{n-1}(x)| &\le K^{n-1} M \frac{|x - x_0|^{n-1}}{(n - 1)!} \\
&\le K^{n-1} M \frac{(b - a)^{n-1}}{(n - 1)!}
\end{aligned}
$$

for every x in the interval and $n = 1, 2, \ldots$. We conclude that each term of the series (10) is less than or equal to the corresponding term of the convergent series of constants

$$M + M + KM(b - a) + K^2 M \frac{(b - a)^2}{2!} + K^3 M \frac{(b - a)^3}{3!} + \cdots,$$

so (3) converges uniformly on the interval $a \leq x \leq b$ to a limit function $y(x)$.

Just as before, the uniformity of the convergence implies that $y(x)$ is a solution of (15) on the whole interval, and all that remains is to show that it is the only such solution. We assume that $\bar{y}(x)$ is also a solution of (15) on the interval. Our strategy is to show that $y_n(x) \to \bar{y}(x)$ for each x as $n \to \infty$; and since we also have $y_n(x) \to y(x)$, it will follow that $\bar{y}(x) = y(x)$. We begin by observing that $\bar{y}(x)$ is continuous and satisfies the equation

$$\bar{y}(x) = y_0 + \int_{x_0}^{x} f[t, \bar{y}(t)]\, dt.$$

If $A = \max |\bar{y}(x) - y_0|$, then for $x_0 \leq x \leq b$ we see that

$$|\bar{y}(x) - y_1(x)| = \left| \int_{x_0}^{x} \{f[t, \bar{y}(t)] - f[t, y_0(t)]\}\, dt \right|$$

$$\leq \int_{x_0}^{x} |f[t, \bar{y}(t)] - f[t, y_0(t)]|\, dt$$

$$\leq K \int_{x_0}^{x} |\bar{y}(t) - y_0|\, dt$$

$$\leq KA(x - x_0),$$

$$|\bar{y}(x) - y_2(x)| = \left| \int_{x_0}^{x} \{f[t, \bar{y}(t)] - f[t, y_1(t)]\}\, dt \right|$$

$$\leq K \int_{x_0}^{x} |\bar{y}(t) - y_1(t)|\, dt$$

$$\leq K^2 A \int_{x_0}^{x} (t - x_0)\, dt = K^2 A \frac{(x - x_0)^2}{2},$$

and in general

$$|\bar{y}(x) - y_n(x)| \leq K^n A \frac{(x - x_0)^n}{n!}.$$

A similar result holds for $a \leq x \leq x_0$, so for any x in the interval we have

$$|\bar{y}(x) - y_n(x)| \leq K^n A \frac{|x - x_0|^n}{n!} \leq K^n A \frac{(b - a)^n}{n!}.$$

Since the right side of this approaches zero as $n \to \infty$, we conclude that $\bar{y}(x) = y(x)$ for every x in the interval, and the proof is complete.

PROBLEMS

1. Let (x_0, y_0) be an arbitrary point in the plane and consider the initial value problem

$$y' = y^2, \qquad y(x_0) = y_0.$$

Explain why Theorem A guarantees that this problem has a unique solution on some interval $|x - x_0| \leq h$. Since $f(x,y) = y^2$ and $\partial f/\partial y = 2y$ are continuous on the entire plane, it is tempting to conclude that this solution is valid for all x. By considering the solutions through the points $(0,0)$ and $(0,1)$, show that this conclusion is sometimes true and sometimes false, and that therefore the inference is not legitimate.

2. Show that $f(x,y) = y^{1/2}$
   (a) does not satisfy a Lipschitz condition on the rectangle $|x| \leq 1$ and $0 \leq y \leq 1$;
   (b) does satisfy a Lipschitz condition on the rectangle $|x| \leq 1$ and $c \leq y \leq d$, where $0 < c < d$.

3. Show that $f(x,y) = x^2|y|$ satisfies a Lipschitz condition on the rectangle $|x| \leq 1$ and $|y| \leq 1$ but that $\partial f/\partial y$ fails to exist at many points of this rectangle.

4. Show that $f(x,y) = xy^2$
   (a) satisfies a Lipschitz condition on any rectangle $a \leq x \leq b$ and $c \leq y \leq d$;
   (b) does not satisfy a Lipschitz condition on any strip $a \leq x \leq b$ and $-\infty < y < \infty$.

5. Show that $f(x,y) = xy$
   (a) satisfies a Lipschitz condition on any rectangle $a \leq x \leq b$ and $c \leq y \leq d$;
   (b) satisfies a Lipschitz condition on any strip $a \leq x \leq b$ and $-\infty < y < \infty$;
   (c) does not satisfy a Lipschitz condition on the entire plane.

6. Consider the initial value problem

$$y' = |y|, \qquad y(x_0) = y_0.$$

   (a) For what points (x_0, y_0) does Theorem A imply that this problem has a unique solution on some interval $|x - x_0| \leq h$?
   (b) For what points (x_0, y_0) does this problem actually have a unique solution on some interval $|x - x_0| \leq h$?

7. For what points (x_0, y_0) does Theorem A imply that the initial value problem

$$y' = y|y|, \qquad y(x_0) = y_0$$

has a unique solution on some interval $|x - x_0| \leq h$?

70 SYSTEMS. THE SECOND ORDER LINEAR EQUATION

Picard's method of successive approximations can also be applied to systems of first order equations. Let us consider, for example, the initial value problem consisting of the following pair of first order equations and initial conditions:

$$\begin{cases} \dfrac{dy}{dx} = f(x,y,z), & y(x_0) = y_0, \\[2mm] \dfrac{dz}{dx} = g(x,y,z), & z(x_0) = z_0, \end{cases} \tag{1}$$

where the right sides are continuous functions in some region of xyz space that contains the point (x_0,y_0,z_0). We use the differential notation here in order to emphasize that x is the independent variable. A solution of such a system is of course a pair of functions $y = y(x)$ and $z = z(x)$ which together satisfy the conditions imposed by (1) on some interval containing the point x_0. As in the case of a single first order equation, it is apparent that the system (1) is equivalent to the system of integral equations

$$\begin{cases} y(x) = y_0 + \displaystyle\int_{x_0}^{x} f[t,y(t),z(t)]\,dt, \\[3mm] z(x) = z_0 + \displaystyle\int_{x_0}^{x} g[t,y(t),z(t)]\,dt, \end{cases} \tag{2}$$

in the sense that the solutions of (1)—if any exist—are precisely the continuous solutions of (2). If we attempt to solve (2) by successive approximations beginning with the constant functions

$$y_0(x) = y_0 \quad \text{and} \quad z_0(x) = z_0,$$

then the Picard method proceeds exactly as before. At the first stage we have

$$\begin{cases} y_1(x) = y_0 + \displaystyle\int_{x_0}^{x} f[t,y_0(t),z_0(t)]\,dt, \\[3mm] z_1(x) = z_0 + \displaystyle\int_{x_0}^{x} g[t,y_0(t),z_0(t)]\,dt; \end{cases}$$

at the second stage we have

$$\begin{cases} y_2(x) = y_0 + \displaystyle\int_{x_0}^{x} f[t,y_1(t),z_1(t)]\,dt, \\[3mm] z_2(x) = z_0 + \displaystyle\int_{x_0}^{x} g[t,y_1(t),z_1(t)]\,dt; \end{cases}$$

and so on. This procedure generates two sequences of functions $y_n(x)$ and

$z_n(x)$; and under suitable hypotheses, the arguments of Theorem 69-A can easily be adapted to prove that these sequences converge to a solution of (1) which exists and is unique on some interval $|x - x_0| \leq h$.

We now specialize to a linear system, in which the functions $f(x,y,z)$ and $g(x,y,z)$ in (1) are linear functions of y and z. That is, we consider an initial value problem of the form

$$\begin{cases} \dfrac{dy}{dx} = p_1(x)y + q_1(x)z + r_1(x), & y(x_0) = y_0, \\[2mm] \dfrac{dz}{dx} = p_2(x)y + q_2(x)z + r_2(x), & z(x_0) = z_0, \end{cases} \tag{3}$$

where the six functions $p_i(x)$, $q_i(x)$, and $r_j(x)$ are continuous on an interval $a \leq x \leq b$ and x_0 is a point in this interval. Since each of these functions is bounded for $a \leq x \leq b$, there exists a constant K such that $|p_i(x)| \leq K$ and $|q_i(x)| \leq K$ for $i = 1, 2$. It is now easy to see that the functions on the right sides of the differential equations in (3) satisfy Lipschitz conditions of the form

$$|f(x,y_1,z_1) - f(x,y_2,z_2)| \leq K(|y_1 - y_2| + |z_1 - z_2|)$$

and

$$|g(x,y_1,z_1) - g(x,y_2,z_2)| \leq K(|y_1 - y_2| + |z_1 - z_2|).$$

Just as in the proof of Theorem 69-B, these conditions can be used to show that (3) has a unique solution on the whole interval $a \leq x \leq b$. Again we spare the reader the details.

These remarks about systems make it possible to give a simple proof of the following basic theorem, which we stated at the beginning of Chapter 3 and which has played an unobtrusive but crucial role in all of our work on second order linear equations.

Theorem A. *Let $P(x)$, $Q(x)$, and $R(x)$ be continuous functions on an interval $a \leq x \leq b$. If x_0 is any point in this interval, and y_0 and y_0' are any numbers whatever, then the initial value problem*

$$\frac{d^2y}{dx^2} + P(x)\frac{dy}{dx} + Q(x)y = R(x), \quad y(x_0) = y_0 \quad \text{and} \quad y'(x_0) = y_0', \tag{4}$$

has one and only one solution $y = y(x)$ on the interval $a \leq x \leq b$.

Proof. If we introduce the variable $z = dy/dx$, then it is clear that every solution of (4) yields a solution of the linear system

$$\begin{cases} \dfrac{dy}{dx} = z, & y(x_0) = y_0, \\[2mm] \dfrac{dz}{dx} = -P(x)z - Q(x)y + R(x), & z(x_0) = y_0', \end{cases} \tag{5}$$

and conversely. We have seen that (5) has a unique solution on the interval $a \leq x \leq b$, so the same is true of (4).

PROBLEM

1. Solve the following initial value problem by Picard's method, and compare the result with the exact solution:

$$\begin{cases} \dfrac{dy}{dx} = z, & y(0) = 1, \\[2mm] \dfrac{dz}{dx} = -y, & z(0) = 0. \end{cases}$$

CHAPTER
14

NUMERICAL
METHODS

BY JOHN S. ROBERTSON

Department of Mathematical Sciences,
U.S. Military Academy,
West Point, New York 10996–1786

71 INTRODUCTION

Despite the broad range of powerful analytical tools presented through-out this book, many occasions cry out for the application of numerical methods for solving ordinary differential equations. For example, an exact solution may be unavailable, or may be of little practical value.[1] This situation occurs when power series solutions to linear second order equations are constructed. In general, the series are rather good approximations near the initial condition, but the Taylor expansions can soon require prohibitively many terms should the solution be required at some large distance from that point. For large systems of equations, an exact solution may exist (in vector form) but the subsequent algebraic manipulations may be overwhelming. Furthermore, numerical solutions

---

[1] For a detailed historical account of the important role played by the application of numerical methods to differential equations, see Garrett Birkhoff's "Numerical Fluid Dynamics," the 1981 John von Neumann Lecture, published in *SIAM Review,* vol. 25, pp. 1–34 (1983).

556

should not be cast in a light of last resort, for they form the mathematician's petri dish—a crucible in which he can conduct any number of experiments on his differential equation and, by proxy, the very thing he is trying to model.[2]

These numerical methods rely on two fundamental but *distinct* approximations. First, a differential equation is replaced with a difference equation and the role played by a continuous independent variable is then assumed by a discrete one. For this approach to be of any use, it is important to understand the conditions under which the solution to the difference equation is close to, that is, converges to, the solution to the differential equation. Second, in virtually all digital computers in use today, the real-number line is approximated by a large but finite subset of rational numbers. Limiting oneself to only a finite range of rationals can have unobvious, but crucial, consequences in certain cases—the errors made by the machine may indeed be catastrophic. At any rate, both of these approximations permit the difference equations to be implemented on an enormous variety of computing hardware. Nevertheless, there are many apocryphal stories told of engineers performing expensive computations on big computers only to obtain nonsense answers. We emphasize here that existence and uniqueness questions, discussed elsewhere in this book, are vitally important and should always be considered first. Beyond these, other problems, such as numerical instability and the existence of *spurious* solutions can cause difficulties. Despite the abundance of well-tuned algorithms for solving ordinary differential equations, the reader should carefully remark the need to be ever-vigilant. Before appealing to the machine for aid, it is always wise to know something about the answer one seeks. That is, the practicing scientist should endeavor to know as much about the solution as is possible. For example, is it bounded? Stable? Periodic? About how big (or small) should the answer be? Careful attention to these issues as discussed in the preceding chapters will stand the reader in good stead for what follows.[3]

---

[2] In 1965, N. J. Zabusky and M. D. Kruskal discovered *solitons* in just this way. By considering a particular version of an equation governing the motion of surface water waves and experimenting with its numerical solution, they deduced the existence of mathematical objects with truly surprising properties. Solitons and the differential equations that govern their behavior have been one of the most intensely studied areas of applied mathematics during the last two decades.

[3] For an excellent historical background on the evolution of numerical methods for differential equations that occurred in the decades surrounding the development of the first digital computers, see Herman H. Goldstine, *The Computer from Pascal to von Neumann*, Princeton University Press, Princeton, 1972.

In order to understand what we mean by a numerical solution of a differential equation, we consider the simple initial-value problem

$$y' = y, \qquad y(0) = 1. \tag{1}$$

The problem has the obvious solution $y = e^x$, and for many theoretical purposes, this is enough. However, in a practical application it might be necessary to know the value of the solution when $x = 0.5$, and the decimal 1.649 is likely to be more useful than the symbol $e^{0.5}$. In contrast to the theoretical solution of (1), a numerical solution can be provided by a table of values for e^x or a pocket calculator. Either way, the number so obtained depended on our knowledge of the formula $y = e^x$.

In this chapter we describe several methods of calculating an approximation numerical solution of the form

$$y' = f(x,y), \qquad y(x_0) = y_0. \tag{2}$$

We shall assume that this problem has a unique solution denoted by $y(x)$. Our methods consist of a computational procedures based solely on the information given by (2), and are completely independent of whether a formula for $y(x)$ is known or not. These numerical methods and others like them are therefore extremely valuable for those initial-value problems that cannot be solved exactly, and also for those having exact formal solutions that are practically intractable.[4]

Let us be a little more specific about the nature of these methods. We shall not approximate the exact solution $y(x)$ for all values of x in some interval, but only for a discrete sequence of points beginning at x_0, say

$$x_0, x_1 = x_0 + h, x_2 = x_1 + h, \ldots, x_n = x_{n-1} + h,$$

where h is a positive number. This means that we want an approximation y_1 to the exact value $y(x_1)$, an approximation y_2 to the exact value $y(x_2)$, and so on. Each numerical method we describe will be a rule for using y_k to compute y_{k+1}.[5] Since we know the initial value $y(x_0) = y_0$ (this is exact), we can apply the rule with $k = 0$ to obtain y_1, with $n = 1$ to obtain y_2, etc. Our general purpose is to apply enough of the details of each method to enable the reader to apply it for himself if the need should ever arise. We avoid details dealing with the plethora of computing machines and programming languages for several reasons.

---

[4] The noted American mathematician R. W. Hamming said that "the purpose of computing is insight, not numbers." Even so, it takes more than insight to build a skyscraper or a space shuttle.

[5] These are so-called *single-step* methods. There are also various multistep methods in which y_{k+1} depends not only on y_k, but possibly on y_{k-1} and earlier terms.

First, those issues are best left to specialized texts in numerical analysis. Second, it is our experience that virtually all students have some familiarity with computing fundamentals and should be able to write programs where appropriate to perform the calculations required by the exercises in this chapter. As to the means, that is better left to the student and his teacher. Third, advances in computing continue at a dizzying pace, and we see no need to burden this book with nonmathematical details that might well be obsolete in only a few short yerars.

We shall illustrate our methods by applying them to the simple problem

$$y' = x + y, \qquad y(0) = 1, \tag{3}$$

which we call our *benchmark* problem. This differential equation in (3) is clearly linear, and the exact solution is easily found to be

$$y = 2e^x - x - 1. \tag{4}$$

We have chosen (3) as our benchmark problem for two reasons. First, it is so simple that a numerical method can be applied to it by hand without obscuring the main steps by a morass of computations. Second, the exact solution (4) can easily be evaluated for various x's with the aid of a pocket calculator, so we have a means of judging the accuracy of the approximate solutions produced by our numerical methods.

PROBLEM

1. Have you encountered any examples in other courses where either the textbook or the instructor referred to numerical solutions of ordinary differential equations? Give an example and discuss what you read or heard.

72 THE METHOD OF EULER

If we integrate the differential equation in (2) from x_0 to $x_1 = x_0 + h$, and use the initial condition $y(x_0) = y_0$, we obtain

$$y(x_1) - y(x_0) = \int_{x_0}^{x_1} f(x,y) \, dx$$

or

$$y(x_1) = y_0 + \int_{x_0}^{x_1} f(x,y) \, dx. \tag{5}$$

Since the unknown function $y = y(x)$ occurs under the integral sign in (5), we can go no further without some sort of approximation to this integral. Different types of approximations correspond to various methods for numerically solving (2).

The Euler method is obtained from the simplest way of approximating the integral in (5). It is worth considering because it paves the way for an understanding of other more accurate but more complicated methods. The idea is to obtain y_1—our approximation to $y(x_1)$—by assuming that the integrand $f(x,y)$ in (5) varies so little over the interval $x_0 \leq x \leq x_1$ that only a small error is made by replacing it by its value $f(x_0,y_0)$ at the left endpoint. This is equivalent to replacing the integrand in (5) with its zeroth order Taylor polynomial, that is,

$$f(x,y) = f(x_0,y_0) + R, \tag{6}$$

where

$$R(x) = [f'(\xi,y(\xi)) + f_y(\xi,y(\xi))y'(\xi)](x - x_0),$$

where R is the Taylor remainder term, $f_y = \partial f / \partial y$ and $x_0 < \xi < x$. Noting that $\underline{y'' = f' + f_y y'}$, we substitute (6) into (5) to obtain

$$y_1 = y_0 + hf(x_0,y_0) + \frac{h^2}{2} y''(\xi).$$

We suppose that $h^2 y''(\xi)/2$ is "small" in an appropriate sense and neglect the term. How small is small in general, and more particularly, when this term is small are important issues that will be discussed in more detail later. (See Problem 6, Section 73, for a related discussion.) Neglecting this term, we have

$$y_1 = y_0 + hf(x_0,y_0), \tag{7}$$

We now continue and obtain y_2 from y_1 in the same way, by the formula $y_2 = y_1 + hf(x,y)$; and in general we have

$$y_{k+1} = y_k + hf(x_k,y_k). \tag{4}$$

for $k = 0, 1, \ldots, n$. The geometric meaning of these formulas is shown in Fig. 101, where the smooth curve is the unknown exact solution which

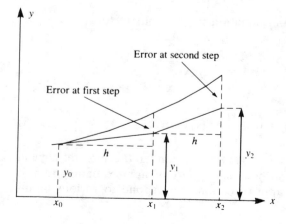

FIGURE 101

is being approximated by the piecewise-linear curve generated constructed from (8). To understand this figure, remember that $f(x_0, y_0)$ is the slope of the tangent line to the curve at the initial point (x_0, y_0). The point y_1 is found by constructing a line segment beginning at (x_0, y_0) with that slope and marching it in the positive x direction a distance of h. That point becomes the second approximation to the solution. The figure indicates the vertical distance between the solution and the approximation as the error at the first stage. An important quantity derived from this, is the *total relative error* $\bar{E}_n$ at the nth step, defined to be

$$\bar{E}_n = \frac{|y(x_n) + y_n|}{|y(x_n)|}. \tag{5}$$

This quantity is often expressed as a percentage, providing a comfortable way to gauge how accurately the numerical solution is performing. Now, using (x_1, y_1) the process is repeated again to obtain the next point at (x_2, y_2), also shown in the figure. The geometric realization of the Euler method suggests that error can build up rather quickly, which is, in general, true.

We illustrate the Euler method by applying it to the benchmark problem (3). We approximate the solution at the points $x_n = 0.2, 0.4, 0.6, 0.8,$ and 1.0 by using intervals of length $h = 0.2$. It is convenient to arrange the calculations as shown in Table 1. In the first line of this table, the initial condition $y = 1$ when $x = 0$ determines the slope $y' = x + y = 1.00$. Since $h = 0.2$ and $y_1 = y_0 + hf(x_0, y_0)$, the next value is given by $1.00 + 0.2(1.00) = 1.20$. This approximation is shifted to the y_n in the second line and the process is repeated to find y_2, which turns out to be 1.48. In the table (and most remaining examples), we retain five figures after the decimal point, and the resulting approximate value of $y(1)$ is 2.97664. The exact value found from (4) is 3.43656, so the error is about 13 percent. If we carry out a similar calculation with $h = 0.1$, then the resulting approximation for $y(1)$ is 3.18748, and the error is reduced to about 7 percent, roughly half of what it was in the first instance. Table 2

TABLE 1
Tabulated values for exact and numerical solutions to (3) with $h = 0.2$

x_n	y_n	Exact	$\bar{E}_n$ (%)
0.0	1.00000	1.00000	0.0
0.2	1.20000	1.24281	3.4
0.4	1.48000	1.58365	6.5
0.6	1.85600	2.04424	9.2
0.8	2.34720	2.65108	11.5
1.0	2.97664	3.43656	13.4

TABLE 2
Tabulated values for exact and numerical solutions to (3) with $h = 0.1$

x_n	y_n	Exact	$\bar{E}_n$ (%)
0.0	1.00000	1.00000	0.0
0.1	1.10000	1.11034	0.9
0.2	1.22000	1.24281	1.8
0.3	1.36200	1.39972	2.7
0.4	1.52820	1.58365	3.5
0.5	1.72102	1.79744	4.3
0.6	1.94312	2.04424	4.9
0.7	2.19743	2.32751	5.6
0.8	2.48718	2.65108	6.2
0.9	2.81590	3.01921	6.7
1.0	3.18748	3.43656	7.2

displays the intermediate results of the Euler method for the benchmark problem in this case.

We can therefore improve the accuracy of the method by taking smaller values of h, but at the expense of more computational work. Even so, after a certain point, reducing the step size will only make errors *worse* as will be discussed in the next section.

PROBLEMS

For the following problems, use the Euler method with $h = 0.1, 0.05,$ and 0.01 to estimate the solution at $x = 1$. Compare your results to the *exact* solution in each instance and discuss how well (or badly!) the Euler method performs.

1. $y' = 2x + 2y,\ y(0) = 1$.
2. $y' = 1/y,\ y(0) = 1$.
3. $y' = e^y,\ y(0) = 0$.
4. $y' = y - \sin x,\ y(0) = -1$.
5. $y' = (x + y - 1)^2,\ y(0) = 0$.
6. This problem illustrates the danger in blindly applying numerical methods. Employ the Euler method to the following initial value problem:

$$y' = \sec^2 x, \qquad y(0) = 0.$$

Use a step size of $h = 0.1$ and determine the numerical solution at $x = 1$. Explain why the initial value problem has no solution at $x = 1$.
7. Refer to Fig. 101. From geometric arguments, for what kind of exact solutions might the Euler method give precise results? Do these results depend on h in any way? Construct two distinct examples to illustrate your ideas.
8. The ordinary differential equation

$$y' = y(1 - y^2),$$

possesses two equilibrium solutions: $\phi_1 = 0$, which is unstable, and $\phi_2 = 1$, which is stable. With the initial condition $y(0) = 0.1$, predict what *should* happen to the solution. Then, with $h = 0.1$, use the Euler method to march the solution out until $x = 3$. What happens to the numerical solution?

73 ERRORS

The notion of *error* is of crucial importance in the study of numerical methods and we will give the idea some special consideration here. We mentioned in the previous section that reducing the step size in the Euler method can be very costly. This occurs for two reasons. First, the number of computations is directly proportional to the number of steps taken. Thus, raising the accuracy raises the computational cost. Secondly, a phenomenon known as round-off error can become important. This is a result of any computer's ability to represent only a finite subset of rational numbers.

> **Example.** Consider the benchmark problem (3). Let us examine what happens if h is made too small. Let us suppose that our calculator has nine decimal digits of precision. Let $h = 10^{-10}$, a very small step size that would seem to yield very accurate answers. Applying the Euler method and computing the first step, we find that the calculator obtains
>
> $$ y_1 = y_0 + hf(x_0, y_0) = 1 + 10^{-10} = 1! \tag{10} $$
>
> The last equality in (10) is not a misprint. Because of its limited precision ability, the calculator represents y_1 as exactly 1. Unfortunately, the same thing will happen to y_2 as well. In this instance, the Euler method would predict a *constant* solution to the test problem, and round-off error has produced a numerical disaster. A detailed analysis of round-off error is beyond the scope of this text.[6] As a result, we will concentrate exclusively on *discretization* error in the rest of this chapter, assuming that round-off error is always negligible.[7]

The local discretization error at the nth step is defined to be $\epsilon_n = y(x_n) - y_n$. (This assumes that y_n is exactly correct.) As shown in the previous section, for the Euler method, this quantity is given by

$$ \epsilon_k = \frac{y''(\xi)h^2}{2}, \tag{11} $$

---

[6] But see Chapter 1 of R. L. Burden and J. D. Faires *Numerical Analysis*, 4th ed., PWS-Kent, Boston, 1989, for a very thorough discussion.

[7] *Caveat computer.*

where $x_{k-1} < \xi < x_k$. First, note that on the interval $x_0 < x < x_n$, the quantity $y''(x)$ is bounded by a positive constant M which is independent of h. Thus, $|\epsilon_k| \leq Mh^2/2$. Reducing the step size by a factor of 2 reduces the error bound on the local discretization error by a factor of 4, for example.

Unfortunately, the story is a bit more complicated than this, since there is nothing to prevent these local errors from accumulating as many steps are taken. This leads to the notion of *total* discretization error at the nth step, E_n. To estimate this quantity, note that, as the numerical solution is marched from x_0 to x_n, n steps are taken, and $n = (x_n - x_0)/h$. Assuming the worst case, that is, that local errors always add together and never cancel, a heuristic bound for the total error can be obtained:

$$|E_n| \leq n \frac{Mh^2}{2} = (x_n - x_0) \frac{Mh}{2}.$$

So, for the Euler method, the total discretization error is never greater than some constant times the step size.

To illustrate these ideas, let us estimate the discretization errors associated with the benchmark problem (3). First, note that $y'' = 2e^x$. It is easy to see that on $0 \leq x \leq 1$, this quantity assumes its largest value at $x = 1$. Thus, $|\epsilon_n| \leq eh^2$. The total error is bounded as well, with $|E_n| \leq eh$. Referring to Table 1 in Section 72, with $h = 0.2$, the total discretization error at $x = 1$ is 0.46 (rounded to two decimal places). The error bound is $e(0.2) = 0.54$, and, as expected, the total error is less than the bound. With $h = 0.1$, the appropriate numbers can be obtained from Table 2 in Section 72. The total error is 0.25 while the error bound is 0.27.

We close this section with some practical advice. Since, in many problems of concern, the exact solution is not available for calculating an error bound, how does one know when h is "small enough?" One way used in practice is to calculate the numerical solution several times, successively halving the step size h. When the results no longer change within the precision desired, it is a good, but not infallible, bet that h is small enough. By the same token, how can one check to see whether h is "too small," that is, that round-off error is not creeping into the problem. One technique is to repeat a calculation using *extended precision* arithmetic. Most programming languages and most computers support this capability. When re-calculated with extended precision, if the numerical results change in any substantial way, it is almost a sure thing that serious round-off errors are occurring. Nevertheless, this test is not foolproof, for it is always possible that the errors will not be visibly manifested even at extended precision. Never forget that, as powerful as computers and numerical methods are, they must be used with care.

PROBLEMS

For the following problems, use the exact solution, together with step sizes $h = 0.2$ and 0.1 to estimate the total discretization error that occurs with the Euler method at $x = 1$.

1. $y' = 2x + 2y$, $y(0) = 1$.
2. $y' = 1/y$, $y(0) = 1$.
3. $y' = e^y$, $y(0) = 0$.
4. $y' = y - \sin x$, $y(0) = -1$.
5. $y' = (x + y - 1)^2$, $y(0) = 0$.
6. Consider the problem $y' = \sin 3\pi x$, with $y(0) = 0$. Determine the exact solution and sketch the graph on the interval $0 \le x \le 1$. Use the Euler method with $h = 0.2$ and $h = 0.1$ and sketch those results on the same axes. Discuss. Now, use the results in this section to calculate a step size sufficient to guarantee a total error of 0.01 at $x = 1$. Apply the Euler method with *this* step size, and compare with the exact solution. Why is this step size so small?

74 AN IMPROVEMENT TO EULER

Errors of this magnitude (13 and 7 percent) are obviously unsatisfactory. They can be reduced considerably by using much smaller values of h, but this can have its hazards as discussed in Section 73 and a better approach is to develop more accurate methods. For example, it is not unreasonable to expect an improvement if we approximate the integrand (5) by the average of its values at the left and right endpoints of the interval, that is, by $\frac{1}{2}[f(x_0,y_0) + f(x_1,y(x_1))]$. This is equivalent to using the *trapezoidal* rule for approximating the definite integral in (5). Making the substitution, we get

$$y_1 = y_0 + \frac{h}{2}[f(x_0,y_0) + f(x_1,y(x_1))]. \tag{12}$$

The difficulty with (12) is that $y(x_1)$ is unknown. However, if we replace $y(x_1)$ by its approximate value as found by the simpler Euler method, which we denote by $z_1 = y_0 + hf(x_0,y_0)$, then (12) assumes the usable form

$$y_1 = y_0 + \frac{h}{2}[f(x_0,y_0) + f(x_1,z_1)]. \tag{13}$$

More generally,

$$y_{k+1} = y_n + \frac{h}{2}[f(x_k,y_k) + f(x_{k+1},z_{k+1})], \tag{14}$$

where

$$z_{k+1} = y_k + hf(x_k,y_k). \tag{15}$$

This method, usually called the improved Euler method or Heun's[8] method, first *predicts*, then *corrects* an estimate for y_k; it is a simple example of a class of numerical techniques called *predictor–corrector methods*. The local truncation error for this method can be shown to be $\epsilon_k = -y'''(\xi)h^3/12$ with $x_k \le \xi \le x_k$; as a result, the total truncation error is proportional to h^2, and we expect more accuracy for the same step size.

One way to visualize the improved Euler method is depicted in Fig. 102. First, the point at (x_1, z_1) is predicted using the Euler method. This point is used to estimate the slope of the solution curve at x_1. This is then averaged with the original slope estimate at (x_0, y_0) to make a better prediction of the solution, namely (x_1, y_1).

To see just how much improvement is obtained, let us apply (14) and (15) to our benchmark problem (3) with a step size of $h = 0.2$. These formulas become

$$z_{k+1} = y_k + 0.2(x_k + y_k),$$

and

$$y_{k+1} = y_k + 0.1[(x_k + y_k) + (x_{k+1} + z_{k+1})].$$

To begin the calculations we set $k = 0$ and use the initial values $x_0 = 0.0$ and $y_0 = 1.0000$ to write

$$z_1 = 1.000 + 0.2(0.0 + 1.000) = 1.200$$

and

$$y_1 = 1.000 + 0.1[(0.0 + 1.000) + (0.2 + 1.2000)] = 1.240.$$

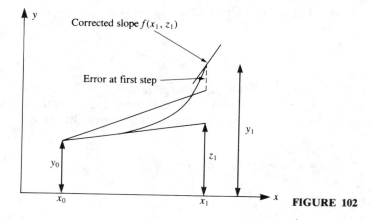

FIGURE 102

---

[8] Karl Heun (1859–1929) was a contemporary of C. Runge and R. Kutta (q.v.). He made contributions to classical mechanics, the theory of special functions, and Gaussian quadrature methods.

TABLE 1
Tabulated values for exact and numerical solutions to (3) with $h = 0.2$ using the improved Euler method

x_n	y_n	Exact	$\bar{E}_n$ (%)
0.0	1.00000	1.00000	0.00
0.2	1.24000	1.24281	0.23
0.4	1.57680	1.58365	0.43
0.6	2.03170	2.04424	0.61
0.8	2.63067	2.65108	0.77
1.0	3.40542	3.43656	0.91

Table 1 shows the approximate values of the solution obtained at the points $x_n = 0.2$, 0.4, 0.8, and 1.0 by continuing this process. The resulting approximate value for $y(1)$ is 3.40542. The error with this method is therefore about 1 percent, which is a substantial improvement over the result obtained with the Euler method and the same step size.

With a smaller step size, results are even better. Table 2 displays the results of applying the improved Euler method to (3) using a step size of $h = 0.1$. The relative error at $x = 1.0$ has been decreased to about 0.2 percent, roughly a fourth of that found previously. Since the total discretization error is proportional h^2, halving the step size leads to the result indicated above.

Clearly, there is a substantial improvement in the accuracy of the improved Euler method at a rather modest increase in the complexity of the formula. Suppose, however, that even more accuracy is desired.

TABLE 2
Tabulated values for exact and numerical solutions to (3) with $h = 0.1$ using the improved Euler method.

x_n	y_n	Exact	$\bar{E}_n$ (%)
0.0	1.00000	1.00000	0.0
0.1	1.11000	1.11034	0.0
0.2	1.24205	1.24281	0.1
0.3	1.39847	1.39972	0.1
0.4	1.58180	1.58365	0.1
0.5	1.79489	1.79744	0.1
0.6	2.04086	2.04424	0.2
0.7	2.32315	2.32751	0.2
0.8	2.64558	2.65108	0.2
0.9	3.01236	3.01921	0.2
1.0	3.42816	3.43656	0.2

Decreasing the step size will work, though, as with the Euler method, it takes longer and will eventually produce unacceptably large errors. There are two main directions in which the strategy of increasing accuracy can be pursued. Perhaps the most natural one is to consider more accurate approximations to the integrand in (5). There are two fundamental ways in which this can be done: by using a polynomial approximant for $f(x,y)$ in the interval $[x_0, x_1]$ or by subdividing the interval. The latter method gives rise to the Runge–Kutta methods, which will be described in the next section. The former approach leads to the multiterm Taylor methods, one of which we briefly describe below.

First, we determine the first order Taylor polynomial for $f(x,y)$ about the point $x = x_0$:

$$f(x,y) = f(x_0, y_0) + [f'(x,y) + f_y(x,y)y'](x - x_0).$$

We then substitute this into (5) to obtain the three-term Taylor scheme:

$$y_{k+1} = y_k + hf(x_0, y_0) + \frac{h^2}{2} y''(x_0), \qquad (16)$$

where we have used the fact that $y'' = [f(x,y)]'$. The local truncation error is $\epsilon_k = y'''(\xi)h^3/12$ where $x_0 \le \xi x_n$. The total truncation error is proportional to h^2. Consequently, (16) is expected to perform comparably to (14).

Table 3 displays the results of applying (16) to (3) with $h = 0.1$. At $x = 1$, this method produces results identical (to the number of decimal places shown) to those obtained with the improved Euler method.

TABLE 3
Tabulated values for exact and numerical solutions to (3) with $h = 0.1$ using the three-term Taylor method

x_n	y_n	Exact	$\bar{E}_n$ (%)
0.0	1.00000	1.00000	0.0
0.1	1.11000	1.11034	0.0
0.2	1.24205	1.24281	0.1
0.3	1.39847	1.39972	0.1
0.4	1.58180	1.58365	0.1
0.5	1.79489	1.79744	0.1
0.6	2.04086	2.04424	0.2
0.7	2.32315	2.32751	0.2
0.8	2.64558	2.65108	0.2
0.9	3.01236	3.01921	0.2
1.0	3.42816	3.43656	0.2

Obviously, better accuracy can be obtained by retaining more terms in the Taylor series (see Problem 8). The drawback to this approach comes from the need to evaluate higher-order derivatives of $f(x,y)$. These derivatives can become unwieldy in a hurry, slowing down the calculation time for a given problem significantly. Even more, $f(x,y)$ may not be available in analytical form. For example, it could consist of discrete experimental data or itself might be the result of a numerical computation. As such, higher order derivative calculations are likely to be so inaccurate as to nullify any gain that might exist in principle. Thus, multiterm Taylor methods are seldom used in practice. There exist much better ways to gain the accuracy needed with far less computational cost, as will be discussed in the next section.

PROBLEMS

For the following problems, use the improved Euler method with $h = 0.1$, 0.05, and 0.01 to estimate the solution at $x = 1$. Compare your results to the *exact* solution and the results obtained with the Euler method in Section 72.

1. $y' = 2x + 2y$, $y(0) = 1$.
2. $y' = 1/y$, $y(0) = 1$.
3. $y' = e^y$, $y(0) = 0$.
4. $y' = y - \sin x$, $y(0) = -1$.
5. $y' = (x + y - 1)^2$, $y(0) = 0$.
6. Think of some examples for which the three-term Taylor method might work better than the improved Euler method. In each instance, describe why and, if possible, use a computer or calculator to illustrate the problem.
7. Think of some examples for which the three-term Taylor method might work poorly. In each instance, describe the source of difficulty. If possible, use a computer or calculator to illustrate the problem.
8. Derive an expression for the *four-term* Taylor method. Apply it to the benchmark problem (3) with a step size of $h = 0.1$ and calculate the solution out to $x = 1$. Is any accuracy gained over the three-term Taylor method?

75 HIGHER ORDER METHODS

As with the improved Euler methods discussed in Section 74, the Runge–Kutta[9] methods can be derived from (5) by using a different

---

[9] Carl Runge (1856–1927) was professor of applied mathematics at Göttingen from 1904 to 1925. He is known for his work on the Zeeman effect and for his discovery of a theorem that foreshadowed the famous Thue–Siegel–Roth theorem in Diophantine equations. He also taught Hilbert to ski. M. W. Kutta (1867–1944), another German applied mathematician, is remembered for his contribution to the Kutta–Joukowski theory of airfoil lift in aerodynamics.

approximation for the integral. Let us consider *Simpson's rule*. In this instance,

$$\int_{x_0}^{x_1} f(x,y)\, dx = \frac{1}{6}[f(x_0,y_0) + 4f(x_{1/2},y(x_{1/2})) + f(x_1,y(x_1))], \quad (17)$$

where $x_{1/2} = x_0 + h/2$. A rigorous derivation of the fourth order Runge–Kutta method is beyond the scope of this chapter. Rather than simply state the results, we give here an *intuitive* development of this extremely important scheme for solving ordinary differential equations.[10]

In much the same way as we applied the other integration formulas, we must make estimates of both $y_{1/2}$ and y_1. The first estimate of $y_{1/2}$ is obtained from Euler's method:

$$y_{1/2} = y_0 + \frac{m_1}{2}, \quad (18)$$

where $m_1 = hf(x_0,y_0)$. The factor of 1/2 is necessary since the step size from x_0 to $x_{1/2}$ is $h/2$. To correct this estimate of $y_{1/2}$, we calculate it again in the following way:

$$y_{1/2} = y_0 + \frac{m_2}{2}, \quad (19)$$

where now $m_2 = hf(x_0 + h/2, y_0 + m_1/2)$. Now, to predict y_1 we use this latter estimate for $y_{1/2}$ and the Euler method:

$$y_1 = y_{1/2} + \frac{m_3}{2}, \quad (20)$$

where now $m_3 = hf(x_0 + h/2, y_0 + m_2/2)$. Finally, we let $m_4 = hf(x + h, y_0 + m_3)$. The Runge–Kutta method is then obtained from substituting each of these estimates into (17) to obtain

$$y_1 = y_0 + \tfrac{1}{6}(m_1 + 2m_2 + 2m_3 + m_4). \quad (21)$$

As with all previous methods, this one can be extended to any number of mesh points in the natural way. At each step, first compute the four

---

[10] It is worth noting that more than one fourth order Runge–Kutta formula can be derived. See B. Carnahan, H. A. Luther, and J. O. Wilkes, *Applied Numerical Methods*, Wiley, New York, 1969, pp. 361–363, for a short, but interesting, historical discussion of this point.

numbers $m_1, \ldots, m_4$:

$$m_1 = hf(x_k, y_k),$$

$$m_2 = hf\left(x_k + \frac{h}{2}, y_k + \frac{m_1}{2}\right),$$

$$m_3 = hf\left(x_k + \frac{h}{2}, y_k + \frac{m_2}{2}\right),$$

$$m_4 = hf(x_k + h, y_k + m_3).$$

Then, y_{k+1} is given by

$$y_{k+1} = y_k + \tfrac{1}{6}(m_1 + 2m_2 + 2m_3 + m_4). \tag{22}$$

This powerful method is capable of giving accurate results without taking h so small that computational labor becomes excessive or that numerical round-off becomes a serious problem. The local truncation error is $\epsilon_k = -y^v(\xi)h^5/180$ where $x_0 \leq x \leq x_n$ and the total truncation error is proportional to h^4. This is one reason for its remarkable accuracy.

We now apply (22) to approximate $y(1)$ in our benchmarks problem (3). With $h = 1$, so that only a *single* step is required, we have

$$m_1 = 1(0 + 1) = 1,$$

$$m_2 = 1(0 + 0.5 + 1 + 0.5) = 2,$$

$$m_3 = 1(0 + 0.5 + 1 + 1) = 2.5,$$

$$m_4 = 1(0 + 1 + 1 + 2.5) = 4.5,$$

so that

$$y_1 = 1 + \tfrac{1}{6}(1 + 4 + 5 + 4.5) = 3.417.$$

This approximation is even better than the improved Euler method with $h = 0.2$! In Table 1, we show the result of applying the Runge–Kutta method to our benchmark problem with $h = 0.2$. Note especially that

TABLE 1
Tabulated values for exact and numerical solutions to (3) with $h = 0.2$ using the Runge–Kutta method

x_n	y_n	Exact	$\bar{E}_n$ (%)
0.0	1.00000	1.00000	0.00000
0.2	1.24280	1.24281	0.00044
0.4	1.58364	1.58365	0.00085
0.6	2.04421	2.04424	0.00125
0.8	2.65104	2.65108	0.00152
1.0	3.43650	3.43656	0.00179

our approximate value for $y(1)$ is 3.43650, which agrees with the exact value to four figures after the decimal point. The relative error is much smaller, in this case less than 0.2%. Halving the step size produces even better results, as shown in Table 2. With $h = 0.1$, the exact and computed solutions agree exactly to the number of decimal places shown, and the relative error at the end of the calculation is now less than 0.02 percent, a very nice result indeed!

PROBLEMS

For the following problems, use the Runge–Kutta method with $h = 0.1$, 0.05, and 0.01 to estimate the solution at $x = 1$. Compare your results to the *exact* solution and the results obtained with both the Euler method in Section 72 and the improved Euler method in Section 74.

1. $y' = 2x + 2y$, $y(0) = 1$.
2. $y' = 1/y$, $y(0) = 1$.
3. $y' = e^y$, $y(0) = 0$.
4. $y' = y - \sin x$, $y(0) = -1$.
5. $y' = (x + y - 1)^2$, $y(0) = 0$.
6. Are there any other numerical integration rules that could be used to generate methods as accurate as the Runge–Kutta method or more so? Find one and attempt to work out the steps necessary for an algorithm. Check your results against the benchmark problem and discuss your findings.
7. Use the Runge–Kutta method with $h = 0.2$ and solve the following equation

$$t^2 y'' - 3ty' + 3y = 1, \qquad y(1) = 0, \qquad y'(1) = 0.$$

Determine the exact solution and compare your results. Does the differential equation possess a solution at $t = 0$? How might the Runge–Kutta method be employed to compute the solution there?

TABLE 2
Tabulated values for exact and numerical solutions to (3) with $h = 0.1$ using the Runge–Kutta method

x_n	y_n	Exact	$\bar{E}_n$ (%)
0.0	1.00000	1.00000	0.00000
0.1	1.11034	1.11034	0.00002
0.2	1.24281	1.24281	0.00003
0.3	1.39972	1.39972	0.00004
0.4	1.58365	1.58365	0.00006
0.5	1.79744	1.79744	0.00007
0.6	2.04424	2.04424	0.00008
0.7	2.32750	2.32751	0.00009
0.8	2.65108	2.65108	0.00010
0.9	3.01920	3.01921	0.00011
1.0	3.43656	3.43656	0.00012

76 SYSTEMS

Heretofore our numerical methods have been employed against first order initial-value problems. It should be clear that many important physical problems are modeled by second and higher order equations (such as vibrating mechanical systems), or even directly as systems of equations (such as predator–prey systems). It is therefore natural to seek ways in which our methods can be extended to treat these types of problems.

Since $d^2y/dt^2 = f(t,y,dy/dt)$ can be transformed into the system of first order equations $dy/dt = x$ and $dx/dt = f(t,y,x)$, it is customary to transform *all* higher order differential equations into systems of first order equations. In this section we will discuss formulas that explicitly treat systems of two first order equations, but the results can be generalized to more equations with relative ease. It should be noted that serious scientific and engineering applications, employing models composed of complicated systems of differential equations, are almost always solved with methods (albeit with a bit more sophistication) very much like the ones we will describe here.

Our objective is to formulate methods for generating numerical solutions to the following system of equations:

$$x' = f(t,x,y), \tag{23}$$

$$y' = g(t,x,y), \tag{24}$$

with initial conditions

$$x(t_0) = x_0, \qquad y(t_0) = y_0. \tag{25}$$

We assume, of course, that the functions f and g are sufficiently smooth so that unique solutions to (24), (24), and (25) exist.[11] As in the previous sections, we seek to construct approximate solutions x_n and y_n to the system at the points $t = t_0$, $t_1 = t_0 + h, \ldots, t_n = t_0 + nh$.

The Euler method takes on an entirely analogous form for this case and is given below:

$$x_{k+1} = x_k + hf(t_k,x_k,y_k), \tag{26}$$

$$y_{k+1} = y_k + hg(t_k,x_k,y_k), \tag{27}$$

where $k = 0, 1, \ldots, n - 1$. The expression for the local truncation error is more complicated for the Euler method in this instance, but it remains true that the total discretization error is proportional to h.

---

[11] See Chapter 11.

Consider the following linear, second order, nonhomogeneous differential equation:

$$\frac{dy^2}{dt^2} + 4y = \cos t \tag{28}$$

with initial conditions $y(0) = y'(0) = 0$. Equation (28) can be thought of as a model for an undamped spring–mass subject to a sinusoidal exterior driving force. At time $t = 0$, the mass lies at its equilibrium position with no initial velocity. The exact solution to (28) is

$$y = \tfrac{1}{3}(\cos t - \cos 2t).$$

Cast into system form, we first let $y' = x$. Then

$$x' = -4y + \cos t, \tag{29}$$

$$y' = x, \tag{30}$$

with initial conditions $x(0) = y(0) = 0$. Table 1 contains the tabulated results[12] for this system on the interval $0 \leq t \leq 1$ using the Euler method with $h = 0.1$ Note that the relative error for y starts out extremely large, decreases to a rather small value, and then begins to increase again. See Problem 5 for a discussion of this phenomenon.

TABLE 1
Tabulated values for exact and numerical solutions to (29) and (30) with $h = 0.1$ using the Euler method

t_n	x_n	y_n	Exact x	Exact y	E_n for y (%)
0.0	0.00000	0.00000	0.00000	0.00000	—
0.1	0.10000	0.00000	0.09917	0.00498	100
0.2	0.19950	0.01000	0.19339	0.01967	49
0.3	0.29351	0.02995	0.27792	0.04333	31
0.4	0.37706	0.05930	0.34843	0.07478	21
0.5	0.44545	0.09701	0.40117	0.11243	14
0.6	0.49440	0.14155	0.43315	0.15433	8.3
0.7	0.52032	0.19099	0.44223	0.19829	3.7
0.8	0.52040	0.24302	0.42726	0.24197	0.4
0.9	0.49286	0.29506	0.38812	0.28294	4.3
1.0	0.43700	0.34435	0.32571	0.31882	8.0

---

[12] This tabulation should convince anyone (should such convincing be needed) trying such a calculation by hand that there is nothing like a computer, together with a good programming language, for accomplishing such a task. Imagine what it was like in the old days (pre-World War II), when virtually all engineering computations were done with a pencil, paper, and perhaps a desk calculator.

The Runge–Kutta method for this system is

$$x_{k+1} = x_k + \tfrac{1}{6}(\mu_{k1} + \mu_{k2} + \mu_{k3} + \mu_{k4}), \tag{31}$$

$$y_{k+1} = y_k + \tfrac{1}{6}(v_{k1} + v_{k2} + v_{k3} + v_{k4}), \tag{32}$$

where

$$
\begin{aligned}
\mu_{k1} &= hf(t_k, x_k, y_k), \\
v_{k1} &= hg(t_k, x_k, y_k), \\
\mu_{k2} &= hf\left(t_k + \frac{h}{2},\, x_k + \frac{\mu_{k1}}{2},\, y_k + \frac{v_{k1}}{2}\right), \\
v_{k2} &= hg\left(t_k + \frac{h}{2},\, x_k + \frac{\mu_{k1}}{2},\, y_k + \frac{v_{k1}}{2}\right), \\
\mu_{k3} &= hf\left(t_k + \frac{h}{2},\, x_k + \frac{\mu_{k2}}{2},\, y_k + \frac{v_{k2}}{2}\right), \\
v_{k3} &= hg\left(t_k + \frac{h}{2},\, x_k + \frac{\mu_{k2}}{2},\, y_k + \frac{v_{k2}}{2}\right), \\
\mu_{k4} &= hf(t_k + h,\, x_k + \mu_{k3},\, y_k + v_{k3}), \\
v_{k4} &= hg(t_k + h,\, x_k + \mu_{k3},\, y_k + v_{k3}).
\end{aligned}
\tag{33}
$$

The total discretization error for this more general Runge–Kutta method remains proportional to h^4. The numerical solution of (29) and (30) with a step size of $h = 0.1$ is displayed in Table 2. Note that the relative error is significantly smaller than that seen with the Euler method as shown in Table 1, and furthermore, the relative error does not exhibit the same degree of fluctuation as that case.

TABLE 2
Tabulated values for exact and numerical solutions to (29) and (30) with $h = 0.1$ using the Runge–Kutta method

t_n	x_n	y_n	Exact x	Exact y	E_n for y (%)
0.0	0.00000	0.00000	0.00000	0.00000	—
0.1	0.09917	0.00498	0.09917	0.00498	0.0006
0.2	0.19339	0.01967	0.19339	0.01967	0.0018
0.3	0.27792	0.04333	0.27792	0.04333	0.0022
0.4	0.34843	0.07478	0.34843	0.07478	0.0023
0.5	0.40117	0.11242	0.40117	0.11243	0.0024
0.6	0.43314	0.15432	0.43315	0.15433	0.0024
0.7	0.44223	0.19829	0.44223	0.19829	0.0024
0.8	0.42726	0.24196	0.42726	0.24197	0.0023
0.9	0.38813	0.28293	0.38812	0.28294	0.0022
1.0	0.32571	0.31881	0.32571	0.31882	0.0021

PROBLEMS

1. Use the Euler method, with step size $h = 0.2$ to evaluate the solution to $y'' - y = 0$, $y(0) = 0$, $y'(1) = 0$ at $t = 0.2$ and $t = 0.4$. Compare your results to the exact solution.

2. Use the Euler method, with step size $h = 0.1$ to evaluate the solution to the following system of equations at $t = 0.5$:

$$x' = y,$$
$$y' = x(1 - x),$$

with $x(0) = y(0) = 1$.

3. Use the Runge–Kutta method (and a computer!) to evaluate the solution to $y'' - y(1 - y)y' + y = 0$, $y(0) = 1$ and $y'(0) = 1$, at $t = 1$. Use step sizes of 0.5, 0.2, and 0.1.

4. Generalize the formulation of the Euler method to a system of *three* first order ordinary differential equations.

5. Using the results listed in Table 1, sketch the graph of y_n and y versus t_n. Explain the fluctuation in the relative error. Does the same error behavior occur for x_n and x? Why does the Runge–Kutta error (see Table 2) not behave this way?

NUMERICAL TABLES

Table 1. Trigonometric functions

ANGLE					ANGLE				
DEGREE	RADIAN	SINE	COSINE	TANGENT	DEGREE	RADIAN	SINE	COSINE	TANGENT
0°	0.000	0.000	1.000	0.000					
1°	0.017	0.017	1.000	0.017	46°	0.803	0.719	0.695	1.036
2°	0.035	0.035	0.999	0.035	47°	0.820	0.731	0.682	1.072
3°	0.052	0.052	0.999	0.052	48°	0.838	0.743	0.669	1.111
4°	0.070	0.070	0.998	0.070	49°	0.855	0.755	0.656	1.150
5°	0.087	0.087	0.996	0.087	50°	0.873	0.766	0.643	1.192
6°	0.105	0.105	0.995	0.105	51°	0.890	0.777	0.629	1.235
7°	0.122	0.122	0.993	0.123	52°	0.908	0.788	0.616	1.280
8°	0.140	0.139	0.990	0.141	53°	0.925	0.799	0.602	1.327
9°	0.157	0.156	0.988	0.158	54°	0.942	0.809	0.588	1.376
10°	0.175	0.174	0.985	0.176	55°	0.960	0.819	0.574	1.428
11°	0.192	0.191	0.982	0.194	56°	0.977	0.829	0.559	1.483
12°	0.209	0.208	0.978	0.213	57°	0.995	0.839	0.545	1.540
13°	0.227	0.225	0.974	0.231	58°	1.012	0.848	0.530	1.600
14°	0.244	0.242	0.970	0.249	59°	1.030	0.857	0.515	1.664
15°	0.262	0.259	0.966	0.268	60°	1.047	0.866	0.500	1.732
16°	0.279	0.276	0.961	0.287	61°	1.065	0.875	0.485	1.804
17°	0.297	0.292	0.956	0.306	62°	1.082	0.883	0.469	1.881
18°	0.314	0.309	0.951	0.325	63°	1.100	0.891	0.454	1.963
19°	0.332	0.326	0.946	0.344	64°	1.117	0.899	0.438	2.050
20°	0.349	0.342	0.940	0.364	65°	1.134	0.906	0.423	2.145
21°	0.367	0.358	0.934	0.384	66°	1.152	0.914	0.407	2.246
22°	0.384	0.375	0.927	0.404	67°	1.169	0.921	0.391	2.356
23°	0.401	0.391	0.921	0.424	68°	1.187	0.927	0.375	2.475
24°	0.419	0.407	0.914	0.445	69°	1.204	0.934	0.358	2.605
25°	0.436	0.423	0.906	0.466	70°	1.222	0.940	0.342	2.748
26°	0.454	0.438	0.899	0.488	71°	1.239	0.946	0.326	2.904
27°	0.471	0.454	0.891	0.510	72°	1.257	0.951	0.309	3.078
28°	0.489	0.469	0.883	0.532	73°	1.274	0.956	0.292	3.271
29°	0.506	0.485	0.875	0.554	74°	1.292	0.961	0.276	3.487
30°	0.524	0.500	0.866	0.577	75°	1.309	0.966	0.259	3.732
31°	0.541	0.515	0.857	0.601	76°	1.326	0.970	0.242	4.011
32°	0.559	0.530	0.848	0.625	77°	1.344	0.974	0.225	4.332
33°	0.576	0.545	0.839	0.649	78°	1.361	0.978	0.208	4.705
34°	0.593	0.559	0.829	0.675	79°	1.379	0.982	0.191	5.145
35°	0.611	0.574	0.819	0.700	80°	1.396	0.985	0.174	5.671
36°	0.628	0.588	0.809	0.727	81°	1.414	0.988	0.156	6.314
37°	0.646	0.602	0.799	0.754	82°	1.431	0.990	0.139	7.115
38°	0.663	0.616	0.788	0.781	83°	1.449	0.993	0.122	8.144
39°	0.681	0.629	0.777	0.810	84°	1.466	0.995	0.105	9.514
40°	0.698	0.643	0.766	0.839	85°	1.484	0.996	0.087	11.43
41°	0.716	0.656	0.755	0.869	86°	1.501	0.998	0.070	14.30
42°	0.733	0.669	0.743	0.900	87°	1.518	0.999	0.052	19.08
43°	0.750	0.682	0.731	0.933	88°	1.536	0.999	0.035	28.64
44°	0.768	0.695	0.719	0.966	89°	1.553	1.000	0.017	57.29
45°	0.785	0.707	0.707	1.000	90°	1.571	1.000	0.000	

Table 2. Exponential functions

x	e^x	e^{-x}	x	e^x	e^{-x}
0.00	1.0000	1.0000	2.5	12.182	0.0821
0.05	1.0513	0.9512	2.6	13.464	0.0743
0.10	1.1052	0.9048	2.7	14.880	0.0672
0.15	1.1618	0.8607	2.8	16.445	0.0608
0.20	1.2214	0.8187	2.9	18.174	0.0550
0.25	1.2840	0.7788	3.0	20.086	0.0498
0.30	1.3499	0.7408	3.1	22.198	0.0450
0.35	1.4191	0.7047	3.2	24.533	0.0408
0.40	1.4918	0.6703	3.3	27.113	0.0369
0.45	1.5683	0.6376	3.4	29.964	0.0334
0.50	1.6487	0.6065	3.5	33.115	0.0302
0.55	1.7333	0.5769	3.6	36.598	0.0273
0.60	1.8221	0.5488	3.7	40.447	0.0247
0.65	1.9155	0.5220	3.8	44.701	0.0224
0.70	2.0138	0.4966	3.9	49.402	0.0202
0.75	2.1170	0.4724	4.0	54.598	0.0183
0.80	2.2255	0.4493	4.1	60.340	0.0166
0.85	2.3396	0.4274	4.2	66.686	0.0150
0.90	2.4596	0.4066	4.3	73.700	0.0136
0.95	2.5857	0.3867	4.4	81.451	0.0123
1.0	2.7183	0.3679	4.5	90.017	0.0111
1.1	3.0042	0.3329	4.6	99.484	0.0101
1.2	3.3201	0.3012	4.7	109.95	0.0091
1.3	3.6693	0.2725	4.8	121.51	0.0082
1.4	4.0552	0.2466	4.9	134.29	0.0074
1.5	4.4817	0.2231	5	148.41	0.0067
1.6	4.9530	0.2019	6	403.43	0.0025
1.7	5.4739	0.1827	7	1096.6	0.0009
1.8	6.0496	0.1653	8	2981.0	0.0003
1.9	6.6859	0.1496	9	8103.1	0.0001
2.0	7.3891	0.1353	10	22026	0.00005
2.1	8.1662	0.1225			
2.2	9.0250	0.1108			
2.3	9.9742	0.1003			
2.4	11.023	0.0907			

Table 3. Natural logarithms ($\ln x = \log_e x$)

This table contains logarithms of numbers from 1 to 10 to the base e. To obtain the natural logarithms of other numbers use the formulas:

$$\ln (10^r x) = \ln x + \ln 10^r \qquad \ln \left(\frac{x}{10^r} \right) = \ln x - \ln 10^r$$

$\ln 10 = 2.302585 \qquad \ln 10^2 = 4.605170 \qquad \ln 10^3 = 6.907755$
$\ln 10^4 = 9.210340 \qquad \ln 10^5 = 11.512925 \qquad \ln 10^6 = 13.815511$

x	0	1	2	3	4	5	6	7	8	9
1.0	0.0 0000	0995	1980	2956	3922	4879	5827	6766	7696	8618
1.1	0.0 9531	*0436	*1333	*2222	*3103	*3976	*4842	*5700	*6551	*7395
1.2	0.1 8232	9062	9885	*0701	*1511	*2314	*3111	*3902	*4686	*5464
1.3	0.2 6236	7003	7763	8518	9267	*0010	*0748	*1481	*2208	*2930
1.4	0.3 3647	4359	5066	5767	6464	7156	7844	8526	9204	9878
1.5	0.4 0547	1211	1871	2527	3178	3825	4469	5108	5742	6373
1.6	0.4 7000	7623	8243	8858	9470	*0078	*0682	*1282	*1879	*2473
1.7	0.5 3063	3649	4232	4812	5389	5962	6531	7098	7661	8222
1.8	0.5 8779	9333	9884	*0432	*0977	*1519	*2078	*2594	*3127	*3658
1.9	0.6 4185	4710	5233	5752	6269	6783	7294	7803	8310	8813
2.0	0.6 9315	9813	*0310	*0804	*1295	*1784	*2271	*2755	*3237	*3716
2.1	0.7 4194	4669	5142	5612	6081	6547	7011	7473	7932	8390
2.2	0.7 8846	9299	9751	*0200	*0648	*1093	*1536	*1978	*2418	*2855
2.3	0.8 3291	3725	4157	4587	5015	5442	5866	6289	6710	7129
2.4	0.8 7547	7963	8377	8789	9200	9609	*0016	*0422	*0826	*1228
2.5	0.9 1629	2028	2426	2822	3216	3609	4001	4391	4779	5166
2.6	0.9 5551	5935	6317	6698	7078	7456	7833	8208	8582	8954
2.7	0.9 9325	9695	*0063	*0430	*0796	*1160	*1523	*1885	*2245	*2604
2.8	1.0 2962	3318	3674	4028	4380	4732	5082	5431	5779	6126
2.9	1.0 6471	6815	7158	7500	7841	8181	8519	8856	9192	9527
3.0	1.0 9861	*0194	*0526	*0856	*1186	*1514	*1841	*2168	*2493	*2817
3.1	1.1 3140	3462	3783	4103	4422	4740	5057	5373	5688	6002
3.2	1.1 6315	6627	6938	7248	7557	7865	8173	8479	8784	9089
3.3	1.1 9392	9695	9996	*0297	*0597	*0896	*1194	*1491	*1788	*2083
3.4	1.2 2378	2671	2964	3256	3547	3837	4127	4415	4703	4990
3.5	1.2 5276	5562	5846	6130	6413	6695	6976	7257	7536	7815
3.6	1.2 8093	8371	8647	8923	9198	9473	9746	*0019	*0291	*0563
3.7	1.3 0833	1103	1372	1641	1909	2176	2442	2708	2972	3237
3.8	1.3 3500	3763	4025	4286	4547	4807	5067	5325	5584	5841
3.9	1.3 6098	6354	6609	6864	7118	7372	7624	7877	8128	8379
4.0	1.3 8629	8879	9128	9377	9624	9872	*0118	*0364	*0610	*0854
4.1	1.4 1099	1342	1585	1828	2070	2311	2552	2792	3031	3270
4.2	1.4 3508	3746	3984	4220	4456	4692	4927	5161	5395	5629
4.3	1.4 5862	6094	6326	6557	6787	7018	7247	7476	7705	7933
4.4	1.4 8160	8387	8614	8840	9065	9290	9515	9739	9962	*0185
4.5	1.5 0408	0630	0851	1072	1293	1513	1732	1951	2170	2388
4.6	1.5 2606	2823	3039	3256	3471	3687	3902	4116	4330	4543
4.7	1.5 4756	4969	5181	5393	5604	5814	6025	6235	6444	6653
4.8	1.5 6862	7070	7277	7485	7691	7898	8104	8309	8515	8719
4.9	1.5 8924	9127	9331	9534	9737	9939	*0141	*0342	*0543	*0744
5.0	1.6 0944	1144	1343	1542	1741	1939	2137	2334	2531	2728
x	0	1	2	3	4	5	6	7	8	9

Note: The * indicates that the first two digits are those at the beginning of the next row.

Table 3. Natural logarithms (ln x = log$_e$ x) (*Cont.*)

x	0	1	2	3	4	5	6	7	8	9
5.0	1.6 0944	1144	1343	1542	1741	1939	2137	2334	2531	2728
5.1	1.6 2924	3120	3315	3511	3705	3900	4094	4287	4481	4673
5.2	1.6 4866	5058	5250	5441	5632	5823	6013	6203	6393	6582
5.3	1.6 6771	6959	7147	7335	7523	7710	7896	8083	8269	8455
5.4	1.6 8640	8825	9010	9194	9378	9562	9745	9928	*0111	*0293
5.5	1.7 0475	0656	0838	1019	1199	1380	1560	1740	1919	2098
5.6	1.7 2277	2455	2633	2811	2988	3166	3342	3519	3695	3871
5.7	1.7 4047	4222	4397	4572	4746	4920	5094	5267	5440	5613
5.8	1.7 5786	5958	6130	6302	6473	6644	6815	6985	7156	7326
5.9	1.7 7495	7665	7843	8002	8171	8339	8507	8675	8842	9009
6.0	1.7 9176	9342	9509	9675	9840	*0006	*0171	*0336	*0500	*0665
6.1	1.8 0829	0993	1156	1319	1482	1645	1808	1970	2132	2294
6.2	1.8 2455	2616	2777	2938	3098	3258	3418	3578	3737	3896
6.3	1.8 4055	4214	4372	4530	4688	4845	5003	5160	5317	5473
6.4	1.8 5630	5786	5942	6097	6253	6408	6563	6718	6872	7026
6.5	1.8 7180	7334	7487	7641	7794	7947	8099	8251	8403	8555
6.6	1.8 8707	8858	9010	9160	9311	9462	9612	9762	9912	*0061
6.7	1.9 0211	0360	0509	0658	0806	0954	1102	1250	1398	1545
6.8	1.9 1692	1839	1986	2132	2279	2425	2571	2716	2862	3007
6.9	1.9 3152	3297	3442	3586	3730	3874	4018	4162	4305	4448
7.0	1.9 4591	4734	4876	5019	5161	5303	5445	5586	5727	5869
7.1	1.9 6009	6150	6291	6431	6571	6711	6851	6991	7130	7269
7.2	1.9 7408	7547	7685	7824	7962	8100	8238	8376	8513	8650
7.3	1.9 8787	8924	9061	9198	9334	9470	9606	9742	9877	*0013
7.4	2.0 0148	0283	0418	0553	0687	0821	0956	1089	1223	1357
7.5	2.0 1490	1624	1757	1890	2022	2155	2287	2419	2551	2683
7.6	2.0 2815	2946	3078	3209	3340	3471	3601	3732	3862	3992
7.7	2.0 4122	4252	4381	4511	4640	4769	4898	5027	5156	5284
7.8	2.0 5412	5540	5668	5796	5924	6051	6179	6306	6433	6560
7.9	2.0 6686	6813	6939	7065	7191	7317	7443	7568	7694	7819
8.0	2.0 7944	8069	8194	8318	8443	8567	8691	8815	8939	9063
8.1	2.0 9186	9310	9433	9556	9679	9802	9924	*0047	*0169	*0291
8.2	2.1 0413	0535	0657	0779	0900	1021	1142	1263	1384	1505
8.3	2.1 1626	1746	1866	1986	2106	2226	2346	2465	2585	2704
8.4	2.1 2823	2942	3061	3180	3298	3417	3535	3653	3771	3889
8.5	2.1 4007	4124	4242	4359	4476	4593	4710	4827	4943	5060
8.6	2.1 5176	5292	5409	5524	5640	5756	5871	5987	6102	6217
8.7	2.1 6332	6447	6562	6677	6791	6905	7020	7134	7248	7361
8.8	2.1 7475	7589	7702	7816	7929	8042	8155	8267	8380	8493
8.9	2.1 8605	8717	8830	8942	9054	9165	9277	9389	9500	9611
9.0	2.1 9722	9834	9944	*0055	*0166	*0276	*0387	*0497	*0607	*0717
9.1	2.2 0827	0937	1047	1157	1266	1375	1485	1594	1703	1812
9.2	2.2 1920	2029	2138	2246	2354	2462	2570	2678	2786	2894
9.3	2.2 3001	3109	3216	3324	3431	3538	3645	3751	3858	3965
9.4	2.2 4071	4177	4284	4390	4496	4601	4707	4813	4918	5024
9.5	2.2 5129	5234	5339	5444	5549	5654	5759	5863	5968	6072
9.6	2.2 6176	6280	6384	6488	6592	6696	6799	6903	7006	7109
9.7	2.2 7213	7316	7419	7521	7624	7727	7829	7932	8034	8136
9.8	2.2 8238	8340	8442	8544	8646	8747	8849	8950	9051	9152
9.9	2.2 9253	9354	9455	9556	9657	9757	9858	9958	*0058	*0158
10.0	2.3 0259	0358	0458	0558	0658	0757	0857	0956	1055	1154
x	0	1	2	3	4	5	6	7	8	9

Table 4. Common logarithms ($\log_{10} x$)

x	0	1	2	3	4	5	6	7	8	9
10	0000	0043	0086	0128	0170	0212	0253	0294	0334	0374
11	0414	0453	0492	0531	0569	0607	0645	0682	0719	0755
12	0792	0828	0864	0899	0934	0969	1004	1038	1072	1106
13	1139	1173	1206	1239	1271	1303	1335	1367	1399	1430
14	1461	1492	1523	1553	1584	1614	1644	1673	1703	1732
15	1761	1790	1818	1847	1875	1903	1931	1959	1987	2014
16	2041	2068	2095	2122	2148	2175	2201	2227	2253	2279
17	2304	2330	2355	2380	2405	2430	2455	2480	2504	2529
18	2553	2577	2601	2625	2648	2672	2695	2718	2742	2765
19	2788	2810	2833	2856	2878	2900	2923	2945	2967	2989
20	3010	3032	3054	3075	3096	3118	3139	3160	3181	3201
21	3222	3243	3263	3284	3304	3324	3345	3365	3385	3404
22	3424	3444	3464	3483	3502	3522	3541	3560	3579	3598
23	3617	3636	3655	3674	3692	3711	3729	3747	3766	3784
24	3802	3820	3838	3856	3874	3892	3909	3927	3945	3962
25	3979	3997	4014	4031	4048	4065	4082	4099	4116	4133
26	4150	4166	4183	4200	4216	4232	4249	4265	4281	4298
27	4314	4330	4346	4362	4378	4393	4409	4425	4440	4456
28	4472	4487	4502	4518	4533	4548	4564	4579	4594	4609
29	4624	4639	4654	4669	4683	4698	4713	4728	4742	4757
30	4771	4786	4800	4814	4829	4843	4857	4871	4886	4900
31	4914	4928	4942	4955	4969	4983	4997	5011	5024	5038
32	5051	5065	5079	5092	5105	5119	5132	5145	5159	5172
33	5185	5198	5211	5224	5237	5250	5263	5276	5289	5302
34	5315	5328	5340	5353	5366	5378	5391	5403	5416	5428
35	5441	5453	5465	5478	5490	5502	5514	5527	5539	5551
36	5563	5575	5587	5599	5611	5623	5635	5647	5658	5670
37	5682	5694	5705	5717	5729	5740	5752	5763	5775	5786
38	5798	5809	5821	5832	5843	5855	5866	5877	5888	5899
39	5911	5922	5933	5944	5955	5966	5977	5988	5999	6010
40	6021	6031	6042	6053	6064	6075	6085	6096	6107	6117
41	6128	6138	6149	6160	6170	6180	6191	6201	6212	6222
42	6232	6243	6253	6263	6274	6284	6294	6304	6314	6325
43	6335	6345	6355	6365	6375	6385	6395	6405	6415	6425
44	6435	6444	6454	6464	6474	6484	6493	6503	6513	6522
45	6532	6542	6551	6561	6571	6580	6590	6599	6609	6618
46	6628	6637	6646	6656	6665	6675	6684	6693	6702	6712
47	6721	6730	6739	6749	6758	6767	6776	6785	6794	6803
48	6812	6821	6830	6839	6848	6857	6866	6875	6884	6893
49	6902	6911	6920	6928	6937	6946	6955	6964	6972	6981
50	6990	6998	7007	7016	7024	7033	7042	7050	7059	7067
51	7076	7084	7093	7101	7110	7118	7126	7135	7143	7152
52	7160	7168	7177	7185	7193	7202	7210	7218	7226	7235
53	7243	7251	7259	7267	7275	7284	7292	7300	7308	7316
54	7324	7332	7340	7348	7356	7364	7372	7380	7388	7396
55	7404	7412	7419	7427	7435	7443	7451	7459	7466	7474
56	7482	7490	7497	7505	7513	7520	7528	7536	7543	7551
57	7559	7566	7574	7582	7589	7597	7604	7612	7619	7627
58	7634	7642	7649	7657	7664	7672	7679	7686	7694	7701
59	7709	7716	7723	7731	7738	7745	7752	7760	7767	7774

Table 4. Common logarithms ($\log_{10} x$) (*Cont.*)

x	0	1	2	3	4	5	6	7	8	9
60	7782	7789	7796	7803	7810	7818	7825	7832	7839	7846
61	7853	7860	7868	7875	7882	7889	7896	7903	7910	7917
62	7924	7931	7938	7945	7952	7959	7966	7973	7980	7987
63	7993	8000	8007	8014	8021	8028	8035	8041	8048	8055
64	8062	8069	8075	8082	8089	8096	8102	8109	8116	8122
65	8129	8136	8142	8149	8156	8162	8169	8176	8182	8189
66	8195	8202	8209	8215	8222	8228	8235	8241	8248	8254
67	8261	8267	8274	8280	8287	8293	8299	8306	8312	8319
68	8325	8331	8338	8344	8351	8357	8363	8370	8376	8382
69	8388	8395	8401	8407	8414	8420	8426	8432	8439	8445
70	8451	8457	8463	8470	8476	8482	8488	8494	8500	8506
71	8513	8519	8525	8531	8537	8543	8549	8555	8561	8567
72	8573	8579	8585	8591	8597	8603	8609	8615	8621	8627
73	8633	8639	8645	8651	8657	8663	8669	8675	8681	8686
74	8692	8698	8704	8710	8716	8722	8727	8733	8739	8745
75	8751	8756	8762	8768	8774	8779	8785	8791	8797	8802
76	8808	8814	8820	8825	8831	8837	8842	8848	8854	8859
77	8865	8871	8876	8882	8887	8893	8899	8904	8910	8915
78	8921	8927	8932	8938	8943	8949	8954	8960	8965	8971
79	8976	8982	8987	8993	8998	9004	9009	9015	9020	9025
80	9031	9036	9042	9047	9053	9058	9063	9069	9074	9079
81	9085	9090	9096	9101	9106	9112	9117	9122	9128	9133
82	9138	9143	9149	9154	9159	9165	9170	9175	9180	9186
83	9191	9196	9201	9206	9212	9217	9222	9227	9232	9238
84	9243	9248	9253	9258	9263	9269	9274	9279	9284	9289
85	9294	9299	9304	9309	9315	9320	9325	9330	9335	9340
86	9345	9350	9355	9360	9365	9370	9375	9380	9385	9390
87	9395	9400	9405	9410	9415	9420	9425	9430	9435	9440
88	9445	9450	9455	9460	9465	9469	9474	9479	9484	9489
89	9494	9499	9504	9509	9513	9518	9523	9528	9533	9538
90	9542	9547	9552	9557	9562	9566	9571	9576	9581	9586
91	9590	9595	9600	9605	9609	9614	9619	9624	9628	9633
92	9638	9643	9647	9652	9657	9661	9666	9671	9675	9680
93	9685	9689	9694	9699	9703	9708	9713	9717	9722	9727
94	9731	9736	9741	9745	9750	9754	9759	9763	9768	9773
95	9777	9782	9786	9791	9795	9800	9805	9809	9814	9818
96	9823	9827	9832	9836	9841	9845	9850	9854	9859	9863
97	9868	9872	9877	9881	9886	9890	9894	9899	9903	9908
98	9912	9917	9921	9926	9930	9934	9939	9943	9948	9952
99	9956	9961	9965	9969	9974	9978	9983	9987	9991	9996

Note: Decimal points are omitted in this table; the entries

	0	1	2
10	0000	0043	0086

mean that $\log_{10}(1.00) = 0.0000$, $\log_{10}(1.01) = 0.0043$, and $\log_{10}(1.02) = 0.0086$ (to four-decimal-place accuracy).

Table 5. Powers and roots

x	x^2	$\sqrt{x}$	x^3	$\sqrt[3]{x}$	x	x^2	$\sqrt{x}$	x^3	$\sqrt[3]{x}$
1	1	1.000	1	1.000	51	2,601	7.141	132,651	3.708
2	4	1.414	8	1.260	52	2,704	7.211	140,608	3.733
3	9	1.732	27	1.442	53	2,809	7.280	148,877	3.756
4	16	2.000	64	1.587	54	2,916	7.348	157,464	3.780
5	25	2.236	125	1.710	55	3,025	7.416	166,375	3.803
6	36	2.449	216	1.817	56	3,136	7.483	175,616	3.826
7	49	2.646	343	1.913	57	3,249	7.550	185,193	3.849
8	64	2.828	512	2.000	58	3,364	7.616	195,112	3.871
9	81	3.000	729	2.080	59	3,481	7.681	205,379	3.893
10	100	3.162	1,000	2.154	60	3,600	7.746	216,000	3.915
11	121	3.317	1,331	2.224	61	3,721	7.810	226,981	3.936
12	144	3.464	1,728	2.289	62	3,844	7.874	238,328	3.958
13	169	3.606	2,197	2.351	63	3,969	7.937	250,047	3.979
14	196	3.742	2,744	2.410	64	4,096	8.000	262,144	4.000
15	225	3.873	3,375	2.466	65	4,225	8.062	274,625	4.021
16	256	4.000	4,096	2.520	66	4,356	8.124	287,496	4.041
17	289	4.123	4,913	2.571	67	4,489	8.185	300,763	4.062
18	324	4.243	5,832	2.621	68	4,624	8.246	314,432	4.082
19	361	4.359	6,859	2.668	69	4,761	8.307	328,509	4.102
20	400	4.472	8,000	2.714	70	4,900	8.367	343,000	4.121
21	441	4.583	9,261	2.759	71	5,041	8.426	357,911	4.141
22	484	4.690	10,648	2.802	72	5,184	8.485	373,248	4.160
23	529	4.796	12,167	2.844	73	5,329	8.544	389,017	4.179
24	576	4.899	13,824	2.884	74	5,476	8.602	405,224	4.198
25	625	5.000	15,625	2.924	75	5,625	8.660	421,875	4.217
26	676	5.099	17,576	2.962	76	5,776	8.718	438,976	4.236
27	729	5.196	19,683	3.000	77	5,929	8.775	456,533	4.254
28	784	5.292	21,952	3.037	78	6,084	8.832	474,552	4.273
29	841	5.385	24,389	3.072	79	6,241	8.888	493,039	4.291
30	900	5.477	27,000	3.107	80	6,400	8.944	512,000	4.309
31	961	5.568	29,791	3.141	81	6,561	9.000	531,441	4.327
32	1,024	5.657	32,768	3.175	82	6,724	9.055	551,368	4.344
33	1,089	5.745	35,937	3.208	83	6,889	9.110	571,787	4.362
34	1,156	5.831	39,304	3.240	84	7,056	9.165	592,704	4.380
35	1,225	5.916	42,875	3.271	85	7,225	9.220	614,125	4.397
36	1,296	6.000	46,656	3.302	86	7,396	9.274	636,056	4.414
37	1,369	6.083	50,653	3.332	87	7,569	9.327	658,503	4.431
38	1,444	6.164	54,872	3.362	88	7,744	9.381	681,472	4.448
39	1,521	6.245	59,319	3.391	89	7,921	9.434	704,969	4.465
40	1,600	6.325	64,000	3.420	90	8,100	9.487	729,000	4.481
41	1,681	6.403	68,921	3.448	91	8,281	9.539	753,571	4.498
42	1,764	6.481	74,088	3.476	92	8,464	9.592	778,688	4.514
43	1,849	6.557	79,507	3.503	93	8,649	9.644	804,357	4.531
44	1,936	6.633	85,184	3.530	94	8,836	9.695	830,584	4.547
45	2,025	6.708	91,125	3.557	95	9,025	9.747	857,375	4.563
46	2,116	6.782	97,336	3.583	96	9,216	9.798	884,736	4.579
47	2,209	6.856	103,823	3.609	97	9,409	9.849	912,673	4.595
48	2,304	6.928	110,592	3.634	98	9,604	9.899	941,192	4.610
49	2,401	7.000	117,649	3.659	99	9,801	9.950	970,299	4.626
50	2,500	7.071	125,000	3.684	100	10,000	10.000	1,000,000	4.642

Table 6. Factorials

n	n!	n	n!	n	n!
0	1.00000 00000 E00	20	2.43290 20082 E18	35	1.03331 47966 E40
1	1.00000 00000 E00	21	5.10909 42172 E19	36	3.71993 32679 E41
2	2.00000 00000 E00	22	1.12400 07278 E21	37	1.37637 53091 E43
3	6.00000 00000 E00	23	2.58520 16739 E22	38	5.23022 61747 E44
4	2.40000 00000 E01	24	6.20448 40173 E23	39	2.03978 82081 E46
5	1.20000 00000 E02	25	1.55112 10043 E25	40	8.15915 28325 E47
6	7.20000 00000 E02	26	4.03291 46113 E26	41	3.34525 26613 E49
7	5.04000 00000 E03	27	1.08888 69450 E28	42	1.40500 61178 E51
8	4.03200 00000 E04	28	3.04888 34461 E29	43	6.04152 63063 E52
9	3.62880 00000 E05	29	8.84176 19937 E30	44	2.65827 15748 E54
10	3.62880 00000 E06	30	2.65252 85981 E32	45	1.19622 22087 E56
11	3.99168 00000 E07	31	8.22283 86542 E33	46	5.50262 21598 E57
12	4.79001 60000 E08	32	2.63130 83693 E35	47	2.58623 24151 E59
13	6.22702 08000 E09	33	8.68331 76188 E36	48	1.24139 15593 E61
14	8.71782 91200 E10	34	2.95232 79904 E38	49	6.08281 86403 E62
15	1.30767 43680 E12			50	3.04140 93202 E64
16	2.09227 89888 E13				
17	3.55687 42810 E14				
18	6.40237 37057 E15				
19	1.21645 10041 E17				

Note: Values are given in scientific notation with the exponent denoted by E; for example, 2.65252 85981 E32 denotes $2.6525285981 \times 10^{32}$.

ANSWERS

SECTION 2, p. 9

2. (a) $y = \frac{1}{3}e^{3x} - \frac{1}{2}x^2 + c$;

   (b) $y = \log x + c$;

   (c) $y = \frac{1}{2}e^{x^2} + c$;

   (d) $y = x \sin^{-1} x + \sqrt{1 - x^2} + c$;

   (e) $y = x - \log(1 + x) + c$;

   (f) $y = \frac{1}{2}\log(1 + x^2) + c$;

   (g) $y = \frac{1}{6}\log\left[\frac{x^2 - x + 1}{(x + 1)^2}\right] + \frac{1}{\sqrt{3}}\tan^{-1}\left[\frac{2x - 1}{\sqrt{3}}\right] + c$;

   (h) $y = \frac{1}{2}(\tan^{-1} x)^2 + c$;

   (i) $x = c(y - 1)e^y$;
   (j) $x^{-4} + y^{-4} = c$;

   (k) $\sin y = cxe^{-x^2}$;

   (l) $y = ce^{x^2}$;
   (m) $x^3 + 3\cos y = c$;

   (n) $y = -\log(\csc x + \cot x) + c$;

   (o) $y = c \cos x$;
   (p) $y = c \sec x$;

   (q) $y = \dfrac{c - x}{1 + cx}$;

   (r) $y = e^{cx}$.

585

3. (a) $y = xe^x - e^x + 3$;

(d) $y = \dfrac{1}{2} \log \left[\dfrac{3x - 3}{x + 1} \right]$;

(b) $y = \sin^2 x + 1$;

(e) $y = \dfrac{1}{8} \log \left[\dfrac{4 - x^2}{3x^2} \right]$;

(c) $y = x \log x - x$;

(f) $y = \dfrac{1}{4} \log \left[(x + 1)^2 (x^2 + 1)^3 \right] - \dfrac{1}{2} \tan^{-1} x + 1$.

4. (a) $3e^{2y} = 2e^{3x} + 1$;

(d) $2 \sin 3x \cos 2y = 1$;

(b) $y = x^2 + \log x$;

(e) $2y + 1 = e^x(\sin x + \cos x)$;

(c) $\tan^{-1} x + e^y = 1$;

(f) $\log x(y + 1) = y - x + 1$.

8. $m = 1, 1/2, -2; y = c_1 e^x + c_2 e^{x/2} + c_3 e^{-2x}$.

SECTION 3, p. 16

1. (a) $x^2 - y^2 = c$;

(c) $r = c(1 - \cos \theta)$;

(b) $x^2 + 2y^2 = c^2$;

(d) $y^2 = -2x + c$.

2. (a) $x^2 + 4y^2 = c^2$;

(b) $x^2 + ny^2 = c^2$.

The orthogonal trajectories are ellipses, and are more and more elongated in the x direction as n is taken to be larger and larger.

3. $r = 2c \cos \theta$.

4. $r = c/(1 + \cos \theta)$.

5. $y^2 = 2xy \dfrac{dy}{dx} + y^2 \left(\dfrac{dy}{dx} \right)^2$; the family is *self-orthogonal* in the sense that when a curve in the family intersects another curve in the family, it is orthogonal to it.

6. (a) $xy = c$;

(e) $x^2 + 2y^2 = c^2$;

(b) $y^2 = \pm 2x + c$;

(f) $y^2 = \pm x^2 + c$;

(c) $y = ce^{\pm x}$;

(g) $\theta = 0$ or $r = 2c \sin \theta$;

(d) $y^2 = cx$;

(h) $\theta = \theta_0$ or $r = ce^{k\theta}$.

7. $y = cx^2$.

8. $xy = ce^{kx}$.

9. The intersections of the cylinders $xy = c$ with the saddle surface $z = y^2 - x^2$.

10. (a) $(xy' - y)^2 = x^2(x^2 - y^2)$;

(d) $y + y'^2 = xy'$;

(b) $(x^2 - y^2 - 1)y' = 2xy$;

(e) $(y - xy')^2 = 1 + y'^2$.

(c) $(x - y)^2(1 + y'^2) = (x + yy')^2$;

SECTION 4, p. 24

2. (a) $T = \dfrac{100 \log 2}{r}$ years;

(b) about 6.93 percent.

3. (a) $A = D\left(\dfrac{e^{kt} - 1}{k} \right)$;

(c) $1866.

(b) $5986;

4. (a) $A = \dfrac{W}{k} + \left(P - \dfrac{W}{k}\right)e^{kt}$;   (c) $T = \dfrac{1}{k}\log\dfrac{W}{W - W_0}$ years;

    (b) $W_0 = kP$;   (d) about 13.86 years.

5. If $x = x(t)$ is his wealth at time t, and $t = 0$ one year ago, then $x = 20/(2 - t)$. Thus, in 6 months $x = 40$ million dollars, and at the end of 1 year (as $t \to 2$) x becomes infinite.

6. At about 10:11 P.M.

7. 3531.

8. In the year A.D. 2076; 6.6 billion.

9. (b) About 15.2 grams.

10. $x = \dfrac{x_0 x_1}{x_0 + (x_1 - x_0)e^{-kx_1 t}}$; when $x = \dfrac{1}{2}x_1$.

12. About 35.35 percent; about 3.125 percent.

13. About 133 days.

14. About 13.53 percent.

16. If $B = A$, then

$$x = \frac{kA^2 abt}{kAabt + 1};$$

and if $B < A$, then

$$x = \frac{AB(1 - e^{-k(A-B)abt})}{A - Be^{-k(A-B)abt}}.$$

The first formula is the limit of the second formula as $B \to A$; students should prove this by using l'Hospital's rule.

17. $\dfrac{1 + (A/x)}{1 + (A/x_0)} = \left[\dfrac{1 + (A/x_1)}{1 + (A/x_0)}\right]^{t/t_1}$.

18. $x = \dfrac{x_0}{x_0 + (1 - x_0)e^{-kt}}$.

19. $40\log 2 \cong 27.72$ minutes.

20. $2\log 2 \cong 1.39$ hours.

21. No later than 36 minutes after the smoking starts.

22. 40 feet.

23. $\dfrac{9}{25} I_0$ and $\dfrac{81}{625} I_0$; $I_0\left(\dfrac{3}{5}\right)^{10/3}$.

25. $\dfrac{\log 5}{\log 2} - 1$ hours.

26. 60°.

27. 16°.

28. At 6 A.M.

29. (a) About 3330 years (1380 B.C.);   (c) about 10,510 years;

    (b) about 3850 years (1900 B.C.);   (d) about 7010 years.

SECTION 5, p. 33

1. $v = \sqrt{\dfrac{g}{c}\dfrac{1 - e^{(-2\sqrt{gc})t}}{1 + e^{(-2\sqrt{gc})t}}}$; the terminal velocity is $\sqrt{\dfrac{g}{c}}$.

2. 2 miles.

3. 256 feet; when $t = 4$, $t = 8$. $v_0^2/2g$; when $t = v_0/g$, $2v_0/g$.

7. $\sqrt{1.5gR}$; $\sqrt{2gR}$.

8. $\sqrt{gR}$, which is approximately 5 miles/second.

MISCELLANEOUS PROBLEMS FOR CHAPTER 1, p. 44

1. $(\sqrt{5} - 1)$ hours before noon.

2. $r = (2 - t)/8$; one more month.

3. After $100 \log 2$ minutes.

4. $100(\sqrt{2} - 1)$ minutes.

5. The intersections of the cylinders $x = cy^4$ with $4x^2 + y^2 + 4z^2 = 36$.

7. $\dfrac{14R^{5/2}}{15r^2\sqrt{2g}}$ seconds.

8. The shape of the surface obtained by revolving $y = cx^4$ about the y-axis.

9. $25h$.

12. $\sqrt{\dfrac{4}{g}} \log (4 + \sqrt{15})$ seconds.

13. $\dfrac{dT}{d\theta} = \mu T$; $T = T_0 e^{\mu\theta}$.

14. $r = r_0 e^{\pi r_0^2 ax/2L}$.

15. The President.

16. Go 2 miles toward the origin and then move outward along one of the spirals $r = e^{\pm\theta/\sqrt{3}}$.

17. $r = \dfrac{a}{\sqrt{2}} e^{-\theta}$; total distance $= a$.

SECTION 7, p. 49

1. (a) $y^2 = x^2 + cx^4$; (e) $y = x \log (\log cx^2)$;
  (b) $y = cx^2(x + y)$; (f) $x^2 - 2xy - y^2 = c$;
  (c) $y = x \tan cx^3$; (g) $y = cx^3 - x$;
  (d) $\cos (y/x) + \log cx = 0$;
  (h) $y\sqrt{x^2 + y^2} + x^2 \log (y + \sqrt{x^2 + y^2}) - 3x^2 \log x + y^2 = cx^2$;
  (i) $y = cx^2/(1 - cx)$; (j) $y^3 = x^3 \log cx^3$.

2. $x^2 + y^2 = cy$.

3. (a) $x + y = \tan (x + c)$; (b) $\tan (x - y + 1) = x + c$.

4. (b) $z = dx + ey$.

5. (a) $\tan^{-1}\left(\dfrac{y+5}{x-1}\right) = \log \sqrt{(x-1)^2 + (y+5)^2} + c$;

(b) $y - x = 5 \log(x + y - 1) + c$;

(c) $\log[(y-x)^2 + (x-1)^2] + 2 \tan^{-1}\left(\dfrac{y-x}{x-1}\right) = c$;

(d) $(x + 2y)(x - 2y - 4)^3 = c$;

(e) $(2x - y + 3)^4 = c(x + 1)^3$.

6. (a) $n = -1/2$, $x = ce^{xy^2}$;

(b) $n = 3/4$, $2 + 5xy^2 = cx^{5/2}$;

(c) $n = -1$, $x = cye^{xy}$.

11. (a) $r = ce^{\theta}$ (in polar coordinates);

(b) $r = ce^{-\theta}$;

(c) $x^2 - y^2 = c$.

SECTION 8, p. 53

1. $xy + \log y^2 = c$.

2. Not exact.

3. $4xy - x^4 + y^4 = c$.

4. Not exact.

5. $xy + \sin xy = c$.

6. Not exact.

7. $xe^y + \sin x \cos y = c$.

8. $\cos \dfrac{x}{y} = c \qquad$ or $\qquad \dfrac{x}{y} = c$.

9. Not exact.

10. $x^2 y^3 + y \sin x = c$.

11. $\log \dfrac{1 + xy}{1 - xy} - 2x = c$.

12. $x^2 y^4 + x \sin y = c$.

13. $\log\left(\dfrac{1 + xy}{1 - xy}\right) + x^2 = c$.

14. $3x^2 + 2(x^2 - y)^{3/2} = c$.

15. Not exact.

16. $xe^{y^2} + \csc y \cot x = c$.

17. $x - y^2 \cos^2 x = c$.

18. $x^2 + y^2 = c^2$.

19. $x^3(1 + \log y) - y^2 = c$.

20. $-y + y^2 - x^2 = c(x + y)$ or $x + y^2 - x^2 = c(x + y)$.

21. $x^2 y^2 (4y^2 - x^2) = c$.

22. (a) $n = 3$, $x^2 y^2 + 2x^3 y = c$;

(b) $n = 1$, $x^2 + e^{2xy} = c$.

SECTION 9, p. 59

2. (a) $\mu = \dfrac{1}{y^4}$, $x^2 - y^2 = cy^3$;

   (b) $\mu = \dfrac{1}{x}$, $2xy - \log x^2 - y^2 = c$;

   (c) $\mu = \dfrac{1}{(xy)^3}$, $3x^2y^4 = 1 + cx^2y^2$;

   (d) $\mu = \sin y$, $e^x \sin y + y^2 = c$;

   (e) $\mu = xe^x$, $x^2e^x \sin y = c$;

   (f) $\mu = \dfrac{1}{(xy)^2}$, $1 + xy^3 = cxy$;

   (g) $\mu = x^2$, $4x^3y^2 + x^4 = c$;

   (h) $\mu = y$, $xy^2 - e^y(y^2 - 2y + 2) = c$;

   (i) $\mu = \dfrac{1}{y}$, $x \log y - x^2 + y = c$;

   (j) $\mu = e^{xy}$, $e^{xy}(x + y) = c$;

   (k) $\mu = e^{x^2/2}$, $e^{x^2/2}(y^3 + x^2 - 2) = c$.

3. When $(\partial M/\partial y - \partial N/\partial x)/(N - M)$ is a function $g(z)$ of $z = x + y$.

4. (a) $-\dfrac{x}{y} = -\dfrac{1}{y} + y + c$;   (h) $2\sqrt{xy} = y + c$;

   (b) $\log \dfrac{x}{y} = \dfrac{1}{3}y^3 + c$;   (i) $-\dfrac{1}{xy} - \log x + y = c$;

   (c) $\tan^{-1}\dfrac{x}{y} = -\dfrac{1}{4}x^4 + c$;   (j) $3x + x^3y^4 + cy = 0$;

   (d) $\log \sqrt{x^2 + y^2} = \tan^{-1}\dfrac{x}{y} + c$;   (k) $x(y^5 + cy) = 4$;

   (e) $\tan^{-1}\dfrac{3y}{x} = 3x + c$;   (l) $y = x/(x^2 + c)$;

   (f) $y = x/(x + c)$;   (m) $xy + x \cos x = \sin x + c$.

   (g) $y = 2x^2 + 3 + cx$;

5. $x^2 \cos(y/x^2) + y \sin(y/x^2) = cx^3$.

6. $r = c/(1 - \cos\theta)$, a parabola.

SECTION 10, p. 61

2. (a) $y = x^4 + cx^3$;

   (b) $y = e^{-x} \tan^{-1} e^x + ce^{-x}$;

   (c) $y = (1 + x^2)^{-1} \log(\sin x) + c(1 + x^2)^{-1}$;

(d) $y = x^2e^{-x} + x^2 - 2x + 2 + ce^{-x}$;

(e) $y = x^2 \csc x + c \csc x$;

(f) $y = -x^3 + cx^2$;

(g) $xy \sin x = \sin x - x \cos x + c$;

(h) $y = 3x^2e^{x^2} + ce^{x^2}$;

(i) $y = (x^3 + c)/\log x$;

(j) $y = x^2(1 + ce^{1/x})$.

3. (a) $\dfrac{1}{y^2} = -x^4 + cx^2$;

(b) $y^3 = 3 \sin x + 9x^{-1} \cos x - 18x^{-2} \sin x - 18x^{-3} \cos x + cx^{-3}$;

(c) $1 + xy \log x = cxy$.

4. (a) $xy^2 = e^y + c$;

(b) $x = ye^y + cy$;

(c) $1 = x^2(y + ce^y)$;

(d) $2xf(y)^3 = f(y)^2 + c$.

5. (a) $x = y - 2 + ce^{-y}$;

(b) $3x + y^2 = c\sqrt{y}$.

7. $\log y = 2x^2 + cx$.

8. $y = \tan x - \sec x$.

9. $x = (10 - t) - \dfrac{8}{10^4}(10 - t)^4,\ 0 \le t \le 10$.

10. (a) 45 pounds;

(b) after $\dfrac{40}{3}(3 - \sqrt{3}) \cong 16.9$ minutes.

11. (a) If $k_2 \ne k_1$, $y = \dfrac{k_1 x_0}{k_2 - k_1}(e^{-k_1 t} - e^{-k_2 t})$; and if $k_2 = k_1$, $y = k_1 x_0 t e^{-k_1 t}$.

(b) About 66 days.

SECTION 11, p. 65

1. (a) $y^2 = c_1 x + c_2$;

(b) $x^2 + (y - c_2)^2 = c_1^2$;

(c) $y = c_1 e^{kx} + c_2 e^{-kx}$;

(d) $y = -\dfrac{1}{2}x^2 - c_1 x - c_1^2 \log (x - c_1) + c_2$;

(e) $2\sqrt{c_1 y - 1} = \pm c_1 x + c_2$;

(f) $y = c_2 e^{c_1 x}$;

(g) $y = x^2 + c_1 \log x + c_2$.

2. (a) $y = 1$ or $3y + x^3 = 3$;

(b) $2y - 3 = 8ye^{3x/2}$;

(c) $y = -\log (2e^{-x} - 1)$.

3. (a) $y = -\log [\cos (x + c_1)] + c_2$;

(b) $y = \log (c_1 e^x + e^{-x}) + c_2$.

4. $T = 2\pi\sqrt{R/g} \cong 89$ minutes.

5. $s = s_0 \cos \sqrt{g/4a}\, t$, period $= 4\pi\sqrt{a/g}$.

SECTION 12, p. 71

2. $T_0 y'' = w(s)\sqrt{1 + (y')^2} + L(x)$.

3. A parabola.

5. $y = c(e^{ax} + e^{-ax})$, where the bottom of the curtain is on the x-axis and the lowest point of the cord is on the y-axis.

6. A horizontal straight line or a catenary.

8. (a) $y = \dfrac{c}{2}\left[\dfrac{1}{1+k}\left(\dfrac{x}{c}\right)^{1+k} - \dfrac{1}{1-k}\left(\dfrac{x}{c}\right)^{1-k}\right] + \dfrac{ck}{1-k^2}$,

so the distance the rabbit runs is $ck/(1 - k^2)$.

(b) $y = \dfrac{1}{2}\left[\dfrac{x^2 - c^2}{2c} - c\log\dfrac{x}{c}\right]$,

and the dog can get closer than $c/2 + \epsilon$ for any $\epsilon > 0$ but not as close as $c/2$.

9. $y = \dfrac{1}{2}\left(\dfrac{x^{k+1}}{c^k} - \dfrac{c^k}{x^{k-1}}\right)$.

If $a > b$ $(k > 1)$, then $y \to -\infty$ as $x \to 0$ and the boat will never land. If $a = b (k = 1)$, then $y \to -c/2$ as $x \to 0$ and the boat will land at $(0, -c/2)$. If $a < b$ $(k < 1)$, then $y \to 0$ as $x \to 0$ and the boat will land at the origin.

SECTION 13, p. 74

2. (a) $I = \dfrac{E_0}{R - kL}e^{-kt} + \left(I_0 - \dfrac{E_0}{R - kL}\right)e^{-Rt/L}$;

(b) $I = \dfrac{E_0}{\sqrt{R^2 + L^2\omega^2}}\sin(\omega t - \alpha) + \left(I_0 + \dfrac{E_0 L\omega}{R^2 + L^2\omega^2}\right)e^{-Rt/L}$,

where $\tan\alpha = L\omega/R$.

4. (a) $Q = E_0 C(1 - e^{-t/RC})$;

(b) case 1, $RC = 1$,

$Q = E_0 Cte^{-t}$; case 2, $RC \neq 1$,

$Q = \dfrac{E_0 C}{RC - 1}[e^{-t/RC} - e^{-t}]$;

(c) $Q = \dfrac{E_0 C}{R^2 C^2\omega^2 + 1}[RC\omega\sin\omega t + \cos\omega t - e^{-t/RC}]$.

5. $Q = Q_0\cos(t/\sqrt{LC})$, $I = (-Q_0/\sqrt{LC})\sin(t/\sqrt{LC})$.

MISCELLANEOUS PROBLEMS FOR CHAPTER 2, p. 75

1. $y = c_2 e^{c_1 x}$.

2. $xy = \log y + c$.

3. $3 \tan^{-1} \dfrac{y + 1}{x - 1} = \log\left[(y + 1)^2 + (x - 1)^2\right] + c.$

4. $y\sqrt{x^2 + y^2} + x^2 \log(y + \sqrt{x^2 + y^2}) + y^2 = 3x^2 \log x + cx^2.$

5. $3y = 2x^2 + cx^2 y^3.$

6. $-\dfrac{1}{2x^2 y^2} = \log \dfrac{y}{x} + c.$

7. $y^2 = c_2 e^{2x} + c_1.$

8. $xy = x \sin x + \cos x + c.$

9. $y = x \log y + cx.$

10. $ye^x - x^2 y^3 = c.$

11. $c_1 \tan^{-1} c_1 x = y + c_2.$

12. $y = x^2 + cx.$

13. $y = x \sin x + 2 \cos x - 2x^{-1} \sin x + cx^{-1}.$

14. $(3x + 2y) + \log(3x + 2y)^2 + x = c.$

15. $x \cos(x + y) = c.$

16. $y = \dfrac{1}{2}(\log x)^2 + c_1 \log x + c_2.$

17. $ye^{xy} + \sin x = c.$

18. $(x - y) \log(x - y) = c - y.$

19. $y = xe^{-x^2} + ce^{-x^2}.$

20. $x^2 y^2 - 2x^3 y - x^4 = c.$

21. $y = x^4(1 + x^2)^{-1} + c(1 + x^2)^{-1}.$

22. $e^x \sin y + \cos xy = c.$

23. $y = c_1 \log(x + \sqrt{1 + x^2}) + c_2.$

24. $2xe^y + x^2 + y^2 - 2x^2 y = c.$

25. $2xe^x e^{-y} + y^2 = c.$

26. $y^4 - x^4 \log x^4 = cx^4.$

27. $3y \cos^3 x = 3 \sin x - \sin^3 x + c.$

28. $y = x(cx^2 - 1)/(cx^2 + 1).$

29. $1 + e^{(x/y)^2} = cy.$

30. $(5y + 4)^2 - 4(5y + 4)(5x + 2) - (5x + 2)^2 = c.$

31. $x^3 \log y = c.$

32. $y^2 \log \dfrac{5x}{x + 3} - 3 \cos y = c.$

33. $x = c(x + y)^2.$

34. $\log x - \dfrac{1}{xy} = c.$

35. $y = \dfrac{1}{2}x^2 - \dfrac{c_1}{2} \log(x^2 + c_1) + c_2.$

36. $x^3 y - xy^3 = c.$

37. $4x^2 y = (x^2 + 1)^3 + c(x^2 + 1).$

38. $3(y - 1)^2 + 4(y - 1)(x + 1) + 3(x + 1)^2 = c.$

39. $xe^{x^2y} = c.$

40. $x^3e^y - x^2 + \cos y = c.$

41. $x = c_1y - \log c_2 y.$

42. $xy(x + y)^2 = c.$

43. $y = x \tan(\log cx).$

44. $\dfrac{1}{y} = 1 + \log x + cx.$

45. $[\cos y][\log(5x + 15)] + \log y = c.$

46. $c_1 y^2 = c_1 x + \log(c_1 x - 1) + c_2.$

47. $xye^x - e^x = c.$

48. $y = x^2/(c - x).$

49. $y^3 = 3(c_2 - x - c_1 y).$

50. $x = \csc y[\log(\sec y) + c].$

51. When $t = 25.$

52. $ce^{2x^5/5} = \dfrac{y - x}{y + x}.$

53. (a) $\dfrac{dz}{dt} + [s(t) + I]z = -Iz^2.$

(b) $y = 1 + z,$ where

$$\frac{1}{z} = e^{((1/2)at^2 + It)}\left[I\int e^{-((1/2)at^2 + It)}\, dt + c\right].$$

55. Burnout velocity $= b \log\left(1 + \dfrac{m_2}{m_1}\right) - \dfrac{gm_2}{a};$

burnout height $= -\dfrac{gm_2^2}{2a^2} + \dfrac{bm_2}{a} + \dfrac{bm_1}{a}\log\dfrac{m_1}{m_1 + m_2}.$

59. (a) If the constant acceleration due to the constant gravitational field is denoted by A, then

$$v = c\left(\frac{1 - e^{-2At/c}}{1 + e^{-2At/c}}\right).$$

SECTION 14, p. 86

1. (a) $y = c_1 + c_2 x^2;$
(b) $a = 1,\ y = c_1 + c_2 x^2 + x^3.$

2. $y = c_1 + c_2 \log x.$

3. (a) $y = c_1 e^{-x} + c_2 e^{2x};$
(b) $y = c_1 e^{-x} + c_2 e^{2x} - 2x + 1.$

4. (a) $y = 1/(2x);$
(b) $y = -3x;$

(c) $y = -\dfrac{1}{3}\sin x$.

5. (a) $y = c_1 x + c_2 + e^x$;   (b) $y = c_1 + c_2 e^{2x} - 2x$;

(c) $y = c_1 e^x + c_2 e^{-x} - \dfrac{1}{2}\sin x$;   (d) $y = c_1 x + c_2 e^x$;

(e) $y = c_1 + c_2 e^{-2x} + 2e^x$.

6. (a) $x^2 y'' - 2xy' + 2y = 0$;   (e) $(1 - x\cot x)y'' - xy' + y = 0$;

(b) $y'' - k^2 y = 0$;   (f) $y'' - 2y' + y = 0$;

(c) $y'' + k^2 y = 0$;   (g) $y'' + 2y' - 3y = 0$;

(d) $y'' + 2y' = 0$;   (h) $x^2 y'' + xy' - y = 0$.

SECTION 15, p. 91

2. $y = x + 2x^2$.

3. $y = -3e^x + 2e^{2x}$.

5. $y_1 = x^2$, $y_2 = x^{-1}$, $y = 3x^2 - 2x^{-1}$.

6. (a) $y = 6e^x + 2e^{-2x}$;   (c) $y = 4e^{-2x} - 3e^{-3x}$;

(b) $y = 0$;   (d) $y = e^{-2} - e^{-x}$.

7. (a) $y = $ a constant or $y = \log(x + c_1) + c_2$.

11. (a) $u = e^{-\frac{1}{2}\int P\,dx}$, $v'' + \left(Q - \dfrac{1}{2}P' - \dfrac{1}{4}P^2\right)v = 0$.

(b) $y = (c_1 x + c_2)e^{-x^2/2}$.

SECTION 16, p. 94

2. (a) $y_2 = -\cos x$, $y = c_1 \sin x + c_2 \cos x$;

(b) $y_2 = -\dfrac{1}{2}e^{-x}$, $y = c_1 e^x + c_2 e^{-x}$.

3. $y_2 = -\dfrac{1}{2}x^{-2}$, $y = c_1 + c_2 x^{-2}$.

4. $y_2 = -\dfrac{1}{4}x^{-2}$, $y = c_1 x^2 + c_2 x^{-2}$.

5. $y = c_1 x + c_2\left[\dfrac{x}{2}\log\left(\dfrac{1+x}{1-x}\right) - 1\right]$.

6. $y = c_1 x^{-1/2}\sin x + c_2 x^{-1/2}\cos x$.

7. (a) $y = c_1 x + c_2 e^x$;

(b) $y = c_1 x + c_2 x^{-2}$;

(c) $y = c_1 x + c_2 x e^x$.

8. $y = c_1 x + c_2 x \displaystyle\int x^{-2} e^{\int x f(x)\,dx}\,dx$.

9. $y = c_1 e^x + c_2 x^2 e^x$.

10. (a) $y_1 = e^x$, $y_2 = e^x \int x^n e^{-x}\, dx$.

   (b) $y = c_1 e^x + c_2(x + 1)$, $y = c_1 e^x + c_2(x^2 + 2x + 2)$,
   $y = c_1 e^x + c_2(x^3 + 3x^2 + 6x + 6)$.

11. $y = c_1 e^x + c_2 e^x \int e^{[-2x + \int f(x)\, dx]}\, dx$.

SECTION 17, p. 97

1. (a) $y = c_1 e^{2x} + c_2 e^{-3x}$;

   (b) $y = c_1 e^{-x} + c_2 x e^{-x}$;

   (c) $y = c_1 \cos 2\sqrt{2}x + c_2 \sin 2\sqrt{2}x$;

   (d) $y = e^x(c_1 \cos \sqrt{3}x + c_2 \sin \sqrt{3}x)$;

   (e) $y = c_1 e^{2x} + c_2 x e^{2x}$;

   (f) $y = c_1 e^{5x} + c_2 e^{4x}$;

   (g) $y = e^{-x/2}\left(c_1 \cos \frac{1}{2}\sqrt{5}x + c_2 \sin \frac{1}{2}\sqrt{5}x\right)$;

   (h) $y = c_1 e^{3x/2} + c_2 x e^{3x/2}$;

   (i) $y = c_1 + c_2 e^{-x}$;

   (j) $y = e^{3x}(c_1 \cos 4x + c_2 \sin 4x)$;

   (k) $y = c_1 e^{-5x/2} + c_2 x e^{-5x/2}$;

   (l) $y = e^{-x}(c_1 \cos \sqrt{2}x + c_2 \sin \sqrt{2}x)$;

   (m) $y = c_1 e^{2x} + c_2 e^{-2x}$;

   (n) $y = e^x\left(c_1 \cos \frac{1}{2}\sqrt{3}x + c_2 \sin \frac{1}{2}\sqrt{3}x\right)$;

   (o) $y = c_1 e^{x/2} + c_2 e^{-x}$;

   (p) $y = c_1 e^{x/4} + c_2 x e^{x/4}$;

   (q) $y = e^{-2x}(c_1 \cos x + c_2 \sin x)$;

   (r) $y = c_1 e^x + c_2 e^{-5x}$.

2. (a) $y = e^{3x-1}$;

   (b) $y = e^x + 2e^{5x}$;

   (c) $y = 5x e^{3x}$;

   (d) $y = e^{-2x}(\cos x + 2 \sin x)$;

   (e) $y = e^{(-2+\sqrt{2})x} - 2e^{(-2-\sqrt{2})x}$;

   (f) $y = \frac{9}{5}e^{x-1} + \frac{1}{5}e^{-9(x-1)}$.

5. (a) $y = x^{-1}[c_1 \cos (\log x^3) + c_2 \sin (\log x^3)]$;

   (b) $y = c_1 x^{-2} + c_2 x^{-2} \log x$;

   (c) $y = c_1 x^3 + c_2 x^{-4}$;

   (d) $y = c_1 x^{3/2} + c_2 x^{-1/2}$;

   (e) $y = c_1 x^2 + c_2 x^2 \log x$;

   (f) $y = c_1 x^2 + c_2 x^{-3}$;

   (g) $y = x^{-1/2}\left[c_1 \cos \left(\frac{1}{2}\sqrt{11} \log x\right) + c_2 \sin \left(\frac{1}{2}\sqrt{11} \log x\right)\right]$;

   (h) $y = c_1 x^{\sqrt{2}} + c_2 x^{-\sqrt{2}}$;

   (i) $y = c_1 x^4 + c_2 x^{-4}$.

7. (a) $y = e^{-x^2/4}\left(c_1 \cos \frac{1}{4}\sqrt{3}x^2 + c_2 \sin \frac{1}{4}\sqrt{3}x^2\right)$;

   (b) not possible.

SECTION 18, p. 103

1. (a) $y = c_1 e^{2x} + c_2 e^{-5x} + \frac{1}{3} e^{4x}$;

(b) $y = c_1 \sin 2x + c_2 \cos 2x + \sin x$;

(c) $y = c_1 e^{-5x} + c_2 x e^{-5x} + 7x^2 e^{-5x}$;

(d) $y = e^x(c_1 \cos 2x + c_2 \sin 2x) + 2 + 4x + 5x^2$;

(e) $y = c_1 e^{3x} + c_2 e^{-2x} - 4x e^{-2x}$;

(f) $y = c_1 e^x + c_2 e^{2x} + 2 \sin 2x + 3 \cos 2x$;

(g) $y = c_1 \sin x + c_2 \cos x + x \sin x$;

(h) $y = c_1 + c_2 e^{2x} + 2x - 3x^2$;

(i) $y = c_1 e^x + c_2 x e^x + 3x^2 e^x$;

(j) $y = e^x(c_1 \cos x + c_2 \sin x) - \frac{1}{2} x e^x \cos x$;

(k) $y = c_1 + c_2 e^{-x} + 2x^5 - 10x^4 + 40x^3 - 120x^2 + 242x$.

2. $y = c_1 \sin kx + c_2 \cos kx + \dfrac{\sin bx}{k^2 - b^2}$ unless $\quad b = k, \quad$ in $\quad$ which $\quad$ case

$\quad y = c_1 \sin kx + c_2 \cos kx - \dfrac{x \cos kx}{2k}$.

3. (a) $y = c_1 \sin 2x + c_2 \cos 2x + x \sin 2x + 2 \cos x - 1 - x + 2x^2$;

(b) $y = c_1 \sin 3x + c_2 \cos 3x - \dfrac{1}{3} x \cos 3x + \dfrac{1}{2} \sin x - 2 e^{-2x} + 3x^3 - 2x$.

SECTION 19, p. 106

1. $y_p = 2x + 4$.

2. $y_p = -\dfrac{1}{4} e^{-x}$.

3. (a) $y_p = -\dfrac{1}{4} \cos 2x \log(\sec 2x + \tan 2x)$;

(b) $y_p = \dfrac{1}{2} x^2 e^{-x} \log x - \dfrac{3}{4} x^2 e^{-x}$;

(c) $y_p = -e^{-x}(8x^2 + 4x + 1)$;

(d) $y_p = \dfrac{1}{2} x e^{-x} \sin 2x + \dfrac{1}{4} e^{-x} \cos 2x \log(\cos 2x)$;

(e) $y_p = \dfrac{1}{10} e^{-3x}$;

(f) $y_p = e^x \log(1 + e^{-x}) - e^x + e^{2x} \log(1 + e^{-x})$.

4. (a) $y_p = x \sin x + \cos x \log(\cos x)$;

(b) $y_p = \cos x \log(\csc x + \cot x) - 2$;

(c) $y_p = \dfrac{1}{2} \cos x \log (\sec x + \tan x) - \dfrac{1}{2} \sin x \log (\csc x + \cot x)$;

(d) $y_p = \dfrac{1}{4} (x^2 \sin x + x \cos x - \sin x)$;

(e) $y_p = -\cos x \log (\sec x + \tan x)$;
(f) $y_p = x \cos x - \sin x - \sin x \log (\cos x)$;
(g) $y_p = -\sin x \log (\csc x + \cot x) - \cos x \log (\sec x + \tan x)$.

5. (b) $y_p(x) = \dfrac{1}{k} \displaystyle\int_0^x f(t) \sin k(x - t)\, dt$.

6. (a) $y = c_1 x + c_2(x^2 + 1) + \dfrac{1}{6} x^4 - \dfrac{1}{2} x^2$;

(b) $y = c_1 e^x + c_2 x^{-1} - x - 1 - \dfrac{1}{3} x^2$;

(c) $y = c_1 x + c_2 e^x + x^2 + 1$;

(d) $y = c_1 e^x + c_2(x + 1) + \dfrac{1}{2} e^{2x}(x - 1)$;

(e) $y = c_1 x + c_2 x^2 - x e^{-x} - (x^2 + x) \displaystyle\int \dfrac{e^{-x}}{x}\, dx$, where this integral is not an elementary function.

SECTION 20, p. 113

1. The frequency is $\dfrac{1}{2\pi} \sqrt{\dfrac{k}{M} - \dfrac{c^2}{2M^2}}$ when $\dfrac{k}{M} - \dfrac{c^2}{2M^2}$ is positive, which is more restrictive than the condition that $\dfrac{k}{M} - \dfrac{c^2}{4M^2} > 0$.

3. $2\pi\sqrt{2r/3g}$ seconds.

4. About 574 pounds.

5. The round trip time is $2\pi\sqrt{R/g}$ seconds, where R is the radius of the earth; this is approximately 90 minutes. The greatest speed is approximately $0.074L$ miles/minute or $4.43L$ miles/hour.

6. $x = \dfrac{1}{2} \cos 4t + \dfrac{1}{4} \sin 4t - t \cos 4t$.

SECTION 21, p. 122

1. (a) $2\sqrt{2}$ years.
(b) $3\sqrt{3}$ years.
(c) 125 years.

2. (a) About 0.39 astronomical units or 36,000,000 miles.
(b) About 29.5 years.

SECTION 22, p. 127

1. $y = c_1 + c_2 e^x + c_3 e^{2x}$.

2. $y = c_1 e^x + e^x (c_2 \cos x + c_3 \sin x)$.

3. $y = c_1 e^x + e^{-x/2} \left(c_2 \cos \frac{1}{2} \sqrt{3} x + c_3 \sin \frac{1}{2} \sqrt{3} x \right)$.

4. $y = c_1 e^{-x} + e^{x/2} \left(c_2 \cos \frac{1}{2} \sqrt{3} x + c_3 \sin \frac{1}{2} \sqrt{3} x \right)$.

5. $y = (c_1 + c_2 x + c_3 x^2) e^{-x}$.

6. $y = (c_1 + c_2 x + c_3 x^2 + c_4 x^3) e^{-x}$.

7. $y = c_1 e^x + c_2 e^{-x} + c_3 \cos x + c_4 \sin x$.

8. $y = c_1 \cos x + c_2 \sin x + c_3 \cos 2x + c_4 \sin 2x$.

9. $y = (c_1 + c_2 x) e^{ax} + (c_3 + c_4 x) e^{-ax}$.

10. $y = (c_1 + c_2 x) \cos ax + (c_3 + c_4 x) \sin ax$.

11. $y = (c_1 + c_2 x) e^{-x} + c_3 \cos x + c_4 \sin x$.

12. $y = (c_1 + c_2 x) e^x + e^{-2x} (c_3 \cos x + c_4 \sin x)$.

13. $y = c_1 e^x + c_2 e^{2x} + c_3 e^{3x}$.

14. $y = c_1 e^{2x} + (c_2 + c_3 x + c_4 x^2) e^{-x}$.

15. $y = (c_1 + c_2 x) e^{2x} + (c_3 + c_4 x) e^{-2x} + c_5 e^{6x}$.

17. $\dfrac{d^4 x_1}{dt^4} + \left[\dfrac{k_1 + k_3}{m_1} + \dfrac{k_2 + k_3}{m_2} \right] \dfrac{d^2 x_1}{dt^2} + \left[\left(\dfrac{k_1 + k_3}{m_1} \right) \left(\dfrac{k_2 + k_3}{m_2} \right) - \dfrac{k_3^2}{m_1 m_2} \right] x_1 = 0$.

18. $x_1 = c_1 \cos \sqrt{\dfrac{k}{m}} t + c_2 \sin \sqrt{\dfrac{k}{m}} t + c_3 \cos \sqrt{\dfrac{3k}{m}} t + c_4 \sin \sqrt{\dfrac{3k}{m}} t$;

$\dfrac{1}{2\pi} \sqrt{\dfrac{k}{m}}$ and $\dfrac{1}{2\pi} \sqrt{\dfrac{3k}{m}}$.

19. $y = c_1 x^3 + c_2 x^2 + c_3 x + c_4 + \sin x + x^4$.

20. $y = c_1 + c_2 e^x + c_3 e^{2x} + 5x + 7e^{3x}$.

21. $y = \dfrac{9}{2} e^x - \dfrac{1}{2} e^{-x} - x$.

22. (a) $y = c_1 + c_2 x + c_3 x^{-1}$;

(b) $y = c_1 x + c_2 x^2 + c_3 x^{-1}$;

(c) $y = c_1 x + c_2 \cos (\log x) + c_3 \sin (\log x)$.

SECTION 23, p. 135

1. $y = \left(\dfrac{1}{4} x - \dfrac{1}{16} \right) e^{2x}$.

2. $y = \dfrac{1}{27} (9x^2 - 24x + 26) e^{2x}$.

3. $y = \dfrac{1}{2} x^5 e^{-2x}$.

4. $y = \frac{1}{2}x^2 e^x$.

5. $y = \left(-\frac{1}{2}x - \frac{1}{4}\right)e^{-x}$.

6. $y = \frac{1}{2}e^{5x}$.

7. $y = x^3 - 6x - 5$.

8. $y = 2x^3 + 9x^2 + 40x + 73$.

9. $y = x^4 - 48x^2 + 384$.

10. $y = -\frac{1}{60}x^5 - \frac{1}{3}x^3 - 2x$.

11. $y = -x^{10} - 151{,}200x^4$.

12. $y = x^4 + 4x^3 + 24x^2 + 69x + 117$.

13. $y = x^4 - 12x^2 + 24$.

14. $y = -2x^3 - 5x^2 - 10x - 10$.

15. $y = \frac{3}{4}x^4 - \frac{10}{3}x^3 + \frac{21}{2}x^2 - 21x + 21$.

16. $y = -e^{2x}(x^3 + 6x)$.

17. $y = \frac{1}{8}e^{2x}(4x^3 - 2x^2 - 18x - 25)$.

18. $y = 2e^{-2x}(x^2 + 4x + 6) + \frac{1}{3}e^{2x}$.

19. $y = -2x^2$.

20. $y = x^3 - 1$.

21. $y = -2x^2$.

22. $y = \frac{1}{2}x(\log x - 1)$.

23. $y = \frac{1}{2}x^2 + 2x + 1$.

24. $y = \frac{1}{48}(4x^3 - 6x^2 + 6x - 3)$.

25. (a) $y = \left(\frac{1}{6}x^3 + c_1x^2 + c_2x + c_3\right)e^{2x}$;

    (b) $y = (2x^3 + c_1x^2 + c_2x + c_3)e^{-x}$;

    (c) $y = (-\sin x + c_1x + c_2)e^{2x}$.

SECTION 24, p. 161

3. $u'' + \left(1 + \frac{1 - 4p^2}{4x^2}\right)u = 0$.

SECTION 25, p. 164

3. If $f(x) \geq 0$ and $k > 0$, then every solution of the equation $y'' + [f(x) + k]y = 0$ has an infinite number of positive zeros.

SECTION 26, p. 171

6. $\sum_{n=1}^{\infty} nx^{n-1}$.

SECTION 27, p. 175

1. (a) $y = a_0\left(1 + x^2 + \dfrac{x^4}{2!} + \dfrac{x^6}{3!} + \dfrac{x^8}{4!} + \cdots\right) = a_0 e^{x^2}$;

(b) $y = a_0 - (a_0 - 1)x + \dfrac{(a_0 - 1)}{2!}x^2 - \dfrac{(a_0 - 1)}{3!}x^3 + \cdots$

$\qquad = 1 + (a_0 - 1)\left(1 - x + \dfrac{x^2}{2!} - \dfrac{x^3}{3!} + \cdots\right) = 1 + (a_0 - 1)e^{-x}$.

2. (a) $y = a_1 x$, no discrepancies;

(b) $y = 0$, $y = ce^{-1/x}$, the latter being analytic at $x = 0$ only when $c = 0$.

3. $\sin^{-1} x = x + \sum_{n=1}^{\infty} \dfrac{1 \cdot 3 \cdots (2n-1)}{2 \cdot 4 \cdots (2n)} \dfrac{x^{2n+1}}{2n+1}$.

5. $y = \dfrac{x^2}{2!} - \dfrac{x^3}{3!} + \dfrac{x^4}{4!} - \cdots$

$\qquad = \left(1 - x + \dfrac{x^2}{2!} - \dfrac{x^3}{3!} + \dfrac{x^4}{4!} - \cdots\right) + x - 1 = e^{-x} + x - 1$.

SECTION 28, p. 182

1. $y = a_0\left(1 + x^2 - \dfrac{1}{3}x^4 + \dfrac{1}{5}x^6 - \dfrac{1}{7}x^8 + \cdots\right) + a_1 x$

$\qquad = a_0(1 + x\tan^{-1} x) + a_1 x$.

2. (a) $y_1(x) = 1 - \dfrac{x^2}{2} + \dfrac{x^4}{2 \cdot 4} - \dfrac{x^6}{2 \cdot 4 \cdot 6} + \cdots$,

$\qquad y_2(x) = x - \dfrac{x^3}{3} + \dfrac{x^5}{3 \cdot 5} - \dfrac{x^7}{3 \cdot 5 \cdot 7} + \cdots$.

3. $a_{n+2} = -\dfrac{(n+1)a_{n+1} - a_{n-1}}{(n+1)(n+2)}$.

(a) $y_1(x) = 1 + \dfrac{x^3}{2 \cdot 3} - \dfrac{x^4}{2 \cdot 3 \cdot 4} + \dfrac{x^5}{2 \cdot 3 \cdot 4 \cdot 5} + \cdots$;

(b) $y_2(x) = x - \dfrac{x^2}{2} + \dfrac{x^3}{2 \cdot 3} + \dfrac{x^4}{2 \cdot 3 \cdot 4} - \dfrac{4x^5}{2 \cdot 3 \cdot 4 \cdot 5} + \cdots$.

4. (c) $a_{n+2} = -\dfrac{p - n}{(n + 1)(n + 2)} a_n,$

$$w(x) = a_0\left[1 - \frac{p}{2!}x^2 + \frac{p(p - 2)}{4!}x^4 - \cdots\right]$$

$$+ a_1\left[x - \frac{(p - 1)}{3!}x^3 + \frac{(p - 1)(p - 3)}{5!}x^5 - \cdots\right].$$

5. (b) $y(x) = a_0\left[1 + \displaystyle\sum_{n=1}^{\infty} \frac{(-1)^n x^{3n}}{2 \cdot 5 \cdot 8 \cdots (3n - 1)3^n n!}\right]$

$$+ a_1\left[x + \sum_{n=1}^{\infty} \frac{(-1)^n x^{3n+1}}{4 \cdot 7 \cdot 10 \cdots (3n + 1)3^n n!}\right].$$

(c) $y(x) = a_0\left[1 + \displaystyle\sum_{n=1}^{\infty} \frac{x^{3n}}{2 \cdot 5 \cdot 8 \cdots (3n - 1)3^n n!}\right]$

$$+ a_1\left[-x - \sum_{n=1}^{\infty} \frac{x^{3n+1}}{4 \cdot 7 \cdot 10 \cdots (3n + 1)3^n n!}\right].$$

6. (a) $y_1(x) = 1 - \dfrac{p \cdot p}{2!}x^2 + \dfrac{p(p - 2)p(p + 2)}{4!}x^4 - \cdots,$

$$y_2(x) = x - \frac{(p - 1)(p + 1)}{3!}x^3$$

$$+ \frac{(p - 1)(p - 3)(p + 1)(p + 3)}{5!}x^5 - \cdots.$$

SECTION 29, p. 191

1. (a) $x = 0$ irregular, $x = 1$ regular;
  (b) $x = 0$ and $x = 1$ regular, $x = -1$ irregular;
  (c) $x = 0$ irregular;

  (d) $x = 0$ and $x = -\dfrac{1}{3}$ regular.

2. (a) ordinary point;
  (b) ordinary point;
  (c) regular singular point;
  (d) regular singular point;
  (e) irregular singular point.

3. (a) $m(m - 1) - 2m + 2 = 0$, $m_1 = 2$, $m_2 = 1$;

  (b) $m(m - 1) - \dfrac{5}{4}m + \dfrac{1}{2} = 0$, $m_1 = 2$, $m_2 = \dfrac{1}{4}$.

4. (a) $y_1(x) = x^{1/2}\left(1 - \dfrac{x}{3!} + \dfrac{x^2}{5!} - \cdots\right) = \sin\sqrt{x},$

$$y_2(x) = 1 - \frac{x}{2!} + \frac{x^2}{4!} - \cdots = \cos\sqrt{x};$$

(b) $y_1(x) = \sum_{n=0}^{\infty} \frac{x^n}{1 \cdot 3 \cdot 5 \cdots (2n+1)}$,

$\qquad y_2(x) = x^{-1/2} \sum_{n=0}^{\infty} \frac{x^n}{2^n n!} = x^{-1/2} e^{x/2}$;

(c) $y_1(x) = x^{1/2}\left(1 - \frac{7}{6}x + \frac{21}{40}x^2 + \cdots\right)$,

$\qquad y_2(x) = 1 - 3x + 2x^2 + \cdots$;

(d) $y_1(x) = x\left(1 + \frac{1}{5}x + \frac{1}{70}x^2 + \cdots\right)$,

$\qquad y_2(x) = x^{-1/2}\left(1 - x - \frac{1}{2}x^2 + \cdots\right)$.

6. (b) $y_2(x) = -xe^{1/x}$.

SECTION 30, p. 198

1. $y = x^2(1 - 4x + 4x^2 + \cdots)$.
2. $y = c_1 x^{1/2} e^x + c_2 x^{1/2} e^x \log x$.

3. (a) $y_1 = 1 - \frac{x^2}{3!} + \frac{x^4}{5!} - \cdots = x^{-1} \sin x$,

$\qquad y_2 = x^{-1}\left(1 - \frac{x^2}{2!} + \frac{x^4}{4!} - \cdots\right) = x^{-1} \cos x$;

(b) $y_1 = x^2\left(1 + \frac{1}{2}x + \frac{1}{20}x^2 - \frac{1}{60}x^3 + \cdots\right)$,

$\qquad y_2 = x^{-1}\left(1 + \frac{1}{2}x + \frac{1}{2}x^2 - \frac{1}{8}x^4 + \cdots\right)$;

(c) $y_1 = x^2\left(1 - \frac{x^4}{3!} + \frac{x^8}{5!} - \cdots\right) = \sin x^2$,

$\qquad y_2 = 1 - \frac{x^4}{2!} + \frac{x^8}{4!} - \cdots = \cos x^2$.

4. $y = x\left(1 - \frac{x^2}{2^2 2!} + \frac{x^4}{2^4 2! 3!} - \cdots\right)$.

5. $y_1 = x^{1/2}\left(1 - \frac{x^2}{3!} + \frac{x^4}{5!} - \cdots\right) = x^{-1/2} \sin x$,

$\qquad y = x^{-1/2}\left(1 - \frac{x^2}{2!} + \frac{x^4}{4!} - \cdots\right) = x^{-1/2} \cos x$.

SECTION 31, p. 203

2. (a) $y = c_1 F\left(2, -1, \dfrac{3}{2}, x\right) + c_2 x^{-1/2} F\left(\dfrac{3}{2}, -\dfrac{3}{2}, \dfrac{1}{2}, x\right)$

$\qquad = c_1\left(1 - \dfrac{4}{3}x\right) + c_2 x^{-1/2} F\left(\dfrac{3}{2}, -\dfrac{3}{2}, \dfrac{1}{2}, x\right);$

(b) $y = c_1 F\left(\dfrac{1}{2}, 1, \dfrac{1}{2}, -x\right) + c_2(-x)^{1/2} F\left(1, \dfrac{3}{2}, \dfrac{3}{2}, -x\right)$

$\qquad = c_1\left(\dfrac{1}{1 + x}\right) + c_2\left[\dfrac{(-x)^{1/2}}{1 + x}\right];$

(c) $y = c_1 F\left(2, 2, \dfrac{1}{2}, \dfrac{x + 1}{2}\right) + c_2\left(\dfrac{x + 1}{2}\right)^{1/2} F\left(\dfrac{5}{2}, \dfrac{5}{2}, \dfrac{3}{2}, \dfrac{x + 1}{2}\right);$

(d) $y = c_1 F\left(1, 1, \dfrac{14}{5}, \dfrac{3 - x}{5}\right) + c_2\left(\dfrac{3 - x}{5}\right)^{-9/5} F\left(-\dfrac{4}{5}, -\dfrac{4}{5}, -\dfrac{4}{5}, \dfrac{3 - x}{5}\right).$

4. (a) $y = c_1 F(p, 1, p, x) + c_2 x^{1-p} F(1, 2 - p, 2 - p, x);$

(b) $y = c_1\left(\dfrac{1}{1 - x}\right) + c_2\left(\dfrac{x^{1-p}}{1 - x}\right);$

(c) $y = c_1\left(\dfrac{1}{1 - x}\right) + c_2\left(\dfrac{\log x}{1 - x}\right).$

5. $y = c_1 F\left(1, -1, -\dfrac{1}{2}, 1 - e^x\right) + c_2(1 - e^x)^{3/2} F\left(\dfrac{5}{2}, \dfrac{1}{2}, \dfrac{5}{2}, 1 - e^x\right).$

SECTION 32, p. 207

1. (a) A regular singular point with exponents $p + 1$ and $-p$.

(b) An irregular singular point.

SECTION 33, p. 256

1. $\dfrac{3\pi}{4} + \displaystyle\sum_1^\infty \dfrac{(-1)^{n+1}\cos(2n - 1)x - \sin(2n - 1)x + \sin 2(2n - 1)x}{2n - 1}.$

2. $\dfrac{1}{4} + \dfrac{1}{\pi}\displaystyle\sum_1^\infty \dfrac{(-1)^{n+1}\cos(2n - 1)x + \sin(2n - 1)x + \sin 2(2n - 1)x}{2n - 1}.$

3. $\dfrac{1}{\pi} - \dfrac{2}{\pi}\displaystyle\sum_1^\infty \dfrac{\cos 2nx}{4n^2 - 1} + \dfrac{1}{2}\sin x.$

4. $\dfrac{1}{2}\cos x + \dfrac{4}{\pi}\displaystyle\sum_1^\infty \dfrac{n \sin 2nx}{4n^2 - 1}.$

5. (a) $\pi;$ \qquad\qquad\qquad\qquad (c) $\cos x;$

(b) $\sin x;$ \qquad\qquad\qquad (d) $\pi + \sin x + \cos x.$

Notice that any finite trigonometric series is automatically the Fourier series of its sum.

6. (a) $\dfrac{4a}{\pi}\left(\sin x + \dfrac{\sin 3x}{3} + \dfrac{\sin 5x}{5} + \cdots\right);$

(b) $\dfrac{4}{\pi}\left(\sin x + \dfrac{\sin 3x}{3} + \dfrac{\sin 5x}{5} + \cdots\right);$

(c) $\sin x + \dfrac{\sin 3x}{3} + \dfrac{\sin 5x}{5} + \cdots;$

(d) $\dfrac{1}{2} + \dfrac{6}{\pi}\left(\sin x + \dfrac{\sin 3x}{3} + \dfrac{\sin 5x}{5} + \cdots\right);$

(e) $\dfrac{3}{2} + \dfrac{2}{\pi}\left(\sin x + \dfrac{\sin 3x}{3} + \dfrac{\sin 5x}{5} + \cdots\right).$

7. After forming the suggested series, continue by subtracting from the series in Problem 1, then dividing by π.

SECTION 34, p. 263

2. $f(x) = \dfrac{\pi}{4} - \dfrac{2}{\pi}\sum_{1}^{\infty}\dfrac{\cos(2n-1)x}{(2n-1)^2} - \sum_{1}^{\infty}(-1)^{n+1}\dfrac{\sin nx}{n}.$

3. $f(x) = -\dfrac{\pi}{4} - \dfrac{2}{\pi}\sum_{1}^{\infty}\dfrac{\cos(2n-1)x}{(2n-1)^2} + 3\sum_{1}^{\infty}\dfrac{\sin(2n-1)x}{2n-1} - \sum_{1}^{\infty}\dfrac{\sin 2nx}{2n}.$

In each case, $\sum\dfrac{1}{(2n-1)^2} = \dfrac{\pi^2}{8}.$

5. $f(x) = \dfrac{\sinh \pi}{\pi}\left[1 + 2\sum_{1}^{\infty}\dfrac{(-1)^n}{n^2+1}\cos nx - 2\sum_{1}^{\infty}\dfrac{(-1)^n n}{n^2+1}\sin nx\right].$

SECTION 35, p. 269

1. Even, odd, neither, odd, even, even, neither, odd.

4. $1 - \dfrac{1}{3} + \dfrac{1}{5} - \cdots = \dfrac{\pi}{4}$ (this concrete sum, familiar to us from elementary calculus, provides strong emphasis for the very remarkable nature of the sine series we are considering: as x varies continuously between 0 and π, each term of the series changes in value, but these changes are so delicately interrelated that the sum of all these variable quantities is constantly equal to $\dfrac{\pi}{4}$—astounding!); $\dfrac{\pi}{4}$.

5. $f(x) = \dfrac{2}{\pi} - \dfrac{4}{\pi}\sum_{1}^{\infty}(-1)^n\dfrac{\cos nx}{4n^2-1}.$

6. $\sin x; \dfrac{2}{\pi} - \dfrac{4}{\pi}\sum_{1}^{\infty}\dfrac{\cos 2nx}{4n^2-1}, 0 \le x \le \pi.$

7. $f(x) = \dfrac{4}{\pi}\sum_{1}^{\infty}\dfrac{\cos(2n-1)x}{(2n-1)^2}.$

8. (a) $\pi - x = \pi + 2 \sum_{1}^{\infty} (-1)^n \dfrac{\sin nx}{n}$;

  (b) $\pi - x = \dfrac{\pi}{2} + \dfrac{4}{\pi} \sum_{1}^{\infty} \dfrac{\cos (2n - 1)x}{(2n - 1)^2}$;

  (c) $\pi - x = 2 \sum_{1}^{\infty} \dfrac{\sin nx}{n}$.

10. (b) $x^2 = 2\pi \sum_{1}^{\infty} (-1)^{n+1} \dfrac{\sin nx}{n} - \dfrac{8}{\pi} \sum_{1}^{\infty} \dfrac{\sin (2n - 1)x}{(2n - 1)^3}$, $\qquad 0 \le x < \pi$.

15. $\sin^2 x = \dfrac{1}{\pi} \sum_{1}^{\infty} \left[\dfrac{2}{2n - 1} - \dfrac{1}{2n + 1} - \dfrac{1}{2n - 3} \right] \sin (2n - 1)x, \qquad 0 \le x \le \pi;$

  $\cos^2 x = \dfrac{1}{\pi} \sum_{1}^{\infty} \left[\dfrac{2}{2n - 1} + \dfrac{1}{2n + 1} + \dfrac{1}{2n - 3} \right] \sin (2n - 1)x, \qquad 0 < x < \pi.$

SECTION 36, p. 274

1. $f(x) = \dfrac{12}{\pi} \sum_{1}^{\infty} \dfrac{1}{2n - 1} \sin (2n - 1) \dfrac{\pi x}{2}.$

2. (a) $f(x) = \dfrac{1}{2} + \dfrac{4}{\pi^2} \sum_{1}^{\infty} \dfrac{\cos (2n - 1)\pi x}{(2n - 1)^2}$;

  (b) $f(x) = 1 - \dfrac{8}{\pi^2} \sum_{1}^{\infty} \dfrac{1}{(2n - 1)^2} \cos (2n - 1) \dfrac{\pi x}{2}.$

4. $f(x) = \dfrac{1}{\pi^2} \sum_{1}^{\infty} \dfrac{\cos 2n\pi x}{n^2}.$

5. $f(x) = 1 + \dfrac{4}{\pi} \sum_{1}^{\infty} \dfrac{(-1)^{n+1}}{2n - 1} \cos (2n - 1) \dfrac{\pi x}{2}.$

6. $\cos \pi x = \cos \pi x.$

7. $f(x) = \dfrac{2}{\pi^2} \sum_{1}^{\infty} \dfrac{\cos 2(2n - 1)\pi x}{(2n - 1)^2}.$

SECTION 38, p. 291

4. $b_1 = \dfrac{4}{\pi}, b_2 = 0, b_3 = \dfrac{4}{3\pi}, b_4 = 0, b_5 = \dfrac{4}{5\pi}.$

5. $b_1 = 2, b_2 = -1, b_3 = \dfrac{2}{3}.$

SECTION 40, p. 308

1. (a) $\lambda_n = 4n^2, y_n(x) = \sin 2nx$;

  (b) $\lambda_n = \dfrac{n^2}{4}, y_n(x) = \sin \dfrac{1}{2}nx$;

(c) $\lambda_n = n^2\pi^2$, $y_n(x) = \sin n\pi x$;

(d) $\lambda_n = \dfrac{n^2\pi^2}{L^2}$, $y_n(x) = \sin \dfrac{n\pi x}{L}$;

(e) $\lambda_n = \dfrac{n^2\pi^2}{4L^2}$, $y_n(x) = \sin \dfrac{n\pi(x+L)}{2L}$;

(f) $\lambda_n = \dfrac{n^2\pi^2}{(b-a)^2}$, $y_n(x) = \sin \dfrac{n\pi(x-a)}{b-a}$.

5. (a) $y(x,t) = \dfrac{8c}{\pi^2} \sum_1^\infty (-1)^{n+1} \dfrac{\sin(2n-1)x \cos(2n-1)at}{(2n-1)^2}$;

(b) $y(x,t) = \dfrac{8}{\pi^2} \sum_1^\infty \dfrac{\sin(2n-1)x \cos(2n-1)at}{(2n-1)^3}$;

(c) $y(x,t) = \dfrac{2}{\pi} \sum_1^\infty \dfrac{1}{n^2} \left[\sin \dfrac{n\pi}{4} + \sin \dfrac{3n\pi}{4} \right] \sin nx \cos nat$.

SECTION 41, p. 316

2. $w(x,t) = \sum_1^\infty b_n e^{-n^2a^2t} \sin nx + g(x)$,

where $g(x) = w_1 + \dfrac{1}{\pi}(w_2 - w_1)x$ and $b_n = \dfrac{2}{\pi} \int_0^\pi [f(x) - g(x)] \sin nx\, dx$.

4. $w(x,t) = e^{-ct} \sum_1^\infty b_n e^{-n^2a^2t} \sin nx$,

where $b_n = \dfrac{2}{\pi} \int_0^\pi f(x) \sin nx\, dx$.

5. $\left. \dfrac{\partial w}{\partial x} \right]_{x=0} = 0$, $\left. \dfrac{\partial w}{\partial x} \right]_{x=\pi} = 0$; $w(x,t) = 100$.

6. $w(x,t) = \dfrac{1}{2}a_0 + \sum_1^\infty a_n e^{-n^2a^2t} \cos nx$, where

$a_n = \dfrac{2}{\pi} \int_0^\pi f(x) \cos nx\, dx$ for $n = 0, 1, 2, \ldots$.

7. $w(x,y) = \sum_1^\infty b_n e^{-ny} \sin nx$, where $b_n = \dfrac{2}{\pi} \int_0^\pi f(x) \sin nx\, dx$.

SECTION 42, p. 322

2. (a) $w(r,\theta) = \dfrac{2}{\pi} - \dfrac{4}{\pi} \sum_1^\infty (-1)^n \dfrac{r^n \cos n\theta}{4n^2 - 1}$;

(b) $w(r,\theta) = 2\left(r \sin\theta - \dfrac{1}{2}r^2 \sin 2\theta + \dfrac{1}{3}r^3 \sin 3\theta - \cdots \right)$;

(c) $w(r, \theta) = \dfrac{1}{\pi} - \dfrac{2}{\pi} \displaystyle\sum_1^\infty r^{2n} \dfrac{\cos 2n\theta}{4n^2 - 1} + \dfrac{1}{2} r \sin \theta;$

(d) $w(r, \theta) = \dfrac{1}{2} + \dfrac{2}{\pi} \left(r \sin \theta + \dfrac{1}{3} r^3 \sin 3\theta + \dfrac{1}{5} r^5 \sin 5\theta + \cdots \right);$

(e) $w(r, \theta) = \dfrac{\pi^2}{12} + \displaystyle\sum_1^\infty (-1)^n \dfrac{r^n \cos n\theta}{n^2}.$

SECTION 43, p. 329

2. $n = 1, y = -\dfrac{1}{2} c_1 x^{-1} + c_2 x.$

3. (a) $(1 - x^2)\mu'' - 2x\mu' + p(p + 1)\mu = 0;$
   (b) $x^2\mu'' + 3x\mu' + (1 + x^2 - p^2)\mu = 0;$
   (c) $(1 - x^2)\mu'' - 3x\mu' + (p^2 - 1)\mu = 0;$
   (d) $\mu'' + 2x\mu' + (2 + 2p)\mu = 0;$
   (e) $\mu'' + x\mu = 0;$
   (f) $x\mu'' + (1 + x)\mu' + (1 + p)\mu = 0.$

4. $\mu = x, y = x^4 e^{x^2}\left[c_1 \displaystyle\int \dfrac{e^{-x^2}}{x^5} dx + c_2 \right].$

6. (b) Legendre's and Airy's.

8. (a) $[(1 - x^2)y']' + p(p + 1)y = 0;$

   (b) $[xy']' + \left(x - \dfrac{p^2}{x} \right)y = 0;$

   (c) $[\sqrt{1 - x^2}\, y']' + \dfrac{p^2}{\sqrt{1 - x^2}} y = 0;$

   (d) $[e^{-x^2}y']' + 2pe^{-x^2}y = 0;$
   (e) $[y']' + xy = 0;$
   (f) $[xe^{-x}y']' + pe^{-x}y = 0.$

10. (b) $\lambda_0 = 0, y_0(x) = 1; \lambda_n = n^2$ for $n = 1, 2, 3, \ldots$, and the eigenfunctions corresponding to each of these λ_n are $\cos nx$ and $\sin nx$.

SECTION 44, p. 340

2. (c) $P_2(x) = \dfrac{1}{2}(3x^2 - 1),$

   $P_3(x) = \dfrac{1}{2}(5x^3 - 3x),$

   $P_4(x) = \dfrac{1}{8}(35x^4 - 30x^2 + 3),$

   $P_5(x) = \dfrac{1}{8}(63x^5 - 70x^3 + 15x).$

SECTION 45, p. 347

4. (a) $f(x) = \frac{1}{4} P_0(x) + \frac{1}{2} P_1(x) + \frac{5}{16} P_2(x) + \cdots$;

(b) $f(x) = \frac{1}{2} (e - e^{-1}) P_0(x) + 3e^{-1} P_1(x) + \frac{1}{2} (5e - 35e^{-1}) P_2(x) + \cdots$.

SECTION 46, p. 356

7. $y = x^{-c} [c_1 J_p(ax^b) + c_2 J_{-p}(ax^b)]$ if p is not an integer;
$y = x^{-c} [c_1 J_p(ax^b) + c_2 Y_p(ax^b)]$ in all cases.

SECTION 47, p. 363

3. $J_2(x) = \frac{2}{x} J_1(x) - J_0(x)$;

$J_3(x) = \left(\frac{8}{x^2} - 1\right) J_1(x) - \frac{4}{x} J_0(x)$;

$J_4(x) = \left(\frac{48}{x^3} - \frac{8}{x}\right) J_1(x) - \left(\frac{24}{x^2} - 1\right) J_0(x)$.

SECTION 48, p. 384

3. $L[\sin^2 ax] = \frac{1}{2} \left(\frac{1}{p} - \frac{p}{p^2 + 4a^2}\right)$ and $L[\cos^2 ax] = \frac{1}{2} \left(\frac{1}{p} + \frac{p}{p^2 + 4a^2}\right)$;

the sum of these transforms is the transform of 1 $(=1/p)$.

4. (a) $\frac{10}{p}$;

(b) $\frac{5!}{p^6} + \frac{p}{p^2 + 4}$;

(c) $\frac{2}{p - 3} - \frac{5}{p^2 + 25}$;

(d) $\frac{4}{p^2 + 4} + \frac{2}{p + 1}$;

(e) $\frac{6!}{p^7}$.

5. (a) $5x^3$;
(b) $2e^{-3x}$;
(c) $2x^2 + 3 \sin 2x$;

(d) $1 - e^{-x}$;
(e) $x - \sin x$.

SECTION 49, p. 388

2. (a) $\frac{1}{pe^{ap}}$;

(b) $\frac{1}{p(e^p - 1)}$;

(c) $\frac{e^p - 1 - p}{p^2(e^p - 1)}$;

(d) $\frac{1 + e^{-\pi p}}{p^2 + 1}$.

SECTION 50, p. 394

1. (a) $\dfrac{5!}{(p+2)^6}$;

   (c) $\dfrac{p-3}{(p-3)^2+4}$.

  (b) $\dfrac{1}{p+1} - \dfrac{2!}{(p+1)^3}$;

2. (a) $2e^{-2x}\sin 3x$;

   (c) $e^{-x}\cos 2x + e^{-x}\sin 2x$.

  (b) $2e^{-3x}x^3$;

3. (a) $y(x) = -e^{-x} + e^{2x}$;

  (b) $y(x) = 3xe^{2x}$;

  (c) $y(x) = 1 - e^{-x}\cos x$;

  (d) $y(x) = -5 + 6x - 3x^2 + x^3 + 5e^{-x}$;

  (e) $y(x) = e^{-x}\sin 2x + e^{-x}\sin x$.

4. $y(x) = y_0 e^{ax} + (y_0' - ay_0)xe^{ax}$.

5. $1 - e^{-x}$.

6. $y = \dfrac{3}{2}e^{-2x}\sin x + \dfrac{1}{2}e^{-2x}\cos x - \dfrac{1}{2}e^{-x}$.

SECTION 51, p. 397

1. $L^{-1}\left[\dfrac{1}{(p^2+a^2)^2}\right] = \dfrac{1}{2a^2}\left(\dfrac{\sin ax}{a} - x\cos ax\right)$.

2. (a) $\dfrac{6ap^2 - 2a^3}{(p^2+a^2)^3}$;

   (b) $\dfrac{3}{4p^2}\sqrt{\dfrac{\pi}{p}}$.

3. (a) $y(x) = cx^2e^x$;

   (b) $y(x) = xe^{-x}$.

5. (a) $\log\dfrac{b}{a}$;

   (b) $\tan^{-1}\dfrac{b}{a}$.

8. (b) $\dfrac{1}{p(1+e^{-p})}$.

SECTION 52, p. 404

2. (a) $y(x) = \cos x$;

   (c) $y(x) = e^{-x}(x-1)^2$;

  (b) $y(x) = e^{2x}$;

   (d) $y(x) = -2\sin x + 4\sin 2x$.

4. $y = cx$.

SECTION 53, p. 410

2. (a) $\dfrac{1}{a}(1 - \cos at)$;

   (c) $\dfrac{1}{a^2}(e^{at} - 1 - at)$;

  (b) $\dfrac{1}{a-b}(e^{at} - e^{bt})$;

   (d) $\dfrac{1}{a^2-b^2}(a\sin bt - b\sin at)$.

4. (a) $y = \dfrac{1}{6}e^{3t} + \dfrac{5}{6}e^{-3t} - e^{-2t}$;

(b) $y = -\dfrac{1}{6}t - \dfrac{1}{36} - \dfrac{1}{45}e^{-3t} + \dfrac{1}{20}e^{2t}$;

(c) $y = 2e^{t} - \dfrac{1}{3}t^{3} - t^{2} - 2t - 2$.

5. (a) $y(t) = \displaystyle\int_{0}^{t} f(\tau)[e^{-(t-\tau)} - e^{-2(t-\tau)}]\, d\tau$;

(b) $\dfrac{1}{20}e^{3t} - \dfrac{1}{4}e^{-t} + \dfrac{1}{5}e^{-2t}$ and $\dfrac{1}{2}t - \dfrac{3}{4} + e^{-t} - \dfrac{1}{4}e^{2t}$.

7. $A(t) = \dfrac{1}{k}\left(1 - \cos\sqrt{\dfrac{k}{M}}\,t\right)$, $h(t) = \dfrac{1}{\sqrt{Mk}}\sin\sqrt{\dfrac{k}{M}}\,t$,

$x(t) = \dfrac{1}{\sqrt{Mk}}\displaystyle\int_{0}^{t} f(\tau)\sin\sqrt{\dfrac{k}{M}}\,(t-\tau)\, d\tau$.

8. (a) $I(t) = \dfrac{E_0}{R}[1 - e^{-Rt/L}]$;

(b) $I(t) = \dfrac{E_0}{L}e^{-Rt/L}$;

(c) $I(t) = \dfrac{E_0}{\sqrt{R^2 + L^2\omega^2}}\sin(\omega t - \alpha) + \dfrac{E_0 L\omega}{R^2 + L^2\omega^2}e^{-Rt/L}$,

where $\tan\alpha = L\omega/R$.

SECTION 54, p. 420

1. (a) $\dfrac{dy}{dx} = z$ (b) $\dfrac{dy}{dx} = z$

$\dfrac{dz}{dx} = xy + x^2 z$; $\dfrac{dz}{dx} = w$

$\dfrac{dw}{dx} = w - x^2 z^2$.

2. $\dfrac{dx}{dt} = v_x$

$\dfrac{dv_x}{dt} = \dfrac{f(t,x,y)}{m}$

$\dfrac{dy}{dt} = v_y$

$\dfrac{dv_y}{dt} = \dfrac{g(t,x,y)}{m}$.

SECTION 55, p. 425

5. (b) $\begin{cases} x = c_1 e^{4t} + c_2 e^{-2t} \\ y = c_1 e^{4t} - c_2 e^{-2t}; \end{cases}$

(c) $\begin{cases} x = 3e^{4t} + 2e^{-2t} \\ y = 3e^{4t} - 2e^{-2t}. \end{cases}$

6. (b) $\begin{cases} x = 2c_1 e^{4t} + c_2 e^{-t} \\ y = 3c_1 e^{4t} - c_2 e^{-t}; \end{cases}$

(c) $\begin{cases} x = 2c_1 e^{4t} + c_2 e^{-t} + 3t - 2 \\ y = 3c_1 e^{4t} - c_2 e^{-t} - 2t + 3. \end{cases}$

8. $\begin{cases} x = c_1 e^t + c_2 t e^t \\ y = c_2 e^t. \end{cases}$

9. (a) $\begin{cases} x = c_1 e^t \\ y = c_2 e^t. \end{cases}$

SECTION 56, p. 433

1. (a) $\begin{cases} x = 2c_1 e^{-t} + c_2 e^t \\ y = c_1 e^{-t} + c_2 e^t; \end{cases}$

(b) $\begin{cases} x = e^{3t}(2c_1 \cos 3t + 2c_2 \sin 3t) \\ y = e^{3t}[c_1(\cos 3t + 3 \sin 3t) + c_2(\sin 3t - 3 \cos 3t)]; \end{cases}$

(c) $\begin{cases} x = -2c_1 e^{3t} + c_2(1 + 2t)e^{3t} \\ y = c_1 e^{3t} - c_2 t e^{3t}; \end{cases}$

(d) $\begin{cases} x = 3c_1 + c_2 e^{-2t} \\ y = 4c_1 + 2c_2 e^{-2t}; \end{cases}$

(e) $\begin{cases} x = c_1 e^{2t} \\ y = c_2 e^{3t}; \end{cases}$

(f) $\begin{cases} x = c_1 e^{-3t} + c_2(1 - t)e^{-3t} \\ y = -c_1 e^{-3t} + c_2 t e^{-3t}; \end{cases}$

(g) $\begin{cases} x = 2c_1 e^{10t} + 3c_2 e^{3t} \\ y = c_1 e^{10t} - 2c_2 e^{3t}; \end{cases}$

(h) $\begin{cases} x = e^{3t}(c_1 \cos 2t + c_2 \sin 2t) \\ y = e^{3t}[c_1(\sin 2t - \cos 2t) - c_2(\sin 2t + \cos 2t)]. \end{cases}$

5. (b) $\begin{cases} x = 3t + 2 \\ y = 2t - 1. \end{cases}$

SECTION 57, p. 439

1. $x \dfrac{d^2 x}{dt^2} = (dx^2 - cx) \dfrac{dx}{dt} + (acx^2 - adx^3) + \left(\dfrac{dx}{dt} \right)^2.$

2. The fox curve is concave up whenever the rabbit curve is rising.

SECTION 58, p. 445

2. Put $c = t_1 - t_2$ and use uniqueness.

3. They are the same except that the directions of all paths are reversed in passing from one to the other.

4. (a) Every point is a critical point, and there are no paths.

(b) Every point on the y-axis is a critical point, and the paths are horizontal half-lines directed out to the left and right from the y-axis.

(c) There are no critical points, and the paths are straight lines with slope 2 directed up to the right.

(d) The point $(0,0)$ is the only critical point, and the paths are half-lines of all possible slopes directed in toward the origin.

5. For equations (1) and (2), they are $(0,0)$, $(\pm\pi,0)$, $(\pm2\pi,0)$, $(\pm3\pi,0)$, . . . ; and for equation (3), $(0,0)$ is the only critical point.

6. (a) $(-2,0)$, $(0,0)$, $(1,0)$; (b) $(2,2)$, $(3,3)$.

7. $\begin{cases} x = c_1e^t \\ y = c_1e^t + e^t + c_2. \end{cases}$

SECTION 59, p. 454

1. (a) (i) The critical points are the points on the x-axis;

  (ii) $dy/dx = 2xy/(x^2 + 1)$;

  (iii) $y = c(x^2 + 1)$.

 (b) (i) $(0,0)$;

  (ii) $dy/dx = -x/y$;

  (iii) $x^2 + y^2 = c^2$.

 (c) (i) There are no critical points;

  (ii) $dy/dx = \cos x$;

  (iii) $y = \sin x + c$.

 (d) (i) The critical points are the points on the y-axis;

  (ii) $dy/dx = -2xy^2$;

  (iii) $y = 1/(x^2 + c)$ and $y = 0$.

2. (a) (i) $\begin{cases} x = c_1e^t \\ y = c_2e^{-t}; \end{cases}$ (iii) $xy = c$;

  (ii) $dy/dx = -y/x$; (iv) unstable.

 (b) (i) $\begin{cases} x = c_1e^{-t} \\ y = c_2e^{-2t}; \end{cases}$ (iii) $y = cx^2$;

  (ii) $dy/dx = 2y/x$; (iv) asymptotically stable.

 (c) (i) $\begin{cases} x = 2c_1\cos 2t + 2c_2\sin 2t \\ y = -c_1\sin 2t + c_2\cos 2t; \end{cases}$

  (ii) $\dfrac{dy}{dx} = \dfrac{-x}{4y}$;

(iii) $\dfrac{x^2}{4c^2} + \dfrac{y^2}{c^2} = 1;$

(iv) stable but not asymptotically stable.

SECTION 60, p. 464

1. (a) Unstable node;
   (b) Asymptotically stable spiral;
   (c) Unstable saddle point;
   (d) Stable but not asymptotically stable center;
   (e) Asymptotically stable node;
   (f) The critical point is not isolated;
   (g) Unstable spiral.
3. (c) The critical point is $(-3,2)$, the transformed system is

$$\begin{cases} \dfrac{d\bar{x}}{dt} = 2\bar{x} - 2\bar{y} \\[2mm] \dfrac{d\bar{y}}{dt} = 11\bar{x} - 8\bar{y}, \end{cases}$$

and the critical point is an asymptotically stable node.
4. (a) $m^2 + 2bm + a^2 = 0;\, p = 2b,\, q = a^2.$
   (b) (i) A stable but not asymptotically stable center; the mass oscillates; the displacement x and velocity $y = dx/dt$ are periodic functions of time.
       (ii) An asymptotically stable spiral; the mass executes damped oscillations; x and $dx/dt \to 0$ through smaller and smaller oscillations.
       (iii) An asymptotically stable node; the mass does not oscillate; x and $dx/dt \to 0$ without oscillating.
       (iv) The same as (iii).
5. $a_2 x^2 - 2a_1 xy - b_1 y^2 = c.$

SECTION 61, p. 470

1. (a) Neither;   (c) Neither;
   (b) Positive definite;   (d) Negative definite.

SECTION 62, p. 479

2. $\dfrac{dy}{dx} = \dfrac{2x^2 y - y^3}{x^3 - 2xy^2}.$

3. Put $D = -pq = (a_1 + b_2)(a_1 b_2 - a_2 b_1) > 0.$
4. No conclusion can be drawn about the stability properties of the nonlinear system (4) at $(0,0)$ when the related linear system (3) has a center at $(0,0)$.
5. (a) Unstable spiral;   (b) Asymptotically stable node.
6. The critical point $(0,0)$ is unstable if $\mu > 0$ and asymptotically stable if $\mu < 0$.

SECTION 63, p. 485

1. If $f(0) = 0$ and $xf(x) < 0$ for $x \neq 0$, the critical point is an unstable saddle point.

3. $y^2 - x^2 + x^4 = 2E$; $(-\sqrt{2}/2,0)$ is a center; $(0,0)$ is a saddle point; and $(\sqrt{2}/2,0)$ is a center.

4. When $z = F(x)$ has a maximum, the critical point is a saddle point; when it has a minimum, the critical point is a center; and when it has a point of inflection, the critical point is a cusp.

SECTION 64, p. 492

2. (a) $\begin{cases} \dfrac{dr}{dt} = r(4 - r^2) \\[2mm] \dfrac{d\theta}{dt} = -4; \end{cases}$

(c) $\begin{cases} x = \dfrac{2\cos 4(t + t_0)}{\sqrt{1 + ce^{-8t}}} \\[4mm] y = \dfrac{-2\sin 4(t + t_0)}{\sqrt{1 + ce^{-8t}}}, \end{cases}$ $\begin{cases} x = 2\cos 4t \\ y = -2\sin 4t. \end{cases}$

4. (a) A periodic solution (Liénard's theorem);
(b) No periodic solution (Theorem B);
(c) No periodic solution (Theorem A);
(d) No periodic solution (Theorem B);
(e) A periodic solution (Liénard's theorem).

SECTION 66, p. 513

1. (a) $(x - c_2)^2 + y^2 = c_1^2$;
(b) $y = c_1 \sin(x - c_2)$.

2. $y = \dfrac{1}{4}(x^2 - x)$.

4. (a) $c_1 = r\cos(\theta - c_2)$; (b) Same as (a).

SECTION 67, p. 523

3. (a) $x = \dfrac{ad}{a^2 + b^2 + c^2}$, $y = \dfrac{bd}{a^2 + b^2 + c^2}$,

$z = \dfrac{cd}{a^2 + b^2 + c^2}.$

5. The catenary $y + \lambda = c_1 \cosh\left(\dfrac{x - c_2}{c_1}\right)$.

SECTION 68, p. 542

1. $y = \dfrac{1}{1 - x} = 1 + x + x^2 + \cdots, |x| < 1;$

$y_1(x) = 1 + x, \qquad y_2(x) = 1 + x + x^2 + \dfrac{1}{3}x^3,$

$y_3(x) = 1 + x + x^2 + x^3 + \dfrac{2}{3}x^4 + \dfrac{1}{3}x^5 + \dfrac{1}{9}x^6 + \dfrac{1}{63}x^7.$

2. $y = e^{x^2} - 1;$

$y_1(x) = x^2, \qquad y_2(x) = x^2 + \dfrac{x^4}{2},$

$y_3(x) = x^2 + \dfrac{x^4}{2} + \dfrac{x^6}{2 \cdot 3},$

$y_4(x) = x^2 + \dfrac{x^4}{2} + \dfrac{x^6}{2 \cdot 3} + \dfrac{x^8}{2 \cdot 3 \cdot 4}.$

3. (a) $y_n(x) = \dfrac{x^2}{2!} + \dfrac{x^3}{3!} + \cdots + \dfrac{x^{n+1}}{(n + 1)!} + e^x \to (e^x - x - 1) + e^x;$

(b) $y_n(x) = 1 + x + 2\left[\dfrac{x^2}{2!} + \dfrac{x^3}{3!} + \dfrac{x^4}{4!} + \cdots + \dfrac{x^{n+1}}{(n + 1)!}\right]$

$\to 1 + x + 2(e^x - x - 1);$

(c) $y_1(x) = (\sin x - x) + 1 + x + \dfrac{x^2}{2!},$

$y_2(x) = -\left(\cos x - 1 + \dfrac{x^2}{2!}\right) + 1 + x + (x^2) + \dfrac{x^3}{3!},$

$y_3(x) = -\left(\sin x - x + \dfrac{x^3}{3!}\right) + 1 + x + \left(x^2 + \dfrac{x^3}{3}\right) + \dfrac{x^4}{4!},$

$y_4(x) = \left(\cos x - 1 + \dfrac{x^2}{2!} - \dfrac{x^4}{4!}\right) + 1 + x + \left(x^2 + \dfrac{x^3}{3} + \dfrac{x^4}{3 \cdot 4}\right) + \dfrac{x^5}{5!}.$

SECTION 69, p. 552

6. (a) All points (x_0, y_0) with $y_0 \neq 0;$
(b) All points (x_0, y_0), since $f(x, y) = |y|$ satisfies a Lipschitz condition on every rectangle.

7. All points $(x_0, y_0).$

SECTION 70, p. 555

1. $\begin{cases} y = \cos x \\ z = -\sin x. \end{cases}$

INDEX